Numerical and Evolutionary Optimization 2024

Numerical and Evolutionary Optimization 2024

Guest Editors

Marcela Quiroz-Castellanos
Oliver Cuate
Leonardo Trujillo
Oliver Schütze

MDPI

Basel • Beijing • Wuhan • Barcelona • Belgrade • Novi Sad • Cluj • Manchester

Guest Editors
Marcela Quiroz-Castellanos
Instituto de Investigaciones en Inteligencia Artificial
Universidad Veracruzana
Xalapa
Mexico

Oliver Cuate
Escuela Superior de Física y Matemáticas
Instituto Politécnico Nacional
Mexico City
Mexico

Leonardo Trujillo
Departamento de Ingeniería en Electrónica y Eléctrica
Instituto Tecnológico de Tijuana
Tijuana
Mexico

Oliver Schütze
Depto de Computacion
Cinvestav
Mexico City
Mexico

Editorial Office
MDPI AG
Grosspeteranlage 5
4052 Basel, Switzerland

This is a reprint of the Special Issue, published open access by the journal *Mathematical and Computational Applications* (ISSN 2297-8747), freely accessible at: https://www.mdpi.com/journal/mca/special_issues/JNB5544SAC.

For citation purposes, cite each article independently as indicated on the article page online and as indicated below:

Lastname, A.A.; Lastname, B.B. Article Title. *Journal Name* **Year**, *Volume Number*, Page Range.

ISBN 978-3-7258-5575-9 (Hbk)
ISBN 978-3-7258-5576-6 (PDF)
https://doi.org/10.3390/books978-3-7258-5576-6

Contents

Preface

This volume is a reprint of the Special Issue that was inspired by the 11th International Workshop on Numerical and Evolutionary Optimization (NEO 2024). The event was held from 03 to 06 September 2024 in Mexico City, Mexico, and was hosted by the Cinvestav.

This volume consists of 17 papers. The order of the presented chapters is organized chronologically by the publication of the respective research papers in *Mathematical and Computational Applications (MCA)*.

We warmly thank all participants of the NEO 2024 as well as all authors who submitted a work to this Special Issue. We hope that this issue can be a contemporary reference regarding the field of numerical and evolutionary optimization and their exciting applications.

We finally use this opportunity to thank our institutions and the funding sources that made this event possible. In particular, we thank the Cinvestav, the Instituto Politécnico Nacional, the TecNM and the TecNM/IT Tijuana, and the Universidad Veracruzana. Further, we acknowledge support from the CONAHCYT (SECIHTI) projects CBF2023-2024-1463 and CF-2023-I-724, the TecNM projects 21802.25-P, 22056.25-P, and 22959.25-P, and the Instituto Politécnico Nacional projects SIP-20240570 and SIP-20251029. Finally, we thank MDPI and MCA for giving us the opportunity to edit the Special Issue and this volume.

Marcela Quiroz-Castellanos, Oliver Cuate, Leonardo Trujillo, and Oliver Schütze
Guest Editors

Mathematical and Computational Applications

MDPI

Editorial

Numerical and Evolutionary Optimization 2024

Marcela Quiroz-Castellanos [1], Oliver Cuate [2], Leonardo Trujillo [3] and Oliver Schütze [4,*]

1 Instituto de Investigaciones en Inteligencia Artificial, Universidad Veracruzana, Xalapa 91000, Mexico
2 Escuela Superior de Física y Matemáticas, Instituto Politécnico Nacional, Mexico City 07738, Mexico
3 Departamento de Ingeniería en Electrónica y Eléctrica, Tecnológico Nacional de México/Instituto Tecnológico de Tijuana, Calzada Tecnológico SN, Tomas Aquino, Tijuana 22414, Mexico
4 Departamento de Computacion, Cinvestav, Mexico City 07360, Mexico
* Correspondence: schuetze@cs.cinvestav.mx

This Special Issue was inspired by the 11th International Workshop on Numerical and Evolutionary Optimization (NEO 2024), held from 3 to 6 September 2024 in Mexico City, Mexico, and hosted by Cinvestav. Solving real-world scientific and engineering problems has always been a challenge, and the complexity of these tasks has increased in recent years as more sources of data and information have been continuously developed. Thus, the design and analysis of powerful search and optimization techniques is of great importance. Two well-established fields that focus on this task are (i) traditional numerical optimization techniques and (ii) bio-inspired metaheuristic methods. Both of these general approaches have unique strengths and weaknesses, allowing researchers to solve some challenging problems while failing to solve others. The goal of the NEO workshop series is to gather experts from both fields to discuss, compare, and merge these complementary perspectives. Collaborative work allows researchers to maximize the strengths and minimize the weaknesses of both paradigms. NEO also intends to help researchers in these fields to understand and tackle real-world problems like pattern recognition, routing, energy, lines of production, prediction, and modeling, among others.

This Special Issue consists of 17 research papers that we will summarize below. The papers are presented chronologically in terms of their publication in *Mathematical and Computational Applications* (*MCA*).

In [1], Alvarado-Méndez et al. propose an alternative methodology for simultaneous parameter and actuator disturbance estimation for a general class of nonlinear systems. To this end, the authors develop an $H\infty$-adaptive nonlinear observer of a class of Lipschitz nonlinear systems with disturbances. The objective is to estimate parameters and monitor the performance of nonlinear processes with model uncertainties. The behavior of the observer is analyzed in the presence of disturbances using Lyapunov stability theory and using an $H\infty$ performance criterion. Numerical simulations are carried out to demonstrate the applicability of this observer on a semi-active car suspension.

In [2], Barradas-Palmeros et al. present a multi-objective feature selection framework that significantly reduces computational costs by integrating fixed and incremental sampling-fraction strategies with memory into a permutation-based Differential Evolution algorithm. Based on the DE-FS^{PM} approach, and adopting the GDE3 selection mechanism, the proposal avoids redundant fitness evaluations and reduces both running time and function evaluations during the search. The experimental results demonstrate the effectiveness of this method in single-objective optimization and suggest that cost-reduction strategies are only partially sustained for multi-objective feature selection.

In [3], Serrano-Cruz et al. develop a mathematical model of the human cardiovascular system to simulate both normal and pathological conditions within the systemic circulation.

Their novel approach reformulates the cardiovascular system into quadratic normal form through coordinate transformation. The latter allows for robust state estimation via sliding mode observers despite linear unobservability, which is particularly valuable for fault detection in scenarios where traditional observability conditions fail. The simulation results demonstrate strong agreement with established data, validating the model's accuracy in representing cardiovascular dynamics.

In [4], Cavita-Huerta et al. present an interesting new approach for classifying physical activities that exclusively uses accelerometry data processed through artificial neural networks (ANNs). The methodology involves data acquisition, preprocessing, feature extraction, and the application of deep learning algorithms to accurately identify activity patterns. A major innovation in this study is the incorporation of a new feature derived from the radius of curvature. The findings demonstrate the potential of this model to improve the precision and reliability of physical activity recognition in wearable health-monitoring systems.

In [5], González-Cárdenas et al. present a trajectory-tracking controller for an inverted pendulum system on a self-balancing differential drive platform. The system modeling is described by considering approximations of the swing angles. A discrete convex representation of the system is obtained via the nonlinear sector technique, considering the nonlinearities associated with the nonholonomic constraint. A discrete parallel distributed compensation controller is designed through an alternative method due to the presence of uncontrollable points that hinder finding a solution for the entire polytope. Finally, the results of simulations and an experiment using a prototype illustrate the effectiveness of the proposal.

In [6], Molina-Pérez and Rojas-López address the problem of enhancing image quality. Since two common goals in this context (improving the contrast and maintaining fine details) conflict with each other, the authors propose a multi-objective optimization approach that combines sigmoid transformation and unsharp masking–highboost filtering with the NSGA-II algorithm. A posterior preference articulation method identifies three representative outcomes from the obtained Pareto front: one maximizing contrast, one maximizing detail, and a balanced "knee point" solution. The method is tested on diverse image types, including medical and natural scenes, showing significant improvements over the original images in both contrast and detail.

In [7], García-Morales et al. propose two new metaheuristics to solve the Internet Shopping Optimization Problem with Sensitive Prices. The first approach is a memetic algorithm that integrates an evolutionary search with an improved local search and adaptive parameter adjustment, while the second one is an enhanced Particle Swarm Optimization algorithm incorporating a diversification technique and adaptive control parameters. Both methods are tested against the Branch and Bound (B&B) algorithm on nine problem instances of varying sizes, showing that the proposed algorithms perform similarly and outperform B&B.

In [8], Pérez-Pérez et al. study the problem of detecting faults and leaks in water pipelines. These systems are notoriously difficult to inspect, with leaks and faults leading to the production of large amounts of waste and inefficiency. Given the scarcity of fresh water in many places around the world, the development of efficient methods to detect faults in water pipelines is of immense importance. The authors present another hybrid approach that combines neuro-fuzzy systems with Kalman filters to achieve very high precision and a low false positive rate under a variety of fault conditions.

In [9], Guadarrama-Estrada et al. present a new approach to design a generalized dynamic observer (GDO) in order to detect and isolate attack patterns that compromise the functionality of cyber–physical systems. The considered attack patterns include denial-of-service (DoS), false data injection (FDI), and random data injection (RDI) attacks. To model an attacker's behavior and enhance the effectiveness of the attack patterns, Markovian logic is employed. A three-tank interconnected system, modeled under the discrete Takagi–Sugeno representation, is used as a case study to validate the proposed approach.

In [10], Contreras Ortiz et al. evaluate vision transformers for an industrial inspection problem, detecting defects in industrial welding. The study considers different versions of the problem, showing that vision transformers outperform convolutional networks and are nearly 30% more effective in a multiclass classification task. The study shows how transformer models can be applied to real-world quality control and inspection tasks.

In [11], Amador-Larrea et al. introduce adaptive mutation strategies for a grouping genetic algorithm, GGA-CGT, applied to the One-Dimensional Bin Packing Problem (1D-BPP). These strategies dynamically control the level of change that will be introduced to each solution by using feedback on population diversity, enabling better exploration. A performance comparison of this algorithm to the base algorithm on selected benchmark problems indicates an increase in the detection of optimal solutions and a severe reduction in the average proportion of individuals with equal fitness (from over 50% to less than 1%), enhancing diversity and avoiding local optima. The adaptive strategies are particularly effective in problem instances with larger item weights. These findings demonstrate the potential of adaptive mechanisms to improve genetic algorithms, offering a robust strategy for tackling complex optimization problems.

In [12], Purata-Aldaz et al. propose MASIP, a heuristic-based approach designed to construct and optimize investment portfolios using principles from the Markowitz and Sharpe models. The methodology performs three key tasks: selecting candidate stocks for an initial portfolio, forecasting asset values over short- and medium-term horizons, and optimizing the portfolio using the Sharpe ratio. The methodology incorporates a dynamic rebalancing process to enhance portfolio performance over time. A comparison on the S&P 500 data shows that MASIP is highly competitive compared to traditional methods for this problem class.

In [13], Marquez-Zepeda et al. present a novel method to monitor and predict air quality, using machine learning tools and Internet of Things (IoT) technologies. The approach is useful for indoor settings, such as offices or classrooms. These technologies can allow for the development of "smart" buildings and "cities", which proactively respond to environmental issues that can affect human health and behavior.

In [14], Sánchez Márquez et al. propose two evolutionary strategies based on the generalized Mallows model for the numerical treatment of the Mixed No-Idle Permutation Flow Shop Scheduling Problem (MNPFSSP). The proposed approaches are compared with algorithms previously used to address the studied problem. Statistical tests of the experimental results show that one of these methods, ES-GMMc, achieves reductions in execution time, especially in instances of large problems, where the shortest computing times are obtained in 23 of 30 instances, without affecting the quality of the solutions.

In [15], Gamboa et al. address a medical problem using a nonlinear third-order mathematical model described by ODEs. In particular, they study the dynamics of insulin and of β-cells, and the concentration of glucose, using a Thau observer, and conduct an analysis to study the models' dynamic bounds. The study represents a crucial initial step towards the potential development of new treatments for diabetes using a digital twin approach, as the method can describe how insulin levels develop over time at various glucose concentrations.

In [16], Goñi et al. study the effective design of content distribution networks over cloud computing platforms. A bio-inspired evolutionary multi-objective optimization approach is applied as a viable alternative to solve realistic problems where exact optimization methods are not applicable. Ad hoc representation and search operators are applied to optimize relevant metrics from the perspective of both system administrators and users. The numerical results indicate that the obtained solutions can provide different options for content distribution network design, enabling fast configuration that fulfills specific quality-of-service demands.

Finally, in [17], Frausto Solís et al. present TAE Predict, a methodology for multivariate time series forecasting of climate variables in the context of climate change. The method incorporates feature selection with an ensemble approach for prediction. The ensemble combines several machine learning models using metaheuristic optimization, and the work analyzes meteorological data from several cities in Mexico. The results are encouraging, showing how metaheuristic optimization and machine learning methods can be effectively hybridized to solve real-world problems.

We thank the NEO 2024 participants and the authors who submitted studies to this Special Issue. We hope that it can serve as a contemporary reference for the use and applications of numerical and evolutionary optimization.

Conflicts of Interest: The authors declare no conflicts of interest.

References

1. Alvarado-Méndez, P.E.; Astorga-Zaragoza, C.M.; Osorio-Gordillo, G.L.; Aguilera-González, A.; Vargas-Méndez, R.; Reyes-Reyes, J. H∞ State and Parameter Estimation for Lipschitz Nonlinear Systems. *Math. Comput. Appl.* **2024**, *29*, 51. [CrossRef]
2. Barradas-Palmeros, J.A.; Mezura-Montes, E.; Rivera-López, R.; Acosta-Mesa, H.G.; Márquez-Grajales, A. Computational Cost Reduction in Multi-Objective Feature Selection Using Permutational-Based Differential Evolution. *Math. Comput. Appl.* **2024**, *29*, 56. [CrossRef]
3. Serrano-Cruz, D.A.; Boutat-Baddas, L.; Darouach, M.; Astorga-Zaragoza, C.M.; Guerrero Ramírez, G.V. Modeling of the Human Cardiovascular System: Implementing a Sliding Mode Observer for Fault Detection and Isolation. *Math. Comput. Appl.* **2024**, *29*, 57. [CrossRef]
4. Cavita-Huerta, E.; Reyes-Reyes, J.; Romero-Ugalde, H.M.; Osorio-Gordillo, G.L.; Escobar-Jiménez, R.F.; Alvarado-Martínez, V.M. Human Activity Recognition from Accelerometry, Based on a Radius of Curvature Feature. *Math. Comput. Appl.* **2024**, *29*, 80. [CrossRef]
5. González-Cárdenas, Y.; López-Estrada, F.R.; Estrada-Manzo, V.; Dominguez-Zenteno, J.; López-Pérez, M. Design and Implementation of a Discrete-PDC Controller for Stabilization of an Inverted Pendulum on a Self-Balancing Car Using a Convex Approach. *Math. Comput. Appl.* **2024**, *29*, 83. [CrossRef]
6. Molina-Pérez, D.; Rojas-López, A.G. Resolving Contrast and Detail Trade-Offs in Image Processing with Multi-Objective Optimization. *Math. Comput. Appl.* **2024**, *29*, 104. [CrossRef]
7. García-Morales, M.A.; Brambila-Hernández, J.A.; Fraire-Huacuja, H.J.; Frausto, J.; Cruz, L.; Gómez, C.; Peña-Ramos, A. New Metaheuristics to Solve the Internet Shopping Optimization Problem with Sensitive Prices. *Math. Comput. Appl.* **2024**, *29*, 119. [CrossRef]
8. Pérez-Pérez, E.J.; González-Baldizón, Y.; Fragoso-Mandujano, J.A.; Guzmán-Rabasa, J.A.; Santos-Ruiz, I. Data-Driven Fault Diagnosis in Water Pipelines Based on Neuro-Fuzzy Zonotopic Kalman Filters. *Math. Comput. Appl.* **2025**, *30*, 2. [CrossRef]
9. Guadarrama-Estrada, A.R.; Osorio-Gordillo, G.L.; Vargas-Méndez, R.A.; Reyes-Reyes, J.; Astorga-Zaragoza, C.M. Cyber–Physical System Attack Detection and Isolation: A Takagi–Sugeno Approach. *Math. Comput. Appl.* **2025**, *30*, 12. [CrossRef]
10. Contreras Ortiz, A.; Santiago, R.R.; Hernandez, D.E.; Lopez-Montiel, M. Multiclass Evaluation of Vision Transformers for Industrial Welding Defect Detection. *Math. Comput. Appl.* **2025**, *30*, 24. [CrossRef]
11. Amador-Larrea, S.; Quiroz-Castellanos, M.; Ramos-Figueroa, O. An Experimental Study of Strategies to Control Diversity in Grouping Mutation Operators: An Improvement to the Adaptive Mutation Operator for the GGA-CGT for the Bin Packing Problem. *Math. Comput. Appl.* **2025**, *30*, 31. [CrossRef]
12. Purata-Aldaz, J.; Frausto-Solís, J.; Castilla-Valdez, G.; González-Barbosa, J.; Sánchez Hernández, J.P. MASIP: A Methodology for Assets Selection in Investment Portfolios. *Math. Comput. Appl.* **2025**, *30*, 34. [CrossRef]
13. Marquez-Zepeda, M.J.; Santos-Ruiz, I.; Pérez-Pérez, E.J.; Navarro-Díaz, A.; Delgado-Aguiñaga, J.A. Internet-of-Things-Based CO2 Monitoring and Forecasting System for Indoor Air Quality Management. *Math. Comput. Appl.* **2025**, *30*, 36. [CrossRef]

14. Sánchez Márquez, E.M.; Pérez-Rodríguez, R.; Ornelas-Rodriguez, M.; Puga-Soberanes, H.J. An Evolutionary Strategy Based on the Generalized Mallows Model Applied to the Mixed No-Idle Permutation Flow Shop Scheduling Problem. *Math. Comput. Appl.* **2025**, *30*, 39. [CrossRef]
15. Gamboa, D.; Galicia, T.C.; Campos, P.J. Thau Observer for Insulin Estimation Considering the Effect of Beta-Cell Dynamics for a Diabetes Mellitus Model. *Math. Comput. Appl.* **2025**, *30*, 43. [CrossRef]
16. Goñi, G.; Nesmachnow, S.; Rossit, D.; Moreno-Bernal, P.; Tchernykh, A. Bio-Inspired Multiobjective Optimization for Designing Content Distribution Networks. *Math. Comput. Appl.* **2025**, *30*, 45. [CrossRef]
17. Frausto Solís, J.; Estrada-Patiño, E.; Ponce Flores, M.; Sánchez-Hernández, J.P.; Castilla-Valdez, G.; González-Barbosa, J. TAE Predict: An Ensemble Methodology for Multivariate Time Series Forecasting of Climate Variables in the Context of Climate Change. *Math. Comput. Appl.* **2025**, *30*, 46. [CrossRef]

Mathematical and Computational Applications

MDPI

Article

$\mathcal{H}_\infty$ State and Parameter Estimation for Lipschitz Nonlinear Systems

Pedro Eusebio Alvarado-Méndez [1], Carlos M. Astorga-Zaragoza [1,*], Gloria L. Osorio-Gordillo [1], Adriana Aguilera-González [2], Rodolfo Vargas-Méndez [1] and Juan Reyes-Reyes [1]

1 Tecnológico Nacional de México/CENIDET, Interior Internado Palmira S/N, Cuernavaca 62490, Morelos, Mexico; d17ce052@cenidet.tecnm.mx (P.E.A.-M.); gloria.og@cenidet.tecnm.mx (G.L.O.-G.); rodolfo.vm@cenidet.tecnm.mx (R.V.-M.); juan.rr@cenidet.tecnm.mx (J.R.-R.)

2 ESTIA Institute of Technology, University of Bordeaux, F-64210 Bidart, France; a.aguilera-gonzalez@estia.fr

* Correspondence: carlos.az@cenidet.tecnm.mx

Abstract: A $\mathcal{H}_\infty$ robust adaptive nonlinear observer for state and parameter estimation of a class of Lipschitz nonlinear systems with disturbances is presented in this work. The objective is to estimate parameters and monitor the performance of nonlinear processes with model uncertainties. The behavior of the observer in the presence of disturbances is analyzed using Lyapunov stability theory and by considering an $\mathcal{H}_\infty$ performance criterion. Numerical simulations were carried out to demonstrate the applicability of this observer for a semi-active car suspension. The adaptive observer performed well in estimating the tire rigidity (as an unknown parameter) and induced disturbances representing damage to the damper. The main contribution is the proposal of an alternative methodology for simultaneous parameter and actuator disturbance estimation for a more general class of nonlinear systems.

Keywords: adaptive observer; nonlinear system; Lipschitz nonlinearities

1. Introduction

Observer design for nonlinear systems satisfying the Lipschitz condition has been the subject of constant research, because these systems have the particularity to represent a wide class of real processes. The Lipschitz property of nonlinear systems was initially used by [1] for observer design, by providing a sufficient conditions to guarantee the asymptotic stability of the observation error. Although studies have been conducted on the design of observers for Lipschitz nonlinear systems, this problem is still insufficiently explored, see, e.g., [2–6]. In [2], the authors presented $\mathcal{H}_\infty$ observers for nonlinear Lipschitz systems using an LPV approach. The observer gains were computed by solving a set of LMI and the observer was evaluated for a neural mass model. In [3], the authors presented an observer-based controller design for stabilizing Lipschitz nonlinear systems with parameter uncertainties and perturbation inputs. The observer-based controller was evaluated for different numerical cases. In [4], the authors presented a generalized observer for nonlinear uncertain descriptor systems satisfying the one-sided Lipschitz condition. Perturbations affecting both inputs and outputs were considered. The goal of the proposed approach was to attenuate the effects of these perturbations. Observers for one-sided Lipschitz nonlinear systems with disturbances and limited communication resources in communication networks were treated in [5], and finally a nonlinear $\mathcal{H}_\infty$ proportional derivative observer for one-sided Lipschitz singular systems with disturbances was designed and tested through simulation for a DC motor in [6].

As described in the previous paragraphs, there have been many works that dealt with the observation and control of various types of Lipschitz nonlinear systems. However, simultaneous state and parameter estimation approaches for this type of system, for monitoring purposes, have not been fully addressed. Process monitoring is typically oriented towards verifying the behavior of certain important state variables of the process.

However, there are faults, disturbances, or unknown inputs that can affect the estimation process, causing dysfunctions or inaccuracies in the control, stabilization, or monitoring of the process.

Parameter estimation techniques could play a crucial role in addressing this issue by continuously updating the model parameters based on the observed data, thereby enhancing the accuracy of monitoring systems. Parameters can vary over time due to system deterioration, among other factors. By accurately estimating them, it becomes possible to better track the process behavior and to early detect anomalies or faults. Integrating parameter estimation approaches into process monitoring systems could potentially reduce the reliance on human operators for fault detection, leading to more reliable and automated monitoring processes. This, in turn, could improve system safety, efficiency, and reliability in various industrial applications. Further research and development in this area could contribute significantly to advancing the field of process monitoring and control.

There are several methods used to estimate process parameters in order to better characterize the systems and to adequately estimate and monitor process variables. Among these methods, adaptive observers have the particularity of being able to estimate state variables and/or one or several parameters of the system, e.g., [6–10]. For instance, in [7], an adaptive observer for estimating unknown parameters by separating measurable states from the non measurable ones was presented. One disadvantage of this observer is that the unknown parameter must be included in the equation of measurable states. Another example was provided in [8], where a descriptor adaptive observer was synthesized for fault estimation in uncertain nonlinear systems. This observer was designed using the $\mathcal{H}_\infty$ approach and Lyapunov stability criteria. The observer was tested on a robotic arm simultaneously affected by actuator and sensor faults. An actuator fault diagnosis and reconfiguration system based on an H_∞ observer was proposed in [9] for a vehicle steering system. Although the proposed approach is interesting, the system requires that all states be measurable, which is not always feasible in practice. On the other hand, a fuzzy adaptive observer for fault and disturbance estimation for Takagi–Sugeno fuzzy systems is presented in [10]. While the Takagi–Sugeno approach is a consistent method for addressing nonlinear problems, algorithms to compute the observer gains by solving a set of LMIs can become complicated for systems with a large number of nonlinearities. Other adaptive observers, prioritizing convergence time, can be found in [11,12].

One of the latent challenges in implementing adaptive observers is considering situations or unforeseen phenomena that may occur in practice, such as disturbances, abrupt or incipient parameter variations, or sensor/actuator faults, in the design process. Work that addressed these kinds of problems was presented in [13–18]. In [13], the authors designed an adaptive observer to estimate the state vector and the unknown parameter, as well as an output feedback controller. They considered uncertainties in the sensors, unknown growth rate, and stochastic disturbances. The gains are adaptively adjusted to account for sensor sensitivity, which is treated as an unknown continuous function. Another interesting study was presented in [14], where the authors proposed an adaptive observer with variable gains to design a fault-tolerant control mechanism for sensor bias faults in the active suspension of vehicles. The approach was specifically developed for this case study and may not be applicable to other practical cases. In [18], an adaptive observer was developed to estimate the uncertainties in linear systems. Other applications of adaptive observers include bioreactors [19], polymerization reactors [20], fuel cells [21], heat exchangers [22], distillation plants [23], induction motors [24], nuclear reactors [25], and reaction-diffusion systems [26], among others.

In this paper, a robust adaptive nonlinear observer $\mathcal{H}_\infty$ for a class of Lipschitz nonlinear systems is presented. The behavior of the observer in the presence of disturbances is analyzed using Lyapunov stability theory and with an $\mathcal{H}_\infty$ strategy successfully employed in previous approaches, as in [27,28]. This work distinguishes itself through several key advancements compared to prior research: (i) It extends the applicability of the methodology to a broader class of nonlinear processes with uncertain models affected by unknown inputs or disturbances, facilitating the estimation of both process variables and unknown parameters; (ii) by incorporating the H_∞ criteria into the design, the observer demonstrates enhanced resilience against undesired disturbances, ensuring a more robust performance; (iii) the simplicity of computing observer gains, eliminating the necessity to solve additional differential equations typically associated with Kalman observers (or filters as presented in [29,30]); (iv) unlike high-gain observers, there is no need for a coordinate transformation in the observer design process, streamlining the implementation and reducing complexity.

From a theoretical perspective, prior research has not specifically addressed adaptive observers for nonlinear Lipschitz systems with unknown parameters, particularly those affected by disturbances. This study employs an $\mathcal{H}_\infty$ approach to attenuate the impact of these unknown disturbances. These important results are summarized in Theorem 1. The applicability of the proposed approach is demonstrated in the performance monitoring of the semi-active suspension of a car.

2. Preliminaries

2.1. Notation

In this article, I_n and 0_n denote the n-dimensional identity and zero matrices, respectively, $\|\cdot\|$ and $\|\cdot\|_{\mathcal{L}_2}$ denote the Euclidean and the $\mathcal{L}_2$ norm, respectively, i.e.,

$$\|x\|_2 = (|x_1|^2 + \cdots + |x_n|^2)^{\frac{1}{2}} = (x^T x)^{\frac{1}{2}}$$

and

$$\|\eta\|_{\mathcal{L}_2} = \sqrt{\int_0^\infty \eta^T(t)\eta(t)dt} < \infty$$

$\mathcal{L}_2$ is the space of piecewise continuous, square-integrable functions. $S > 0$ is a symmetric positive definite matrix, whereas $S \geq 0$ is a symmetric positive semi-definite matrix. $T < 0$ is a symmetric negative definite matrix, whereas $T \leq 0$ is a symmetric negative semi-definite matrix. A variable with a hat $\hat{x}$ denotes the estimated value of x. A^T and A^{-1} denote the transpose and inverse of matrix A, respectively.

sup is the supremum of a set, i.e., the least upper bound in a set, for example

$$\sup\{x \in \mathbb{R} | 0 < x < 10\} = \sup\{x \in \mathbb{R} | 0 \leq x \leq 10\} = 10$$

min is the smallest value of a set, for example

$$\min\{-5 \leq x \leq 5\} = 5$$

$C^\perp$ denotes the orthogonal projection on to $null(C)$, the kernel or null space of matrix C.

2.2. Problem Formulation

Consider the following nonlinear system:

$$\mathcal{S} : \begin{cases} \dot{x}(t) = Ax(t) + \Psi(y,u) + \Phi(x,\theta,u) + N\eta(t) \\ y(t) = Cx(t) \end{cases} \tag{1}$$

with

$$\Phi(x,\theta,u) = \Phi_1(x,u) + B\Phi_2(x,u)\theta(t) \tag{2}$$

where $x(t) \in \mathbb{R}^n$ is the state vector, $\theta(t) \in \mathbb{R}^q$ is the unknown parameter vector, $u(t) \in \mathbb{R}^m$ is the input, $y(t) \in \mathbb{R}^p$ is the output of the system, $\eta(t) \in \mathbb{R}^r$ is a bounded disturbance vector; $\Phi(x,\theta,u) \in \mathbb{R}^n$ is a nonlinear function depending on states $x(t)$, unknown parameters $\theta(t)$, and inputs $u(t)$. This function can be decomposed as is shown in Equation (2). $\Psi(y,u) \in \mathbb{R}^n$ is a nonlinear function depending on outputs and inputs. Finally, $A \in \mathbb{R}^{n\times n}$, $B \in \mathbb{R}^{n\times l}$, $C \in \mathbb{R}^{p\times n}$, and $N \in \mathbb{R}^{n\times l}$ are constant matrices of appropriate dimensions.

Assumption 1. *Inputs $u(t)$, outputs $y(t)$, the parameter vector $\theta(t)$, and the disturbance $\eta(t)$ are assumed to be bounded.*

Assumption 1 implies that controlled variables u, measurements y, and parameters θ are limited by the actuators, sensors, or physical limitations of the process.

Assumption 2. *It is assumed that the nonlinear function $\Phi(x,\theta,u)$ satisfies the Lipschitz condition with respect to state variables for bounded values of $u(t)$ and $\theta(t)$, i.e.,*

$$\|\Phi(x,\theta,u) - \Phi(\hat{x},\theta,u)\| \leq \gamma \|(x - \hat{x})\| \tag{3}$$

where γ is the Lipschitz constant of function Φ.

As described previously, the nonlinear function $\Phi(x,\theta,u)$ can be decomposed into two terms: $\Phi_1(x,u)$ and $B\Phi_2(x,u)\theta(t)$, where the second term is affine to the parameter vector $\theta(t)$.

Assumption 3. *Functions $\Phi_1(x,u)$ and $\Phi_2(x,u)$ are also Lipschitz functions with regards to $x(t)$ and bounded inputs $u(t)$.*

Assumptions 2 and 3 imply that the dynamics of a real system can be represented by differential equations involving uniform continuity (Lipschitz) functions. This property guarantees the existence and uniqueness of the solution of differential equations to an initial value problem. Indeed, this is the key feature to exploit in the design process.

Lemma 1 ([31]). *Let $\mathcal{M}$ and $\mathcal{N}$ be two constant matrices of appropriate dimensions. Then, the following inequality*

$$\mathcal{M}^T\mathcal{N} + \mathcal{N}^T\mathcal{M} \leq \alpha\mathcal{M}^T\mathcal{M} + \frac{1}{\alpha}\mathcal{N}^T\mathcal{N}$$

holds for any scalar $\alpha > 0$.

Consider now the following adaptive nonlinear observer:

$$\mathcal{O}: \begin{cases} \dot{\hat{x}}(t) = A\hat{x}(t) + \Psi(y,u) + \Phi_1(\hat{x},u) + B\Phi_2(\hat{x},u)\hat{\theta}(t) + L(y(t) - C\hat{x}(t)) \\ \dot{\hat{\theta}}(t) = \Gamma\Phi_2^T(\hat{x},u)H(y(t) - C\hat{x}(t)), \text{ with } \Gamma > 0, \\ \hat{y}(t) = C\hat{x}(t) \end{cases} \tag{4}$$

where $\hat{x}(t)$ is the estimate of the state vector, $\hat{\theta}(t)$ is the parameter estimation vector, and $\Gamma \in \mathbb{R}^{q\times q}$ is a positive definite matrix. Matrices L and H must be selected in such a way that the convergence of the observer is guaranteed.

Consider the following errors

$$e_x(t) = x(t) - \hat{x}(t) \tag{5}$$

$$e_\theta(t) = \theta(t) - \hat{\theta}(t) \tag{6}$$

where $e_x(t)$ represents the state estimation error, and $e_\theta(t)$ represents the parameter estimation error.

The derivate of Equation (5) is

$$\dot{e}_x(t) = \dot{x}(t) - \dot{\hat{x}}(t) \tag{7}$$

$$\begin{aligned} &= Ax + \Psi(y,u) + \Phi_1(x,u) + B\Phi_2(x,u)\theta + N\eta - \\ &\quad A\hat{x} - \Psi(y,u) - \Phi_1(\hat{x},u) - B\Phi_2(\hat{x},u)\hat{\theta} - L(y - C\hat{x}) \end{aligned} \tag{8}$$

By adding and subtracting the term $B\Phi_2(\hat{x},u)\theta(t)$, we obtain

$$\dot{e}_x(t) = (A - LC)e_x(t) + \Phi_1(x,u) + B\Phi_2(x,u)\theta(t) - \Phi_1(\hat{x},u) - B\Phi_2(\hat{x},u)\theta(t) + B\Phi_2(\hat{x},u)\theta(t) - B\Phi_2(\hat{x},u)\hat{\theta}(t) + N\eta(t) \quad (9)$$

By taking into account the consideration marked in a box as

$$e_\Phi(t) = \Phi_1(x,u) + B\Phi_2(x,u)\theta(t) - \Phi_1(\hat{x},u) - B\Phi_2(\hat{x},u)\theta(t) \quad (10)$$
$$= \Phi(x,\theta,u) - \Phi(\hat{x},\theta,u) \quad (11)$$

Equation (9) becomes

$$\dot{e}_x(t) = (A - LC)e_x(t) + e_\Phi(t) + B\Phi_2(\hat{x},u)e_\theta(t) + N\eta(t). \quad (12)$$

By considering that $\theta(t)$ is a constant parameter, i.e., $\dot{\theta}(t) = 0$, then

$$\dot{e}_\theta(t) = \dot{\theta}(t) - \dot{\hat{\theta}}(t)$$
$$= -\Gamma\Phi_2(\hat{x},u)^T HCe_x(t) \quad (13)$$

Considering the Lipschitz condition of Equation (3), presented in [32], a condition is proposed that ensures the stability of the observer:

$$e_\Phi^T(t)Qe_\Phi(t) \leq e_x^T(t)Re_x(t) \quad (14)$$

where Q and R are two positive definite symmetric matrices.

Equations (12) and (13) are written in matrix form as

$$\underbrace{\begin{bmatrix}\dot{e}_x(t)\\ \dot{e}_\theta(t)\end{bmatrix}}_{\dot{\delta}(t)} \& = \underbrace{\begin{bmatrix}A - LC & B\Phi_2(\hat{x},u)\\ -\Gamma\Phi_2(\hat{x},u)^T HC & 0_q\end{bmatrix}}_{\mathbb{A}} \underbrace{\begin{bmatrix}e_x(t)\\ e_\theta(t)\end{bmatrix}}_{\delta(t)} + \underbrace{\begin{bmatrix}I_n\\ 0_{q\times n}\end{bmatrix}}_{\mathbb{B}} e_\Phi(t) + \underbrace{\begin{bmatrix}N\\ 0_{q\times l}\end{bmatrix}}_{\mathbb{N}} \eta(t) \quad (15)$$

and from Equations (1) and (4) we obtain

$$r(t) = Cx(t) - C\hat{x}(t)$$
$$= \underbrace{\begin{bmatrix}C & 0_{p\times q}\end{bmatrix}}_{\mathbb{C}} \underbrace{\begin{bmatrix}e_x(t)\\ e_\theta(t)\end{bmatrix}}_{\delta(t)} \quad (16)$$

where $r(t) = y(t) - \hat{y}(t)$ is the output error estimation.

The problem is to propose an adaptive observer for the class of Lipschitz nonlinear systems given in Equations (1) and (2), in order to simultaneously estimate the process variables $x(t)$ and the parameter vector $\theta(t)$, and so that the worst case estimation error energy over all bounded energy disturbances $\eta(t)$ is minimized, i.e.,

1. for $\eta(t) = 0$, the errors $e_x(t) = x(t) - \hat{x}(t)$ and $e_\theta(t) = \theta(t) - \hat{\theta}(t)$ converge asymptotically to zero.
2. for $\eta \neq 0$ we solve the min $\sup_{\eta\in\mathcal{L}_2-\{0\}} \frac{||r(t)||_{\mathcal{L}_2}}{||\eta(t)||_{\mathcal{L}_2}}$.

3. H_∞ Adaptive Observer Design

In this section, the H_∞ observer design is presented. The following theorem gives the sufficient conditions for Equation (15) to be stable and $||r(t)||_{\mathcal{L}_2} < \lambda||\eta(t)||_{\mathcal{L}_2}$ for $\eta(t) \neq 0$.

Stability of the Observer

This section is devoted to the stability analysis of Equation (15). The following theorem gives the conditions for the stability in a set of LMIs.

Theorem 1. *There exists an observer having the form given in Equation* (4) *for the nonlinear system* (1) *such that the dynamic error of Equation* (15) *is stable and* $||r(t)||_{\mathcal{L}_2} < \beta||\eta(t)||_{\mathcal{L}_2}$*, if there exists positive definite matrices P, R, and Q, and positive scalar* β *such that the following LMI is satisfied:*

$$\begin{bmatrix} PA - SC + A^T P - C^T S^T + C^T C + R & PN & P \\ N^T P & -\bar{\beta} I_{n\times l} & 0_n \\ P & 0_{n\times l} & Q \end{bmatrix} \leq 0 \quad (17)$$

where the observer gain L is solved as $L = P^{-1}S$*, and the observer matrix H is obtained as* $H = B^T P C^{-1}$.

Proof. Consider the following Lyapunov candidate function:

$$V(t) = \delta^T(t) X \delta(t) > 0 \quad (18)$$

where

$$X = \begin{bmatrix} P & 0_{n\times q} \\ 0_{q\times n} & \Gamma^{-1} \end{bmatrix} > 0 \quad (19)$$

the derivative of $V(t)$ along the solution of (15) is given by

$$\dot{V}(t) = \dot{\delta}(t)^T X \delta(t) + \delta(t)^T X \dot{\delta}(t) \quad (20)$$

$$= \delta^T(t)(\mathbb{A}^T X + X\mathbb{A})\delta(t) + \delta^T(t) X \mathbb{N} \eta(t) + \eta^T(t) \mathbb{N}^T X \delta(t) + \delta^T(t) X \mathbb{B} e_\Phi(t) + e_\Phi^T(t) \mathbb{B} X \delta(t) \quad (21)$$

by replacing matrices $\mathbb{A}$, $\mathbb{B}$, $\mathbb{N}$ from Equation (15) and X from Equation (19) then

$$\dot{V}(t) = e_x^T(t) P(A - LC) e_x(t) + e_x^T(t) P e_\Phi(t) + e_x^T(t) P B \Phi_2(\hat{x}, u) e_\theta + e_x^T(t) P N \eta(t) + e_x^T(t)(A - LC)^T P e_x(t) + e_\Phi^T(t) P e_x(t) + e_\theta^T(t) \Phi_2(\hat{x}, u)^T B^T P e_x(t) + \eta^T(t) N^T P e_x(t) - e_\theta^T(t) \Phi_2(\hat{x}, u)^T H C e_x(t) - e_x^T(t) C^T H^T \Phi_2(\hat{x}, u) e_\theta(t) \quad (22)$$

Note that if the equality $B^T P C^\perp = 0$ is satisfied, this implies that there exists matrices H and L, such that $B^T P = HC$ [33], where $C^\perp$ represents an orthogonal projection onto $null(C)$. With this consideration, the above inequality can be simplified as follows:

$$\dot{V}(t) = e_x^T(t) P(A - LC) e_x(t) + e_x^T(t) P e_\Phi(t) + e_x^T(t) P N \eta(t) + e_x^T(t)(A - LC)^T P e_x(t) + e_\Phi^T(t) P e_x(t) + \eta^T(t) N^T P e_x(t)$$

$$= e_x^T(t)[(A - LC)^T P + P(A - LC)] e_x(t) + 2 e_x^T(t) P e_\Phi(t) + e_x^T(t) P N \eta(t) + \eta^T(t) N^T P e_x(t) \quad (23)$$

There exists an scalar $\beta > 0$ such that

$$\dot{V}(t) < \beta^2 \eta^T(t) \eta(t) - r^T(t) r(t) \quad (24)$$

by integrating the two sides of this inequality we obtain

$$\int_0^\infty \dot{V}(\tau) d\tau < \int_0^\infty \beta^2 \eta^T(\tau) \eta(\tau) d\tau - \int_0^\infty r^T(\tau) r(\tau) d\tau$$

or equivalently $V(\infty) - V(0) < \beta^2 ||\eta(t)||_2^2 - ||r(t)||_2^2$. Under zero initial conditions, we obtain

$$V(\infty) < \beta^2 ||\eta(t)||_2^2 - ||r(t)||_2^2$$

which leads to $||r(t)||_2^2 < \beta^2||\eta(t)||_2^2$. From Equation (24), we can deduce

$$\dot{V}(t) + r^T(t)r(t) - \beta^2\eta^T(t)\eta(t) < 0 \tag{25}$$

By replacing $\dot{V}(t)$ from Equation (23) and $r(t)$ from Equation (16), we obtain

$$\begin{aligned} &e_x^T(t)[(A-LC)^TP + P(A-LC)]e_x(t) + 2e_x^T(t)Pe_\Phi(t) \\ &+ e_x^T(t)PN\eta(t) + \eta^T(t)N^TPe_x(t) + e_x^T(t)C^TCe_x(t) - \beta^2\eta^T(t)\eta(t) < 0 \end{aligned} \tag{26}$$

By applying the following equivalence in the framed term, we obtain

$$2e_x^T(t)Pe_\Phi(t) = 2e_x^T(t)PQ^{-1/2}Q^{1/2}e_\Phi(t) \tag{27}$$

By using Lemma 1, we can obtain the following inequality from Equation (27):

$$2e_x^T(t)PQ^{-1/2}Q^{1/2}e_\Phi(t) \leq e_x^T(t)PQ^{-1}Pe_x(t) + e_\Phi^T(t)Qe_\Phi(t) \tag{28}$$

Now, by using the condition given in Equation (14) in the framed expression, we obtain the following inequality from (26)

$$\begin{aligned} &e_x^T(t)[(A-LC)^TP + P(A-LC)]e_x(t) + e_x^T(t)PQ^{-1}Pe_x(t) + e_x^T(t)Re_x(t) \\ &+ e_x^TPN(t)\eta(t) + \eta^T(t)N^TPe_x(t) + e_x^T(t)C^TCe_x(t) - \beta^2\eta^T(t)\eta(t) \leq 0 \end{aligned} \tag{29}$$

This can be written in matrix form as

$$\begin{bmatrix} e_x(t) \\ \eta(t) \end{bmatrix}^T \Omega \begin{bmatrix} e_x(t) \\ \eta(t) \end{bmatrix} \leq 0 \tag{30}$$

where

$$\Omega = \begin{bmatrix} P(A-LC) + (A-LC)^TP + C^TC + PQ^{-1}P + R & PN \\ N^TP & -\beta^2 I_{n\times l} \end{bmatrix}.$$

If $\Omega \leq 0$, the index performance given in (25) is verified. By using the Schur complement, we obtain

$$\Omega = \begin{bmatrix} P(A-LC) + (A-LC)^TP + C^TC + R & PN & P \\ N^TP & -\overline{\beta} I_{n\times l} & 0_n \\ P & 0_{n\times l} & Q \end{bmatrix} \leq 0$$

where $\overline{\beta} = \beta^2$. By simplifying the therm $S = PL$, we obtain

$$\Omega = \begin{bmatrix} PA - SC + A^TP - C^TS^T + C^TC + R & PN & P \\ N^TP & -\overline{\beta} I_{n\times l} & 0_n \\ P & 0_{n\times l} & Q \end{bmatrix} \leq 0 \tag{31}$$

By solving the LMI (31), the observer gains L and H can be easily obtained, as stated in the theorem. □

4. Application to a Semi-Active Automotive Suspension

A semi-active suspension composed by a magnetorheological (MR) damper is represented in Figure 1. The system is represented by the following mathematical model [34]:

$$m_s\ddot{z}_s(t) = -k_s(z_s(t) - z_{us}(t)) - F_{MR}(t) \tag{32}$$
$$m_{us}\ddot{z}_{us}(t) = k_s(z_s(t) - z_{us}(t)) - k_t(z_{us}(t) - z_r(t)) + F_{MR}(t) \tag{33}$$

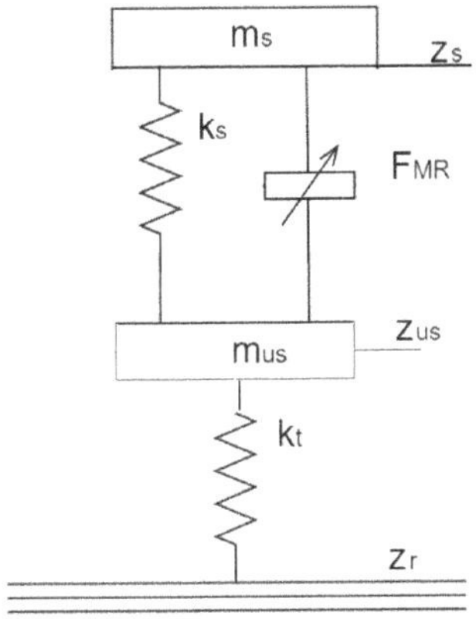

Figure 1. Semi-active suspension diagram.

The semi-active damping force ($F_{MR}(t)$), with the inclusion of a manipulation signal (electric current) is represented as follows:

$$F_{MR}(t) = If_c\rho(t) + b_1\dot{z}_{def}(t) + b_2 z_{def}(t) + \eta(t) \tag{34}$$

where ρ is the nonlinear part representing the hysteresis of the force provided by the magnetorheological damper [35]. Such non-linearity is described by:

$$\rho(t) = tanh(a_1\dot{z}_{def}(t) + a_2 z_{def}(t)) \tag{35}$$

The nomenclature of the parameters and variables of the model is described in Table 1.

Table 1. List of parameters and variables of the mathematical model.

Parameter	Description	Value	Units
a_1, a_2	Pre-effort zone of F_{MR}	37.8, 22.15	(Ns)/m
b_1, b_2	Post-effort zone of F_{MR}	2830.86, −7897.21	(Ns)/m
f_c	Damping force	600.95	N/A
I	Electric current	2	A
k_s	Spring stiffness coefficient	86,378	N/m
k_t	Tire stiffness coefficient	260,000	N/m
m_s, m_{us}	Suspended mass and Unsprung (tire) mass	470, 110	kg
Variable	Description	Role	Units
z_{def}	Vertical damper position	output y_3	m
$\dot{z}_{def}$	Vertical damper speed	$\dot{y}_3$	m/s
z_r	Road profile	input	m
z_s, z_{us}	Vertical displacement of m_s, m_{us}	outputs y_1 and y_2	m
$\dot{z}_s, \dot{z}_{us}$	Vertical speed of m_s, m_{us}	state x_2 and x_4	m/s
$\ddot{z}_s, \ddot{z}_{us}$	Vertical acceleration of m_s, m_{us}	$\dot{x}_2$ and $\dot{x}_4$	m^2/s
ρ	Shock absorber hysteresis	nonlinear function	
F_{MR}	Force MR	damping force	N

The measured outputs are $y_1(t) = z_s(t)$, $y_2(t) = z_{us}(t)$ and $y_3(t) = z_s(t) - z_{us}(t) = z_{def}(t)$. In addition, consider the following change of variables: $x_1(t) = z_s(t)$, $x_2(t) = \dot{z}_s(t) = \dot{x}_1(t)$, $\dot{x}_2(t) = \ddot{z}_s(t)$, $x_3(t) = z_{us}(t)$, $x_4(t) = \dot{z}_{us}(t) = \dot{x}_3(t)$ and $\dot{x}_4(t) = \ddot{z}_{us}(t)$.

The model given by Equations (32) and (33) can be rewritten as follows:

$$\begin{aligned}
\dot{x}_1(t) &= x_2(t) \\
\dot{x}_2(t) &= -\frac{b_2+k_s}{m_s}x_1(t) - \frac{b_1}{m_s}x_2(t) + \frac{b_2+k_s}{m_s}x_3(t) \\
&\quad + \frac{b_1}{m_s}x_4(t) - \frac{f_c\rho}{m_s}I(t) - \frac{1}{m_s}\eta(t) \\
\dot{x}_3(t) &= x_4(t) \\
\dot{x}_4(t) &= \frac{b_2+k_s}{m_{us}}x_1(t) + \frac{b_1}{m_{us}}x_2(t) - \frac{b_2+k_s+k_t}{m_{us}}x_3(t) \\
&\quad - \frac{b_1}{m_{us}}x_4(t) + \frac{f_c\rho}{m_{us}}I(t) + \frac{1}{m_{us}}\eta(t) \\
&\quad + \frac{k_t}{m_{us}}z_r(t)
\end{aligned} \tag{36}$$

The state, the output, and the input vectors are $x(t) = [x_1(t)\ x_2(t)\ x_3(t)\ x_4(t)]^T$, $y(t) = [y_1(t)\ y_2(t)\ y_3(t)]^T$, $u(t) = [\eta(t)\ I(t)\ z_r(t)]^T$, where $x_1(t)$ and $x_2(t)$ are the vertical chassis position and the vertical chassis speed, $x_3(t)$ and $x_4(t)$ are the vertical tire position and the vertical tire speed, $y_1(t)$ is the vertical chassis position, $y_2(t)$ is the vertical tire position and $y_3(t)$ is the vertical damper position, $z_r(t)$ is the road profile, and $\eta(t)$ represents a disturbance in the damper. This disturbance occurs when driving on a road with potholes and bumps, as well as due to excess luggage or passengers getting into the car. A damaged shock absorber causes an imbalance in the chassis and increases the undesirable pitching and rolling motion of the car.

Equation (36) can be represented in the form of system (1):

$$\begin{aligned}
\underbrace{\begin{bmatrix}\dot{x}_1\\ \dot{x}_2\\ \dot{x}_3\\ \dot{x}_4\end{bmatrix}}_{\dot{x}} &= \underbrace{\begin{bmatrix} 0 & 1 & 0 & 0 \\ -\frac{b_2+k_s}{m_s} & -\frac{b_1}{m_s} & \frac{b_2+k_s}{m_s} & \frac{b_1}{m_s} \\ 0 & 0 & 0 & 1 \\ \frac{b_2+k_s}{m_{us}} & \frac{b_1}{m_{us}} & -\frac{b_2+k_s}{m_{us}} & -\frac{b_1}{m_{us}} \end{bmatrix}}_{A} \underbrace{\begin{bmatrix}x_1\\ x_2\\ x_3\\ x_4\end{bmatrix}}_{x} \\
&+ \underbrace{\begin{bmatrix} 0 \\ -\frac{f_c\rho}{m_s}I \\ 0 \\ \frac{f_c\rho}{m_{us}}I \end{bmatrix}}_{\Psi(y,u)} + \underbrace{\underbrace{\begin{bmatrix}0\\0\\0\\1\end{bmatrix}}_{B} \underbrace{\begin{bmatrix}\frac{z_r - x_3}{m_{us}}\end{bmatrix}}_{\Phi_2(x_3,u)} \underbrace{\theta}_{k_t}}_{\Phi(x,u,\theta)} + \underbrace{\begin{bmatrix} 0 \\ -\frac{1}{m_s} \\ 0 \\ \frac{1}{m_{us}} \end{bmatrix}}_{N} \eta
\end{aligned} \tag{37}$$

$$\underbrace{\begin{bmatrix}y_1\\ y_2\\ y_3\end{bmatrix}}_{y} = \underbrace{\begin{bmatrix} 1 & 0 & 0 & 0 \\ 0 & 0 & 1 & 0 \\ 1 & 0 & -1 & 0 \end{bmatrix}}_{C} \underbrace{\begin{bmatrix}x_1\\ x_2\\ x_3\\ x_4\end{bmatrix}}_{x} \tag{38}$$

It can be seen in Equation (38) that function $\Phi(x,u,\theta)$ is:

$$\Phi(x,u,\theta) = \begin{bmatrix} \phi_1(x,u,\theta) \\ \phi_2(x,u,\theta) \\ \phi_3(x,u,\theta) \\ \phi_4(x,u,\theta) \end{bmatrix} = \begin{bmatrix} 0 \\ 0 \\ 0 \\ \frac{z_r - x_3}{m_{us}}\theta \end{bmatrix}$$

In order to verify Assumption 2, the Lipschitz constant γ of function $\Phi(x,u,\theta)$ is computed as follows (see Lemma 3.1 in [36]):

$$\gamma = \left\| \frac{\partial \Phi(x,u,\theta)}{\partial x} \right\|_1 = \left\| \begin{matrix} 0 & 0 & 0 & 0 \\ 0 & 0 & 0 & 0 \\ 0 & 0 & 0 & 0 \\ 0 & 0 & -\frac{1}{m_{us}} & 0 \end{matrix} \right\|_1 = \frac{1}{m_{us}}$$

The nonlinear function $\Phi(x,u,\theta)$ satisfies the Lipschitz condition with respect to the state variables [7]. Assumption 3 is also verified. The Lipschitz constant of function $\Phi_2(x_3,u) \in \mathbb{R}$ is the same as the Lipschitz constant of function $\Phi(x,u,\theta)$, i.e.,:

$$\left\| \frac{\partial \Phi_2(x_3,u)}{\partial x_3} \right\|_1 = \frac{1}{m_{us}}$$

Therefore, the observer (4) is used to simultaneously estimate the state variables and the unknown parameter θ.

By considering that the parameter to be estimated is the spring stiffness coefficient k_t, then the observer (4) for system (37) is

$$\underbrace{\begin{bmatrix} \dot{\hat{x}}_1 \\ \dot{\hat{x}}_2 \\ \dot{\hat{x}}_3 \\ \dot{\hat{x}}_4 \end{bmatrix}}_{\dot{\hat{x}}} = \underbrace{\begin{bmatrix} 0 & 1 & 0 & 0 \\ -\frac{b_2+k_s}{m_s} & -\frac{b_1}{m_s} & \frac{b_2+k_s}{m_s} & \frac{b_1}{m_s} \\ 0 & 0 & 0 & 1 \\ \frac{b_2+k_s}{m_{us}} & \frac{b_1}{m_{us}} & -\frac{b_2+k_s}{m_{us}} & -\frac{b_1}{m_{us}} \end{bmatrix}}_{A} \underbrace{\begin{bmatrix} \hat{x}_1 \\ \hat{x}_2 \\ \hat{x}_3 \\ \hat{x}_4 \end{bmatrix}}_{\hat{x}}$$

$$+ \underbrace{\begin{bmatrix} 0 \\ -\frac{f_c\,\rho}{m_s} I \\ 0 \\ \frac{f_c\,\rho}{m_{us}} I \end{bmatrix}}_{\Psi(y,u)} + \underbrace{\underbrace{\begin{bmatrix} 0 \\ 0 \\ 0 \\ 1 \end{bmatrix}}_{B} \underbrace{\left[\frac{z_r - \hat{x}_3}{m_{us}} \right]}_{\Phi_2(\hat{x}_3,u)} \underbrace{\hat{\theta}}_{k_t}}_{\Phi(\hat{x},u,\theta)} + L(y - C\hat{x})$$

$$\dot{\hat{\theta}} = \Gamma \left[\frac{z_r - \hat{x}_3}{m_{us}} \right]^T H(y - C\hat{x})$$

The matrices L and H are the gains of the observer, and they must be selected to guarantee the estimation convergence of the estimated states and parameters. The gain $\Gamma > 0$ is a positive scalar. In this case, $\Gamma = 125$ is chosen, because this value allows an adequate convergence time for the observer.

5. Simulation Results

To evaluate the performance of the proposed observer, the behavior of the suspension was analyzed when a disturbance occurs in the damping force $\eta(t)$, affecting the position of the piston, causing poor vehicle comfort, and risk of rollover due to disturbances on the road $z_r(t)$. The parameters presented in Table 1 were considered to estimate the unknown parameter k_t, which represents the stiffness of the tire. The simulation was implemented using MATLAB, with a simulation time of 65 s. This time frame was chosen to ensure the system stabilized adequately before any subsequent disturbances or inputs could affect it again. The first-order Euler method was used to integrate the differential equations, with an integration step of 1ms. The initial conditions of the system and the observer were $x(0) = [0\ 0\ 0\ 0]^T$ and $\hat{x}(0) = [0.1\ 0.1\ 0.1\ 0.1]^T$. The electric current was $I(t) = 2$ A.

A road profile was assumed starting as a straight path, and then it passed through two consecutive speed bumps and finally it continued with its path $z_r(t)$ as shown in Figure 2. This road profile was considered as an input $z_r(t)$. It can be appreciated that each bump on the road exerted a vertical force on the vehicle during 3 s, affecting the vertical positions x_1, x_3 and z_{def}.

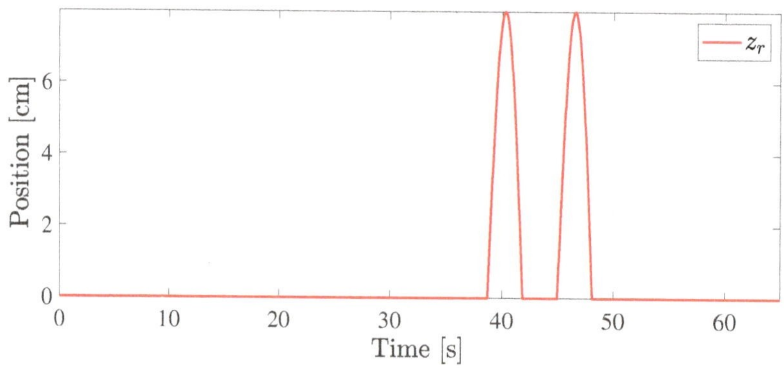

Figure 2. Road profile z_r: a straight path and two speed bumps.

The system had a disturbance in the actuator $\eta(t)$. The disturbance profile, shown in Figure 3, corresponds to around 15% of the shock absorber's operating range, thereby affecting the comfort and safety of passengers. This disturbance influences the position of the shock absorber piston gradually, diminishing its ability to dampen the vehicle oscillations resulting from road irregularities.

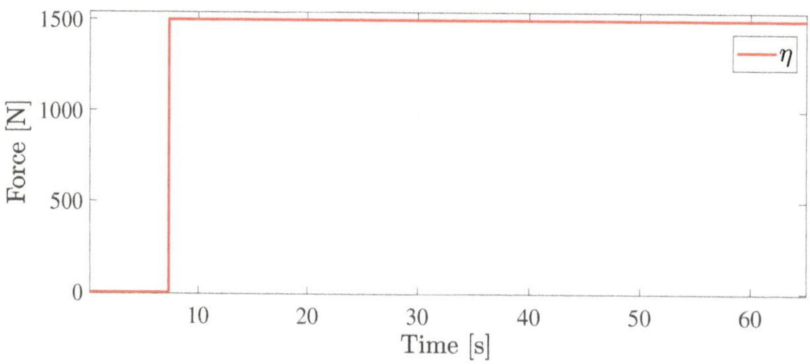

Figure 3. Disturbance η affecting the semi-active damping force $F_{MR}(t)$ (see Equation (34)).

The observer gains were obtained by solving the LMI presented in Equation (17) using the MATLAB toolbox YALMIP:

$$L = \begin{bmatrix} -0.9627 & -0.9677 & 0.0050 \\ 1.3959 & -1.0688 & 2.4647 \\ -0.6003 & -0.5931 & -0.0072 \\ -1.0688 & 1.3799 & -2.4487 \end{bmatrix} \tag{39}$$

$$H = \begin{bmatrix} -0.3696 & -0.3708 & 0.0012 \end{bmatrix} \tag{40}$$

As shown in Figure 4, the unknown parameter $\theta = k_t$ was adequately estimated with appropriate time convergence.

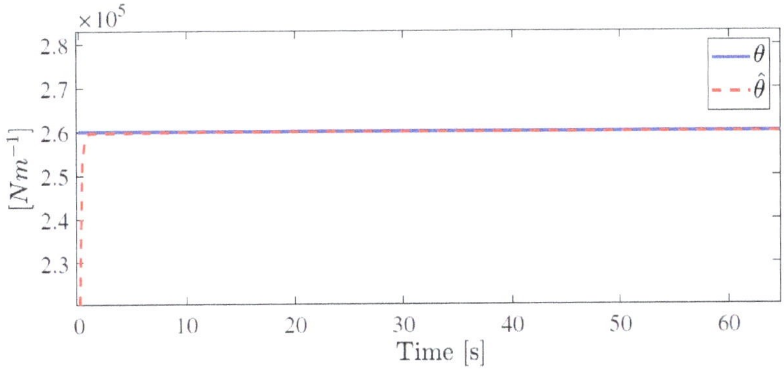

Figure 4. Simulated wheel stiffness coefficient θ (solid line) and its estimation value $\hat{\theta}$ (dotted line).

Once the parameter k_t had been estimated, the observer was able to estimate the position of the chassis x_1 and its estimated value $\hat{x}_1$ (Figure 5). The effect of the disturbance η was observed at $t = 8$ s. This harmed the comfort and safety of passengers. The observer attenuated the effect of the disturbance by minimizing the oscillation of $\hat{x}_1$. It can be seen that the disturbance caused an alteration in the behavior of the shock absorber when the vehicle went over speed bumps on the road.

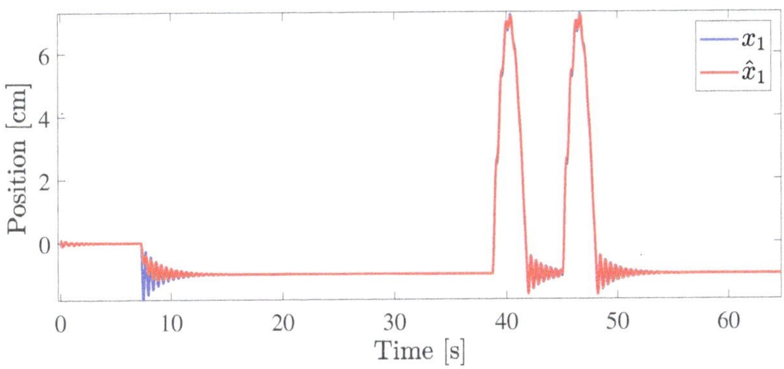

Figure 5. Chassis vertical position x_1 (blue line) and its estimated value $\hat{x}_1$ (red line).

The position of the tire x_3 and its estimated value $\hat{x}_3$ are shown in Figure 6.

It can be seen that the effect of the disturbance η on the damper caused oscillations, which forced the tire to follow the path over speed bumps on the road. Once again, the observer attenuated the oscillation of $\hat{x}_3$, obtaining an adequate estimation of the output despite the presence of the disturbance.

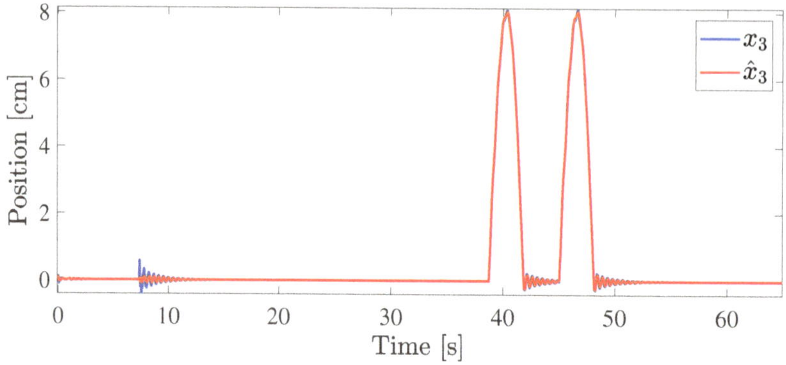

Figure 6. Tire vertical position x_3 (blue line) and its estimated value $\hat{x}_3$ (red line).

Knowing the positions of the chassis and the tire, the vertical position of the shock absorber could be calculated (as seen in Figure 7). One can observe the effect of the disturbance on the actuator when the tire passed over the road profile, allowing us to minimize the disturbance to our observer.

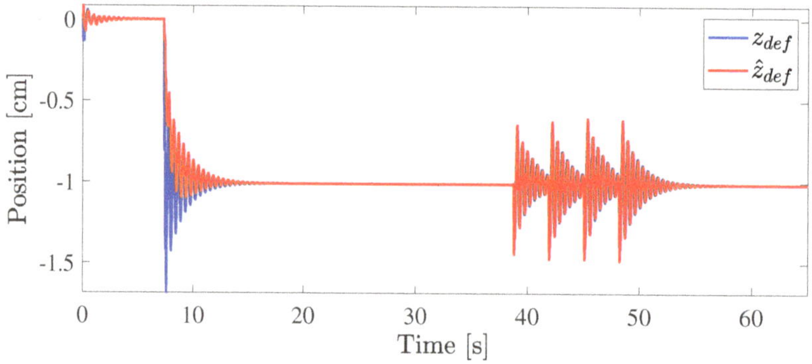

Figure 7. Vertical damper position z_{def} (blue line) and its estimated value $\hat{z}_{def}$ (red line).

The behavior of the suspension was affected by the disturbance η. The greater the force that disturbed the shock absorber, the greater its deformation. The observer estimated the tire's stiffness coefficient k_t to monitor the deterioration of the tire and managed to attenuate the disturbance.

6. Conclusions

An $\mathcal{H}_\infty$ adaptive observer was presented for processes that can be modeled as nonlinear Lipschitz systems. The proposed conditions under a passivity constraint were employed to deal with nonlinear systems with certain unknown parameters. The proposed observer was able to simultaneously estimate unknown states and parameters, even in the presence of disturbances. The main advantage of this observer is that it can be applied to a wider class of systems with unknown parameters. Moreover, by incorporating the $\mathcal{H}_\infty$ criteria into the design, the observer demonstrated enhanced resilience against undesired disturbances, while ensuring robust performance. Unlike high-gain observers, the observer design process does not necessitate a coordinate transformation, thereby streamlining implementation and reducing complexity. A semi-active car suspension was used to test the performance of the proposed observer. Thanks to the $\mathcal{H}_\infty$ approach, the effect of disturbances or unknown inputs could be attenuated, allowing for better monitoring of the systems. The simplicity of computing the observer gains was demonstrated, eliminating the need to solve the additional differential equations usually associated with Kalman observers. It is well known that Kalman observers (or filters as presented in [29,30]) require additional differential equations to recursively compute the observer gain, incorporating the predicted covariance matrix to estimate the accuracy of state estimates (e.g., [37]). In contrast, the proposed observer employs fixed-value observer gains, which are computed offline once, by solving the LMIs provided in Theorem 1. The simulation results demonstrated the effectiveness of the proposed approach in dealing with a practical system. As future work, we expect to apply the $\mathcal{H}_\infty$ approach at the output to supervise the operation of dynamic systems in the presence of sensor disturbances.

Future work will focus on developing adaptive observers for non-Lipschitz nonlinear systems, such as those involving dry friction. This approach aims to broaden the scope and address more realistic scenarios.

Author Contributions: Conceptualization, P.E.A.-M. and G.L.O.-G.; methodology, P.E.A.-M., G.L.O.-G. and C.M.A.-Z.; software, P.E.A.-M., G.L.O.-G., R.V.-M. and J.R.-R.; validation, P.E.A.-M., G.L.O.-G. and R.V.-M.; formal analysis, P.E.A.-M., G.L.O.-G. and A.A.-G.; investigation, P.E.A.-M., G.L.O.-G. and C.M.A.-Z.; writing—original draft preparation, P.E.A.-M., G.L.O.-G. and R.V.-M.; writing—review and editing, P.E.A.-M., G.L.O.-G., R.V.-M., A.A.-G. and J.R.-R.; supervision, G.L.O.-G., A.A.-G. and C.M.A.-Z. All authors have read and agreed to the published version of the manuscript.

Funding: This research received no external funding.

Data Availability Statement: Data are contained within the article.

Acknowledgments: The authors acknowledge CONAHCYT for supporting Pedro Eusebio Alvarado Méndez through a Ph.D. Scholarship.

Conflicts of Interest: The authors declare no conflicts of interest.

References

1. Thau, F. Observing the state of non-linear dynamic systems. *Int. J. Control* **1973**, *17*, 471–479. [CrossRef]
2. Zemouche, A.; Rajamani, R.; Trinh, H.; Zasadzinski, M. A new LMI based $\mathcal{H}_\infty$ observer design method for Lipschitz nonlinear systems. In Proceedings of the 2016 European Control Conference (ECC), Aalborg, Denmark, 29 June–1 July 2016; pp. 2011–2016.
3. Zemouche, A.; Rajamani, R.; Kheloufi, H.; Bedouhene, F. Robust observer-based stabilization of Lipschitz nonlinear uncertain systems via LMIs-discussions and new design procedure. *Int. J. Robust Nonlinear Control* **2017**, *27*, 1915–1939. [CrossRef]
4. Shaheen, B.; Nazir, M.S.; Rehan, M.; Ahmad, S. Robust generalized observer design for uncertain one-sided Lipschitz systems. *Appl. Math. Comput.* **2020**, *365*, 124588. [CrossRef]
5. Wang, X.; Park, J.H. State-Based Dynamic Event-Triggered Observer for One-Sided Lipschitz Nonlinear Systems with Disturbances. *IEEE Trans. Circuits Syst. II Express Briefs* **2022**, *69*, 2326–2330. [CrossRef]
6. Mu, Y.; Zhang, H.; Yan, Y.; Wu, Z. A design framework of nonlinear $\mathcal{H}_\infty$ PD observer for one-sided Lipschitz singular systems with disturbances. *IEEE Trans. Circuits Syst. II Express Briefs* **2022**, *69*, 3304–3308. [CrossRef]
7. Besançon, G. Remarks on nonlinear adaptive observer design. *Syst. Control Lett.* **2000**, *41*, 271–280. [CrossRef]
8. Zhang, J.; Swain, A.K.; Nguang, S.K. Robust $\mathcal{H}_\infty$ adaptive descriptor observer design for fault estimation of uncertain nonlinear systems. *J. Frankl. Inst.* **2014**, *351*, 5162–5181. [CrossRef]
9. Wang, H.; Wang, Q.; Zhang, H.; Han, J. H-Infinity Observer for Vehicle Steering System with Uncertain Parameters and Actuator Fault. *Actuators* **2022**, *11*, 43. [CrossRef]
10. Mu, Y.; Zhang, H.; Ren, H.; Cai, Y. Fuzzy adaptive observer-based fault and disturbance reconstructions for TS fuzzy systems. *IEEE Trans. Circuits Syst. II Express Briefs* **2021**, *68*, 2453–2457.
11. Xiong, X.; Pal, A.K.; Liu, Z.; Kamal, S.; Huang, R.; Lou, Y. Discrete-time adaptive super-twisting observer with predefined arbitrary convergence time. *IEEE Trans. Circuits Syst. II Express Briefs* **2020**, *68*, 2057–2061. [CrossRef]
12. Qin, Q.; Gao, G.; Zhong, J. Finite-Time Adaptive Extended State Observer-Based Dynamic Sliding Mode Control for Hybrid Robots. *IEEE Trans. Circuits Syst. II Express Briefs* **2022**, *69*, 3784–3788. [CrossRef]
13. Li, W.; Yao, X.; Krstic, M. Adaptive-gain observer-based stabilization of stochastic strict-feedback systems with sensor uncertainty. *Automatica* **2020**, *120*, 109112. [CrossRef]
14. Yan, S.; Sun, W.; Yu, X.; Gao, H. Adaptive Sensor Fault Accommodation for Vehicle Active Suspensions via Partial Measurement Information. *IEEE Trans. Cybern.* **2021**, *52*, 12290–12301. [CrossRef] [PubMed]
15. Bzioui, S.; Channa, R. An Adaptive Observer Design for Nonlinear Systems Affected by Unknown Disturbance with Simultaneous Actuator and Sensor Faults. Application to a CSTR. *Biointerface Res. Appl. Chem.* **2021**, *12*, 4847–4856.

16. Bonargent, T.; Menard, T.; Gehan, O.; Pigeon, E. Adaptive observer design for a class of Lipschitz nonlinear systems with multirate outputs and uncertainties: Application to attitude estimation with gyro bias. *Int. J. Robust Nonlinear Control* **2021**, *31*, 3137–3162. [CrossRef]
17. Zheng, Y.; Liu, Y.; Song, R.; Ma, X.; Li, Y. Adaptive neural control for mobile manipulator systems based on adaptive state observer. *Neurocomputing* **2022**, *489*, 504–520. [CrossRef]
18. Sleiman, M.; Bouyekhf, R.; Al Chami, Z.; El Moudni, A. Uncertainty observer and stabilization for transportation network with constraints. *J. Appl. Math. Comput.* **2022**, *68*, 1107–1133. [CrossRef]
19. Perrier, M.; De Azevedo, S.F.; Ferreira, E.; Dochain, D. Tuning of observer-based estimators: Theory and application to the on-line estimation of kinetic parameters. *Control Eng. Pract.* **2000**, *8*, 377–388. [CrossRef]
20. Astorga, C.M.; Othman, N.; Othman, S.; Hammouri, H.; McKenna, T.F. Nonlinear continuous–discrete observers: Application to emulsion polymerization reactors. *Control Eng. Pract.* **2002**, *10*, 3–13. [CrossRef]
21. Arcak, M.; Gorgun, H.; Pedersen, L.M.; Varigonda, S. A nonlinear observer design for fuel cell hydrogen estimation. *IEEE Trans. Control Syst. Technol.* **2004**, *12*, 101–110. [CrossRef]
22. Astorga-Zaragoza, C.M.; Zavala-Río, A.; Alvarado, V.; Méndez, R.M.; Reyes-Reyes, J. Performance monitoring of heat exchangers via adaptive observers. *Measurement* **2007**, *40*, 392–405. [CrossRef]
23. Ramos-Hernández, E.; Astorga-Zaragoza, C.; Reyes, J.R.; Ramırez-Rasgado, F.; Osorio-Gordillo, G.; Ruiz-Acosta, S. Estimation of Process Variables in a Steam Distillation Plant. Congreso Nacional de Control Automático. 2023. Available online: https://revistadigital.amca.mx/wp-content/uploads/2023/12/0103.pdf (accessed on 29 May 2024)
24. Farza, M.; M'saad, M.; Menard, T.; Ltaief, A.; Maatoug, T. Adaptive observer design for a class of nonlinear systems. Application to speed sensorless induction motor. *Automatica* **2018**, *90*, 239–247. [CrossRef]
25. Dong, Z.; Liu, M.; Guo, Z.; Huang, X.; Zhang, Y.; Zhang, Z. Adaptive state-observer for monitoring flexible nuclear reactors. *Energy* **2019**, *171*, 893–909. [CrossRef]
26. Zhong, J.; Feng, Y.; Chen, X.; Zeng, C. Observer-based piecewise control of reaction–diffusion systems with the non-collocated output feedback. *J. Appl. Math. Comput.* **2023**, *69*, 4187–4211. [CrossRef]
27. de Jesús Rubio, J.; Lughofer, E.; Pieper, J.; Cruz, P.; Martinez, D.I.; Ochoa, G.; Islas, M.A.; Garcia, E. Adapting H-infinity controller for the desired reference tracking of the sphere position in the Maglev process. *Inf. Sci.* **2021**, *569*, 669–686. [CrossRef]
28. Asad, M.; Rehan, M.; Ahn, C.K.; Tufail, M.; Basit, A. Distributed H_∞ State and Parameter Estimation over Wireless Sensor Networks under Energy Constraints. *IEEE Trans. Netw. Sci. Eng.* **2024**, *11*, 2976–2988. [CrossRef]
29. Chen, C.; Sun, F.; Xiong, R.; He, H. A Novel Dual H Infinity Filters Based Battery Parameter and State Estimation Approach for Electric Vehicles Application. *Energy Procedia* **2016**, *103*, 375–380. [CrossRef]
30. Gong, X.; Suh, J.; Lin, C. A novel method for identifying inertial parameters of electric vehicles based on the dual H infinity filter. *Veh. Syst. Dyn.* **2019**, *58*, 28–48. [CrossRef]
31. Xu, S. Robust H_∞ filtering for a class of discrete-time uncertain nonlinear systems with state delay. *IEEE Trans. Circuits Syst. I Fundam. Theory Appl.* **2002**, *49*, 1853–1859.
32. Ekramian, M.; Hosseinnia, S.; Sheikholeslam, F. Observer design for non-linear systems based on a generalised Lipschitz condition. *IET Control Theory Appl.* **2011**, *5*, 1813–1818. [CrossRef]
33. Ekramian, M.; Sheikholeslam, F.; Hosseinnia, S.; Yazdanpanah, M.J. Adaptive state observer for Lipschitz nonlinear systems. *Syst. Control Lett.* **2013**, *62*, 319–323. [CrossRef]
34. Guo, S.; Yang, S.; Pan, C. Dynamic modeling of magnetorheological damper behaviors. *J. Intell. Mater. Syst. Struct.* **2006**, *17*, 3–14. [CrossRef]
35. Tudon-Martinez, J.C.; Morales-Menéndez, R.; Ramirez-Mendoza, R.; Sename, O.; Dugard, L. Fault tolerant control in a semi-active suspension. *IFAC Proc. Vol.* **2012**, *45*, 1173–1178. [CrossRef]
36. Khalil, H.K.; Grizzle, J.W. *Nonlinear Systems*; Prentice Hall: Upper Saddle River, NJ, USA, 2002; Volume 3. Available online: https://nasim.hormozgan.ac.ir/ostad/UploadedFiles/863740/863740-8707186354456456.pdf (accessed on 29 May 2024).
37. Zhao, J.; Mili, L. A decentralized H-infinity unscented Kalman filter for dynamic state estimation against uncertainties. *IEEE Trans. Smart Grid* **2018**, *10*, 4870–4880. [CrossRef]

Mathematical and Computational Applications

MDPI

Article

Computational Cost Reduction in Multi-Objective Feature Selection Using Permutational-Based Differential Evolution

Jesús-Arnulfo Barradas-Palmeros [1,*], Efrén Mezura-Montes [1], Rafael Rivera-López [2], Hector-Gabriel Acosta-Mesa [1] and Aldo Márquez-Grajales [3]

1 Artificial Intelligence Research Institute, Universidad Veracruzana, Xalapa 91097, Veracruz, Mexico; emezura@uv.mx (E.M.-M.); heacosta@uv.mx (H.-G.A.-M.)

2 Departamento de Sistemas y Computación, Instituto Tecnológico de Veracruz, Veracruz 91897, Veracruz, Mexico; rafael.rl@veracruz.tecnm.mx

3 INFOTEC Center for Research and Innovation in Information and Communication Technologies, Pocitos, Aguascalientes 20326, Aguascalientes, Mexico; li.aldomg@gmail.com

* Correspondence: zs23000652@estudiantes.uv.mx

Abstract: Feature selection is a preprocessing step in machine learning that aims to reduce dimensionality and improve performance. The approaches for feature selection are often classified according to the evaluation of a subset of features as filter, wrapper, and embedded approaches. The high performance of wrapper approaches for feature selection is associated at the same time with the disadvantage of high computational cost. Cost-reduction mechanisms for feature selection have been proposed in the literature, where competitive performance is achieved more efficiently. This work applies the simple and effective resource-saving mechanisms of the fixed and incremental sampling fraction strategies with memory to avoid repeated evaluations in multi-objective permutational-based differential evolution for feature selection. The selected multi-objective approach is an extension of the DE-FS^{PM} algorithm with the selection mechanism of the GDE3 algorithm. The results showed high resource savings, especially in computational time and the number of evaluations required for the search process. Nonetheless, it was also detected that the algorithm's performance was diminished. Therefore, the results reported in the literature on the effectiveness of the strategies for cost reduction in single-objective feature selection were only partially sustained in multi-objective feature selection.

Keywords: feature selection; differential evolution; cost reduction; multi-objective optimization

1. Introduction

Feature selection (FS) is a dimensionality reduction preprocessing task in machine learning (ML) that deals with selecting the most relevant features in a dataset and discarding the noisy and irrelevant ones. By preferring only the features with meaningful information, the learning process complexity is reduced, and an ML algorithm's performance in classification or clustering tasks is expected to be improved [1]. FS is an NP-Hard combinatorial problem with an exponentially growing search space of 2^n possible solutions for a dataset with n features. Given the complexity of the problem, nature-inspired metaheuristics, including Evolutionary Computation (EC) algorithms, constitute popular and effective methods for FS [2].

Dimensionality reduction methods have gained attention given the growing volumes of data with high dimensions generated nowadays in applications from different fields. By reducing the data dimensions, the ML algorithm training process is expected to be more efficient and fruitful [3,4]. An alternative feature reduction approach is feature extraction or feature reduction. It uses an algorithm to generate new features that combine and represent the information from the dataset features. In contrast, FS consists of selecting some of the original features in the dataset. An advantage of FS is that the interpretability of the data is not affected [5,6]. This work focuses on FS.

As presented in [3,7,8], the FS mechanism of assessing and comparing the goodness of a subset of features allows the classification of FS approaches into filter, wrapper, and embedded. Filters are the fastest approach, using an evaluation metric independent of the classification or clustering algorithm. Filters usually focus on the relevance of each feature or maximize the interaction between selected features. Wrapper approaches use an ML algorithm to determine the quality of a feature subset. The features that increase the performance of the ML algorithm are expected to be selected. Wrappers are the most resource-demanding approach, given the required several runs of the ML algorithm. Nonetheless, wrappers generally obtain the highest performance. Finally, embedded approaches incorporate the FS process in the training process of the ML algorithm. Decision trees are an example of an embedded approach that selects the most relevant features for classification in the training process. Embedded approaches are expected to be more complex than filters and faster than wrappers.

Differential Evolution (DE) is an evolutionary algorithm (EA) that is highlighted for its simplicity and robustness when applied to optimization problems [9]. As presented in [10], various works have extended the DE algorithm's classic single-objective form to a multi-objective one. For the FS problem, various multi-objective DE proposals can be found in the literature using two common types of solution representation: a real-value codification with a threshold to determine if a feature is selected and binary adaptations of DE. Binary DE versions require changes in the mutation operator of DE, whereas using a real-value representation with a threshold allows the application of basic operators without changes. Examples of multi-objective DE for feature selection are found in [11,12] for filters with real-value codification, in [13,14] for wrappers with real-value codification, and in [15,16] for wrappers with binary representation. Alternative multi-objective DE approaches for FS can be found in [17] for unsupervised FS and [18] for FS using reinforcement learning.

An alternative and effective solution representation for DE applied for the FS problem is presented in [19] where the permutational-based DE algorithm for FS (DE-FS^{PM}) is proposed. Permutations represent the individuals, and the mutation operator is modified to work with the selected representation. The reported results from the DE-FS^{PM} show that it outperformed other metaheuristic and classic approaches for FS. In this manner, the effectiveness of using the DE adaptation to the permutational space was proved. In [20], the DE-FS^{PM} algorithm is extended to a multi-objective version by combining it with the Generalized Differential Evolution 3 (GDE3) algorithm. The results show that the proposal effectively found a set of solutions that represent a trade-off between the objectives of minimizing the prediction error of a classifier and the number of selected features. Nonetheless, the single objective version of the DE-FS^{PM} algorithm finds subsets with lower classification errors.

Cost-reduction approaches for the FS process are proposed in [21–23], where the search process for the most relevant subset of features is conducted using a fraction of the dataset instances, selected with random sampling, to reduce the evaluation cost. In [21], using the randomly selected group of dataset instances is tested in filter approaches for FS and feature extraction. The authors define a fixed number of instances to be sampled (100, 250, 500, 1500, and 2000) and select a subset of features with a predefined size of ten. Six methods are used with large-scale datasets, resulting in an execution time reduction when using the reduced subset of instances with minimal impact on the performance of the feature reduction method.

In [22], the approach of using random sampling for reducing the number of instances in the search process for feature selection, and therefore the computational cost, is extended for wrapper approaches for FS. The authors proposed three different random sampling-based strategies to reduce the number of dataset instances in the search process: the fixed, the incremental, and the evolving sampling fraction. In addition, the Success History parameter adaptation for DE from [24] was adapted to the FS problem with the DE-FS^{PM}

algorithm. The results were promising in maintaining the performance of the FS procedure, but the resource savings were scarce in computational time.

An extension of the work from [22] is presented in [23]. A memory mechanism to avoid repeated evaluations is added to work with the fixed and incremental sampling fraction strategies applied to the DE-FS^{PM} algorithm. The memory attempts to solve the problem of wasted resources associated with evaluating duplicated individuals in DE presented in [25]. The method could perform similarly to the DE-FS^{PM} original algorithm using the fixed sampling fraction strategy with memory to avoid repeated evaluations. The proposed approach reduced the average of 35.15% of the computational time and detected that an average of 35.35% of the evaluations could be avoided.

This work is based on the future work stated in [22,23], where the resource-saving mechanisms of the sampling strategies and memory to avoid repeated evaluations were proposed. The main contributions of this paper are the following:

- We incorporate sampling strategies and memory to avoid repeated evaluations into the GDE3-based multi-objective version of the DE-FS^{PM} algorithm.
- We introduce two novel proposals: GDE3fix and GDE3inc. The former utilizes the fixed sampling fraction strategy, while the latter incorporates the incremental sampling fraction strategy. Both proposals use memory to avoid repeated evaluations.
- We test the robustness of the cost-reduction mechanisms in multi-objective FS. The main goal was to determine if the results from the single-objective approach, where the computational cost was reduced without diminishing algorithm performance, are maintained in the multi-objective version of the algorithm.
- We thoroughly analyze the effects of the proposals in terms of computational time consumption, the number of evaluations performed by the algorithm, and the number of instances used for an evaluation. Additionally, future work is described, including possible areas for improvement.

The rest of this document is organized into four sections. In Section 2, the details of the multi-objective FS process are introduced. After that, the DE algorithm is described along with its adaptation to the permutational codification of solutions and its extension to multi-objective optimization. Additionally, the cost-reduction mechanisms are presented in detail. Section 3 presents the experimentation details and results. Finally, Sections 4 and 5 include the analysis of the results, the conclusions, and future research directions.

2. Materials and Methods

2.1. Multi-Objective Feature Selection

As presented in [6], the FS process is guided by two main objectives: maximizing the classification accuracy and minimizing the number of selected features. These objectives conflict and different feature subsets can represent trade-offs between obtaining a higher accuracy performance or a smaller subset of features. Usually, the classification accuracy maximization is transformed into the minimization of the classification error. This way, the goal is to minimize both objectives. Given that there are no constraints to the FS problem, in [26], the multi-objective FS problem is modeled following Equation (1). x represents a subset of selected features.

$$minimize\ F(x) = [f_1(x), f_2(x)] \tag{1}$$

$f_1(x)$ represents the classification error and is calculated using Equation (2). In the equation, the True Positive (TP), False Positive (FP), True Negative (TN), and False Negative (FN) values from the confusion matrix are required. A cross-validation (CV) approach is followed in our proposal to calculate the error rate. $f_2(x)$ corresponds to the number of selected features as presented in Equation (3). d is the number of selected features, and n is the total number of features in the dataset. $f_1(x)$ and $f_2(x)$ are in the range of [0, 1].

$$f_1(x) = \frac{FP + FN}{TP + FP + TN + FN} \tag{2}$$

$$f_2(x) = \frac{d}{n} \tag{3}$$

As presented in [27], a multi-objective optimization algorithm returns a set of solutions instead of a single solution as in single-objective optimization. The concept of Pareto optimal solutions is presented in multi-objective optimization. A solution is Pareto optimal if it is not dominated by any other solution. The weak dominance ($\preceq$) is defined in Equation (4), where a solution x_1 weakly dominates x_2 ($x_1 \preceq x_2$) if, and only if, all its objective values are less than or equal to the objective values of x_2. The dominance ($\prec$) of a solution x_1 over x_2 ($x_1 \prec x_2$) is given by Equation (5), adding to the weak dominance the condition that at least one objective value $f_m(x_1)$ must be less than $f_m(x_2)$. In the case of the previously stated modeling of the FS problem, $m = 2$ since the objectives are the classification error and the normalized number of selected features. The Pareto optimal solutions compose the Pareto optimal set. The visualization of the solutions in the Pareto optimal set is the Pareto front.

$$x_1 \preceq x_2 \; iff \; \forall m \in \{1,2,\ldots,M\} : f_m(x_1) \leq f_m(x_2) \tag{4}$$

$$x_1 \prec x_2 \; iff \; x_1 \preceq x_2 \land \exists m \in \{1,2,\ldots,M\} : f_m(x_1) < f_m(x_2) \tag{5}$$

2.2. *Differential Evolution*

DE is a population-based metaheuristic algorithm for continuous optimization proposed in [28]. N individuals constitute the population. A vector of the size of the dimensions of the problem represents an individual (x_i) in the population. The initial population consists of N individuals randomly created. Each individual has an associated fitness value provided by an evaluation using the selected fitness function that will guide the search process. The population is evolved in an iterative process where in each generation g, the mutation and crossing procedures are applied for each target vector x_i to generate the noise (v_i) and trial (u_i) vectors. A user-defined parameter is the maximum number of generations G to evolve the population.

As presented in [29], the basic DE version is called $DE/rand/1/bin$. $rand$ and 1 come from calculating v_i, where Equation (6) is applied. $rand$ refers to the random selection of the r_0, r_1, and r_2 individuals from the population different from themselves and the x_i that is being considered for mutation and crossing. The value 1 is given by the calculation of one vector difference. The scaling factor F is a scalar value defined by the user. bin comes from the binomial distribution associated with the uniform crossover operator presented in Equation (7). The crossover rate CR is another user-defined parameter that controls the probability of choosing $u_{i,j}$ from v_i or x_i. To ensure that u_i is not copied from x_i, a random position J_{rand} is chosen, where $u_{i,j}$ is guaranteed to be $v_{i,j}$. Alternative DE variants have been proposed in the literature, changing the calculation of v_i [9]. Some examples include $DE/best/1/$, $DE/rand/2/$, and $DE/current\text{-}to\text{-}best/$. An alternative crossover mechanism is the exponential crossover (exp).

$$v_i = r_0 + F(r_1 - r_2) \tag{6}$$

$$u_{i,j} = \begin{cases} v_{i,j} & \text{if } (rand_j \leq CR) \; or \; (j = J_{rand}); j = 1,\ldots,|x_i| \\ x_{i,j} & \text{otherwise} \end{cases} \tag{7}$$

After u_i is computed and evaluated, a binary tournament is performed to determine the individual with better fitness value comparing x_i and u_i. The winner is included in the population for the next generation of the evolutionary process. The previous process is an elitist selection mechanism that guarantees that the best solution found in the search process is never discarded. Once the G generations are computed, the process ends, and the individual with the highest fitness value in the population is returned as the best solution for the problem.

2.3. Permutational-Based Differential Evolution for Feature Selection

In [19], the DE-FS^{PM} algorithm is proposed, modifying the DE algorithm to be applied to the FS problem. The first modification is that each individual is represented by a permutation with the indexes of the features in the dataset and a number zero used in the decoding process. The indexes that appear before the zero in the permutation are considered the selected features by the individual. Given the alternative encoding of the individuals, the DE procedure is adapted to the permutational space. The main changes are applied to the v_i calculation. Equations (8) and (9) are used instead of Equation (6).

$$r_1 \leftarrow \mathbf{P}r_2 \tag{8}$$

$$v_i \leftarrow \mathbf{P}_F r_0 \tag{9}$$

Equation (8) presents the calculation of the permutation matrix $\mathbf{P}$ that maps r_1 and r_2. After calculating $\mathbf{P}$, a scaled permutation matrix $\mathbf{P}_F$ is required. $\mathbf{P}_F$ is used to apply some changes to r_0 in Equation (9); the parameter F controls how disruptive the mutation is. The process of calculating $\mathbf{P}_F$ is presented in [30]. For each row i in $\mathbf{P}$, if there is a 0 in the position $\mathbf{P}[i,i]$ and a random number $rand_i$ is greater than F, the row i is swapped with the row j where the position $\mathbf{P}[j,i]$ is a 1. When the swap is produced, a number 1 is included in the diagonal of the matrix. The positions in the diagonal of $\mathbf{P}_F$ with a value of 1 represent no changes to r_0 when Equation (9) is applied. Greater F values will produce little change to $\mathbf{P}$ due to the low probability of $rand_i$ being greater than F. If $F = 1$, no changes will be applied to $\mathbf{P}$. By contrast, smaller F values will allow the setting of more elements in the $\mathbf{P}_F$ diagonal as 1. If $F = 0$, the resulting $\mathbf{P}_F$ will be the diagonal matrix and $v_i = r_0$.

The uniform crossover operator of DE presented in Equation (7) is maintained in the DE-FS^{PM} algorithm. This way, some elements in the permutation of u_i come from v_i and others from x_i. As a result, u_i could possess repeated elements requiring the application of a repair mechanism. The repair mechanism proposed in [19] removes all the repeated elements in u_i; the remaining elements are moved to the left to take the empty spaces left by the removed elements. Finally, the permutation is completed with the missing elements in the order they appear in x_i. The mutation and crossover process of the DE-FS^{PM} algorithm is illustrated with an example in Figure 1.

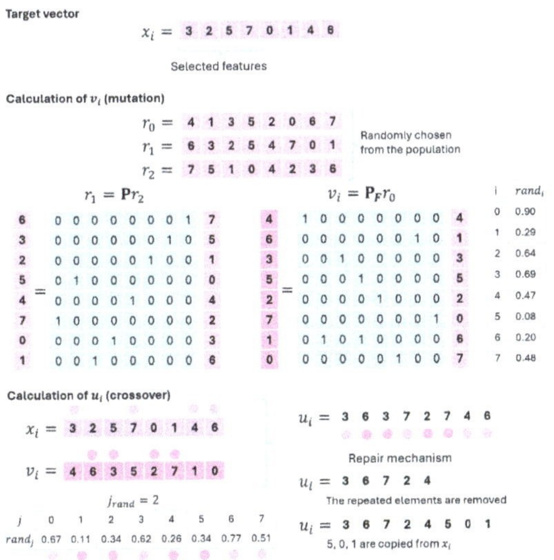

Figure 1. An example of the DE-FS^{PM} algorithm mutation and crossover procedures. The decoding of an individual x_i is presented with an example of calculating u_i and applying the repair mechanism to maintain a valid permutation. The random values required by the process are also shown. The parameters

considered for the example are $F = 0.6$ and $CR = 0.4$. After calculating u_i, it is evaluated, and its associated fitness value is compared with the x_i fitness value. The most fitted one is included in the next-generation population. Different colors are used in the figure to represent the x_i, v_i, and u_i vectors.

The accuracy metric guides the FS process in the DE-FS^{PM} algorithm. A five-fold stratified cross-validation (CV) evaluation using the k-nearest-neighbors (KNN) algorithm is used as the fitness function. To determine the k value for the KNN algorithm, at the beginning of the search process, different values of k from 1 to 20 with a step of 2 are tested using all the dataset features, selecting the k with the highest performance in a ten-fold CV. After selecting k, the dataset is evaluated using all features with a ten-fold stratified CV. The ten-fold stratified CV evaluation is conducted again at the end of the search process with the selected subset of features.

The DE-FS^{PM} algorithm requires a preprocessing step for the datasets used for the feature selection process. First, the missing values are imputed, considering the mean for numerical features and the mode for categorical features. The next step is converting the categorical features into numerical features. Finally, all features are normalized following a min-max normalization.

2.4. Generalized Differential Evolution 3 Algorithm

The GDE3 algorithm was proposed in [31], extending the capabilities of the DE algorithm to deal with problems with M objectives and K constraints. As presented in [10], the changes to the DE algorithm are applied in the selection procedure. Earlier GDE versions considered the Pareto dominance and crowding distance measurement when comparing x_i and u_i. GDE3's population can grow temporally when specific criteria are met (as explained later), keeping both x_i and u_i. At the end of each generation, if the population grew, the Fast-Non-Dominated Sort and Crowding Distance (CD) assignment algorithms from [32] are applied to reduce the population to the size N. The Fast-Non-Dominated Sort assigns the non-dominated solutions to the front number 1. Front number 2 comprises the non-dominated solutions from the rest, and so on until all solutions are included in a front. The fronts formed by the algorithm are progressively added to the population for the next generation until one front does not fit entirely due to a surpass of N. Then, CD is applied to rank the individuals in the front. The ones with the highest CD value are selected.

In [27], it is explained that the changes applied to the DE algorithm in the generalized versions were intended to be as little as possible. As mentioned earlier, the modification is presented in the selection mechanism, and the rest is the same as the $DE/rand/1/bin$ procedure. The GDE3 algorithm is equivalent to the DE original procedure when applied to problems with $M = 1$ and $K = 0$, i.e., problems with one objective and no constraints. To deal with constraint satisfaction problems ($K > 0$), the selection mechanism of GDE3 is based on the concept of constraint-domination ($\prec_c$), a variant of the domination ($\prec$) concept presented in Section 2.1. $x_1 \prec_c x_2$ when:

- x_1 and x_2 are unfeasible, but x_1 violates the constraints less.
- x_1 is feasible and x_2 is unfeasible.
- x_1 and x_2 are feasible and $x_1 \prec x_2$.

The selection mechanism defined in [31] for the GDE3 algorithm consists of the following three cases:

1. When x_i and u_i are infeasible. The one that violates the least number of constraints is selected.
2. If x_i is feasible and u_i is unfeasible, x_i is selected. On the other hand, if u_i is feasible and x_i is unfeasible, u_i is selected.
3. When x_i and u_i are feasible, u_i is selected if $u_i \preceq x_i$; x_i is selected if $x_i \prec u_i$; and both x_i and u_i are selected if $x_i \nprec u_i$ and $u_i \nprec x_i$.

The third case is the only one in which the GDE3 algorithm allows population growth, given that both x_i and u_i can be selected. Since the FS problem presents no constraints ($k = 0$), only the third case in the GDE3 selection mechanism is applied. In [20], the DE-FS^{PM} algorithm is extended to work with two objectives using the optimization framework of the GDE3 algorithm. The objectives considered are the classification error and the number of selected features. The selection mechanism is the only modification to the DE-FS^{PM} algorithm. In our research proposal, the number of selected features is normalized by dividing it by the total number of features in the dataset as presented in Section 2.1. The GDE3-based multi-objective version of the DE-FS^{PM} algorithm is presented in Algorithm 1. As observed in Algorithm 1, the procedure returns the Pareto optimal solutions. Hence, a mechanism must be established to choose one of the returned solutions.

Algorithm 1 The multi-objective $DE - FS^{PM}$ algorithm

Require: $DE - FS^{PM} - GDE3\ (CR,\ F,\ N,\ G)$
  Input: The crossover rate (CR), the scale factor (F), the population size (N), and the number of generations (G).
  Output: Pareto optimal solutions in the current population.

  $\mathbb{X}_0 \leftarrow \emptyset$
  for each $i \in \{1, \ldots, N\}$ **do**
    $x_i \leftarrow$ A permutation chosen at random from the solution space.
    $\mathbb{X}_0 \leftarrow \mathbb{X}_0 \cup \{x_i\}$
  end for
  for each $g \in \{1, \ldots, G\}$ **do**
    $\mathbb{X}_g \leftarrow \emptyset$
    $m \leftarrow 0$
    for each $x_i \in \mathbb{X}_g$ **do**
      $v_i \leftarrow$ Mutated vector using Equations (8) and (9).
      $u_i \leftarrow$ Trial vector calculated using Equation (7) and the repair procedure.
      $\mathbb{X}_g \leftarrow \mathbb{X}_g \cup \begin{cases} \{u_i\} & \text{if } u_i \prec x_i \\ \{x_i\} & \text{otherwise} \end{cases}$
      if $u_i \not\prec x_i \wedge x_i \not\prec u_i$ **then**
        $\mathbb{X}_g \leftarrow \mathbb{X}_g \cup \{u_i\}$
        $m \leftarrow m + 0$
      end if
    end for
    if $m > 0$ **then**
      $FR \leftarrow$ fast-non-dominated-sort$(\mathbb{X}_g)$
      $\mathbb{X}_g \leftarrow \emptyset$
      $j \leftarrow 1$
      while $|\mathbb{X}_g| < N$ **do**
        if $|\mathbb{X}_g| + |FR_j| \leq N$ **then**
          $\mathbb{X}_g \leftarrow \mathbb{X}_g \cup FR_j$
        else
          $CD_j \leftarrow$ Crowding-distance(FR_j)
          $sort(FR_j, CD_j)$ ▷ Sort FR_j in descending CD_j order
          $\mathbb{X}_g \leftarrow \mathbb{X}_g \cup FR_j[0 : N - |\mathbb{X}_g|]$
        end if
        $j \leftarrow j + 1$
      end while
    end if
  end for
  $FR \leftarrow$ fast-non-dominated-sort$(\mathbb{X}_g)$
  return FR_1

2.5. Cost-Reduction Mechanisms for Feature Selection

The considered mechanisms for computational cost reduction are the fixed and incremental sampling fraction strategies from [22] and their incorporation with memory to avoid repeated evaluations from [23]. Both proposals were applied to the DE-FS^{PM} algorithm that considers only one objective in its optimization process. Nonetheless, the mechanisms can also be applied to multi-objective optimization processes. The incorporation of both mechanisms into the multi-objective feature selection process is described next.

2.5.1. Fixed Sampling Fraction with Memory

The FS search process is conducted with a fraction of the dataset instances in the fixed sampling fraction strategy. The user defines two parameters: the initial sampling fraction S and the number of blocks in the search B. S is used at the beginning of the search process to apply random sampling and select part of the dataset instances. At that step, the memory used to avoid repeated evaluations is empty. When an individual is decoded for evaluation, the procedure searches in memory if there is a stored fitness value associated with the subset of features represented by the individual. The stored fitness value is returned if a coincidence is found in memory. Otherwise, the evaluation is performed using the fitness function. In the case of a multi-objective process, the memory stores the value obtained from the assessment of the individual in each of the considered objectives.

The parameter B divides the search process's generations into groups. In the fixed sampling fraction, the memory is reset at the beginning of each block. The previous block is used to control the size of the memory. For example, if G is set as 100, B is set as five. The memory is reset at generations 1, 21, 41, 61, and 81. The memory mechanism is expected to have a more considerable impact on the last blocks of the search process due to the algorithm's convergence, where population diversity diminishes and the probability of finding a duplicate individual increases.

2.5.2. Incremental Sampling Fraction with Memory

When the incremental sampling fraction strategy is applied, the process has the same start conditions as the fixed sampling fraction strategy. The user must also define the parameters S and B. Random sampling is applied at the beginning of the process, considering S to select part of the dataset instances. The memory mechanism is also used to avoid repeated evaluations. The characteristic aspect of the incremental sampling fraction strategy is that, at the beginning of each block of generations, not only is the memory reset, but more instances are proportionally added to the search process. In the last block of generations, all dataset instances are used. This way, fewer instances are considered in early generations, expecting that the algorithm will find promising areas of the search space with less costly evaluations. Then, late generations will likely be able to consider feature subsets with better generalization capabilities and avoid overfitting the selected feature subset to a fraction of dataset instances.

Given that the conditions of the FS problem change when more instances are considered, the population is reevaluated at the beginning of each block. The previous point is a drawback of the strategy due to the extra evaluations. Figure 2 presents a diagram of the multi-objective version of the DE-FS^{PM} algorithm with the resource-saving modifications in the search process. It is seen that the incremental sampling fraction strategy requires the reevaluation process when a new block starts, while the fixed sampling fraction only requires resetting the memory.

Figure 2. Diagram of the multi-objective DE-FS^{PM} algorithm with the cost-reduction mechanisms applied in the search process. S and B are considered 0.5 and 6, respectively, as examples of their effect on managing the dataset instances in the search process.

3. Results

To test the effectiveness of the cost-reduction mechanisms from Section 2.5, the experimentation approach from [23] is adopted. The algorithm proposals are run 30 times for each of the eighteen datasets selected for experimentation. The code was implemented using Python, and the experiments were run in the virtual environments provided by Google Colab Pro. The datasets are selected from the UCI Machine Learning Repository [33], and their details are presented in Table 1.

Table 1. Details of the datasets selected for experimentation.

Dataset	Features	Instances	Classes
Arrhythmia	279	452	16
Audiology	69	226	24
Australian	14	690	2
Cylinder-b	39	540	2
CRX	15	690	2
Dermatology	34	366	6
German-c	20	1000	2
Hill valley	100	1212	2
Ionosphere	34	351	2
M-libras	90	360	15
Musk 1	168	476	2
Parkinsons	22	195	2
Sonar	60	208	2
Soybean	35	683	19
SPECTF	44	267	2
Vehicle	18	846	4
Vote	16	435	2
WDBC	30	569	2

The proposals are identified as:

1. **GDE3**: for the GDE3 version of the DE-FS^{PM} algorithm.
2. **GDE3fix**: for the GDE3 version of the DE-FS^{PM} algorithm using the fixed sampling fraction strategy with memory to avoid repeated evaluations.
3. **GDE3inc**: for the GDE3 version of the DE-FS^{PM} algorithm using the incremental sampling fraction strategy with memory to avoid repeated evaluations.

The parameter configuration is adopted from [19], where the parameters G and N are defined for the DE-FS^{PM} algorithm. F and CR are obtained in the fine-tuning of the GDE3 version of the DE-FS^{PM} algorithm from [20]. Finally, the parameters S and B for the cost-reduction mechanisms are presented in [23]. The details are provided next.

- G: 200 as the maximum number of generations.
- N: 5 times the number of features in the dataset. The value is bound to have at least 200 individuals and at most 450.
- F: 0.8305.
- CR: 0.7049.
- S: 0.6 for GDE3fix and GDE3inc.
- B: 10 for GDE3fix and GDE3inc.

Three criteria were considered when selecting a solution from the resulting Pareto front of each method. First, the solution with the best performance in $f_1(x)$ represents the lowest classification error obtained by a solution. After that, the solution in the knee point of the front is considered a middle point in the trade-off between $f_1(x)$ and $f_2(x)$. The knee point is calculated as the solution in the Pareto front with the smallest Euclidian distance concerning the point (0,0). Finally, the solution with better performance minimizing $f_2(x)$ is selected.

Three key aspects were monitored to assess the computational resources saved by the cost-reduction mechanism: the execution time, the number of evaluations used by the method, and the number of instances avoided in evaluation. The execution time is the time required on average by the FS procedure in the runs of each one of the proposals. The number of evaluations refers to the times an individual is evaluated with $f_1(x)$ and $f_2(x)$. In the GDE3fix and GDE3inc proposals, the memory to avoid repeated evaluations is used to reduce the number of evaluations. Finally, the effect of using the fixed and incremental sampling strategies and the memory mechanism is measured with the sum of the number of instances used while evaluating an individual. Additionally, the hypervolume and spacing indicators for multi-objective optimization are reported.

The results are divided into three subsections of the document. First, in Section 3.1, the proposals are compared in terms of accuracy and the number of selected features. Then, in Section 3.2, the resource savings achieved by the GDE3fix and GDE3inc proposals are presented. Finally, Section 3.3 shows the performance of the proposals measured by the selected multi-objective indicators.

3.1. Accuracy Performance and Number of Selected Features

Table 2 presents the average results of the proposals when the solution selected from the Pareto front is the one with the minimum classification error. For this selection, the priority is given to $f_1(x)$. For comparison, the table includes the accuracy and the number of selected features performance of the data without FS, and the results reported in [23] for the DE-FS^{PM} algorithm. It is seen that the single-objective version of the DE-FS^{PM} algorithm achieves a higher accuracy performance. Nonetheless, in almost all cases, the GDE3-based proposals attained the goal of reducing the data dimensionality and increasing the accuracy of using the dataset without feature selection.

The next comparison point is when the solution is selected from the Pareto front in the knee point. These solutions represent the trade-off between the accuracy performance and the number of selected features. Table 3 presents the results. In this case, the accuracy performance of the dataset without FS is presented as a reference. In all cases, the procedure reduced the dimensionality of the data. However, in most cases, the accuracy results are poorer than the accuracy performance achieved by the method without FS.

Table 2. Accuracy and feature selection results of selecting the solution with the best performance in accuracy from the Pareto front. The best result is marked in bold, and the best result considering only the GDE3, GDE3fix, and GDE3inc proposals is underlined.

Dataset	WoFS		DE-FS(PM)		GDE3		GDE3fix		GDE3inc	
	SF	Acc	SF	Acc	SF	Acc	SF	Acc	SF	Acc
Arrhtythmia	279	58.79	10.9	**75.66**	**10.5**	70.73	16.0	66.37	12.1	66.75
Audiology	69	76.47	**25.1**	**85.63**	42.9	82.43	43.7	79.72	44.9	81.00
Australian	14	85.98	**6.7**	**86.82**	9.5	86.48	**6.7**	86.29	7.6	86.47
Cylinder-b	39	74.51	3.9	**84.19**	**2.2**	83.51	10.2	80.97	2.3	82.65
CRX	15	86.13	9.0	**87.49**	10.8	87.17	**8.0**	86.16	9.0	86.94
Dermatology	34	96.89	**19.1**	**98.35**	25.7	98.02	23.9	97.17	26.6	97.65
German-c	20	75.35	**9.1**	**77.53**	15.4	76.14	12.6	74.34	14.8	75.65
Hill valley	100	64.54	**9.9**	**71.64**	10.5	69.52	14.9	67.24	16.6	68.48
Ionosphere	34	86.97	6.5	**94.82**	6.6	93.45	**5.7**	90.52	8.0	92.13
M-libras	90	86.06	**24.0**	**88.89**	24.9	87.86	39.5	86.23	47.3	87.19
Musk 1	168	85.65	**36.1**	**94.23**	42.4	91.23	52.5	87.98	60.5	89.45
Parkinsons	22	95.84	10.9	**98.84**	**9.6**	98.24	13.3	95.44	12.8	97.39
Sonar	60	86.70	**22.4**	**93.79**	25.2	90.56	25.1	87.00	34.3	89.03
Soybean	35	91.85	**17.3**	**94.91**	21.3	94.30	23.3	92.69	22.6	93.61
SPECTF	44	77.68	**5.5**	**84.22**	6.9	83.00	15.3	79.34	9.2	79.93
Vehicle	18	70.08	**9.9**	**74.31**	11.1	73.19	11.6	71.62	11.9	72.15
Vote	16	93.50	5.2	**96.59**	4.7	96.54	4.3	95.38	**2.5**	95.93
WDBC	30	97.07	18.2	**97.55**	23.5	97.26	**17.6**	96.82	22.5	97.17

Table 3. Accuracy and feature selection results of selecting the solution in the knee point from the Pareto front. The best result is marked in bold, and the best result considering only the GDE3, GDE3fix, and GDE3inc proposals is underlined.

Dataset	WoFS		DE-FS(PM)		GDE3		GDE3fix		GDE3inc	
	SF	Acc	SF	Acc	SF	Acc	SF	Acc	SF	Acc
Arrhtythmia	279	58.79	10.9	**75.66**	9.2	70.54	11.0	66.12	**7.7**	66.81
Audiology	69	76.47	25.1	**85.63**	**9.8**	73.08	11.4	68.08	13.0	72.95
Australian	14	85.98	6.7	**86.82**	**1.0**	85.51	**1.0**	85.51	**1.0**	85.42
Cylinder-b	39	74.51	3.9	**84.19**	**2.0**	83.30	2.2	81.69	2.1	82.41
CRX	15	86.13	9.0	**87.49**	**1.0**	84.57	**1.0**	83.82	**1.0**	84.86
Dermatology	34	96.89	19.1	**98.35**	**4.4**	90.42	4.5	87.98	4.7	90.87
German-c	20	75.35	9.1	**77.53**	**1.8**	71.13	1.9	70.84	1.9	71.69
Hill valley	100	64.54	9.9	**71.64**	6.9	69.46	**5.5**	66.15	6.8	68.11
Ionosphere	34	86.97	6.5	**94.82**	**2.2**	89.17	2.4	87.33	2.7	89.20
M-libras	90	86.06	24.0	**88.89**	**8.5**	86.92	9.6	84.59	8.8	85.44
Musk 1	168	85.65	36.1	**94.23**	12.3	88.69	**11.5**	85.16	12.8	86.91
Parkinsons	22	95.84	10.9	**98.84**	2.0	93.48	**1.9**	89.47	2.1	92.04
Sonar	60	86.70	22.4	**93.79**	5.8	86.32	**5.0**	79.35	5.9	83.19
Soybean	35	91.85	17.3	**94.91**	**5.5**	85.94	5.9	83.55	5.9	83.96
SPECTF	44	77.68	5.5	**84.22**	2.3	83.30	2.3	79.76	**2.0**	79.95
Vehicle	18	70.08	9.9	**74.31**	**2.2**	65.88	2.4	65.82	2.8	66.93
Vote	16	93.50	5.2	**96.59**	**1.0**	95.13	**1.0**	95.63	**1.0**	95.63
WDBC	30	97.07	18.2	**97.55**	2.0	95.31	**1.8**	93.86	2.0	94.79

The third selection strategy was to prioritize the dimensional reduction and select the solution in the Pareto front with better performance in $f_2(x)$. Table 4 presents the results of choosing the solution with the smallest feature subset. The accuracy performance of the dataset without FS is presented as a reference. It is seen that, in this case, the dimensionality reduction is maximum. All the proposals found a subset with only one feature, but the accuracy was severely affected. It is worth noticing the cases where only one feature obtains a performance close to the accuracy using all features, such as the arrhythmia, Australian, cylinder-b, CRX, and vote datasets.

Table 4. Accuracy and feature selection results of selecting the solution with the best performance in the number of selected features from the Pareto front. The best result is marked in bold, and the best result considering only the GDE3, GDE3fix, and GDE3inc proposals is underlined.

Dataset	WoFS		DE-FS(PM)		GDE3		GDE3fix		GDE3inc	
	SF	Acc	SF	Acc	SF	Acc	SF	Acc	SF	Acc
Arrhtythmia	279	58.79	10.9	**75.66**	<u>1</u>	<u>57.64</u>	<u>1</u>	56.04	<u>1</u>	56.42
Audiology	69	76.47	25.1	**85.63**	<u>1</u>	<u>29.13</u>	<u>1</u>	27.68	<u>1</u>	27.57
Australian	14	85.98	6.7	**86.82**	<u>1</u>	<u>85.51</u>	<u>1</u>	<u>85.51</u>	<u>1</u>	85.29
Cylinder-b	39	74.51	3.9	**84.19**	<u>1</u>	<u>74.88</u>	<u>1</u>	74.56	<u>1</u>	74.86
CRX	15	86.13	9.0	**87.49**	<u>1</u>	84.56	<u>1</u>	83.28	<u>1</u>	<u>84.78</u>
Dermatology	34	96.89	19.1	**98.35**	<u>1</u>	<u>47.10</u>	<u>1</u>	43.27	<u>1</u>	46.09
German-c	20	75.35	9.1	**77.53**	<u>1</u>	68.08	<u>1</u>	67.65	<u>1</u>	<u>69.60</u>
Hill valley	100	64.54	9.9	**71.64**	<u>1</u>	<u>55.56</u>	<u>1</u>	53.45	<u>1</u>	54.77
Ionosphere	34	86.97	6.5	**94.82**	<u>1</u>	<u>79.17</u>	<u>1</u>	77.97	<u>1</u>	78.77
M-libras	90	86.06	24.0	**88.89**	<u>1</u>	<u>25.88</u>	<u>1</u>	23.49	<u>1</u>	25.29
Musk 1	168	85.65	36.1	**94.23**	<u>1</u>	<u>63.07</u>	<u>1</u>	61.74	<u>1</u>	62.44
Parkinsons	22	95.84	10.9	**98.84**	<u>1</u>	<u>80.61</u>	<u>1</u>	78.30	<u>1</u>	79.46
Sonar	60	86.70	22.4	**93.79**	<u>1</u>	67.07	<u>1</u>	65.22	<u>1</u>	<u>67.20</u>
Soybean	35	91.85	17.3	**94.91**	<u>1</u>	29.56	<u>1</u>	<u>30.51</u>	<u>1</u>	28.89
SPECTF	44	77.68	5.5	**84.22**	<u>1</u>	78.33	<u>1</u>	78.15	<u>1</u>	<u>78.50</u>
Vehicle	18	70.08	9.9	**74.31**	<u>1</u>	<u>50.43</u>	<u>1</u>	48.79	<u>1</u>	49.58
Vote	16	93.50	5.2	**96.59**	<u>1</u>	95.38	<u>1</u>	95.41	<u>1</u>	<u>95.54</u>
WDBC	30	97.07	18.2	**97.55**	<u>1</u>	<u>91.07</u>	<u>1</u>	90.35	<u>1</u>	90.73

3.2. Computational Cost Reduction

The effects of the cost-reduction mechanisms on the computational time, the number of instances used for evaluation, and the number of evaluations are presented in Tables 5 and 6. Table 5 presents the average execution time of the three proposals with the percentage of reduction obtained using cost-reduction mechanisms. Both cost-reduction proposals could significantly reduce the computational time, but GDE3fix achieved a slightly higher time reduction than GDE3inc.

Table 6 shows that the memory mechanism can avoid around 80% of the evaluations in the search process. In reducing the number of instances used for evaluation, it is seen that GDE3fix achieves a higher reduction than GDE3inc. This behavior is expected given that GDE3inc proportionally incorporates more instances in the search process.

Figure 3 summarizes the effect of incorporating the fixed and incremental sampling fraction strategies with memory to avoid repeated evaluations in the multi-objective version of the DE-FS^{PM} algorithm. It is seen that, despite reducing the required evaluations in a similar percentage, the GDE3inc proposal achieved a higher reduction in the computational cost of the FS procedure and in the number of instances used during evaluation. A higher reduction in the number of instances was expected, given that GDE3fix uses the same amount of instances throughout the FS process, while GDE3inc uses more instances in the final generations.

Table 5. Time reduction obtained when applying the cost-reduction mechanisms to the multi-objective feature selection process of the DE-FS^{PM} algorithm with GDE3. The larger reduction in computational time between the GDE3fix and GDE3inc proposals is marked in bold.

Dataset	GDE3	GDE3fix		GDE3inc	
Arrhythmia	8038.508	2465.115	**69.33%**	2514.494	68.72%
Audiology	3991.253	1035.456	**74.06%**	1170.440	70.67%
Australian	3032.478	233.618	92.30%	228.902	**92.45%**
Cylinder-b	2696.103	429.128	84.08%	364.854	**86.47%**
CRX	2939.885	287.545	90.22%	239.618	**91.85%**
Dermatology	2561.947	544.807	**78.73%**	818.316	68.06%
German-c	4036.628	423.846	**89.50%**	524.601	87.00%
Hill valley	11,660.199	2057.723	**82.35%**	2662.471	77.17%
Ionosphere	2171.877	353.161	**83.74%**	372.972	82.83%
M-libras	5755.865	1765.322	**69.33%**	2640.832	54.12%
Musk 1	7632.818	3085.924	**59.57%**	3704.495	51.47%
Parkinsons	1853.851	299.767	**83.83%**	348.367	81.21%
Sonar	3417.529	782.014	**77.12%**	849.622	75.14%
Soybean	2844.210	820.758	71.14%	700.032	**75.39%**
SPECTF	2627.962	315.105	**88.01%**	331.941	87.37%
Vehicle	3228.412	423.969	86.87%	420.571	**86.97%**
Vote	2278.412	173.897	92.37%	164.088	**92.80%**
WDBC	3075.011	492.209	**83.99%**	518.768	83.13%
Average			**80.92%**		78.49%

Table 6. Percentages of avoided evaluations and reduction in the number of instances used for evaluation. The proposal representing more significant savings in each case is marked in bold.

Dataset	Evaluations		Instances	
	GDE3fix	GDE3inc	GDE3fix	GDE3inc
Arrhythmia	73.05%	**75.94%**	**83.84%**	75.94%
Audiology	**77.64%**	74.12%	**86.59%**	74.12%
Australian	90.91%	**93.56%**	**94.55%**	93.56%
Cylinder-b	83.24%	**88.07%**	**89.94%**	88.07%
CRX	89.43%	**93.00%**	**93.66%**	93.00%
Dermatology	**75.43%**	68.12%	**85.23%**	68.12%
German-c	**86.74%**	86.16%	**92.04%**	86.16%
Hill valley	**77.46%**	74.32%	**86.48%**	74.32%
Ionosphere	84.12%	**84.37%**	**90.46%**	84.37%
M-libras	**69.98%**	57.58%	**81.99%**	57.58%
Musk 1	**51.83%**	51.52%	**71.06%**	51.52%
Parkinsons	**84.35%**	84.11%	**90.61%**	84.11%
Sonar	77.29%	**78.07%**	**86.36%**	78.07%
Soybean	71.12%	**73.27%**	**82.66%**	73.27%
SPECTF	89.09%	**89.84%**	**93.46%**	89.84%
Vehicle	85.66%	**86.55%**	**91.39%**	86.55%
Vote	92.38%	**95.01%**	**95.43%**	95.01%
WDBC	81.10%	**82.55%**	**88.67%**	82.55%
Average	**80.04%**	79.79%	**88.02%**	79.79%

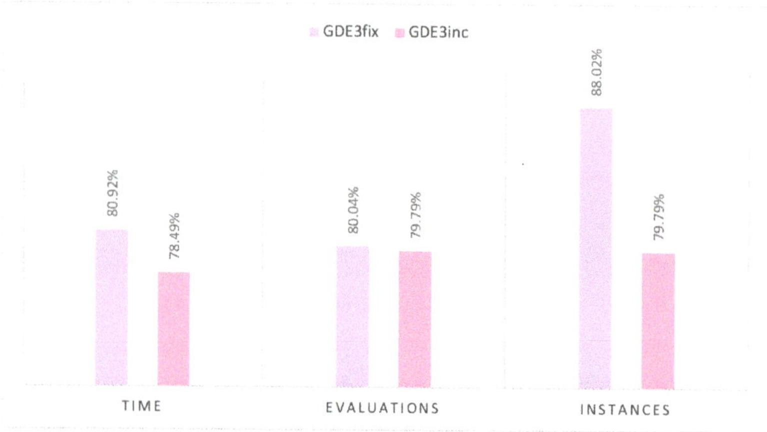

Figure 3. Resource consumption reduction achieved by the proposals GDE3fix and GDE3inc. The plot presents the average reductions concerning the GDE3 procedure in terms of computational time, the number of evaluations, and the number of instances used when evaluating the individuals.

3.3. Multi-Objective Optimization Indicators

Finally, Table 7 presents the performance of the proposals in the hypervolume and spacing indicators for multi-objective optimization. In most cases, the GDE3 proposal achieved a larger hypervolume value, but the GDE3fix and GDE3 proposals presented a better spacing value. The values from the spacing are expected to be small in this problem due to the discrete nature of the Pareto front. The solutions select feature subsets with integer sizes, distributing them in the front.

Table 7. Proposal's results for the hypervolume and spacing indicators. The method with the best performance in each case is marked in bold.

Dataset	Hypervolume			Spacing		
	GDE3	GDE3fix	GDE3inc	GDE3	GDE3fix	GDE3inc
Arrhythmia	**0.7093**	0.6816	0.6660	0.0002	**0.0000**	0.0001
Audiology	**0.7870**	0.7352	0.7422	0.0027	**0.0003**	**0.0003**
Australian	0.8102	**0.8141**	0.8061	0.0005	**0.0002**	0.0004
Cylinder-b	**0.8348**	0.7809	0.7953	0.0006	0.0008	**0.0000**
CRX	0.8167	**0.8201**	0.8105	0.0005	**0.0003**	0.0007
Dermatology	**0.9182**	0.9122	0.9083	0.0030	0.0004	**0.0003**
German-c	0.7282	**0.7301**	0.7130	0.0161	**0.0004**	0.0005
Hill valley	**0.6926**	0.6457	0.6712	0.0003	**0.0001**	**0.0001**
Ionosphere	**0.9086**	0.9028	0.8876	0.0008	**0.0001**	0.0002
M-libras	**0.8663**	0.8203	0.8508	0.0053	**0.0004**	0.0005
Musk 1	**0.9044**	0.8822	0.8852	0.0012	0.0008	**0.0004**
Parkinsons	**0.9366**	0.9208	0.9151	0.0037	**0.0003**	0.0011
Sonar	**0.8955**	0.8830	0.8638	0.0019	**0.0002**	0.0004
Soybean	**0.8707**	0.8531	0.8501	0.0041	**0.0004**	**0.0004**
SPECTF	0.8337	**0.8401**	0.7936	0.0026	0.0009	**0.0002**
Vehicle	**0.6832**	0.6722	0.6618	0.0033	0.0005	**0.0003**
Vote	**0.9048**	0.9028	0.8992	0.0001	0.0001	**0.0000**
WDBC	**0.9442**	0.9436	0.9385	0.0038	0.0007	**0.0003**

4. Discussion

Statistical tests were conducted as suggested in [34] to evaluate if the resource-saving mechanisms could maintain the performance of the GDE3 algorithm. First, the Friedman test was applied with the accuracy results of the GDE3, GDE3fix, and GDE3inc proposals from Table 2, obtaining a p-value of 1.52×10^{-8} that indicates significant differences in the means of the proposals. The post hoc Nemenyi test was then applied to compare the proposals, and the results showed that significant differences were presented in all cases.

An alternative comparison is performed using the hypervolume indicator results to see if the cost-saving mechanisms are affecting the multi-objective search capabilities of the algorithm. The statistical tests were rerun considering the hypervolume values of the GDE3, GDE3fix, and GDE3inc proposals from Table 7. The Friedman test resulted in a p-value of 3.84×10^{-5}, showing significant differences among the proposals. The Nemenyi post hoc test indicated significant differences for the GDE3 and the GDE3fix proposals and for the GDE3 and the GDE3inc procedures.

In contrast with the findings from [23], incorporating the fixed sampling fraction proposal with memory to avoid repeated evaluations resulted in diminishing the performance of the multi-objective FS method. Nonetheless, the fixed and incremental sampling strategies with memory achieved considerably higher resource savings. The time reduction observed in the single-objective approach was reported as 35.16% and 48.61%, respectively, for the fixed and incremental proposals. Whereas in the multi-objective approach, the time reduction was 80.92% and 78.49%, for the fixed and incremental proposals, respectively. The fixed proposal achieved a higher reduction in this case.

An interesting thing to analyze is the number of evaluations the memory mechanism avoids. The reduction of 80% in the number of evaluations clearly indicates that the adaptation of the DE-FS^{PM} algorithm to multi-objective optimization using GDE3 is finding a high number of duplicate individuals. Population diversity and the crowdedness of the solutions are aspects to be considered in future attempts to improve the algorithm's performance.

Another future direction of implementing the cost-reduction mechanisms with the multi-objective version of the DE-FS^{PM} algorithm is to find an adequate compromise between performance and resource savings. As seen in [23], the algorithm's performance is not affected, but the reported resource savings are fewer than the ones found in this work despite using the same configuration of the mechanisms. The previous observation suggests that the cost-reduction mechanisms require a more specific parameter configuration for their application in different FS approaches. Additionally, a comparison among the subsets of features selected by each proposal can provide valuable insights into the differences in the results of the FS process. Applying the FS processes in a real-life application, like the one from [35], will result in a deeper comparison of the results when collaborating with an expert in the field.

5. Conclusions

In this work, fixed and incremental sampling fraction strategies with memory to avoid repeated evaluations were implemented as cost-reduction mechanisms in a multi-objective permutational-based DE approach for FS. The proposed approach for multi-objective FS presented some limitations. The main one is in the method's performance in the accuracy and multi-objective indicators, which were reduced by the use of cost-reduction mechanisms. The proposals presented limited exploration capabilities, evidenced by the number of evaluations they avoid due to the finding of repeated individuals. The results exhibit high savings of computational resources at the expense of reducing the method's performance. The previous observation shows that the findings in the single-objective version of the algorithm, where the fixed sampling fraction proposal achieved the goal of reducing the computational cost of a wrapper approach for FS without diminishing its performance, is not maintained in a multi-objective FS approach.

The multi-objective approach for FS using the permutational version of GDE3 does not achieve accuracy results as high as the single-objective version of the DE-FS^{PM} algorithm. Nevertheless, the advantage of using a multi-objective approach is that more than one solution to the problem is found, allowing the user to choose the most convenient subset of features among the options in the Pareto front. The GDE3-based proposals found solutions with smaller feature subsets and involved different compromises between the accuracy performance and the number of selected features. The previous results are an advantage of the multi-objective approach. However, it requires an additional step in the process that can be performed with an automatic technique or by following the recommendation of an expert in the data field in the context of using the FS technique in a real-life application.

The parameter configuration used affects the algorithm's search capabilities. In DE, the F and CR parameters control the algorithm's exploration/exploitation capabilities. Finding an adequate set of parameters for every dataset in the FS process is a complicated task. A parameter adaptation scheme can help increase the algorithm's search capabilities in future work. Another critical aspect for future improvements is considering mechanisms to improve the population's diversity. Given the high number of repeated individuals found, the procedure appears to suffer from stagnation. Following the previous point, another area to explore in future changes to the algorithm is the mechanism used to reduce the population when it grows with the selection mechanism of the GDE3 algorithm. Finding more effective selection criteria for the FS problem will be helpful.

Future experimentation, in which the resource-saving mechanisms are applied to different single- and multi-objective approaches for FS, will provide more insights into how robust the mechanisms are for computational cost reduction. As presented in this work, finding a balance in the severity of the proposed resource savings is necessary if the goal is to maintain the algorithm's performance. Consequently, additional experimentation focusing on the effect of the parameters S and B will be helpful. A further aspect to be considered in future experimentation is the mechanism used to select a solution among the Pareto optimal ones that the algorithm is returning. A more complex selection procedure can provide options for considering a different trade-off between the objectives than just selecting the solution that performs best in one objective or the knee point in the front.

Author Contributions: Conceptualization, J.-A.B.-P., R.R.-L., E.M.-M., H.-G.A.-M. and A.M.-G.; methodology, R.R.-L. and E.M.-M.; software, J.-A.B.-P.; validation, R.R.-L., E.M.-M. and H.-G.A.-M.; formal analysis, J.-A.B.-P. and A.M.-G.; investigation, J.-A.B.-P.; resources, R.R.-L. and E.M.-M.; writing—original draft preparation, J.-A.B.-P.; writing—review and editing, R.R.-L., E.M.-M., H.-G.A.-M. and A.M.-G.; visualization, J.-A.B.-P. and A.M.-G.; supervision, E.M.-M. and R.R.-L. All authors have read and agreed to the published version of the manuscript.

Funding: This research received no external funding.

Data Availability Statement: The data presented in this study are available in the UCI Machine Learning repository at https://archive.ics.uci.edu/, reference number [33].

Acknowledgments: The first author (CVU 1142850) acknowledges support from the Mexican National Council of Humanities, Science, and Technology (CONAHCYT) with a scholarship to pursue PhD studies at Universidad Veracruzana.

Conflicts of Interest: The authors declare no conflicts of interest.

Abbreviations

The following abbreviations are used in this manuscript:

FS	Feature Selection
DE	Differential Evolution
EA	Evolutionary Algorithm
CV	Cross-validation
GDE3	Generalized Differential Evolution 3
KNN	K-nearest-neighbors
DE-FS^{PM}	Permutational-based Differential Evolution Algorithm for Feature Selection

References

1. Sharma, M.; Kaur, P. A Comprehensive Analysis of Nature-Inspired Meta-Heuristic Techniques for Feature Selection Problem. *Arch. Comput. Methods Eng.* **2021**, *28*, 1103–1127. [CrossRef]
2. Dokeroglu, T.; Deniz, A.; Kiziloz, H.E. A comprehensive survey on recent metaheuristics for feature selection. *Neurocomputing* **2022**, *494*, 269–296. [CrossRef]
3. Abdulwahab, H.M.; Ajitha, S.; Saif, M.A.N. Feature selection techniques in the context of big data: Taxonomy and analysis. *Appl. Intell.* **2022**, *52*, 13568–13613. [CrossRef]
4. Brezočnik, L.; Fister, I.; Podgorelec, V. Swarm Intelligence Algorithms for Feature Selection: A Review. *Appl. Sci.* **2018**, *8*, 1521. [CrossRef]
5. Agrawal, P.; Abutarboush, H.F.; Ganesh, T.; Mohamed, A.W. Metaheuristic Algorithms on Feature Selection: A Survey of One Decade of Research (2009–2019). *IEEE Access* **2021**, *9*, 26766–26791. [CrossRef]
6. Xue, B.; Zhang, M.; Browne, W.N.; Yao, X. A Survey on Evolutionary Computation Approaches to Feature Selection. *IEEE Trans. Evol. Comput.* **2016**, *20*, 606–626. [CrossRef]
7. Dhal, P.; Azad, C. A comprehensive survey on feature selection in the various fields of machine learning. *Appl. Intell.* **2022**, *52*, 4543–4581. [CrossRef]
8. Theng, D.; Bhoyar, K.K. Feature selection techniques for machine learning: A survey of more than two decades of research. *Knowl. Inf. Syst.* **2024**, *66*, 1575–1637. [CrossRef]
9. Ahmad, M.F.; Isa, N.A.M.; Lim, W.H.; Ang, K.M. Differential evolution: A recent review based on state-of-the-art works. *Alex. Eng. J.* **2022**, *61*, 3831–3872. [CrossRef]
10. Mezura-Montes, E.; Reyes-Sierra, M.; Coello, C.A.C. Multi-objective Optimization Using Differential Evolution: A Survey of the State-of-the-Art. In *Advances in Differential Evolution*; Chakraborty, U.K., Ed.; Springer: Berlin/Heidelberg, Germany, 2008; pp. 173–196. [CrossRef]
11. Hancer, E.; Xue, B.; Zhang, M. Differential evolution for filter feature selection based on information theory and feature ranking. *Knowl.-Based Syst.* **2018**, *140*, 103–119. [CrossRef]
12. Hancer, E.; Xue, B.; Zhang, M. An evolutionary filter approach to feature selection in classification for both single- and multi-objective scenarios. *Knowl.-Based Syst.* **2023**, *280*, 111008. [CrossRef]
13. Xue, B.; Fu, W.; Zhang, M. Multi-objective Feature Selection in Classification: A Differential Evolution Approach. In *Proceedings of the Simulated Evolution and Learning*; Dick, G., Browne, W.N., Whigham, P., Zhang, M., Bui, L.T., Ishibuchi, H., Jin, Y., Li, X., Shi, Y., Singh, P., et al., Eds.; Springer: Cham, Switzerland, 2014; pp. 516–528.
14. Wang, P.; Xue, B.; Liang, J.; Zhang, M. Differential Evolution-Based Feature Selection: A Niching-Based Multiobjective Approach. *IEEE Trans. Evol. Comput.* **2023**, *27*, 296–310. [CrossRef]
15. Bidgoli, A.A.; Ebrahimpour-Komleh, H.; Rahnamayan, S. A Novel Multi-objective Binary Differential Evolution Algorithm for Multi-label Feature Selection. In Proceedings of the 2019 IEEE Congress on Evolutionary Computation (CEC), Wellington, New Zealand, 10–13 June 2019; pp. 1588–1595. [CrossRef]
16. Wang, P.; Xue, B.; Liang, J.; Zhang, M. Feature Selection Using Diversity-Based Multi-objective Binary Differential Evolution. *Inf. Sci.* **2023**, *626*, 586–606. [CrossRef]
17. Hancer, E. A new multi-objective differential evolution approach for simultaneous clustering and feature selection. *Eng. Appl. Artif. Intell.* **2020**, *87*, 103307. [CrossRef]
18. Yu, X.; Hu, Z.; Luo, W.; Xue, Y. Reinforcement learning-based multi-objective differential evolution algorithm for feature selection. *Inf. Sci.* **2024**, *661*, 120185. [CrossRef]
19. Rivera-López, R.; Mezura-Montes, E.; Canul-Reich, J.; Cruz-Chávez, M.A. A permutational-based Differential Evolution algorithm for feature subset selection. *Pattern Recognit. Lett.* **2020**, *133*, 86–93. [CrossRef]
20. Mendoza-Mota, J.A. Selección de Atributos con un Enfoque Evolutivo Multiobjetivo. Master's Thesis, Laboratorio Nacional de Informática Avanzada, Xalapa-Enríquez, Mexico, 2021.
21. Malekipirbazari, M.; Aksakalli, V.; Shafqat, W.; Eberhard, A. Performance comparison of feature selection and extraction methods with random instance selection. *Expert Syst. Appl.* **2021**, *179*, 115072. [CrossRef]
22. Barradas-Palmeros, J.A.; Rivera-López, R.; Mezura-Montes, E.; Acosta-Mesa, H.G. Experimental Study of the Instance Sampling Effect on Feature Subset Selection Using Permutational-Based Differential Evolution. In *Proceedings of the Advances in Computational Intelligence, MICAI 2023 International Workshops*; Calvo, H., Martínez-Villaseñor, L., Ponce, H., Zatarain Cabada, R., Montes Rivera, M., Mezura-Montes, E., Eds.; Springer: Cham, Switzerland, 2024; pp. 409–421. [CrossRef]
23. Barradas-Palmeros, J.A.; Mezura-Montes, E.; Rivera-López, R.; Acosta-Mesa, H.G. Computational Cost Reduction in Wrapper Approaches for Feature Selection: A Case of Study Using Permutational-Based Differential Evolution (In press). In Proceedings of the 2024 IEEE Congress on Evolutionary Computation (CEC), Yokohama, Japan, 30 June–5 July 2024.
24. Tanabe, R.; Fukunaga, A. Success-history based parameter adaptation for Differential Evolution. In Proceedings of the 2013 IEEE Congress on Evolutionary Computation, Cancun, Mexico, 20–23 June 2013; pp. 71–78. [CrossRef]
25. Kitamura, T.; Fukunaga, A. Duplicate Individuals in Differential Evolution. In Proceedings of the 2022 IEEE Congress on Evolutionary Computation (CEC), Padua, Italy, 18–23 July 2022; pp. 1–8. [CrossRef]
26. Al-Tashi, Q.; Abdulkadir, S.J.; Rais, H.M.; Mirjalili, S.; Alhussian, H. Approaches to Multi-Objective Feature Selection: A Systematic Literature Review. *IEEE Access* **2020**, *8*, 125076–125096. [CrossRef]

27. Kukkonen, S.; Coello, C.A. Generalized Differential Evolution for Numerical and Evolutionary Optimization. In *NEO 2015: Results of the Numerical and Evolutionary Optimization Workshop NEO 2015 Held at September 23–25 2015 in Tijuana, Mexico*; Schütze, O., Trujillo, L., Legrand, P., Maldonado, Y., Eds.; Springer International Publishing: Cham, Switzerland, 2017; pp. 253–279. [CrossRef]
28. Storn, R.; Price, K. Differential Evolution—A Simple and Efficient Heuristic for global Optimization over Continuous Spaces. *J. Glob. Optim.* **1997**, *11*, 341–359. [CrossRef]
29. Eiben, A.E.; Smith, J.E. *Introduction to Evolutionary Computing*; Springer: Berlin/Heidelberg, Germany, 2015. [CrossRef]
30. Price, K.V.; Storn, R.M.; Lampinen, J.A. *Differential Evolution: A Practical Approach to Global Optimization*; Springer: Berlin/Heidelberg, Germany, 2005. [CrossRef]
31. Kukkonen, S.; Lampinen, J. GDE3: The third evolution step of generalized differential evolution. In Proceedings of the 2005 IEEE Congress on Evolutionary Computation, Edinburgh, UK, 2–5 September 2005; Volume 1, pp. 443–450. [CrossRef]
32. Deb, K.; Pratap, A.; Agarwal, S.; Meyarivan, T. A fast and elitist multiobjective genetic algorithm: NSGA-II. *IEEE Trans. Evol. Comput.* **2002**, *6*, 182–197. [CrossRef]
33. Kelly, M.; Longjohn, R.; Nottingham, K. The UCI Machine Learning Repository. Available online: https://archive.ics.uci.edu (accessed on 25 July 2023).
34. Derrac, J.; García, S.; Molina, D.; Herrera, F. A practical tutorial on the use of nonparametric statistical tests as a methodology for comparing evolutionary and swarm intelligence algorithms. *Swarm Evol. Comput.* **2011**, *1*, 3–18. [CrossRef]
35. Vargas-Moreno, I.; Rodríguez-Landa, J.F.; Acosta-Mesa, H.G.; Fernández-Demeneghi, R.; Oliart-Ros, R.; Hernández Baltazar, D.; Herrera-Meza, S. Effects of Sterculia Apetala Seed Oil on Anxiety-like Behavior and Neuronal Cells in the Hippocampus in Rats. *J. Food Nutr. Res.* **2023**, *11*, 211–222. [CrossRef]

Mathematical and Computational Applications

MDPI

Article

Modeling of the Human Cardiovascular System: Implementing a Sliding Mode Observer for Fault Detection and Isolation

Dulce A. Serrano-Cruz [1,2,*], Latifa Boutat-Baddas [2,*], Mohamed Darouach [2], Carlos M. Astorga-Zaragoza [1] and Gerardo V. Guerrero Ramírez [1]

1 Tecnológico Nacional de México, CENIDET, Cuernavaca C. P. 62490, Mexico; carlos.az@cenidet.tecnm.mx (C.M.A.-Z.); gerardo.gr@cenidet.tecnm.mx (G.V.G.R.)

2 CRAN, UMR 7039, Université de Lorraine, IUT de Longwy, 186, Rue de Lorraine, 54400 Cosnes et Romain, France; mohamed.darouach@univ-lorraine.fr

* Correspondence: alejandra.serrano17ee@cenidet.edu.mx (D.A.S.-C.); latifa.baddas@univ-lorraine.fr (L.B.-B.)

Abstract: This paper presents a mathematical model of the cardiovascular system (CVS) designed to simulate both normal and pathological conditions within the systemic circulation. The model introduces a novel representation of the CVS through a change of coordinates, transforming it into the "quadratic normal form". This model facilitates the implementation of a sliding mode observer (SMO), allowing for the estimation of system states and the detection of anomalies, even though the system is linearly unobservable. The primary focus is on identifying valvular heart diseases, which are significant risk factors for cardiovascular diseases. The model's validity is confirmed through simulations that replicate hemodynamic parameters, aligning with existing literature and experimental data.

Keywords: cardiovascular system; heart diseases; pressure–volume loops; normal form; sliding mode observer; fault detection and isolation; unobservable states

1. Introduction

The World Health Organization (WHO) has reported that cardiovascular diseases (CVDs) continue to be the world's leading cause of mortality [1]. In clinical practice, arterial pressure (AP) is frequently used to evaluate cardiovascular health and disease. However, it has been suggested that aortic pressure (A_o), measured near to the heart, provides more information about health behavior and diseases compared to blood pressure measurements. Nevertheless, the widespread adoption of aortic pressure measurement faces significant challenges due to the invasive and/or inconvenient procedures required, as well as the need for skilled physicians to perform direct measurements, such as cardiac catheterization and carotid artery tonometry [2].

These challenges have spurred research on the cardiovascular system (CVS) from various perspectives, including those that diverge from traditional medical approaches. Among these perspectives are computational and mathematical models, which allow for experiments that are much simpler and less expensive to conduct than in vivo or in vitro heart experiments. Given the complexity of the cardiovascular system, interdisciplinary expertise may be helpful in developing dynamic models that can predict cardiovascular events in patients with heart failure, myocardial infarction, or valvular heart disease. Research on the CVS is ongoing, with significant efforts dedicated to modeling the CVS to better understand its behavior and to develop new, reliable diagnostic techniques [3–6]. Simplified parameter models are particularly noteworthy, as they provide streamlined representations of the primary behaviors of each CVS component [3–12]. Mathematical models have thus emerged as valuable tools, offering simpler and less expensive alternatives to in vitro heart experiments [6,13–15]. For systems described by dynamic models, there are methods for fault localization and detection that rely on creating fault indicators, also known as residuals. These indicators are determined by comparing actual system measurements with estimates made by an observer. Numerous results based on linear models have been published across various contexts [9,16–18]. While observers used to reconstruct system states have been very useful for monitoring and detecting anomalies, the implementation of observers for nonlinear models has not been extensively explored. This is particularly challenging because the design of observers is generally more sensitive when a nonlinear model is required to represent the system's behavior. Currently, there are no straightforward, generic methods for creating observers for all types of nonlinear systems.

This work used a model of the cardiovascular system (CVS) from the study [19], designed to simulate both normal and pathological conditions within the systemic circulation. The model introduces a novel representation of the CVS by transforming it into the "quadratic normal form" through a change of coordinates. This model offers a structured approach to understanding the complexity of this system, aiding the development of clinical decision support systems for cardiovascular diseases (CVDs), and facilitates the implementation of a sliding mode observer (SMO), enabling the estimation of system states and the detection of anomalies, even though the system is linearly unobservable.

This paper is structured as follows: Section 2 provides a brief overview of CVS functionality. In Section 3, the justification for employing electrical analogies in the CVS description is discussed, along with the depiction of the CVS model and a validation of the proposed model against clinical indices and experimental data. Section 4 explores the potential of the quadratic normal form and sliding mode observer design as suitable tools for CVD modeling. Section 5 presents the CVS model designed for anomaly detection, focusing on two types of faults in the mitral and aortic valves. Simulation results are then presented to illustrate how this design can estimate system states for cardiovascular activity monitoring. Finally, Section 6 presents the conclusions.

2. Cardiovascular System Model

This section is dedicated to describing the dynamic behavior of the CV system from both a medical and control theory perspective.

2.1. Anatomy and Physiology of the Cardiac Cycle

The dynamic behavior of the cardiac cycle can be described as a distribution network of blood vessels to supply oxygenated and deoxygenated blood throughout the body, thanks to the heart behaving as a pump and its pressure–volume (PV) loops.

- **Blood circulation pathway**

The path followed by the blood is presented as a closed circuit, starting at the heart, which is responsible for pumping blood. This is illustrated in Figure 1 through a schematic cross-section of the heart, consisting of double atria-ventricular chambers on both sides. Where, the ventricles act as the primary pumps, while the atria serve as preload chambers that regulate the distinct paths of blood circulation. Specifically, the right side of the heart regulates blood flow in to the pulmonary artery, which carries to the lungs, where blood is oxygenated in the lungs and then it returns to the left side of the heart entering through the left atrium. Subsequently, the oxygen-rich blood is pumped by the left ventricle through the aorta, regulating blood circulation to the rest of the body [20]. Additionally, it is important to note that blood flows in one direction only, due to one-way valves being situated between the chambers to prevent reflux, and at the output of the ventricles, called semilunar valves as shown in Figure 1.

- **Cardiac cycle phases**

From a functional point of view, the cardiac cycle is divided into two alternating phases: diastole (dilatation period) and systole (contraction period), which are simplified into four stages as shown in Figure 1:

(1) The first stage is atrial diastole and the beginning of ventricular systole, during which the atria relax while the ventricles contract and the atrioventricular valves close. This increases the pressure inside the ventricles but not enough to open the semilunar valves.

(2) The second stage is ventricular diastole, when the pressure inside the ventricles rapidly decreases, the atrioventricular valves open, and the chambers passively fill due to their relaxation combined with atrial systole, during which the atria contract to fill the ventricles.
(3) The third stage is atrial systole, during which the pressure in the ventricles rises until it exceeds that of the arteries. This leads to the opening of the semilunar valves and the ejection of blood into the pulmonary artery, marking the beginning of systemic circulation.
(4) The last stage marks the end of ventricular systole and the start of the ventricular and atrial diastole. During this phase, the pressure in the ventricles decreases rapidly, and all chambers passively fill due to their relaxation. This transition leads into a new cardiac cycle, beginning with atrial systole.

An alternative method to graphically describe and characterize the cardiac cycle is through the use of a left ventricle (PV) loop. This loop illustrates the relationship between left ventricular pressure (LVP) and left ventricular volume (LVV) across the four stages of the cardiac cycle. It enables the identification of changes in cardiac function, including the factors related to preload and afterload, as well as heart contractility (for more information see [6]).

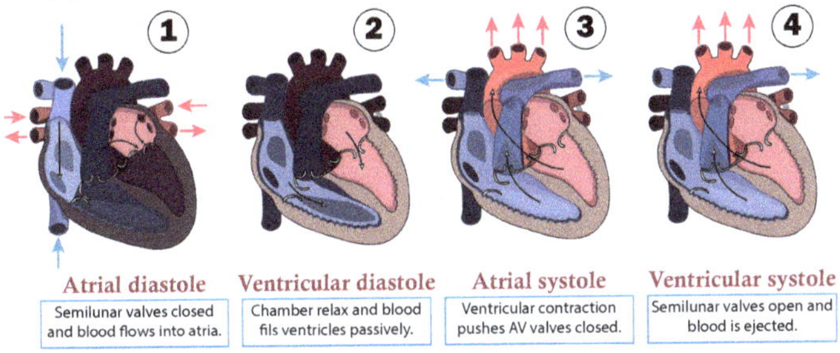

Figure 1. Cardiac cycle of circulatory system.

2.2. Valve Pathologies

Valvular heart diseases are a leading cause of cardiovascular morbidity and mortality worldwide [1,21]. Among the most frequent valve pathologies are those impacting the aortic and mitral valves. These pathologies often result in both stenosis (narrowing) and regurgitation (impaired closure).

Aortic valve stenosis refers to the insufficient opening of the valve during systole, often caused by congenital abnormalities or the progressive buildup of calcium on the valve leaflets with age [6,22,23]. Conversely, a malfunction in the aortic valve closure, known as aortic valve regurgitation, results in backward leakage into the left ventricle during diastole. This condition shares similar causes with aortic valve stenosis [22].

Both aortic stenosis and regurgitation lead to the hypertrophy of the left ventricle in response to increased stress, resulting in the thickening of the left ventricular muscle and the subsequent elevation of left ventricular pressure.

3. Description of the Cardiovascular System Model

This section outlines the equivalences between electrical and hydrodynamic indices. We can simulate the CVS electrically by utilizing the equivalencies between hydrodynamic and electrical indicators. This model addresses the simulation of the system and the contractile activity of the heart once the relationships and equivalencies between an electrical circuit and the behavior defined by a segment of the CVS are established.

3.1. Equivalent Electric Model

The heart is a highly complex system that presents significant challenges for mathematical modeling. In recent years, numerous dynamical state-space models with varying levels of complexity have been developed [10]. The main methodology employed is that of lumped parameter models, which provide simplified descriptions of the predominant behavior of each component involved in the CVS [4,7,10–12].

The model discussed in this work is based on an electrical representation of the CVS, as proposed in [4,7,12]. The choice of this model is motivated by the need for a comprehensive model that can be validated from a medical perspective and is capable of describing cardiovascular phenomena, such as valve pathologies, which are among the primary risk factors for cardiovascular diseases (CVDs). This model primarily targets the left chambers of the heart, assuming that voltages are analogous to pressure and currents are analogous to blood flow. The systemic resistance R_S is the resistance to flow from the descending aorta through the capillary vessels, venous, and pulmonary circulation to reach the left atrium. Left ventricular pressure (LVP) is represented by the voltage across the time-varying contractile capacity $C(t)$, where its capacitance is defined as the inverse of left ventricular elastance $E(t)$. The term $E(t)$ represents the elastance of the heart at time t, which is a function of the pressure. The mitral and aortic valves are represented as ideal diodes, D_A and D_M, in series with resistance R_A and R_M, respectively. The capacitor C_A, represents the elasticity of the ascending aorta, simulating the pressure variations caused by the opening and closing of the aortic valve. Finally, the remaining components model the anatomical characteristics of the circulatory system, including the elasticity represented by C_S, inertia (L_S), and resistance R_C of the descending aorta [6]. The electrical model circuit in Figure 2 has been thoroughly analyzed in [6,20].

The state variables and parameter values of the cardiovascular circuit model shown in Figure 2, as referenced from [6,17,20], are detailed in Tables 1 and 2 below:

Table 1. State variables of the cardiovascular system and their physiological significance of the circuit model shown in Figure 2.

Variables	Abbreviation	Physiological Meaning (Unit)
$x_1(t)$	LVP(t)	Left ventricular pressure (mmHg)
$x_2(t)$	LAP(t)	Left atrial pressure (mmHg)
$x_3(t)$	AP(t)	Descending arterial pressure (mmHg)
$x_4(t)$	$A_o(t)$	Ascending aortic pressure (mmHg)
$x_5(t)$	F(t)	Total aortic flow (mL/s)

Table 2. Parameter values of the CVS circuit model shown in Figure 2.

Parameter	Value	Physiological Meaning
C_S	1.33	Systemic compliance
C_R	4.40	Left atrial compliance
C_R	4.40	Aortic compliance
L_S	0.0005	Inertia of blood in aorta
R_C	0.0398	Characteristic resistance
R_M	0.005	Mitral valve resistance
R_A	0.001	Aortic valve resistance
	Left Ventricle	
E_{max}	2	Maximum volume in diastole
E_{min}	0.06	Minimum volume in diastole
V_0	10	Reference volume at zero pressure (mL)
H_R	75	Heart rate (bpm)
	Elastance	
	1.17	Shape parameter
	0.7	Shape parameter
	1.55	Amplitude
	1.9	Ascending slope of the LV relaxation time
	21.9	Descending slope of the LV relaxation time

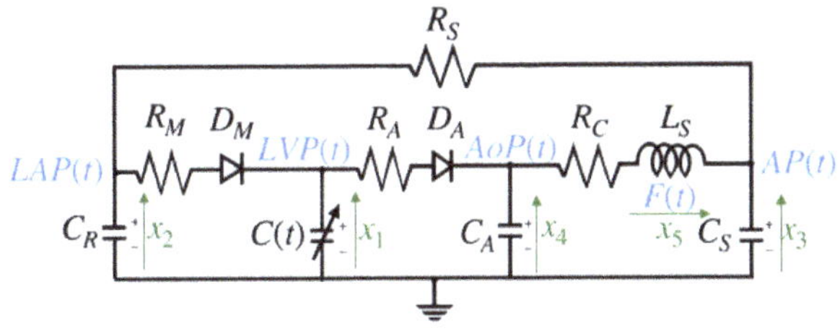

Figure 2. Cardiovascular circuit model.

3.2. Elastance

Elastance, denoted as $E(t)$, relates to the state of contraction of the left ventricle. It represents the relationship between the pressure and volumes of the LV, as defined by the following expression:

$$E(t) = \frac{LVP(t)}{LVV(t) - V_0} = \frac{x_1(t)}{LVV(t) - V_0} \tag{1}$$

where $LVP(t)$ is the left ventricular pressure, $LVV(t) = \frac{x_1(t)}{E(t)} + V_0$ is the left ventricular volume, and V_0 is a reference volume, which corresponds to the theoretical volume in the ventricle at zero pressure. The elastance function $E(t)$ has been addressed in various studies [20]. These studies concur that the definition can be mathematically approximated using an expression where the points at which the left ventricular function reaches its maximum and minimum are identified used the expression:

$$E(t) = (E_{max} - E_{min})E_n(t_n) + E_{min} \tag{2}$$

where E_{max} and E_{min} are constants related to the end-systolic volume (ESV) and end-diastolic volume (EDV), representing the left ventricular volumes at systole and diastole, respectively. The end-systolic pressure–volume relationship (ESPVR) denotes the maximal pressure of the left ventricle.

The elastance function is implemented using a number of mathematical approximations, including $E_n(t_n)$, the so-called "double hill" [24]. In this paper, $E_n(t_n)$ represents the normalized elastance at time t_n. "Normalized" means that it has been adjusted or expanded to fit a specific range, often between 0 and 1 or −1 and 1. The normalized elastance $E_n(t_n)$ is scaled proportionally between E_{min} and E_{max}. Specifically, when $E_n(t_n) = 0$, $E(t)$ equal E_{min}, and when $E_n(t_n) = 1$, $E(t)$ equal E_{max}. In the context of the cardiovascular system, $E_n(t_n)$ describes how elastance dynamically varies over time adjusted to heart rate, and this relationship is expressed by:

$$E_n(t_n) = 1.55\left(\frac{(\frac{tn}{0.7})^{1.9}}{1 + (\frac{tn}{0.7})^{1.9}}\right)\left(\frac{1}{1 + (\frac{tn}{1.17})^{21.9}}\right) \tag{3}$$

where $t_n = t/(0.2 + 0.15\frac{60}{H_R})$, with H_R being the heart rate expressed in beats per minute (bpm). The first term within the brackets describes the ascending segment of the curve, while the subsequent term portrays its descending counterpart. The value 1.55 corresponds to the amplitude of elastance, which is associated with the maximum arterial pressure. Additionally, 1.9 and 21.9 indicate the ascending and descending slopes during the LV relaxation period, respectively, while 0.7 and 1.17 are constants that determine the proportional representation of each curve over the cardiac cycle. Figure 3 illustrates the graphical representations of these curves.

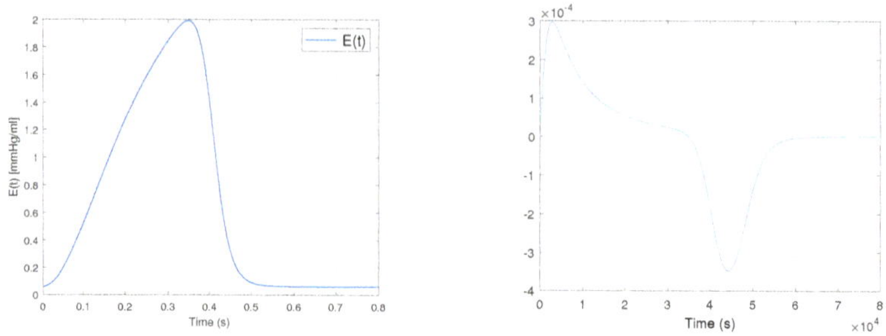

(a) (b)

Figure 3. Plot (**a**) shows the elastance function $E(t) = \frac{1}{C(t)}$ and (**b**) shows the expression $-\frac{\dot{C}(t)}{C(t)}$ for a healthy heart during a single cardiac cycle.

3.3. Presentation of the Mathematical Model of the Cardiovascular System

We consider the CVS model basically given by the following equations [12]:

$$\begin{pmatrix} \dot{x}_1 \\ \dot{x}_2 \\ \dot{x}_3 \\ \dot{x}_4 \\ \dot{x}_5 \end{pmatrix} = \begin{pmatrix} \frac{-\dot{C}(t)}{C(t)} & 0 & 0 & 0 & 0 \\ 0 & -\frac{1}{R_S C_R} & \frac{1}{R_S C_R} & 0 & 0 \\ 0 & \frac{1}{R_S C_S} & -\frac{1}{R_S C_S} & 0 & \frac{1}{C_S} \\ 0 & 0 & 0 & 0 & -\frac{1}{C_A} \\ 0 & 0 & -\frac{1}{L_S} & \frac{1}{L_S} & -\frac{R_C}{L_S} \end{pmatrix} \begin{pmatrix} x_1 \\ x_2 \\ x_3 \\ x_4 \\ x_5 \end{pmatrix} + \begin{pmatrix} \frac{1}{C(t)R_M} & \frac{1}{C(t)R_A} \\ \frac{-1}{C_R R_M} & 0 \\ 0 & 0 \\ 0 & \frac{-1}{C_A R_A} \\ 0 & 0 \end{pmatrix} \begin{pmatrix} D_M(x_2 - x_1) \\ D_A(x_4 - x_1) \end{pmatrix} \tag{4}$$

where $\begin{pmatrix} x_1 \\ x_2 \\ x_3 \\ x_4 \\ x_5 \end{pmatrix}$ represents the state vector of CVS circuit model (see Figure 2 and Tables 1 and 2) and $\begin{pmatrix} D_M \\ D_A \end{pmatrix}$ represents the natural control input sequences of cardiovascular system, with D_M is the state of the mitral valve and D_A is the state of the aortic valve given by:

$$D_M = \begin{cases} 0, & x_2 < x_1 \\ 1, & x_2 \geq x_1 \end{cases}, D_A = \begin{cases} 0, & x_1 < x_4 \\ 1, & x_1 \geq x_4 \end{cases} \tag{5}$$

3.4. Quadratic Normal Form of the CVS [25]

For the remainder of our work of the CVS system, we take the output vector as $y(t) = \begin{pmatrix} y_1 \\ y_2 \end{pmatrix} = \begin{pmatrix} x_5 \\ x_4 \end{pmatrix}$ and the input vector as $u(t) = \begin{pmatrix} u_1 \\ u_2 \\ u_3 \\ u_4 \end{pmatrix} = \begin{pmatrix} D_M \\ D_A \\ D_M \frac{1}{C(t)} \\ D_A \frac{1}{C(t)} \end{pmatrix}$. Given this output, it is easy to demonstrate that x_1 is linearly unobservable in system (4). To proceed with calculating the quadratic normal form, we introduce the following change of coordinates:

$$\begin{aligned} z_1^1 &= x_5 \\ z_2^1 &= -\frac{1}{L_S} x_3 \\ z_3^1 &= -\frac{1}{L_S R_S C_S} x_2 + \frac{1}{L_S R_S C_S} x_3 \\ z_1^2 &= x_4 \\ z_2^2 &= C(t) x_1(t) \end{aligned} \tag{6}$$

which is equivalent to

$$
\begin{array}{lllll}
x_1 & = & \frac{1}{C(t)} z_2^2 & = & \xi_1 \\
x_2 & = & L_S R_S C_S z_3^1 + L_S z_2^1 & = & \xi_2 \\
x_3 & = & -L_S z_2^1 & = & \xi_3 \\
x_4 & = & z_1^2 & = & \xi_4 \\
x_5 & = & z_1^1 & = & \xi_5
\end{array} \quad (7)
$$

We directly obtain the quadratic normal form (QNF) [25,26] of the CVS model:

$$
\begin{cases}
\dot{z}_1^1 = -\frac{R_C}{L_S} z_1^1 + z_2^1 + \frac{1}{L_S} z_1^2 \\
\dot{z}_2^1 = -\frac{1}{L_S C_S} z_1^1 + z_3^1 \\
\dot{z}_3^1 = \frac{1}{L_S R_S C_S^2} z_1^1 - \beta z_3^1 - k_1 z_3^1 u_1 - k_2 z_2^1 u_1 - k_3 z_2^2 u_3 \\
\dot{z}_1^2 = -\frac{1}{C_A} z_1^1 - \frac{1}{C_A R_A} z_1^2 u_2 - \frac{1}{C_A R_A} z_2^2 u_4 \\
\dot{z}_2^2 = \frac{L_S R_S C_S}{R_M} z_3^1 u_1 - \frac{L_S}{R_M} z_2^1 u_1 + \frac{1}{R_A} z_1^2 u_2 - \frac{1}{R_M} z_2^2 u_3 - \frac{1}{R_A} z_2^2 u_4 \\
y_1 = z_1^1 \\
y_2 = z_1^2
\end{cases} \quad (8)
$$

where $\beta = \frac{1}{R_S C_R} + \frac{1}{R_S C_S}$, $k_1 = \frac{1}{C_R R_M}$, $k_2 = \frac{1}{R_S C_S C_R R_M}$ and $k_3 = \frac{1}{L_S R_S C_S C_R R_M}$.

And $u_1 = D_M$, $u_2 = D_A$, $u_3 = D_M \frac{1}{C(t)}$ and $u_4 = D_A \frac{1}{C(t)}$.

Remark 1. *As a result, building on the work in [25], thanks to the quadratic terms $k_3 u_3 z_2^2$ and $\frac{1}{C_A R_A} z_2^2 u_4$, we can recover observability for z_2^2.*

3.5. Validation of the Quadratic Normal Form of the CVS Model

This section presents the validation process for the quadratic normal form of the CVS model obtained, drawing upon previously established validations and incorporating diverse analytical perspectives, as detailed in seminal works such as [6,17]. Initially, the model's accuracy is substantiated by juxtaposing the waveforms of principal variables, as depicted in Figure 4, against empirical data from healthy subjects reported in [6]. Subsequently, the model's robustness is assessed through its responsiveness to alterations in preload and afterload factors, ensuring its consistency and reliability in simulating physiological conditions.

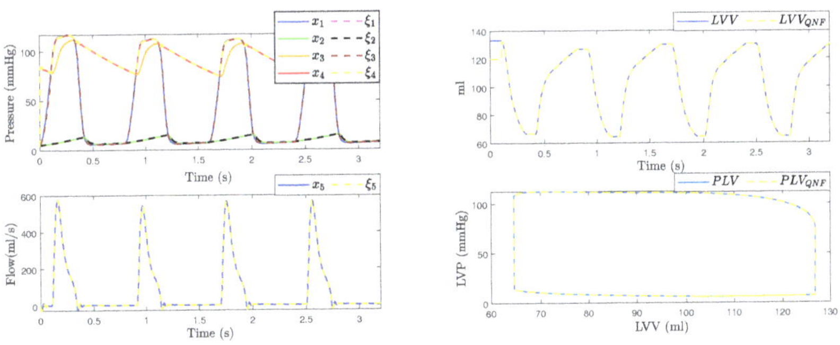

(**a**) (**b**)

Figure 4. Hemodynamic waveforms of the CVS model (4) compared with experimental data presented in [6]. (**a**) Original states $x_i(t)$ and QNF states $\xi_i(t)$ of CVS; (**b**) Left ventricular volume (LVV) and preload volume (PLV) in original and QNF system.

With these results, we can affirm that the quadratic normal form obtained offers a novel alternative for the representation of the classical model presented in the literature.

Additionally, it is crucial to emphasize that a significant advantage of the quadratic normal form is its capacity to enable the design of observers. These observers can estimate the states of the system that are not directly measurable and apply other control theory concepts. An example of such an application, as discussed in this work, is fault detection and estimation.

The validation of the model also involves a dynamic analysis concerning preload and afterload factors. To evaluate this aspect, we analyze the preload and afterload signals generated by both the original and the quadratic normal form of the CVS model. Figure 5b displays the left ventricular pressure data and the corresponding pressure–volume loop obtained from our model using a volume of $V_0 = 10$ mL. It is observed that the dynamics obtained are consistent with those described in [6].

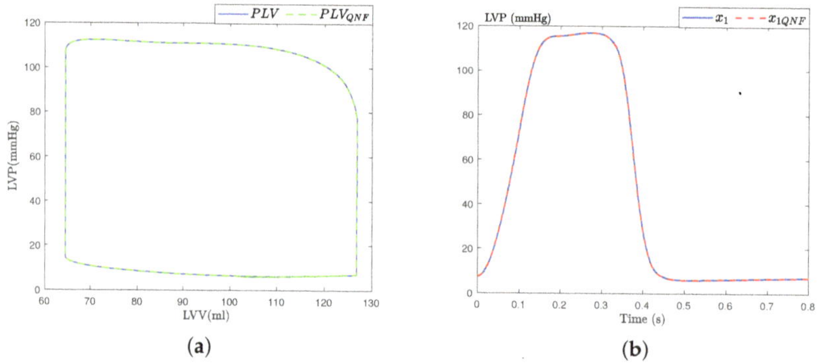

Figure 5. PV loops for state variable x_1: Left ventricular pressure. (**a**) PV-loops; (**b**) Left ventricular pressure.

4. Sliding Mode Observer Design

In this subsection, we will outline the structure of the observer. For a deeper exploration into the design and analysis of observability, readers are encouraged to consult the works cited in [27]. The observer structure presented here accounts for quadratic observability singularities that arise due to state separation or universal input. This methodology is derived from the step-by-step sliding mode approach as detailed in references [28–30]. We assume that the states z_1^1 and z_1^2 are directly measurable, but the others are not. The sliding mode observer is described as follows:

$$\left\{\begin{array}{lcl} \dot{\hat{z}}_1^1 & = & -\frac{R_C}{L_S} z_1^1 + \hat{z}_2^1 + \frac{1}{L_S} z_1^2 + \delta_1^1 sign(z_1^1 - \hat{z}_1^1) \\ \dot{\hat{z}}_2^1 & = & -\frac{1}{L_S C_S} z_1^1 + \hat{z}_3^1 + E_1^1 \delta_2^1 sign(\tilde{z}_2^1 - \hat{z}_2^1) \\ \dot{\hat{z}}_3^1 & = & \frac{1}{L_S R_S C_S^2} z_1^1 - \beta \tilde{z}_3^1 - k_1 \tilde{z}_3^1 u_1 - k_2 \tilde{z}_2^1 u_1 - k_3 \tilde{z}_2^2 u_3 + E_2^1 \delta_3^1 sign(\tilde{z}_3^1 - \hat{z}_3^1) \\ \dot{\hat{z}}_1^2 & = & -\frac{1}{C_A} z_1^1 - \frac{1}{C_A R_A}(z_1^2 u_2 + \hat{z}_2^2 u_4) + \delta_1^2 sign(z_1^2 - \hat{z}_1^2) \\ \dot{\hat{z}}_2^2 & = & -\frac{L_S R_S C_S}{R_M} \tilde{z}_3^1 u_1 - \frac{L_S}{R_M} \tilde{z}_2^1 u_1 - \frac{1}{R_M} \tilde{z}_2^2 u_3 + \frac{1}{R_A} z_1^2 u_2 - \frac{1}{R_A} \tilde{z}_2^2 u_4 + E_1^2 \delta_2^2 sign(\tilde{z}_2^2 - \hat{z}_2^2) \end{array}\right. \tag{9}$$

In system (9), the auxiliary components $\tilde{z}_i^q$ are calculated algebraically as follows:

$$\begin{array}{lcl} \tilde{z}_2^1 & = & \hat{z}_2^1 + \delta_1^1 sign(z_1^1 - \hat{z}_1^1) \\ \tilde{z}_3^1 & = & \hat{z}_3^1 + \delta_2^1 sign(\tilde{z}_2^1 - \hat{z}_2^1) \\ \tilde{z}_2^2 & = & \hat{z}_2^2 + \frac{E_{SM}}{k_3 u_3 + E_{SM} - 1} E_2^1 \delta_3^1 sign(\tilde{z}_3^1 - \hat{z}_3^1) \end{array}$$

with the following conditions:

If $\hat{z}_1^1 = z_1^1$ and $\hat{z}_1^2 = z_1^2$ then $E_1^1 = E_1^2 = 1$ otherwise $E_1^1 = E_1^2 = 0$,
If $\hat{z}_2^1 = z_2^1$ and $E_1^1 = E_1^2 = 1$ then $E_2^1 = 1$ otherwise $E_2^1 = 0$.

And, to ensure observability is not lost near the observability singularity, you must accurately set E_{SA} and E_{SM}, such that:

If u_1 =1 then $E_{SM}=1$ otherwise $E_{SM}=0$, If u_4 =1 then $E_{SA}=1$ otherwise $E_{SA}=0$. Which gives:

$$\begin{array}{rcl} \tilde{z}_3^1 & = & \hat{z}_3^1 + \frac{E_{SM}}{-\frac{L_S R_S C_S}{R_M} u_1 + E_{SM} - 1} \delta_2^2 sign(\tilde{z}_2^2 - \hat{z}_2^2) \\ \tilde{z}_2^2 & = & \hat{z}_2^2 + \frac{C_A R_A E_{SA}}{u_4 + E_{SA} - 1} E_1^2 \delta_1^2 sign(z_1^2 - \hat{z}_1^2) \end{array}$$

Remark 2. *The quality of the estimation of z_2^2 depends on the choice of E_{SA} and E_{SM}. Therefore, it is essential to adjust E_{SA} and E_{SM} within a small neighborhood of the singularity to ensure that the structure without feedback is applied for the minimum amount of time necessary.*

Remark 3. *Since u_1 and u_2 cannot be equal to one at the same time, also $u_3 = u_1 \frac{1}{C(t)}$, and $u_4 = u_2 \frac{1}{C(t)}$ then when $u_3 = 1$ and $u_4 = 0$, we use the quadratic term $k_3 z_2^2 u_3$ to recover the information of z_2^2 in this case*

$$\tilde{z}_2^2 = z_2^2 + \frac{E_{SM}}{k_3 u_3 + E_{SM} - 1} E_2^1 \delta_3^1 sign(\tilde{z}_3^1 - \hat{z}_3^1)$$

and when $u_3 = 0$ and $u_4 = 1$, we use the quadratic term $\frac{1}{C_A R_A} z_2^2 u_4$ to recover the information of z_2^2 in this case

$$\tilde{z}_2^2 = \hat{z}_2^2 + \frac{C_A R_A E_{SA}}{u_4 + E_{SA} - 1} E_1^2 \delta_1^2 sign(z_1^2 - \hat{z}_1^2)$$

Proof. The proof of convergence for the observation error is detailed in our article [19]. In it, the stability and convergence analysis of the observer employs equivalent vector methods [31]. The strategy for ensuring the observer's convergence is implemented step by step across different sliding surfaces. This approach guarantees the convergence of the observation error to zero in three steps and in finite time, in the Lyapunov sense, as further supported by references [19,28–30]. □

5. Model for Anomaly Detection

In this section, we will only present the adaptation of the model (8) by including the following fault vector F, which describes the variations affecting the mitral valve f_m and the aortic valve f_{ao}. In the context of the cardiovascular system, variations affecting the mitral valve D_m and the aortic valve D_a are significant contributors to valvular heart diseases, which are a leading cause of cardiovascular morbidity and mortality. We conceptualize the fault in the mitral valve (f_m) as the nominal value modeled as a percentile addition or subtraction to the input value (1 or 0) defined in Equation (6). Similarly, the fault in the aortic valve (f_{ao}) is considered. Two simple tests were run to evaluate the performance of the proposed FDI methodology: 50% mitral regurgitation, 50% aortic regurgitation. The fault vector is defined by the following equation:

$$F(t) = \begin{bmatrix} f_m \\ f_{ao} \end{bmatrix}, \text{such as: } \left\{ \begin{array}{l} U_1 = D_M + f_m \\ U_2 = D_A + f_{ao} \end{array} \right. \quad (10)$$

where D_M and D_A are the nominal values of the real state of the mitral and aortic valves, respectively, f_m and f_{ao} are the faults corresponding to the mitral and aortic valves, respectively. The faulty model of the CVS have the following form:

$$\left\{ \begin{array}{rcl} \dot{x}_1 & = & \frac{-\dot{C}(t)}{C(t)} x_1 - \frac{U_1}{C(t) R_M}(x_1 - x_2) - \frac{U_2}{C(t) R_A}(x_1 - x_4) \\ \dot{x}_2 & = & \frac{1}{R_S C_R}(x_3 - x_2) + \frac{1}{C_R R_M}(x_1 - x_2) U_1 \\ \dot{x}_3 & = & \frac{1}{R_S C_S}(x_2 - x_3) + \frac{1}{C_S} x_5 \\ \dot{x}_4 & = & -\frac{1}{C_A} x_5 + \frac{1}{C_A R_A}(x_1 - x_4) U_2 \\ \dot{x}_5 & = & -\frac{1}{L_S} x_3 + \frac{1}{L_S} x_4 - \frac{R_C}{L_S} x_5 \end{array} \right. \quad (11)$$

Remark 4. *In observer design, U_1 and U_2 are considered as bounded unknown inputs [29,30].*

The bank of two observers and the residual generator proposed are associated with the SMO designed previously in [19]. In this paper, residual generation is achieved by two single step-by-step sliding mode observers where the faults has been estimated by the observers. In this study, residual generation is obtained through single-step sliding mode observers, which estimate faults and then detect anomalies in the dynamic behavior of the cardiovascular system.

5.1. Residual Generation of Anomalies

The most common valve pathologies are related to the aortic and mitral valves. In both cases, these involve a defect in the closure of the valve, known as valve regurgitation. Aortic valve regurgitation refers to a defect in the valve closure that leads to backward leakage into the left ventricle during diastole. Patients with aortic regurgitation exhibit PV loops with increased amplitude and displacement to the right, indicating that the stroke work is higher, and the pressure–volume area is also increased compared to a healthy case. Similarly, mitral valve pathologies involve leakage during systole from the LV to the left atria (LA).

Based on this information, this section presents the design of the fault detection and isolation (FDI) system, this design is based on the assumption that only one fault can occur at any given time. Therefore, two simulation scenarios were considered for fault detection based on the analysis of the generated waveforms pressure. As shown in Figure 6, the first scenario is mitral regurgitation (f_m), and in scenario 2, the fault is aortic regurgitation (f_{ao}). To meet this requirement effectively, the sliding mode observers presented in [19] are used, which enable precise estimation of sensor measurements.

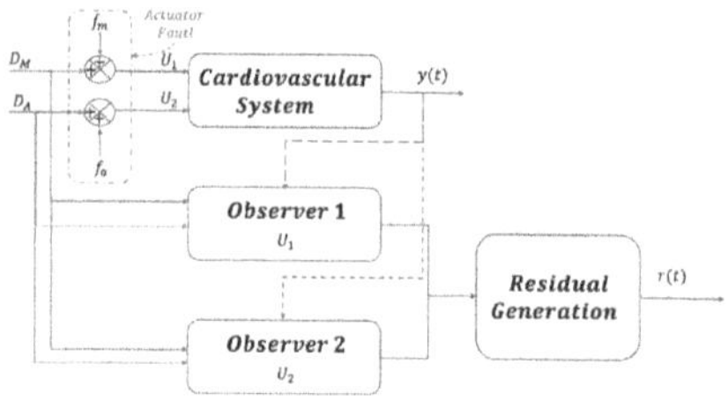

Figure 6. Bank of observers for all actuator faults estimation.

In Figure 6, the $y(t)$ denotes the output vector of CVS, where $y_1 = x_5$ and $y_2 = x_4$. These outputs serve as inputs to the observer, as they are the signals required for the observer to initiate the estimation process.

In this paper, the residuals are defined as the difference between the output variables represents that the most basic residual $r(t)$ can be expressed as:

$$r = y(t) - \hat{y}(t) = (sing(\bar{y}))_{eq} \tag{12}$$

Define $e = x - \hat{x}$ as the state estimation error, where

$$\begin{array}{rcl} \hat{x}_1 & = & \frac{1}{C(t)}\hat{z}_2^2 \\ \hat{x}_2 & = & -L_S\hat{z}_2^1 - L_S R_S C_S \hat{z}_3^1 \\ \hat{x}_3 & = & -L_S\hat{z}_2^1 \\ \hat{x}_4 & = & \hat{z}_1^2 \\ \hat{x}_5 & = & \hat{z}_1^1 \end{array}.$$

The bank of two observers and the residual generator proposed is associated with the SMO designed previously in [19]. In this paper, the residual generation is achieved by the means of two single step-by-step sliding mode observer, where the faults have been estimated by the observers. When there is no fault, i.e., $f_m = 0$ and $f_{ao} = 0$, the error will asymptotically converge to the true state. We also observed that the residuals in the presence of an anomaly in the mitral valve (U_2) are almost the same as the residuals in the aortic valve (U_1). However, the changes in the pressure and flow signals of the system are different. This is why, due to the robustness of the observer, we could implement failure isolation and determine when a failure occurs in the mitral and aortic valves.

Below, Table 3 presents a comprehensive signature for residual generation. This table is instrumental in understanding the nuanced differences in residual patterns, which are key to our failure isolation strategy. Each signature has been meticulously derived to ensure the precise detection and localization of anomalies within the mitral and aortic valves, highlighting the sophisticated nature of our observer's diagnostic capabilities.

Table 3. Signature for the residual generation.

Residuals/Faults	Anomalie on U_1	Anomalie on U_2
$r_1(t)$	0	0
$r_2(t)$	0	0
$r_3(t)$	1	1
$r_4(t)$	0	0
$r_5(t)$	1	1

5.2. Simulation Results

For the initial condition, we refer to those specified in [17], defined as follows:

$$x = [7.4, 5, 85, 82, 0]^T \text{and } \hat{z} = [5, -11 \times 10^4, 5.7143 \times 10^4, 0, 150]^T.$$

Figure 7 shows the hemodynamic waveforms for a healthy individual with a heart rate (HR) of 75 bpm. The waveforms are consistent, as will be explained. The systolic pressure (LAP) and diastolic pressure (LVP) were measured at 117 mmHg and 77 mmHg, respectively. The ascending aorta pressure (AoP), resulting from the opening and closing of the aortic valve and the pressure wave propagation along the aorta, presented a delayed waveform. From the transformation presented in Equation (1), we can derive the estimated states $\hat{x}_1$, $\hat{x}_2$, and $\hat{x}_3$ from the measurable outputs $\hat{x}_4$ and $\hat{x}_5$. Figure 7a presents the states of systems (4) and the observer (9). It demonstrates how the state estimation converges completely for all states within a time frame of 0.2 s.

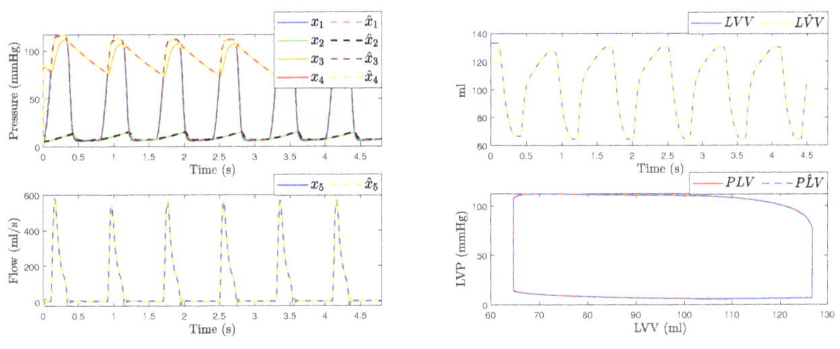

(a) (b)

Figure 7. Hemodynamic waveforms of the CVS model (4) compared with experimental data presented in [6]. (**a**) shows original states $x_i(t)$ and observer states $\hat{x}_i(t)$ of CVS for a normal heart and (**b**) shows left ventricular volume (LVV) and preload volume (PLV) in original and SMO of CVS.

Then, the left ventricle volume and preload volume (LVV) in Figure 7b represent the result of changing afterload conditions. Even with variations in preload and afterload, the relationship between end-systolic pressure and left ventricular volume should be roughly linear if the model functions as predicted. This relationship is known as the end-systolic pressure–volume relationship. By employing an especially built sliding mode observer to estimate the system's state x_1, we were able to determine the left ventricle's volume and preload volume using expression (1). This is because the state x_1 is described by recalling Frank–Starling's law, allowing us to gain more insight into the hemodynamic behavior of the heart in a healthy individual. The conditions were simulated with $E_{\max} = 2$ mmHg/mL, $E_{\min} = 0.05$ mmHg/mL, and $V_0 = 10$ mL.

Remark 5. *These data are compared and confirmed with the results described in [6,7,20], where the aortic pressure and flow waveforms are all consistent with hemodynamics data on healthy individuals.*

In the following discussion, two different fault scenarios are presented: Scenario 1 involves mitral regurgitation, while Scenario 2 involves aortic regurgitation.

5.3. Scenario 1: Mitral Regurgitation

In this scenario, we consider a 50% regurgitation in the mitral valve (i.e., $f_m = 0.5$ when $u_1 = 0$ and $f_{ao} = 0$). The simulation of the mitral valve fault was modeled by adding a binary value (1 or 0) to the input u_1. This modification was introduced at time $t = 2.5$ s, corresponding to the fourth cardiac cycle. Figure 8 illustrates the outcomes following the fault occurrence in the mitral valve. It is observed that, post-fault, the dynamics of the aortic valve, u_3 and u_4 are altered due to their dependence on u_1 and u_2. Figure 8b,c presents the simulated hemodynamic waveforms for an individual with heart failure. The simulation indicates changes in the dynamic system upon fault occurrence, with a decrease in blood flow waveforms and alterations in pressure waveforms.

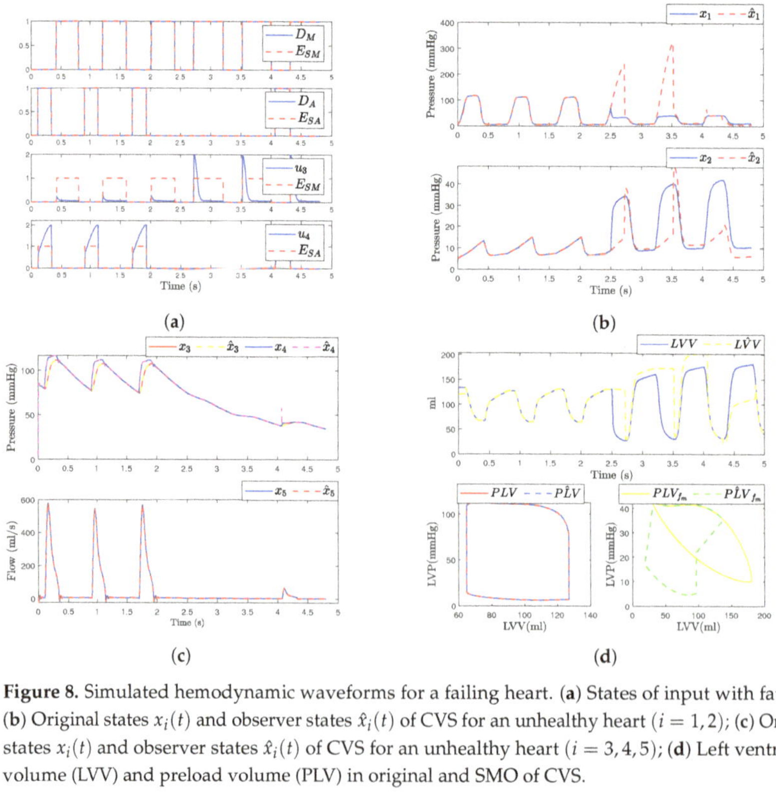

Figure 8. Simulated hemodynamic waveforms for a failing heart. (**a**) States of input with fault f_m; (**b**) Original states $x_i(t)$ and observer states $\hat{x}_i(t)$ of CVS for an unhealthy heart ($i = 1, 2$); (**c**) Original states $x_i(t)$ and observer states $\hat{x}_i(t)$ of CVS for an unhealthy heart ($i = 3, 4, 5$); (**d**) Left ventricular volume (LVV) and preload volume (PLV) in original and SMO of CVS.

Additionally, we show that the sliding mode observer can adapt to the change in the system dynamics. The observer can reconstruct the unobservable state when the failure occurs, but only states x_3, x_4, and x_5 are able to convergent again to the true state, as shown in Figure 8c. Furthermore, assuming that the heart is healthy, we were able to determine the left ventricular volume and preload volume using the response described by expression (1); providing that the heart is healthy, as shown in Figure 8d. Here x_i is the original model and $\hat{x}_i$ is the estimated signal provided by the observer (for $i = 1, 2, 3, 4, 5$). On the other hand, when the failure occurs, the SMO is not able to estimate the volume correctly due to the loss of x_1, as illustrated in the figure showing the failure in the original model PLV_{f_m} and $P\hat{L}V_{f_m}$ (Figure 8d).

5.4. Case 2: Aortic Regurgitation

In this scenario, we consider a 50% regurgitation in the aortic valve (i.e., $f_{ao} = 0.5$ when $u_2 = 0$ and $f_m = 0$). The simulation of fault in the aortic valve was modeled by adding a binary value (1 or 0) to the input u_2. This change was introduced at the time $t = 2.5$ s. Figure 9 represents the results after the fault occurs in the aortic valve. It is evident that, after the fault occurs in the aortic valve, there is also a change in the mitral valve dynamics. Additionally, u_3 and u_4 exhibit altered dynamics due to their dependence on u_1 and u_2. Figure 9 presents the simulation waveforms of the hemodynamics for an individual with aortic regurgitation. We observe changes in the system dynamics when the failure occurs, including variations in blood flow waveforms and an increase in pressure waveforms.

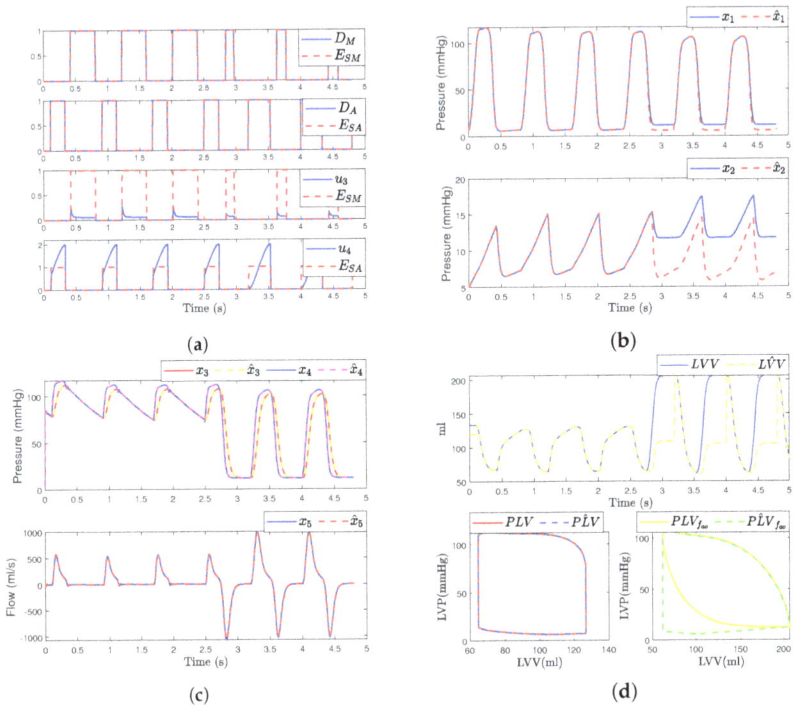

Figure 9. Simulated hemodynamic waveforms for a failing heart. (**a**) Input with fault f_{ao}; (**b**) Original states $x_i(t)$ and observer states $\hat{x}_i(t)$ of CVS for an unhealthy heart ($i = 1, 2$); (**c**) Original states $x_i(t)$ and observer states $\hat{x}_i(t)$ of CVS for an unhealthy heart ($i = 3, 4, 5$); (**d**) Left ventricular volume (LVV) and preload volume (PLV) in original and SMO of CVS.

However, we demonstrate that the SMO has the capacity to adapt to the change in in system dynamics. Furthermore, the observer is able to reconstruct the unobservable state when the failure occurs; however, only the states x_3, x_4, and x_5 can be convergent again to the true state, while states x_1 and x_2 have a constant error, as depicted in Figure 9b. Additionally, we were able to determine the left ventricular volume and preload volume using the response described by expression (1), assuming that the heart is healthy, as illustrated in Figure 9d. Here x_i is the original model and $\hat{x}_i$ is the estimated signal provided by the observer (for $i = 1,2,3,4,5$). On the other hand, when a failure occurs, the SMO is unable to accurately estimate the volume due to the loss of x_1, as shown in the comparison between the original model $PLV_{f_{ao}}$ and the observer-estimated model $P\hat{L}V_{f_{ao}}$ (Figure 9d).

In summary, the results align well with the hemodynamic parameters reported in the existing literature and experimental data, which support the validity of the proposed model and demonstrate its capability to produce results that are comparable to medical data. However, a larger-scale study with a greater number of tests would be necessary to obtain more precise results.

6. Conclusions

This study presents a comprehensive mathematical model of the cardiovascular system capable of simulating both normal and pathological states, specifically focusing on fault detection and isolation. The proposed model, which incorporates electrical analogies, offers a novel representation by transforming the cardiovascular system into a QNF. This form facilitates the design of a sliding mode observer, enhancing the model's ability to estimate system states and detect anomalies such as valvular heart diseases, which are significant risk factors for cardiovascular diseases.

Our results indicate that the SMO can adapt to changes in system dynamics and reconstruct unobservable states when faults occur. The observer successfully estimated the states x_3, x_4, and x_5, while x_1 and x_2 showed persistent errors under fault conditions. The model's validity was affirmed through simulations that replicated hemodynamic parameters that are consistent with the existing literature and experimental data. Additionally, the SMO demonstrated its effectiveness in scenarios of aortic and mitral valve regurgitation by accurately reconstructing the system dynamics post-failure. The results obtained were validated by comparing the data and the simulations presented in [6,7,20].

The findings underscore the potential of the proposed model and observer in clinical decision support, offering a less invasive, economical, and efficient alternative for monitoring cardiovascular health and diagnosing pathologies. However, further studies with larger datasets and a higher number of tests are recommended to refine the model and enhance the precision of the results.

Overall, this work contributes significantly to the field of cardiovascular modeling, providing a robust tool for understanding and managing cardiovascular diseases through advanced fault detection and isolation techniques.

Author Contributions: All co-authors contributed to this work. D.A.S.-C. and L.B.-B. conceived the idea, wrote the original draft, contributed to the investigation and analysis, performed the simulations, edited the manuscript, and contributed to the illustrations. M.D. and C.M.A.-Z. conceived the idea, contributed to the editing and supervised the manuscript. G.V.G.R. provided support in obtaining the data and revised the manuscript. All authors have read and agreed to the published version of the manuscript.

Funding: This research was funded by CONAHCyT, proyect number 320056: Sostenibilidad y Control Automatico.

Data Availability Statement: Experimental data associated with this article will be made available upon request.

Acknowledgments: The authors acknowledge the support provided by CONAHyT and Tecnológico Nacional de México.

Conflicts of Interest: The authors declare no conflicts of interest.

Abbreviations

The following abbreviations are used in this manuscript:

CVD	Cardiovascular disease
WHO	World Health Organization
CVS	Cardiovascular system
LVP	Left ventricular pressure
LV	Left ventricle
LVV	Left ventricular volume
EDV	End-diastolic volume
ESV	End-systolic volume
SMO	Sliding mode observer
FDI	Fault detection and isolation
PV	Pressure–volume
PVA	Pressure–volume area
LA	Left atria
H_R	Heart rate
LAP	Left atrial pressure
AoP	Ascending aorta pressure
F	Total flow

References

1. World Health Organization. Cardiovascular Diseases (CVDs). Available online: https://www.who.int/news-room/fact-sheets/detail/cardiovascular-diseases-(cvds) (accessed on 8 May 2024).
2. Chen, C.; Ting, C.-T.; Nussbacher, A.; Nevo, E.; Kass, D.A.; Pak, P.; Wang, S.-P.; Chang, M.; Yin, F.C.P. Validation of carotid artery tonometry as a means of estimating augmentation index of ascending aortic pressure. *Hypertension* **1996**, *27*, 168–175. [CrossRef]
3. Fónod, R.; Krokavec, D. Actuator fault estimation using neuro-sliding mode observers. In Proceedings of the IEEE 16th International Conference on Intelligent Engineering Systems (INES), Lisbon, Portugal, 13–15 June 2012; pp. 405–410.
4. Ledezma, F.D.; Laleg-Kirati, T.M. Detection of Cardiovascular Anomalies: Hybrid Systems Approach. *IFAC Proc. Vol.* **2012**, *45*, 222–227. [CrossRef]
5. Laleg-Kirati, T.M.; Belkhatir, Z.; Ledezma, F.D. Application of Hybrid Dynamical Theory to the Cardiovascular System. In *Hybrid Dynamical Systems*; Djemai, M., Defoort, M., Eds.; Springer: Cham, Switzerland, 2015; pp. 315–328.
6. Traver, J.; Nuevo-Gallardo, C.; Tejado, I.; Fernández-Portales, J.; Ortega-Morán, J.F.; Pagador, J.B.; Vinagre, B.M. Cardiovascular circulatory system and left carotid model: A fractional approach to disease modeling. *Fractal Fract.* **2022**, *6*, 64. [CrossRef]
7. Diaz Ledezma, F.; Laleg-Kirati, T.M. A first approach on fault detection and isolation for cardiovascular anomalies detection. In Proceedings of the 2015 American Control Conference (ACC), Chicago, IL, USA, 1–3 July 2015; pp. 5788–5793.
8. El Khaloufi, G.; Chaibi, N.; Alaoui, S.; Boumhidi, I.; Driss, E.-J. Design of Observer for Cardiovascular Anomalies Detection: An LMI approach. In Proceedings of the 2022 International Conference on Intelligent Systems and Computer Vision (ISCV), Fez, Morocco, 18–20 May 2022.
9. Ghasemi, Z.; Jeon, W.; Kim, C.-S.; Gupta, A.; Rajamani, R.; Hahn, J.-O. Observer-based deconvolution of deterministic input in coprime multichannel systems with its application to noninvasive central blood pressure monitoring. *J. Dyn. Syst. Meas. Control* **2020**, *142*, 091006. [CrossRef] [PubMed]
10. Ortiz-Rangel, E.; Guerrero-Ramírez, G.V.; García-Beltrán, C.D.; Guerrero-Lara, M.; Adam-Medina, M.; Astorga-Zaragoza, C.M.; Reyes-Reyes, J.; Posada-Gómez, R. Dynamic modeling and simulation of the human cardiovascular system with PDA. *Biomed. Signal Process. Control* **2022**, *71*, 103151. [CrossRef]
11. Laubscher, R.; van der Merwe, J.; Liebenberg, J.; Herbst, P. Dynamic simulation of aortic valve stenosis using a lumped parameter cardiovascular system model with flow regime dependent valve pressure loss characteristics. *Med. Eng. Phys.* **2022**, *106*, 103838. [CrossRef] [PubMed]
12. Simaan, M.A.; Ferreira, A.; Chen, S.; Antaki, J.F.; Galati, D.G. A dynamical state space representation and performance analysis of a feedback-controlled rotary left ventricular assist device. *IEEE Trans. Control. Syst. Technol.* **2018**, *17*, 15–28. [CrossRef]
13. Ferreira, A.; Chen, S.; Simaan, M.A.; Boston, J.R.; Antaki, J.F. A nonlinear state-space model of a combined cardiovascular system and a rotary pump. In Proceedings of the 44th IEEE Conference on Decision and Control, Seville, Spain, 15 December 2005; pp. 897–902.
14. Korakianitis, T.; Shi, Y. A concentrated parameter model for the human cardiovascular system including heart valve dynamics and atrioventricular interaction. *Med. Eng. Phys.* **2006**, *28*, 613–628. [CrossRef] [PubMed]

15. Simaan, M.A. Modeling and control of the heart left ventricle supported with a rotary assist device. In Proceedings of the 47th IEEE Conference on Decision and Control, Cancun, Mexico, 9–11 December 2008; pp. 2656–2661.
16. Astorga-Zaragoza, C. Observer-based monitoring of the cardiovascular system. *IEEE Trans. Circuits Syst. II Express Briefs* **2019**, *67*, 501–505. [CrossRef]
17. Belkhatir, Z.; Laleg-Kirati, T.-M.; Tadjine, M. Residual generator for cardiovascular anomalies detection. In Proceedings of the European Control Conference (ECC), Strasbourg, France, 24–27 June 2014; pp. 1862–1868.
18. Ramírez-Rasgado, F.; Hernández-González, O.; Astorga-Zaragoza, C.M.; Farza, M.; Barreto-Arenas, O.; Guerrero-Sánchez, M.E. Observer-based supervision of the cardiovascular system with delayed measurements. In Proceedings of the Congreso Nacional de Control Automático, Tuxtla Gutiérrez, Mexico, 12–14 October 2022.
19. Serrano-Cruz, D.; Boutat-Baddas, L.; Darouach, M.; Zaragoza, C.M.A.; Ramirez, G.V.G. Sliding mode observer design for fault estimation in cardiovascular system. In Proceedings of the Congreso Nacional de Control Automático, Acapulco, Mexico, 25–27 October 2023.
20. Simaan, M.A. Rotary heart assist devices. In *Springer Handbook of Automation*; Nof, S., Ed.; Springer: Berlin/Heidelberg, Germany, 2009; pp. 1409–1422.
21. Mensah, G.A.; Roth, G.A.; Fuster, V. The global burden of cardiovascular diseases and risk factors: 2020 and beyond. *J. Am. Coll. Cardiol.* **2019**, *74*, 2529–2532. [CrossRef] [PubMed]
22. Aluru, J.S.; Barsouk, A.; Saginala, K.; Rawla, P.; Barsouk, A. Valvular heart disease epidemiology. *Med. Sci.* **2022**, *10*, 32. [CrossRef] [PubMed]
23. Hollenberg, S.M. Valvular Heart Disease in Adults: Etiologies, Classification, and Diagnosis. *FP Essent.* **2017**, *457*, 11–16. [PubMed]
24. Manoliu, V. Modeling cardiovascular hemodynamics in a model with nonlinear parameters. In Proceedings of the E-Health and Bioengineering Conference (EHB), Iasi, Romania, 19–21 November 2015; pp. 1–4.
25. Boutat-Baddas, L.; Boutat, D.; Barbot, J.; Tauleigne, R. Quadratic observability normal form. In Proceedings of the 40th IEEE Conference on Decision and Control (Cat. No.01CH37228), Orlando, FL, USA, 4–7 December 2001; Volume 3, pp. 2942–2947.
26. Boutat-Baddas, L. Analyse des singularités d'observabilité et de détectabilité: Application à la synchronisation des circuits éléctroniques chaotiques. Ph.D. Thesis, Université de Cergy-Pontoise, Cergy, France, 2002.
27. Serrano-Cruz, D.A.; Boutat-Baddas, L.; Darouach, M.; Astorga-Zaragoza, C.-M. Observer design for a nonlinear cardiovascular system. In Proceedings of the 9th International Conference on Systems and Control, Caen, France, 24–26 November 2021; pp. 294–299.
28. Barbot, J.; Boukhobza, T.; Djemai, M. Sliding mode observer for triangular input form. In Proceedings of the 35th IEEE Conference on Decision and Control, Kobe, Japan, 13 December 1996; Volume 2, pp. 1489–1490.
29. Barbot, J.; Floquet, T. Iterative higher order sliding mode observer for nonlinear systems with unknown inputs. *Dyn. Contin. Discret. Impuls. Syst.* **2010**, *17*, 1019–1033.
30. Floquet, T.; Barbot, J.-P. Super Twisting Algorithm-Based Step-by-Step Sliding Mode Observers for Nonlinear Systems with Unknown Inputs. *Int. J. Syst. Sci.* **2007**, *38*, 803–815. [CrossRef]
31. Draženović, B. The invariance conditions in variable structure systems. *Automatica* **1969**, *5*, 287–295. [CrossRef]

Mathematical and Computational Applications

MDPI

Article

Human Activity Recognition from Accelerometry, Based on a Radius of Curvature Feature

Elizabeth Cavita-Huerta, Juan Reyes-Reyes *, Héctor M. Romero-Ugalde, Gloria L. Osorio-Gordillo, Ricardo F. Escobar-Jiménez and Victor M. Alvarado-Martínez

Centro Nacional de Investigación y Desarrollo Tecnológico, Tecnológico Nacional de México, Cuernavaca 62493, Morelos, Mexico; d17ce053@cenidet.tecnm.mx (E.C.-H.); hector@cenidet.edu.mx (H.M.R.-U.); gloria.og@cenidet.tecnm.mx (G.L.O.-G.); ricardo.ej@cenidet.tecnm.mx (R.F.E.-J.); victor.am@cenidet.tecnm.mx (V.M.A.-M.)

* Correspondence: juan.rr@cenidet.tecnm.mx

Abstract: Physical activity recognition using accelerometry is a rapidly advancing field with significant implications for healthcare, sports science, and wearable technology. This research presents an interesting approach for classifying physical activities using solely accelerometry data, signals that were taken from the available "MHEALTH dataset" and processed through artificial neural networks (ANNs). The methodology involves data acquisition, preprocessing, feature extraction, and the application of deep learning algorithms to accurately identify activity patterns. A major innovation in this study is the incorporation of a new feature derived from the radius of curvature. This time-domain feature is computed by segmenting accelerometry signals into windows, conducting double integration to derive positional data, and subsequently estimating a circumference based on the positional data obtained within each window. This characteristic is computed across the three movement planes, providing a robust and comprehensive feature for activity classification. The integration of the radius of curvature into the ANN models significantly enhances their accuracy, achieving over 95%. In comparison with other methodologies, our proposed approach, which utilizes a feedforward neural network (FFNN), demonstrates superior performance. This outperforms previous methods such as logistic regression, which achieved 93%, KNN models with 90%, and the InceptTime model with 88%. The findings demonstrate the potential of this model to improve the precision and reliability of physical activity recognition in wearable health monitoring systems.

Keywords: accelerometry; physical activity; artificial neural networks; radius of curvature; classification models; human activity recognition

1. Introduction

Physical inactivity is alarmingly recognized as the fourth leading risk factor for global mortality [1]. The close association between engaging in physical activities and sports with a range of health benefits is well-documented [2,3]. Consequently, enhancing lifestyles, promoting healthy behaviors, and mitigating chronic diseases have become critical priorities [4]. Furthermore, the integration of sophisticated technology and computing in healthcare plays an increasingly vital role in our daily lives, opening up numerous potential research areas, notably in Human Activity Recognition (HAR) [5,6].

1.1. Human Activity Recognition

HAR predominantly utilizes wearable devices to identify and categorize human physical activities in both controlled and uncontrolled settings. In recent years, HAR has gained prominence due to its extensive applications across various fields such as healthcare, sports science, video game development, and activity tracking [7]. Especially, HAR in healthcare plays a crucial role in monitoring, managing chronic conditions, promoting healthy lifestyles, detecting health issues early, and supporting personalized care, refs. [8–10]. It has the potential to enhance patient outcomes, optimize healthcare delivery, and contribute to advancements in medical research and public health initiatives. It can be implemented using data from different types of sensors; these are generally divided into two main categories: (1) environmental sensors, such as video cameras used in monitoring areas, which are limited by their need for fixed infrastructure [11,12], and (2) integrated sensors located within portable devices that enhance and expand the capabilities of human monitoring systems at any location [13].

Environmental sensors often require costly on-site installation for effective monitoring [14], while vision systems rely on cameras and are frequently perceived as intrusive [15]. The most widely used alternatives, wearable devices, have garnered significant attention from researchers due to their popularity, ease of use, and affordability. These devices are embedded in various gadgets such as smartphones, smartwatches, medical monitoring devices, or fitness bands [16,17].

Various studies have introduced HAR algorithms using signals captured by wearable sensors attached to different body parts, achieving commendable classification performance [18]. However, challenges arise with the discomfort users may experience when wearing sensors on specific body parts and the energy limitations of mobile devices. These sensors are often uncomfortable for users and do not provide a viable long-term solution. Hence, developing HAR algorithms that allow for the use of wearable devices on multiple body parts is an opportunity area [19–22].

One of the most effective sensors in HAR applications is the accelerometer. Given the advancements in integrated circuit technology, it can be worn continuously for days or even weeks, presenting a practical solution to the limitations related to energy consumption and wearer comfort. HAR, with accelerometers, utilizes these sensors to detect and classify human activities by analyzing their movement patterns. Accelerometers directly measure human body movements, responding to factors such as frequency, intensity, and inclination [23]. This allows for the accurate monitoring of physical activities in real-time.

This motion sensor can detect acceleration in the three orthogonal axes of space (x, y and z). Therefore, it is essential to extract features considering each axis independently or in combination. This approach enables the sensor to capture signals accurately from any movement plane of the human body, offering a comprehensive and detailed perspective on movement, as illustrated in Figure 1.

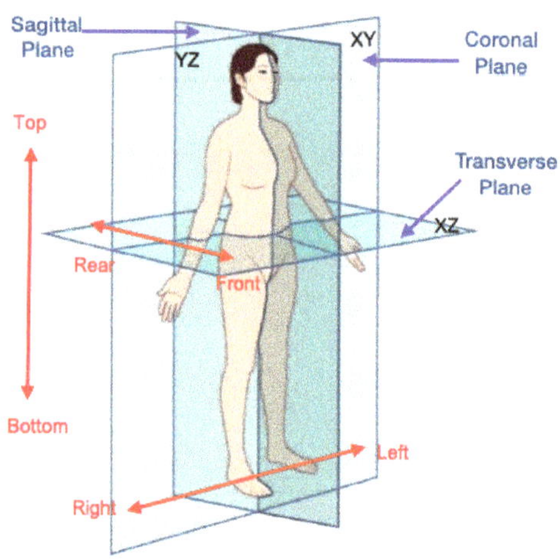

Figure 1. Planes of human movement.

1.2. Sensor Position

When utilizing accelerometers in HAR, it is crucial to consider that the output patterns of accelerometer signals vary depending on the body position. Therefore, alternative approaches have been proposed to develop multiple models capable of classifying physical activity across different accelerometer placements on the body [19–22]. In this way, the user has more freedom of choice, increasing the probability of the success of HAR and enhancing models' performance [24–26].

Analyses such as those presented in [27] show how the classification capability and performance of the model are directly affected by the number and combination of accelerometer positions used. However, this can result in consequences such as increased costs, reduced comfort, and the generation of larger and more complex datasets [26].

HAR systems use machine learning or deep learning models to recognize activities [28]. Deep learning models have bolstered the performance and robustness of HAR [29], accelerating its adoption. These models can decrease the computational work in the data preprocessing and feature extraction phases, while enhancing generalization performance and model reliability [30].

Hence, determining the accelerometer's position facilitates the adaptation of HAR systems to different body configurations. One significant contribution of this work is that detecting the sensor's position allows for the development of specific deep learning models that enhance the precision and reliability of HAR.

1.3. Related Works

This section reviews the main advances and methodologies in HAR, including feature extraction techniques, machine learning models, and innovative approaches. The goal is to provide a clear overview of current trends and identify opportunities for further research and improvement in HAR technology.

Despite their various applications, HAR algorithms still face several challenges, such as: (1) the complexity and diversity of daily activities, (2) variability between subjects for the same activity, (3) computational inefficiency on embedded and portable devices, and (4) difficulty in handling data [31]. HAR systems utilizing wearable inertial sensors, such as accelerometers, utilize temporal signals.

This signal type has not been extensively researched, necessitating innovative approaches to extract valuable features for HAR. In ref. [32], the activity recognition accuracy is improved by incorporating attention into multi-head convolutional neural networks for better feature extraction and selection. In a related work, ref. [33] proposed a feature incremental activity recognition method named FIRF, which evaluates the performance of continuously recognizing new emerging features. These features, extracted from a new sensor, are incrementally added at different time steps. Additionally, ref. [34] proposed the Down-Sampling One-Dimensional Local Binary Pattern method. Their HAR system consists of two stages: firstly, a conversion was applied to the sensor signals to extract statistical features from the newly formed signals, and secondly, classification was performed using these features. In [35], attention mechanisms enhance temporal convolutional networks (TCNs) to better capture temporal dependencies and identify key patterns in activity data. Similarly, ref. [36] improves model performance by combining wearable ambient light sensors with traditional IMUs to detect environmental changes linked to activities, leading to better accuracy. Additionally, ref. [37] explores using a single triaxial accelerometer with features from the time, frequency, and statistical domains to capture human activity dynamics more comprehensively.

The exploration of new features is crucial for developing models that deliver robust real-world results. By integrating innovative characteristics, we can enhance the accuracy of activity recognition systems, leading to the advancement of smarter wearable devices for monitoring and evaluating physical activity. Hence, this study focuses on extracting a new characteristic closely linked to movement dynamics: the radius of curvature. This time-domain feature is computed by segmenting accelerometry signals into windows, conducting double integration to derive positional data, and subsequently estimating a circumference based on the data obtained within each window, across movement planes.

Table 1 provides a comparison between the proposed model and other state-of-the-art approaches in activity classification. This comparison highlights the differences in methodologies and performance metrics across various studies. It is highlighted that the proposed FFNN-based model has a competitive performance that is equal or superior to the most accurate models, which shows that the developed approach is as effective or more in HAR as the state-of-the-art techniques.

Table 1. Performance comparison of the proposed model and state-of-the-art methods.

Authors	Methods	Accuracy
Geravesh, and Rupapara, et al. [38] (2023)	KNN	94%
Hafeez, Alotaibi, et al. [39] (2023)	Logistic Regression (LR)	93%
Jantawong, Ponnipa, et al. [40] (2021)	InceptTime model	88%
Zhang, Haoxi, et al. [32] (2019)	Multi-head Convolutional Attention	95%
Mekruksavanich, Jitpattanakul, et al. [41] (2022)	ResNet-SE model	94–97%
Lohit, Wang, et al. [42] (2019)	Temporal Transformer Networks	78%
Neverova, Natalia, et al. [43] (2016)	Dense Clockwork RNN	93%
Our proposed approach	FFNN	95–97%

1.4. Based Methodology

To design and evaluate any HAR systems, specific steps must traditionally be followed to retrieve activity information from the sensor. These steps are referred to as the activity recognition chain [44]. Firstly, data collection is conducted, focusing on determining the specifics of the data acquisition process. The second step consists of performing data preprocessing using methods to fit the signals and extract the characteristics that are used as input information in the classification models. The third step involves selecting the appropriate learning model and training it. Finally, in the fourth step, the model is evaluated in terms of activity recognition metrics, as shown in Figure 2.

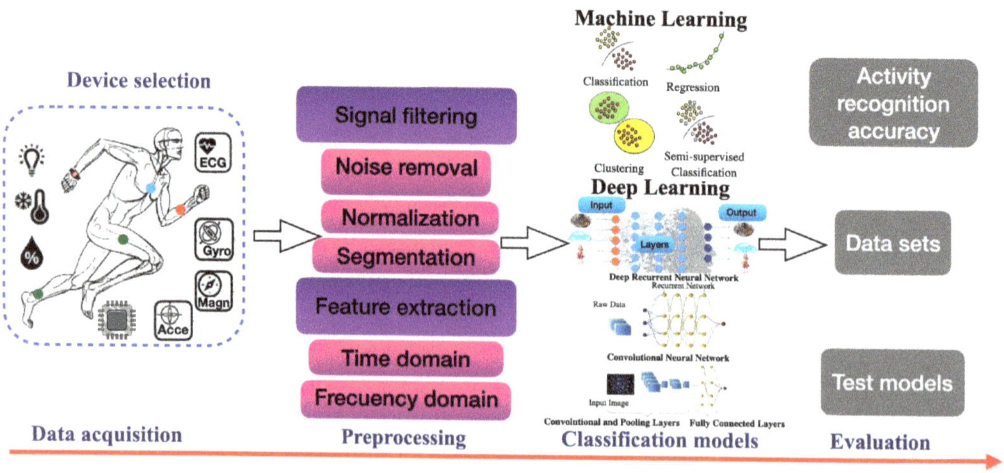

Figure 2. Human activity recognition chain.

As indicated in the HAR chain process, it is necessary to consider important aspects of signal acquisition, such as the sampling frequency. In the context of capturing human body motion, the majority of its energy, specifically 99%, is concentrated below 15 Hz [45]. Therefore, a sampling frequency of over 30 Hz is sufficient to capture interesting information, present in refs. [46–48].

Additionally, the automatic classification of accelerometry data requires a preprocessing phase where various techniques are used to enhance signals, such as filtering, signal segmentation, and feature extraction. Complex feature extraction techniques are needed to improve success, which would lead to an increased data processing power [49].

Within the preprocessing stage, segmentation is performed, where time-series sensor data are segmented before feature extraction; the sliding window technique is a common example of the methodologies used at this stage [50]. Some research also implements an overlap between two consecutive segments; from each sampled window, a vector of features must be obtained [51]. Feature extraction can be classified into two main domains: the time domain and the frequency domain, with the time domain being the most commonly used due to its simplicity of computation [52].

ANNs are an important branch of artificial intelligence, defined as models inspired by the structure and function of the human brain [15]. They have been shown to be effective on a wide range of tasks, including HAR, as they have the ability to learn complex patterns in data and extract discriminative features, resulting in a powerful tool [53].

In accordance with the Table 2, the proposed model relies solely on accelerometers, whereas previous studies use multiple sensors and perform feature extraction in the frequency and time domains, or combinations of these, which increases the processing complexity to achieve high performance. Therefore, this model could serve as a viable alternative for devices with limited resources, as its performance balances the use of multiple sensors and complex transformations. Hence, exploring new features closely related to human movement dynamics could provide enough information to enhance model accuracy.

Table 2. Comparison with previous research.

Authors	Dataset	Sensor	Model	Feature Extraction Domain	Accuracy
Bennasar M. et al. [54] (2022)	WISDIM	Acce	SVM, KNN	Frequency, Time	90.6–93.2%
Gil M. et al. [55] (2020)	PAMAP2	Acce	CNNs	Frequency	89.8–96.6%
Dua N. et al. [56] (2021)	PAMAP2	Acce, Gyro, Mag	Multi-Input CNN-GRU	Time	95.2%
Kutlay M. et al. [57] (2016)	mHealth	Acce, Gyro, Mag	SVM, MLP	Time	91.7%, 83.2%
Cosma G. et al. [58] (2019)	mHealth	Acc, Gyro	KNN	Time, Frequency	47.5–82.3%
Proposed Model	mHealth	Acce	FFNN	Time	95–97%

The main proposal and contribution of this work is the adaptation of well-known temporal features combined with a new feature, which is obtained based on the calculation of the curvature radius from positions obtained in the three planes of motion. This characteristic is closely related to the movement patterns that people perform during physical activity, enabling HAR algorithms to be performed successfully, thus maintaining computational simplicity in the algorithms and reducing the need for complex data transformations. To classify activities, a feedforward neural network (FFNN) was used.

2. Materials and Methods

2.1. Dataset

This work has been developed using the available "MHEALTH dataset" (Mobile Health), created to evaluate human behavior analysis techniques based on multimodal body sensing [46]. To capture the data from accelerometry signals, three triaxial accelerometer sensors were used. Authors recorded accelerations, from sensors placed on the chest, right wrist, and left ankle of each subject, attached using elastic straps to measure the motion experienced by different parts of the body. All sensing activities were recorded at a sampling rate of 50 Hz, which is considered sufficient for capturing human activity. The data were collected in an out-of-laboratory setting with no restrictions on how the activities were to be performed, except that the subjects were required to exert maximum effort.

The dataset includes recordings of body motion from ten volunteers with diverse profiles, collected while they performed 12 physical activities considered common in daily life. Additionally, the labels used serve to identify the activities, as shown in Table 3.

Table 3. Activity Set.

Label	Activity	Duration
L1	Standing	1 min
L2	Sitting and relaxing	1 min
L3	Lying down	1 min
L4	Walking	1 min
L5	Climbing stairs	1 min
L6	forward waist bends	20×
L7	Frontal elevation of arms	20×
L8	Knee-bending (crouching)	20×
L9	Cycling	1 min
L10	Jogging	1 min
L11	Running	1 min
L12	Jump front–back	20×

In the "Duration" column, Nx is the number of repetitions or the duration of the exercises (min).

2.2. System Architecture for HAR Process

The general architecture of the HAR system used in this work is shown in Figure 3. The proposed scheme encompasses data preprocessing and transformation, as well as the use of a Random Forest (RF) classifier model to detect the sensor position on the body; lastly, the activity classification is performed using FFNN models, selected based on the position detected earlier, therefore three different models have been developed to calculate HAR. Given that raw acceleration data from portable sensors experience considerable variations over time, classifying them based only on data from a single point becomes impossible [31]. Therefore, HAR methods are based on a series of data collected over a given time interval [59]; in this work, periods of 20 s are selected to perform the classification task.

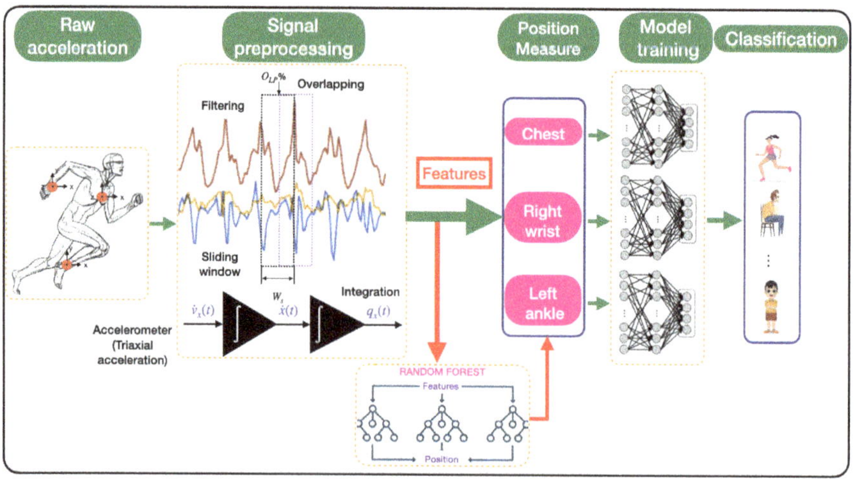

Figure 3. Schematic of the proposed system for HAR process.

In next sections, a detailed study of different stages is presented.

2.3. Preprocessing Data

The raw accelerometry signals are characterized by being affected by signal noise, which makes it difficult to use them in their original state. The presence of noise and its negative effects on the classification models are mitigated in the preprocessing stage of the data, making them suitable for use as inputs to the recognition models [60].

In the preprocessing step, the accelerometer raw signals that were sampled at 50 Hz are filtered to reduce noise. According to [48], an ideal low-pass frequency filter set at a (cutoff frequency of 10 Hz) allows us to capture all relevant acceleration for physical activity while minimizing the impact of noise, which is utilized in this research.

2.4. Data Segmentation

The time-dependent raw acceleration dataset is divided into segments, which can vary in length according to the selected window size, during the data segmentation process. All operations related to signal preprocessing are performed considering each of these segments. Previous studies in the context of activity classification have already examined different configurations in terms of window size and signal segmentation process.

They claim that overlapping windows are more suitable because they can handle transitions more accurately and avoid loss of information between segments [31]. In most of the studies performed for activity classification, the length of the windows varies in the range of 1 to 10 s [49]. Increasing the length of the windows in some cases can improve the recognition accuracy, but it is not always the best alternative. In any case, the window size should be chosen so that each window contains enough samples (at least one activity cycle) to distinguish similar movements [61]. The estimation of the number of windows can be carried out by using the following mathematical expression:

$$K = \left\lfloor \frac{N - M}{R} \right\rfloor + 1 \tag{1}$$

where (K) is the number of windows, (N) the signal length, (M) the window size and (R) the overlap factor.

The following expression is used to calculate the number of overlap samples:

$$X_{OL} = S_R * W_S \left\lfloor \frac{O_{LP}}{100} \right\rfloor \tag{2}$$

where (X_{OL}) is total number of overlap samples, (S_R) sampling rate (Hz), (W_S) window size (s) and (O_{LP}) overlap percentage.

The sliding window approach is the predominant method used in the segmentation step of HAR. In this method, sensor signals are divided into windows of a fixed size. When adjacent windows overlap, this technique is referred to as overlapping sliding window.

In this research, data segmentation is carried out using the sliding window technique, considering a window size of 2 s and an overlap of 50%, shown in Figure 4.

Taking into account the above details, 19 windows are generated within each 20 s time interval considered for classification process, sampled at 50 Hz.

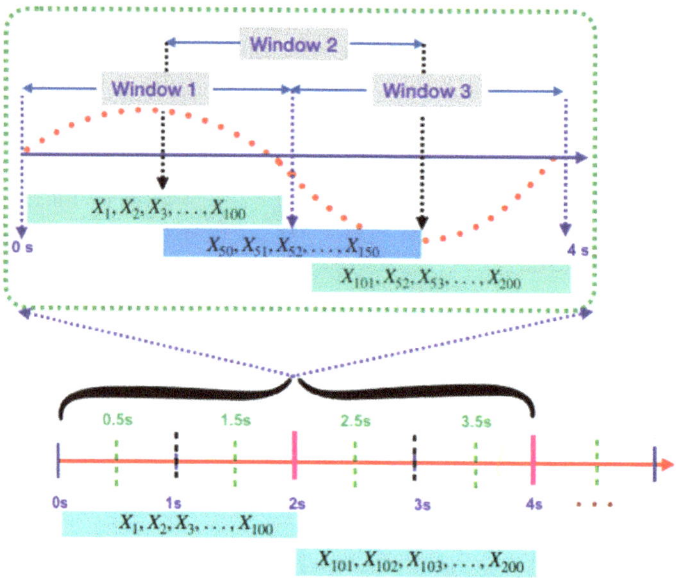

Figure 4. Segmentation process using sliding window technique.

2.5. Feature Extraction

Each time window is transformed into a feature vector, which is the input data in the position detection and activity classification models. In particular, our methodology consists of transforming the accelerometry signals into different types of representations, such as velocity and position, from which the characteristics are extracted. Therefore, to execute this transformation process, double integration using Euler's method is initially performed on the acceleration triaxial signals, considering the following reference frame, shown in Figure 5.

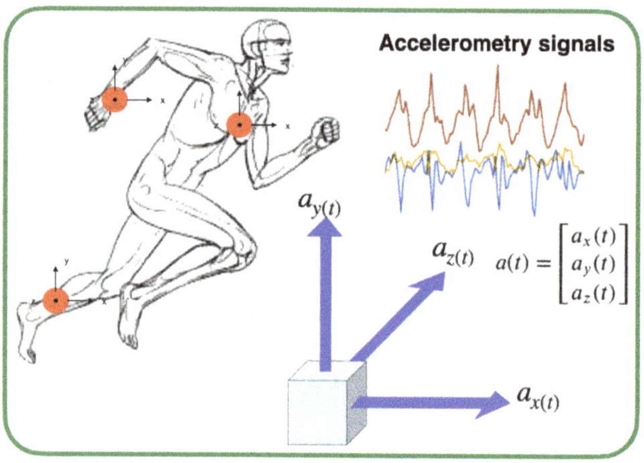

Figure 5. Triaxial accelerometer signals, reference frame.

Therefore, the kinematics of the position, velocity and acceleration can be expressed for each component as follows:

- X-axis component, where $a_x(t), v_x(t), x(t)$ are the instantaneous acceleration, velocity and position, respectively:

$$\Sigma_{Kx}: \begin{cases} \dot{v}_x(t) = a_x(t) \\ \dot{x}(t) = v_x(t) \\ q_x(t) = [v_x(t), \ \ x(t)]^T \end{cases} \tag{3}$$

- Y-axis component, where $a_y(t), v_y(t), y(t)$ are the instantaneous acceleration, velocity and position:

$$\Sigma_{Ky}: \begin{cases} \dot{v}_y(t) = a_y(t) \\ \dot{y}(t) = v_y(t) \\ q_y(t) = [v_y(t), \ \ y(t)]^T \end{cases} \tag{4}$$

- Z-axis component, where $a_z(t), v_z(t), z(t)$ are the instantaneous acceleration, velocity and position:

$$\Sigma_{Kz}: \begin{cases} \dot{v}_z(t) = a_z(t) \\ \dot{z}(t) = v_z(t) \\ q_z(t) = [v_z(t), \ \ z(t)]^T \end{cases} \tag{5}$$

Using the Euler method of double integration in each of the acceleration components, mathematically expressed below and considering initial conditions $x(t_1), v_x(t_1), y(t_1), v_y(t_1), z(t_1), v_z(t_1) = 0$, the following is obtained:

- X-axis component:

$$v_x(t_{k+1}) = v_x(t_k) + \frac{1}{w_s} a_x(t_k) \tag{6}$$

$$x(t_{k+1}) = x(t_k) + \frac{1}{w_s} v_x(t_k); \ k = 0, 2, \ldots w_s \tag{7}$$

- Y-axis component:

$$v_y(t_{k+1}) = v_y(t_k) + \frac{1}{w_s} a_y(t_k) \tag{8}$$

$$y(t_{k+1}) = y(t_k) + \frac{1}{w_s} v_y(t_k); \ k = 0, 2, \ldots w_s \tag{9}$$

- Z-axis component:

$$v_z(t_{k+1}) = v_z(t_k) + \frac{1}{w_s} a_z(t_k) \tag{10}$$

$$z(t_{k+1}) = z(t_k) + \frac{1}{w_s} v_z(t_k); \ k = 0, 2, \ldots w_s \tag{11}$$

where w_s is the window size.

Once positions, velocities, and accelerations in (x, y, z) axes have been obtained in each window, as shown Figure 6, feature extraction is performed.

This study proposes a novel approach to derive a new feature closely related to the dynamics of motion, which varies depending on the activity being performed. To achieve this, we calculate the curvature radius ($\hat{r}$) from the positions obtained for each window using the Euler's double integration method, considering different movement planes.

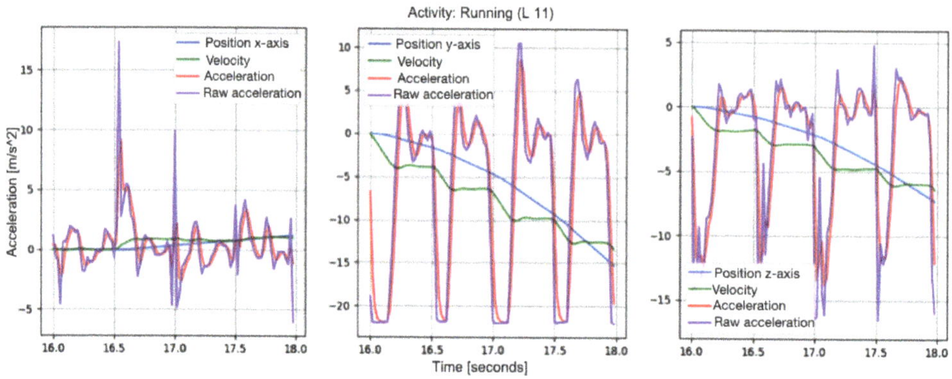

Figure 6. Accelerometry, velocity and position signals per window.

2.6. Proposed Estimating Model for Curvature Radius

Taking into account the general equation of the circumference in a Cartesian plane, where $[h, k]$ are the coordinates of the center of the circle, r is the radius of the circle and $[x, y]$ are the coordinates of the position points on the plane, Figure 7, shows the calculation process in the coronal plane (x, y) axes.

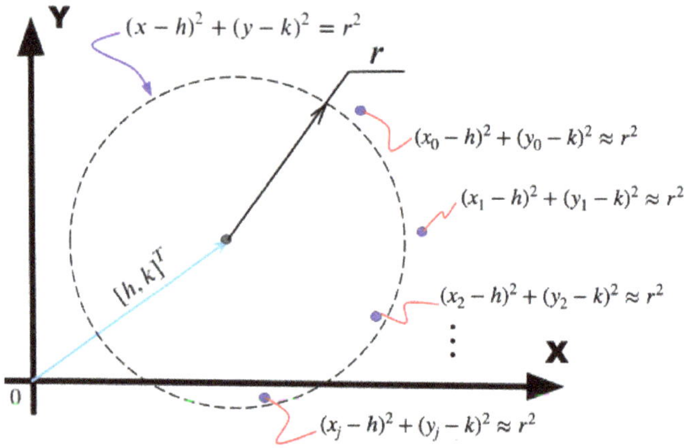

Figure 7. Graphical description of point $[h, k]^T$ and the radius of curvature.

To calculate a point whose coordinate or position vector is denoted by the vector $[h, k]^T$, this point is approximately equidistant to the number of points contained in the selected window size (w_s), considering $[x_0, y_0]^T, \ldots, [x_j, y_j]^T$ at a distance r, the mathematical expression is:

$$(x_j - h)^2 + (y_j - k)^2 = r^2, \quad j = 0, 1, \ldots, w_s; \tag{12}$$

To solve the point $[h, k]^T$, it is necessary to re-express these equations in vector notation, expressed as follows:

$$\begin{bmatrix} x_j - h, & y_j - k \end{bmatrix} \begin{bmatrix} x_j - h \\ y_j - k \end{bmatrix} \cong r^2, \quad i = 0, 1, \ldots, w_s; \tag{13}$$

$$\left(\begin{bmatrix} x_j & y_j \end{bmatrix} - \begin{bmatrix} h, & k \end{bmatrix} \right) \left(\begin{bmatrix} x_j \\ y_j \end{bmatrix} - \begin{bmatrix} h \\ k \end{bmatrix} \right) \cong r^2, \quad j = 0, 1, \ldots, w_s; \tag{14}$$

$$\begin{bmatrix} x_j - h \\ y_j - k \end{bmatrix} = \begin{bmatrix} x_j \\ y_j \end{bmatrix} - \begin{bmatrix} h \\ k \end{bmatrix} \quad (15)$$

The algebraic manipulations generate the following:

$$\begin{bmatrix} x_j & y_j \end{bmatrix}\begin{bmatrix} x_j \\ y_j \end{bmatrix} - \begin{bmatrix} h & k \end{bmatrix}\begin{bmatrix} x_j \\ y_j \end{bmatrix} - \begin{bmatrix} x_j & y_j \end{bmatrix}\begin{bmatrix} h \\ k \end{bmatrix} + \begin{bmatrix} h & k \end{bmatrix}\begin{bmatrix} h \\ k \end{bmatrix} \cong r^2 \quad (16)$$

$$\begin{bmatrix} x_j & y_j \end{bmatrix}\begin{bmatrix} x_j \\ y_j \end{bmatrix} - 2\begin{bmatrix} x_j & y_j \end{bmatrix}\begin{bmatrix} h \\ k \end{bmatrix} \cong r^2 - \begin{bmatrix} h & k \end{bmatrix}\begin{bmatrix} h \\ k \end{bmatrix} \quad (17)$$

So, it is defined that:

$$p_j := \begin{bmatrix} x_j \\ y_j \end{bmatrix} \quad (18)$$

Then:

$$p_j^T p_j - 2p_j^T \begin{bmatrix} h \\ k \end{bmatrix} \cong r^2 - \begin{bmatrix} h & k \end{bmatrix}\begin{bmatrix} h \\ k \end{bmatrix}, \; j = 0, 1, \dots w_s; \quad (19)$$

In summary, we have a number of equations equal to the number of points within a window, and these equations express the same quadratic difference; therefore, we can formulate the equivalences between them.

$$p_0^T p_0 - 2p_0^T \begin{bmatrix} h \\ k \end{bmatrix} \cong p_i^T p_i - 2p_i^T \begin{bmatrix} h \\ k \end{bmatrix}, \; i = 1, 2, \dots w_s. \quad (20)$$

Window size (w_s) equations can be constructed, leaving the affine terms on the left-side:

$$-2\left(p_0^T - p_i^T\right)\begin{bmatrix} h \\ k \end{bmatrix} \cong p_i^T p_i - p_0^T p_0, \; i = 1, 2, \dots w_s; \quad (21)$$

The matrix expression considering the coronal plane is:

$$-2\begin{bmatrix} p_0^T - p_1^T \\ p_0^T - p_2^T \\ \cdot \\ \cdot \\ \cdot \\ p_0^T - p_{w_s}^T \end{bmatrix}\begin{bmatrix} h_c \\ k_c \end{bmatrix} \cong \begin{bmatrix} p_1^T p_1 - p_0^T p_0 \\ p_2^T p_2 - p_0^T p_0 \\ \cdot \\ \cdot \\ \cdot \\ p_{w_s}^T p_{w_s} - p_0^T p_0 \end{bmatrix} \quad (22)$$

Finally, the point $[h_c, k_c]^T$ corresponding to the center of the circle is obtained as follows:

$$\begin{bmatrix} h_c \\ k_c \end{bmatrix} \cong -\frac{1}{2}\begin{bmatrix} p_0^T - p_1^T \\ p_0^T - p_2^T \\ \cdot \\ \cdot \\ \cdot \\ p_0^T - p_{w_s}^T \end{bmatrix}^{+} \cdot \begin{bmatrix} p_1^T p_1 - p_0^T p_0 \\ p_2^T p_2 - p_0^T p_0 \\ \cdot \\ \cdot \\ \cdot \\ p_{w_s}^T p_{w_s} - p_0^T p_0 \end{bmatrix}, \quad (23)$$

where $[\cdot]^+$ is known as the pseudo-inverse of $[\cdot]$.

Once the value of point $[h_c, k_c]^T$ is obtained in the coronal plane, we proceed to calculate the value of the approximately equidistant distance $\hat{r}_c$ (curvature radius), as follows:

$$\hat{r}_c = \frac{1}{w_s}\sum_{j=0}^{w_s}\sqrt{(x_j - h_c)^2 + (y_j - k_c)^2}, \; j = 0, 1, \dots w_s; \quad (24)$$

where $\hat{r}_c$ is the curvature radius, (x_j, y_j) are the coordinates of the position points in the plane, (h_c, k_c) are the coordinates of the center of the circle, considering all of them in the context of calculus in the coronal plane. In this work, the signals are segmented into 2 s periods, which means $w_s = 100$. Figure 8 shows an example of the graphical result after applying our methodology, where the value of $\hat{r}$ depends on the trajectory of points corresponding to the calculated position, according to the analyzed plane.

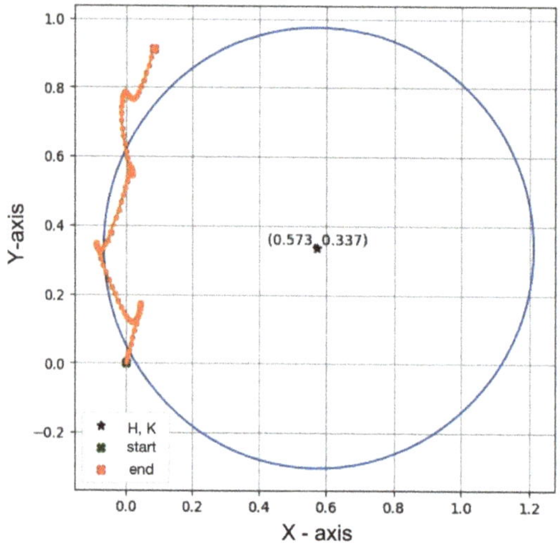

Figure 8. Graphical example of the calculation process for $\hat{r}_c$ in the coronal plane.

Therefore, it can be observed that, from the position trajectory obtained in the coronal plane during a two-second window, an upward movement with slight lateral deviations is recorded. This pattern corresponds directly to the dynamics of a person's running motion. Moreover, the circumference calculated with the proposed algorithm demonstrates its efficiency, since it fits well to most of the points of the obtained trajectory. The value of the calculated radius of curvature ($\hat{r}_c$) is also shown.

2.7. Proposed Feature Vector

Feature selection is a critical aspect in the development of ANN models. By carefully choosing relevant features extracted solely from the time domain, we rationalize the model complexity while ensuring robust classification accuracy in HAR.

Following this approach, the proposed feature vector is composed, considering that the estimation of curvature radius for each of the three planes of motion involves applying the method described (Section 2.6) Additionally, the mean and variance of the acceleration, velocity, and position signals from the three axes are calculated, as shown in Table 4.

Considering the segmentation process (Section 2.4), each processed data slice has 19 windows. Therefore, in the feature extraction process, 19 curvature radii are obtained in each plane, respectively, as shown in the following mathematical expressions.

Table 4. Proposed feature vector.

Features
$\hat{r}_c$ in a coronal plane
$\hat{r}_s$ in a saggital plane
$\hat{r}_t$ in a transverse plane
Mean acceleration in (x, y, z) axes
Mean velocity in (x, y, z) axes
Mean position in (x, y, z) axes
Variance acceleration in (x, y, z) axes
Variance velocity in (x, y, z) axes
Variance position in (x, y, z) axes

- Curvature radius in the saggital plane $\hat{r}_s$, where the point $[h_s, k_s]^T$ corresponding to the center of the circle is obtained as follows:

$$\begin{bmatrix} h_s \\ k_s \end{bmatrix} \cong -\frac{1}{2} \begin{bmatrix} p_0^T - p_1^T \\ p_0^T - p_2^T \\ \cdot \\ \cdot \\ \cdot \\ p_0^T - p_{w_s}^T \end{bmatrix}^+ \cdot \begin{bmatrix} p_1^T p_1 - p_0^T p_0 \\ p_2^T p_2 - p_0^T p_0 \\ \cdot \\ \cdot \\ \cdot \\ p_{w_s}^T p_{w_s} - p_0^T p_0 \end{bmatrix}, p_j := \begin{bmatrix} y_j \\ z_j \end{bmatrix}, \ j = 0, 1, \dots w_s; \tag{25}$$

where $[\cdot]^+$ is known as the pseudo-inverse of $[\cdot]$, (y_j, z_j) are the coordinates of the position points in the plane. Therefore, the curvature radius is obtained as follows:

$$\hat{r}_s = \frac{1}{w_s} \sum_{j=0}^{w_s} \sqrt{(y_j - h_s)^2 + (z_j - k_s)^2}, \ j = 0, 1, \dots w_s; \tag{26}$$

- Curvature radius in the transverse plane $\hat{r}_t$, where the point $[h_t, k_t]^T$ corresponding to the center of the circle is obtained as follows:

$$\begin{bmatrix} h_t \\ k_t \end{bmatrix} \cong -\frac{1}{2} \begin{bmatrix} p_0^T - p_1^T \\ p_0^T - p_2^T \\ \cdot \\ \cdot \\ \cdot \\ p_0^T - p_{w_s}^T \end{bmatrix}^+ \cdot \begin{bmatrix} p_1^T p_1 - p_0^T p_0 \\ p_2^T p_2 - p_0^T p_0 \\ \cdot \\ \cdot \\ \cdot \\ p_{w_s}^T p_{w_s} - p_0^T p_0 \end{bmatrix}, p_j := \begin{bmatrix} x_j \\ z_j \end{bmatrix}, \ j = 0, 1, \dots w_s; \tag{27}$$

where $[\cdot]^+$ is known as the pseudo-inverse of $[\cdot]$, (x_j, z_j) are the coordinates of the position points in the plane. Therefore, the curvature radius is obtained as follows:

$$\hat{r}_t = \frac{1}{w_s} \sum_{j=0}^{w_s} \sqrt{(x_j - h_t)^2 + (z_j - k_t)^2}, \ j = 0, 1, \dots w_s; \tag{28}$$

Another 342 features are also extracted, corresponding to the calculation of the mean and variance of the position, velocity, and acceleration signals along the (x, y, z) axes. Finally, a vector of 399 features per activity is used to input data into both the HAR and RF algorithms.

2.8. Accelerometer Position Detection

The information required to determine the position of the accelerometer on the body is derived from the accelerations, which vary depending on where the sensor is placed. The core concept of on-body positioning involves analyzing acceleration data while the user is engaged in specific activities. Previous research has indicated that incorporating positional information enhances the model's precision in HAR [62].

In this sense, for the automatic detection of the sensor position, an RF classifier containing three possible corresponding classes (chest, right wrist and left ankle) is used; the model is trained from the characteristics obtained in the feature vector proposed in this work, considering that there are N training examples, where each example has D features. The features are represented as a vector $x_i = (xi_1, xi_2, \ldots, xi_D)$, for each example i. The corresponding class labels are represented as a vector Q_i for each example i.

$$D = \{(x_i, Q_i)\}_{i=1}^{N} \tag{29}$$

where x_i are the inputs, the features extracted at (x, y, z), and Q_i outputs class, corresponding to the detected position (chest, right wrist, left ankle).

The classification output of a RF model is determined by counting the class predictions from each tree and selecting the most frequent class as the final prediction.

$$y_{pred} = argmax(\sum_{i=1}^{M} \alpha(Q_i)) \tag{30}$$

where y_{pred} is the predicted output, M is the number of trees and $\alpha(Q_i)$ is an indicator function that returns 1 if the prediction of tree i is equal to class Q, and 0 otherwise.

2.9. Measurement Validation

In this work, the signals from "MHEALTH dataset" are utilized to execute the HAR process, which has been widely used in some research [57,58,63]. The data are divided into two groups for training and testing the ANN models that classify the activities, and to perform an on-body position detection of the accelerometer using the RF algorithm. According to [64], a training and testing dataset split of 70/30 was determined to be the most effective ratio for training and validating machine learning models. Therefore, 70% of the participants' data are used for training the models, specifically, while the remaining 30% are used for testing them. The following metrics were employed to evaluate the final classification performance of our proposed approach in HARL

Accuracy indicates the proportion of correct predictions (positive and negative) over the total predictions made, for each class.

$$Accuracy = \frac{TP + TN}{TP + TN + FP + FN} \tag{31}$$

where TP is the number of true positive instances, TN is the number of true negatives, FP is the number of false positives, and FN is the number of false negatives.

Precision measures the accuracy of positive predictions made by the model.

$$Precision = \frac{TP}{TP + FP} \tag{32}$$

Recall shows the proportion of positive instances that were correctly identified by the model for each class.

$$Recall = \frac{TP}{TP + FN} \tag{33}$$

F1-score is the harmonic mean of precision and recall, providing a metric that balances both measures.

$$F1 - score = 2 * \frac{Precision * Recall}{Precision + Recall} = \frac{2 * TP}{2 * TP + FP + FN} \tag{34}$$

3. Results: Accelerometer Position Detection and Human Activity Recognition

3.1. Detection Position Model

A RF classifier is a supervised machine learning algorithm. Its main job is to build multiple decision trees during training and combine their predictions to obtain a more robust and accurate final prediction. Each decision tree is constructed using a random subset of features and a random subset of training data. For classification, each tree in the forest generates a class prediction and, in the end, a majority vote or averaging of the predictions is used to determine the final class assigned. The RF algorithm was utilized to determine the sensor's location on the body. Given a proposed feature vector (Section 2.7). The performance of the model is shown in the following confusion matrix, Figure 9 from which some important points can be determined.

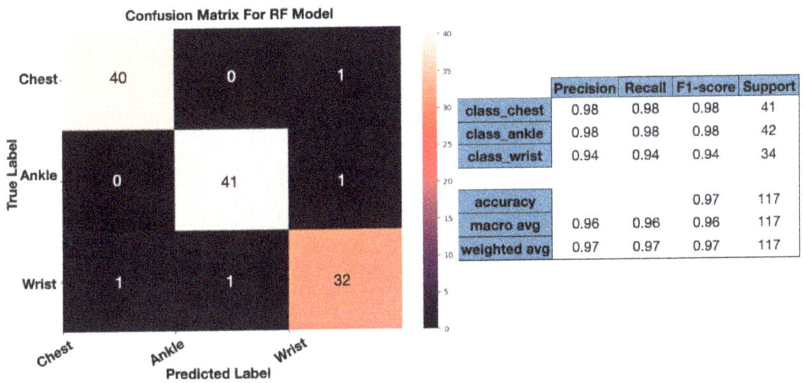

Figure 9. Confusion matrix of RF model.

The results show that the RF model performs excellently overall, with an accuracy, recall, and F1-score above 0.94 for all evaluated classes, achieving an overall accuracy of 97%. Both macro- and weighted average metrics are also high, indicating consistent and robust classification performance across classes.

Additionally, when testing the classification using only the curvature radius as the input feature, an accuracy of 70% is achieved, demonstrating that this feature provides the most useful information to the model, although the model shows a slightly lower performance for the wrist class (94%) due to the high variability in hand movements.

3.2. *Physical Activity Classification Models*

Once the position of the sensor has been determined, the next step is to classify the 12 previously labeled activities, shown in Table 3. A proposed features vector composed of 399 elements is used as the input to train activity recognition models, including curvature radius and statistics such as the mean and variance, with the features shown in Table 4. Therefore, three different neural network model structures of FFNN type were constructed. For any of the three cases, the input vector has the same dimensions, so the difference resides in the parameters of the ANN applied in each case. The results in Figure 10 show the performance of physical activity models.

According to the results from the chest model, the activities with an F1-score of 1, indicating the best performance, are standing still (L1), sitting and relaxing (L2), lying down (L3), walking (L4), forward waist bends (L6), frontal elevation of arms (L7), knee-bending (crouching) (L8), cycling (L9), and jump front–back (L12). This shows perfect precision and recall for these activities. The activities with the lowest performance are climbing stairs (L5), jogging (L10), and running (L11), which have lower F1-scores. Overall, the model has an accuracy of 95%, indicating strong classification ability.

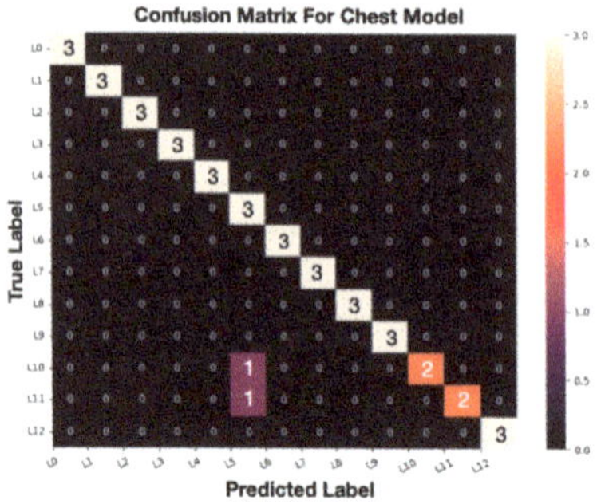

	Precision	Recall	F1-score	Support
class_0	1	1	1	3
class_1	1	1	1	3
class_2	1	1	1	3
class_3	1	1	1	3
class_4	1	1	1	3
class_5	0.60	1	0.75	3
class_6	1	1	1	3
class_7	1	1	1	3
class_8	1	1	1	3
class_9	1	1	1	3
class_10	1	0.67	0.80	3
class_11	1	0.67	0.80	3
class_12	1	1	1	3
accuracy			0.95	39
macro avg	0.97	0.95	0.95	39
weighted avg	0.97	0.95	0.95	39

Chest model

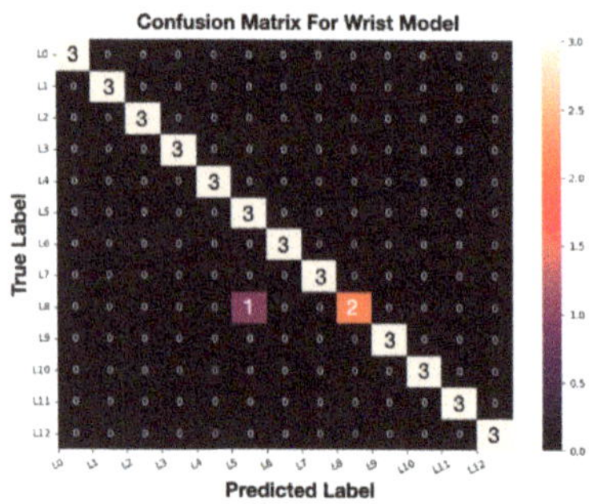

	Precision	Recall	F1-score	Support
class_0	1	1	1	3
class_1	1	1	1	3
class_2	1	1	1	3
class_3	1	1	1	3
class_4	1	1	1	3
class_5	0.75	1	0.86	3
class_6	1	1	1	3
class_7	1	1	1	3
class_8	1	0.67	0.80	3
class_9	1	1	1	3
class_10	1	1	1	3
class_11	1	1	1	3
class_12	1	1	1	3
accuracy			0.97	39
macro avg	0.98	0.97	0.97	39
weighted avg	0.98	0.97	0.97	39

Wrist model.

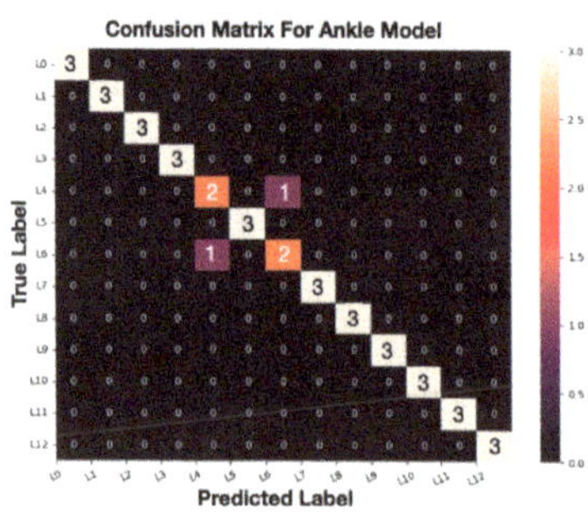

	Precision	Recall	F1-score	Support
class_0	1	1	1	3
class_1	1	1	1	3
class_2	1	1	1	3
class_3	1	1	1	3
class_4	0.67	0.67	0.67	3
class_5	1	1	1	3
class_6	0.67	0.67	0.67	3
class_7	1	1	1	3
class_8	1	1	1	3
class_9	1	1	1	3
class_10	1	1	1	3
class_11	1	1	1	3
class_12	1	1	1	3
accuracy			0.95	39
macro avg	0.95	0.95	0.95	39
weighted avg	0.95	0.95	0.95	39

Ankle model

Figure 10. Performances of the HAR algorithms based on FFNNs when the model is tested.

For the ankle model, activities such as standing still (L1), sitting and relaxing (L2), lying down (L3), climbing stairs (L5), frontal elevation of arms (L7), knee-bending (crouching) (L8), cycling (L9), jogging (L10), running (L11), and jump front–back (L12) have accuracies and recalls of 1, showing perfect classification. Activities like walking (L4) and forward waist bends (L6) have lower accuracies and recalls of 0.67, indicating less accuracy. Overall, the model achieves a 95% accuracy, demonstrating strong performance in most cases.

Finally, in the wrist model, the activities with the best performance, where the model achieves perfect classification without errors, are standing still (L1), sitting and relaxing (L2), lying down (L3), walking (L4), forward waist bends (L6), frontal elevation of arms (L7), cycling (L9), jogging (L10), running (L11), and jump front–back (L12). The accuracy for climbing stairs (L5) and knee-bending (crouching) (L8) indicates that the percentage of times the model predicts the activity is 75% and 67%, respectively, which are the lowest. Overall, this model achieved the highest accuracy, at 97%.

Using only the curvature radius for training the models, the accuracy rates are 72% for the chest, 73% for the ankle, and 77% for the wrist, highlighting the usefulness of this feature.

Leave-One-Subject-Out (LOSO) validation was employed to assess the model's robustness and generalization by evaluating its performance on data from a subject excluded during training. This approach offers a realistic estimate of how well the model generalizes to new individuals, helping to prevent overfitting and ensuring the model identifies general patterns rather than memorizing specific features of the training data.

The LOSO evaluation is showing in Table 5, it reveals a remarkable consistency in the accuracies obtained; the variations between subjects are minimal. This indicates that the model is robust to individual differences in the data. While the chest and wrist sensors show consistently high accuracies, the ankle sensor shows slight variability. This suggests that while the chest and wrist sensors offer greater classification stability, the ankle sensor may be subject to more noise or variability in the data, which influences the accuracy of the model.

Table 5. Leave-One-Subject-Out (LOSO) cross-validation results.

	Accuracy		
Subject	**Chest**	**Wrist**	**Ankle**
Subject 1	96.34%	98.12%	94.72%
Subject 2	96.87%	98.25%	95.21%
Subject 3	97.21%	98.08%	94.85%
Subject 4	96.12%	97.99%	95.67%
Subject 5	97.85%	98.43%	94.18%
Subject 6	96.63%	98.51%	95.73%
Subject 7	95.21%	97.92%	95.49%
Subject 8	97.34%	98.67%	94.92%
Subject 9	97.58%	97.84%	93.67%
Subject 10	97.96%	98.29%	95.26%

4. Conclusions

In conclusion, the results suggest that the proposed features involving curvature radius are suitable and effective for constructing robust models for physical activity classification and position detection. Notably, using curvature radius alone achieves an accuracy of up to 77%. This underscores the significant role of curvature radius in HAR models, providing crucial insights into the dynamics and characteristics of human movements. The RF model performs excellent overall performance, with an overall accuracy of 97%, indicating its ability to correctly classify most positions. The wrist signals model shows the highest accuracy at 97%, but all classification models perform well, with accuracies above 95%.

Discussion and Limitations

Our methodology's advantage lies in its ability to achieve high precision using only a few features; this offers advantages in terms of simplicity and computational efficiency, and unlike methods that employ frequency domain features or combinations of both domains, our approach avoids complex transformations, thus reducing processing and memory requirements.

Although, the calculation of the curvature radius is highly dependent on the dynamics of the movement. In activities with little dynamic variation, the ability of the model to extract useful information may be diminished. The sensitivity of the radius of curvature to the lack of variability in the data may limit the effectiveness of the method, which suggests that future research should address these aspects to further optimize the accuracy and applicability in various physical activity monitoring scenarios.

Despite these challenges, the simplicity and efficiency of our approach provides an attractive balance between performance and computational load, which is beneficial for applications on portable devices with limited resources.

Author Contributions: Conceptualization, E.C.-H., J.R.-R. and H.M.R.-U.; methodology, E.C.-H., J.R.-R. and H.M.R.-U.; software, E.C.-H., H.M.R.-U. and J.R.-R.; validation, E.C.-H., J.R.-R., H.M.R.-U. and G.L.O.-G.; formal analysis, E.C.-H., H.M.R.-U. and J.R.-R.; investigation, E.C.-H., R.F.E.-J., V.M.A.-M. and G.L.O.-G.; writing—original draft preparation, E.C.-H.; writing—review and editing, E.C.-H., H.M.R.-U. and J.R.-R.; supervision, J.R.-R., H.M.R.-U., V.M.A.-M., R.F.E.-J. and G.L.O.-G. All authors have read and agreed to the published version of the manuscript.

Funding: This research received no external funding.

Data Availability Statement: Data are contained within the article.

Acknowledgments: The authors acknowledge CONAHCYT for supporting Elizabeth Cavita Huerta through a Ph.D. Scholarship.

Conflicts of Interest: The authors declare no conflicts of interest.

References

1. Cao, Z.B. Physical activity levels and physical activity recommendations in Japan. In *Physical Activity, Exercise, Sedentary Behavior and Health*; Springer: Tokyo, Japan, 2015; pp. 3–15.
2. Black, N.; Johnston, D.W.; Propper, C.; Shields, M.A. The effect of school sports facilities on physical activity, health and socioeconomic status in adulthood. *Soc. Sci. Med.* **2019**, *220*, 120–128. [CrossRef] [PubMed]
3. Atiq, F.; Mauser-Bunschoten, E.P.; Eikenboom, J.; van Galen, K.P.; Meijer, K.; de Meris, J.; Cnossen, M.H.; Beckers, E.A.; Laros-van Gorkom, B.A.; Nieuwenhuizen, L.; et al. Sports participation and physical activity in patients with von Willebrand disease. *Haemophilia* **2019**, *25*, 101–108. [CrossRef] [PubMed]
4. Afshin, A.; Babalola, D.; Mclean, M.; Yu, Z.; Ma, W.; Chen, C.Y.; Arabi, M.; Mozaffarian, D. Information technology and lifestyle: A systematic evaluation of internet and mobile interventions for improving diet, physical activity, obesity, tobacco, and alcohol use. *J. Am. Heart Assoc.* **2016**, *5*, e003058. [CrossRef] [PubMed]
5. Ungurean, L.; Brezulianu, A. An internet of things framework for remote monitoring of the healthcare parameters. *Adv. Electr. Comput. Eng.* **2017**, *17*, 11–16. [CrossRef]
6. Ramanujam, E.; Perumal, T.; Padmavathi, S. Human activity recognition with smartphone and wearable sensors using deep learning techniques: A review. *IEEE Sens. J.* **2021**, *21*, 13029–13040. [CrossRef]
7. Nweke, H.F.; Teh, Y.W.; Mujtaba, G.; Al-Garadi, M.A. Data fusion and multiple classifier systems for human activity detection and health monitoring: Review and open research directions. *Inf. Fusion* **2019**, *46*, 147–170. [CrossRef]
8. Paraschiakos, S.; de Sá, S.; Okai, J.; Slagboom, E.; Beekman, M.; Knobbe, A. RNNs on Monitoring Physical Activity Energy Expenditure in Older People. *arXiv* **2020**, arXiv:2006.01169. Available online: https://tinyurl.com/cfp7849a (accessed on 6 July 2021).
9. Trost, S.G.; Wong, W.K.; Pfeiffer, K.A.; Zheng, Y. Artificial neural networks to predict activity type and energy expenditure in youth. *Med. Sci. Sport. Exerc.* **2012**, *44*, 1801. [CrossRef]
10. Jang, Y.; Song, Y.; Noh, H.W.; Kim, S. A basic study of activity type detection and energy expenditure estimation for children and youth in daily life using 3-axis accelerometer and 3-stage cascaded artificial neural network. In Proceedings of the 2015 37th Annual International Conference of the IEEE Engineering in Medicine and Biology Society (EMBC), Milan, Italy, 25–29 August 2015; pp. 2860–2863.
11. Dimiccoli, M.; Cartas, A.; Radeva, P. *Activity Recognition from Visual Lifelogs: State of the Art and Future Challenges*; Elsevier Ltd.: Amsterdam, The Netherlands, 2018; pp. 121–134. [CrossRef]
12. Kang, K.H.; Shin, S.H.; Jung, J.; Kim, Y.T. Estimation of a Physical Activity Energy Expenditure with a Patch-Type Sensor Module Using Artificial Neural Network; John Wiley & Sons, Ltd.: Hoboken, NJ, USA, 2019; pp. 1–9. [CrossRef]
13. Zeng, M.; Nguyen, L.T.; Yu, B.; Mengshoel, O.J.; Zhu, J.; Wu, P.; Zhang, J. Convolutional neural networks for human activity recognition using mobile sensors. In Proceedings of the 6th International Conference on Mobile Computing, Applications and Services, Austin, TX, USA, 6–7 November 2014; pp. 197–205.
14. Xu, T.; Zhou, Y.; Zhu, J. New advances and challenges of fall detection systems: A survey. *Appl. Sci.* **2018**, *8*, 418. [CrossRef]
15. Sathyanarayana, S.; Satzoda, R.K.; Sathyanarayana, S.; Thambipillai, S. Vision-based patient monitoring: A comprehensive review of algorithms and technologies. *J. Ambient Intell. Humaniz. Comput.* **2018**, *9*, 225–251. [CrossRef]
16. Sunny, J.T.; George, S.M.; Kizhakkethottam, J.J. Applications and Challenges of Human Activity Recognition using Sensors in a Smart Environment. *IJIRST—Int. J. Innov. Res. Sci. Technol.* **2015**, *2*, 50–57.
17. Bulling, A.; Blanke, U.; Schiele, B. A tutorial on human activity recognition using body-worn inertial sensors. *ACM Comput. Surv. (CSUR)* **2014**, *46*, 1–33. [CrossRef]

18. Attal, F.; Mohammed, S.; Dedabrishvili, M.; Chamroukhi, F.; Oukhellou, L.; Amirat, Y. Physical human activity recognition using wearable sensors. *Sensors* **2015**, *15*, 31314–31338. [CrossRef] [PubMed]
19. Mannini, A.; Sabatini, A.M.; Intille, S.S. Accelerometry-based recognition of the placement sites of a wearable sensor. *Pervasive Mob. Comput.* **2015**, *21*, 62–74. [CrossRef] [PubMed]
20. Fujinami, K.; Kouchi, S. Recognizing a Mobile Phone's Storing Position as a Context of a Device and a User. In *Mobile and Ubiquitous Systems: Computing, Networking, and Services, Proceedings of the International Conference on Mobile and Ubiquitous Systems: Computing, Networking, and Services, Beijing, China, 12–14 December 2012*; Springer: Berlin/Heidelberg, Germany, 2012; pp. 76–88.
21. Durmaz Incel, O. Analysis of movement, orientation and rotation-based sensing for phone placement recognition. *Sensors* **2015**, *15*, 25474–25506. [CrossRef]
22. Kunze, K.; Lukowicz, P. Dealing with sensor displacement in motion-based onbody activity recognition systems. In Proceedings of the 10th International Conference on Ubiquitous Computing, Seoul, Republic of Korea, 21–24 September 2008; pp. 20–29.
23. Garnotel, M.; Simon, C.; Bonnet, S. Physical activity estimation from accelerometry. In Proceedings of the 2019 41st Annual International Conference of the IEEE Engineering in Medicine and Biology Society (EMBC), Berlin, Germany, 23–27 July 2019; pp. 6–10.
24. Kurban, O.C.; Yildirim, T. Daily motion recognition system by a triaxial accelerometer usable in different positions. *IEEE Sens. J.* **2019**, *19*, 7543–7552. [CrossRef]
25. Clevenger, K.A.; Pfeiffer, K.A.; Montoye, A.H. Cross-generational comparability of hip-and wrist-worn ActiGraph GT3X+, wGT3X-BT, and GT9X accelerometers during free-living in adults. *J. Sport. Sci.* **2020**, *38*, 2794–2802. [CrossRef]
26. Bao, L.; Intille, S.S. Activity recognition from user-annotated acceleration data. In *Pervasive Computing, Proceedings of the International Conference on Pervasive Computing, Vienna, Austria, 21–23 April 2004*; Springer: Berlin/Heidelberg, Germany, 2004; pp. 1–17.
27. Altini, M.; Penders, J.; Vullers, R.; Amft, O. Estimating energy expenditure using body-worn accelerometers: A comparison of methods, sensors number and positioning. *IEEE J. Biomed. Health Inform.* **2014**, *19*, 219–226. [CrossRef]
28. Wang, J.; Chen, Y.; Hao, S.; Peng, X.; Hu, L. Deep learning for sensor-based activity recognition: A survey. *Pattern Recognit. Lett.* **2019**, *119*, 3–11. [CrossRef]
29. Sharifani, K.; Amini, M. Machine learning and deep learning: A review of methods and applications. *World Inf. Technol. Eng. J.* **2023**, *10*, 3897–3904.
30. Kanjo, E.; Younis, E.M.; Ang, C.S. Deep learning analysis of mobile physiological, environmental and location sensor data for emotion detection. *Inf. Fusion* **2019**, *49*, 46–56. [CrossRef]
31. Lara, O.D.; Labrador, M.A. A survey on human activity recognition using wearable sensors. *IEEE Commun. Surv. Tutor.* **2012**, *15*, 1192–1209. [CrossRef]
32. Zhang, H.; Xiao, Z.; Wang, J.; Li, F.; Szczerbicki, E. A novel IoT-perceptive human activity recognition (HAR) approach using multihead convolutional attention. *IEEE Internet Things J.* **2019**, *7*, 1072–1080. [CrossRef]
33. Hu, C.; Chen, Y.; Peng, X.; Yu, H.; Gao, C.; Hu, L. A Novel Feature Incremental Learning Method for Sensor-Based Activity Recognition. *IEEE Trans. Knowl. Data Eng.* **2019**, *31*, 1038–1050. [CrossRef]
34. Kuncan, F.; Kaya, Y.; Kuncan, M. A novel approach for activity recognition with down-sampling 1D local binary pattern. *Adv. Electr. Comput. Eng.* **2019**, *19*, 35–44. [CrossRef]
35. Wei, X.; Wang, Z. TCN-attention-HAR: Human activity recognition based on attention mechanism time convolutional network. *Sci. Rep.* **2024**, *14*, 7414. [CrossRef]
36. Ray, L.S.S.; Geißler, D.; Liu, M.; Zhou, B.; Suh, S.; Lukowicz, P. ALS-HAR: Harnessing Wearable Ambient Light Sensors to Enhance IMU-based HAR. *arXiv* **2024**, arXiv:2408.09527.
37. Liandana, M.; Hostiadi, D.P.; Pradipta, G.A. A New Approach for Human Activity Recognition (HAR) Using A Single Triaxial Accelerometer Based on a Combination of Three Feature Subsets. *Int. J. Intell. Eng. Syst.* **2024**, *17*, 235–250.
38. Geravesh, S.; Rupapara, V. Artificial neural networks for human activity recognition using sensor based dataset. *Multimed. Tools Appl.* **2023**, *82*, 14815–14835. [CrossRef]
39. Hafeez, S.; Alotaibi, S.S.; Alazeb, A.; Al Mudawi, N.; Kim, W. Multi-Sensor-Based Action Monitoring and Recognition via Hybrid Descriptors and Logistic Regression. *IEEE Access* **2023**, *11*, 48145–48157. [CrossRef]
40. Jantawong, P.; Jitpattanakul, A.; Mekruksavanich, S. Enhancement of Human Complex Activity Recognition using Wearable Sensors Data with InceptionTime Network. In Proceedings of the 2021 2nd International Conference on Big Data Analytics and Practices (IBDAP), Bangkok, Thailand, 26–27 August 2021.
41. Mekruksavanich, S.; Jitpattanakul, A.; Sitthithakerngkiet, K.; Youplao, P.; Yupapin, P. Resnet-se: Channel attention-based deep residual network for complex activity recognition using wrist-worn wearable sensors. *IEEE Access* **2022**, *10*, 51142–51154. [CrossRef]
42. Lohit, S.; Wang, Q.; Turaga, P. Temporal transformer networks: Joint learning of invariant and discriminative time warping. In Proceedings of the IEEE/CVF Conference on Computer Vision and Pattern Recognition, Long Beach, CA, USA, 15–20 June 2019; pp. 12426–12435.
43. Neverova, N.; Wolf, C.; Lacey, G.; Fridman, L.; Chandra, D.; Barbello, B.; Taylor, G. Learning human identity from motion patterns. *IEEE Access* **2016**, *4*, 1810–1820. [CrossRef]

44. AlShorman, O.; Alshorman, B.; Masadeh, M.S. A review of physical human activity recognition chain using sensors. *Indones. J. Electr. Eng. Inform. (IJEEI)* **2020**, *8*, 560–573.
45. Karantonis, D.M.; Narayanan, M.R.; Mathie, M.; Lovell, N.H.; Celler, B.G. Implementation of a real-time human movement classifier using a triaxial accelerometer for ambulatory monitoring. *IEEE Trans. Inf. Technol. Biomed.* **2006**, *10*, 156–167. [CrossRef]
46. Banos, O.; Garcia, R.; Holgado-Terriza, J.A.; Damas, M.; Pomares, H.; Rojas, I.; Saez, A. mHealthDroid: A novel framework for agile development of mobile health applications. In Proceedings of the International Workshop on Ambient Assisted Living, Belfast, UK, 2–5 December 2014; pp. 2–5.
47. Yang, J. Toward physical activity diary: Motion recognition using simple acceleration features with mobile phones. In Proceedings of the 1st International Workshop on Interactive Multimedia for Consumer Electronics, Beijing, China, 23 October 2009; pp. 1–10.
48. Fridolfsson, J.; Börjesson, M.; Buck, C.; Ekblom, Ö.; Ekblom-Bak, E.; Hunsberger, M.; Lissner, L.; Arvidsson, D. Effects of frequency filtering on intensity and noise in accelerometer-based physical activity measurements. *Sensors* **2019**, *19*, 2186. [CrossRef]
49. Preece, S.J.; Goulermas, J.Y.; Kenney, L.P.; Howard, D. A comparison of feature extraction methods for the classification of dynamic activities from accelerometer data. *IEEE Trans. Biomed. Eng.* **2008**, *56*, 871–879. [CrossRef]
50. Dehghani, A.; Sarbishei, O.; Glatard, T.; Shihab, E. A quantitative comparison of overlapping and non-overlapping sliding windows for human activity recognition using inertial sensors. *Sensors* **2019**, *19*, 5026. [CrossRef]
51. San-Segundo, R.; Montero, J.M.; Barra-Chicote, R.; Fernández, F.; Pardo, J.M. Feature extraction from smartphone inertial signals for human activity segmentation. *Signal Process.* **2016**, *120*, 359–372. [CrossRef]
52. Shoaib, M.; Bosch, S.; Incel, O.D.; Scholten, H.; Havinga, P.J. A survey of online activity recognition using mobile phones. *Sensors* **2015**, *15*, 2059–2085. [CrossRef]
53. Nweke, H.F.; Teh, Y.W.; Al-Garadi, M.A.; Alo, U.R. Deep learning algorithms for human activity recognition using mobile and wearable sensor networks: State of the art and research challenges. *Expert Syst. Appl.* **2018**, *105*, 233–261. [CrossRef]
54. Bennasar, M.; Price, B.A.; Gooch, D.; Bandara, A.K.; Nuseibeh, B. Significant features for human activity recognition using tri-axial accelerometers. *Sensors* **2022**, *22*, 7482. [CrossRef]
55. Gil-Martín, M.; San-Segundo, R.; Fernandez-Martinez, F.; Ferreiros-López, J. Improving physical activity recognition using a new deep learning architecture and post-processing techniques. *Eng. Appl. Artif. Intell.* **2020**, *92*, 103679. [CrossRef]
56. Dua, N.; Singh, S.N.; Semwal, V.B. Multi-input CNN-GRU based human activity recognition using wearable sensors. *Computing* **2021**, *103*, 1461–1478. [CrossRef]
57. Kutlay, M.A.; Gagula-Palalic, S. Application of machine learning in healthcare: Analysis on MHEALTH dataset. *Southeast Eur. J. Soft Comput.* **2016**, *4* . [CrossRef]
58. Cosma, G.; Mcginnity, T.M. Feature extraction and classification using leading eigenvectors: applications to biomedical and multi-modal mHealth data. *IEEE Access* **2019**, *7*, 107400–107412. [CrossRef]
59. Sztyler, T.; Stuckenschmidt, H. On-body localization of wearable devices: An investigation of position-aware activity recognition. In Proceedings of the 2016 IEEE International Conference on Pervasive Computing and Communications (PerCom), Sydney, NSW, Australia, 14–19 March 2016; pp. 1–9.
60. Ciuti, G.; Ricotti, L.; Menciassi, A.; Dario, P. MEMS sensor technologies for human centred applications in healthcare, physical activities, safety and environmental sensing: A review on research activities in Italy. *Sensors* **2015**, *15*, 6441–6468. [CrossRef]
61. Janidarmian, M.; Roshan Fekr, A.; Radecka, K.; Zilic, Z. A comprehensive analysis on wearable acceleration sensors in human activity recognition. *Sensors* **2017**, *17*, 529. [CrossRef]
62. Coskun, D.; Incel, O.D.; Ozgovde, A. Phone position/placement detection using accelerometer: Impact on activity recognition. In Proceedings of the 2015 IEEE Tenth International Conference on Intelligent Sensors, Sensor Networks and Information Processing (ISSNIP), Singapore, 7–9 April 2015; pp. 1–6.
63. O'Halloran, J.; Curry, E. A Comparison of Deep Learning Models in Human Activity Recognition and Behavioural Prediction on the MHEALTH Dataset. In Proceedings of the Irish Conference on Artificial Intelligence and Cognitive Science, Galway, Ireland, 5–6 December 2019; pp. 212–223.
64. Nguyen, Q.H.; Ly, H.B.; Ho, L.S.; Al-Ansari, N.; Le, H.V.; Tran, V.Q.; Prakash, I.; Pham, B.T. Influence of data splitting on performance of machine learning models in prediction of shear strength of soil. *Math. Probl. Eng.* **2021**, *2021*, 4832864. [CrossRef]

Mathematical and Computational Applications

MDPI

Article

Design and Implementation of a Discrete-PDC Controller for Stabilization of an Inverted Pendulum on a Self-Balancing Car Using a Convex Approach

Yasmani González-Cárdenas [1], Francisco-Ronay López-Estrada [1,*], Víctor Estrada-Manzo [2], Joaquin Dominguez-Zenteno [1,*] and Manuel López-Pérez [1]

1 TURIX-Dynamics Diagnosis and Control Group, Tecnológico Nacional de México, Tuxtla Gutierrez 29580, Mexico; yagcardenas@gmail.com (Y.G.-C.); manuel.lp@tuxtla.tecnm.mx (M.L.-P.)

2 Department of Mechatronics, Universidad Politécnica de Pachuca, Zempoala 43830, Mexico; victor_estrada@upp.edu.mx

* Correspondence: frlopez@tuxtla.tecnm.mx (F.-R.L.-E.); joaquin.dz@tuxtla.tecnm.mx (J.D.-Z.)

Abstract: This paper presents a trajectory-tracking controller of an inverted pendulum system on a self-balancing differential drive platform. First, the system modeling is described by considering approximations of the swing angles. Subsequently, a discrete convex representation of the system via the nonlinear sector technique is obtained, which considers the nonlinearities associated with the nonholonomic constraint. The design of a discrete parallel distributed compensation controller is achieved through an alternative method due to the presence of uncontrollable points that avoid finding a solution for the entire polytope. Finally, simulations and experimental results using a prototype illustrate the effectiveness of the proposal.

Keywords: self-balancing inverted pendulum; controller design; TS system; trajectory tracking

1. Introduction

The control of critical systems is an important area of study in robotics, as these systems operate around unstable equilibrium points; once they reach a critical zone, it is impossible to re-stabilize them due to physical limitations in their actuators; a classic example is the inverted pendulum on a self-balancing platform. The main challenge lies in maintaining two coupled joints within the allowable limits of the equilibrium point despite the inevitable presence of measurement noise, disturbances, and model inaccuracies [1,2]. One way to prevent the system from going beyond its bounds is using constrained predictive control strategies [3,4]. However, these strategies come with high computational costs, making them prohibitive for most of the low-cost embedded hardware.

As the balancing platform is essentially a mobile robot with differential traction, movement restrictions arise due to its nonholonomic nature, adding complexity and turning it into a multi-objective problem. Balance must be maintained while following a specific trajectory on the plane, which poses significant challenges in planning and control execution [5]. The literature covers various subjects: hardware, sensing configurations, and control methods for two-wheeled and self-balancing robots [6–8]. For instance, the authors in [1] introduced a low-cost prototype with nested control to regulate the longitudinal displacement and speed via the pitch angle. In [9,10], advanced control algorithms are proposed using adaptive sliding mode and direct fuzzy control for improved velocity tracking and stability. The works [11,12] presented a nonlinear control approach, avoiding linearization around the equilibrium points; a machine-learning-based adaptive fuzzy logic-proportional integral controller for variable payloads in a two-wheeled underactuated-mobile inverted pendulum was developed by [13]. Recently, refs. [14,15] proposed a self-balancing robot with PID control using sensors and complementary filters for practical angular velocity estimation and motor stability; the controller was implemented with Arduino. However, to the best of the authors' knowledge, few works have been dedicated to the study of two-wheeled and self-balancing platforms with a coupled inverted pendulum.

This paper presents the model and control of the inverted pendulum coupled with a two-wheeled and self-balancing differential robot. The model is developed under specific considerations, thus, allowing for simplifications without losing accuracy within the operating region. A discrete parallel distributed compensation (PDC) controller is designed for control and trajectory tracking based on an exact convex model [16]. Experimental results are presented to illustrate the performance of the proposed method.

The rest of the paper is organized as follows: Section 2 outlines the fundamental concepts and the problem to be solved. Section 3 presents the main results related to obtaining the mathematical model and controller design. Section 4 illustrates the performance of the proposed control scheme by means of simulations, while Section 5 validates the control strategy by means of an experimental test. Finally, Section 6 closes the paper with some final remarks and conclusions.

Preliminaries: Throughout the paper, the following notation will be employed: In the case of a matrix $A \in \mathbb{R}^{n\times n}$, A^T and A^{-1} stand for the transpose and the inverse, respectively. Given the complex variable $z \in \mathbb{C}$, $\bar{z}$ represents its complex conjugate. Additionally, the symbol $*$ indicates the transpose of the element in the symmetric position of the matrix.

2. Problem Statement

Consider the inverted pendulum system coupled to a two-wheeled and self-balancing mobile platform with differential traction, as displayed in Figure 1. This system can be modeled as a double inverted pendulum, which includes rotational motion and longitudinal movement. The differential traction mechanism allows the platform to follow a specific path. The control objective for the mobile platform is to track a desired reference signal while the pendulum is stabilized at the vertical position.

Figure 1. Prototype of the inverted pendulum on a self-balancing differential robot.

A free-body diagram of the system can be seen in Figure 2, where J_1 and J_2 represent the joints. In this case, joint J_2 couples the inverted pendulum with the cart and its pivotal, allowing the pendulum to swing while mounted on the mobile platform. The parameters and variables of this system are shown in Table 1.

Table 1. Relationship of parameters for the self-balancing cart model with an inverted pendulum.

Parameter	Units	Description
m_1	[Kg]	Mass of the cart
m_2	[Kg]	Mass of the pendulum
m_w	[Kg]	Mass of the wheels

Table 1. *Cont.*

Parameter	Units	Description
θ_L	[rad]	Angle of the left wheel
θ_R	[rad]	Angle of the right wheel
θ_1	[rad]	Angle of the cart
θ_2	[rad]	Angle of the pendulum
θ_3	[rad]	Angle of the cart position in space
l_1	[m]	Distance to the center of mass in the cart
l_2	[m]	Distance to the center of mass of the pendulum
L_1	[m]	Distance between axle and pendulum
L_2	[m]	Distance of the pendulum
w	[m]	Separation between wheels
r	[m]	Radius of the wheels
I_w	[kg·m^2]	Moment of inertia of the wheels
I_1	[kg·m^2]	Moment of inertia of the cart
I_2	[kg·m^2]	Moment of inertia of the pendulum
g	[m/s^2]	Gravitational acceleration constant
P	[m]	Distance of the displaced point

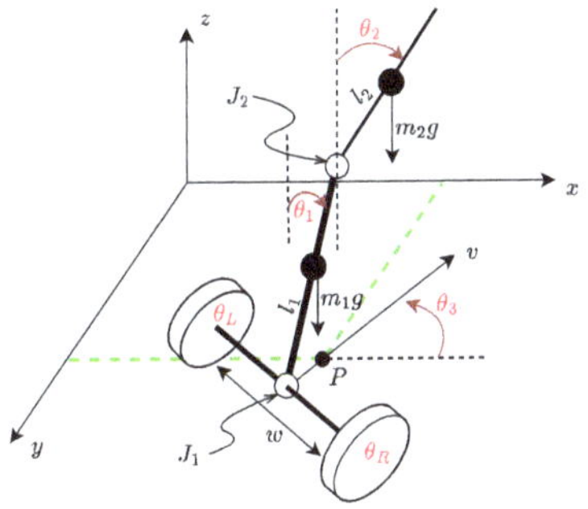

Figure 2. Diagram of a self-balancing car with an inverted pendulum.

Considering the Euler–Lagrange energy approach to model the longitudinal motion, where both wheels rotate together to stabilize the oscillating body. The energy equations for each wheel are defined by the following:

$$\begin{aligned} E_{kL}(t) &= \frac{1}{2}m_w(r\dot{\theta}_L(t))^2 + \frac{1}{2}I_w\dot{\theta}_L^2(t), \\ E_{kR}(t) &= \frac{1}{2}m_w(r\dot{\theta(t)}_R)^2 + \frac{1}{2}I_w\dot{\theta}_R^2(t). \end{aligned} \tag{1}$$

The potential energy for both is zero ($E_{pL}(t) = E_{pR}(t) = 0$) as they remain on the horizontal plane. For the position of the cart ($\theta_3(t) = 0$, $y(t) = 0$), the localization of the center of mass in the plane (x, z) of the joint J_1 (self-balancing) is represented by the following:

$$\begin{aligned} x_1(t) &= \frac{r}{2}(\theta_L(t) + \theta_R(t)) + l_1 \sin(\theta_1(t)), \\ z_1(t) &= l_1 \cos(\theta_1(t)), \end{aligned}$$

and the velocity of the center of mass $v_1(t) = [\dot{x}_1(t) \;\; \dot{z}_1(t)]^T$:

$$\begin{aligned} \dot{x}_1(t) &= \frac{r}{2}(\dot{\theta}_L(t) + \dot{\theta}_R(t)) + \dot{\theta}_1(t)l_1 \cos(\theta_1(t)), \\ \dot{z}_1(t) &= -\dot{\theta}_1(t)l_1 \sin(\theta_1(t)). \end{aligned}$$

As for the kinetic and potential energies at the joint J_1, we have

$$\begin{aligned} E_{k1}(t) &= \frac{1}{2}m_1(v_1(t)v_1^T(t)) + \frac{1}{2}I_1\dot{\theta}_1^2(t), \\ E_{p1}(t) &= m_1 g z_1(t); \end{aligned}$$

which are equivalent to:

$$\begin{aligned} E_{k1}(t) &= \frac{r^2}{8}m_1(\dot{\theta}_L(t)+\dot{\theta}_R(t))^2 + \frac{1}{2}(m_1 l_1^2 + I_1)\dot{\theta}_1^2(t) + \frac{r}{2}m_1 l_1(\dot{\theta}_L(t)+\dot{\theta}_R(t))\dot{\theta}_1(t)\cos(\theta_1(t)), \\ E_{p1}(t) &= m_1 g l_1 \cos(\theta_1(t)). \end{aligned} \quad (2)$$

Similarly, for the joint J_2 (pendulum), its position and velocity in the plane (x,z); $v_2(t) = [\dot{x}_2(t) \;\; \dot{z}_2(t)]^T$ are described by the following expressions:

$$\begin{aligned} x_2(t) =& \frac{r}{2}(\theta_L(t)+\theta_R(t)) + L_1\sin(\theta_1(t)) + l_2\sin(\theta_2(t)), \\ z_2(t) =& L_1\cos(\theta_1(t)) + l_2\cos(\theta_2(t)), \\ \dot{x}_2(t) =& \frac{r}{2}(\dot{\theta}_L(t)+\dot{\theta}_R(t)) + \dot{\theta}_1(t)L_1\cos(\theta_1(t)) + \dot{\theta}_2(t)l_2\cos(\theta_2(t)), \\ \dot{z}_2(t) =& -\dot{\theta}_1(t)L_1\sin(\theta_1(t)) - \dot{\theta}_2(t)l_2\sin(\theta_2(t)). \end{aligned}$$

Their kinetic and potential energies are obtained as follows:

$$\begin{aligned} E_{k2}(t) =& \frac{1}{2}m_2(v_2(t)v_2(t)^T) + \frac{1}{2}I_2\dot{\theta}_2^2(t), \\ E_{p2}(t) =& m_2 g z_2(t). \end{aligned}$$

by replacing $v_2(t)$ and $z_2(t)$, we have:

$$\begin{aligned} E_{k2}(t) =& \frac{r^2}{8}m_2(\dot{\theta}_L(t)+\dot{\theta}_R(t))^2 + \frac{1}{2}m_2 L_1^2\dot{\theta}_1^2(t) + \frac{1}{2}(m_2 l_2^2 + I_2)\dot{\theta}_2^2(t) + \\ & \frac{r}{2}m_2 L_1(\dot{\theta}_L(t)+\dot{\theta}_R(t))\dot{\theta}_1(t)\cos(\theta_1(t)) + \frac{r}{2}m_2 l_2(\dot{\theta}_L(t)+\dot{\theta}_R(t))\dot{\theta}_2(t)\cos(\theta_2(t)) + \\ & m_2 L_1 l_2 \dot{\theta}_1(t)\dot{\theta}_2(t)\cos(\theta_1(t)-\theta_2(t)), \\ E_{p2}(t) =& m_2 g L_1\cos(\theta_1(t)) + m_2 g l_2\cos(\theta_2(t)). \end{aligned} \quad (3)$$

Nonholonomic Model

The above equations do not consider the kinematic part of the nonholonomic inherent in the differential traction vehicle setup. Following [5], the model of the differential robot considering the displaced point is as follows:

$$\begin{aligned} \dot{x}(t) =& \Big(\frac{r}{2}\cos(\theta_3(t)) - \frac{r}{w}P\sin(\theta_3(t))\Big)\omega_R(t) + \Big(\frac{r}{2}\cos(\theta_3(t)) + \frac{r}{w}P\sin(\theta_3(t))\Big)\omega_L(t), \\ \dot{y}(t) =& \Big(\frac{r}{2}\sin(\theta_3(t)) + \frac{r}{w}P\cos(\theta_3(t))\Big)\omega_R(t) + \Big(\frac{r}{2}\sin(\theta_3(t)) - \frac{r}{w}P\cos(\theta_3(t))\Big)\omega_L(t), \\ \dot{\theta}_3(t) =& \frac{r}{w}(\omega_R(t) - \omega_L(t)), \end{aligned} \quad (4)$$

where $\omega_R(t)$ and $\omega_L(t)$ are the wheel velocities. By considering these velocities, the angle θ_3 can be computed in real-time using the following odometry equation:

$$\theta_3(t) = \frac{r}{w}(\theta_R(t) - \theta_L(t)).$$

Note that it is not possible to simultaneously control x, y, and θ_3 due to the nonholonomic constraints; therefore, θ_3 is treated as an exogenous parameter and the dynamic equation associated with θ_3 is removed from (4). Now, let us consider the accelerations at each wheel $\dot{\omega}_R(t)$ and $\dot{\omega}_L(t)$, an augmented system can be formulated with $\ddot{\theta}_L = \dot{\omega}_L(t) = \alpha_L(t)$ and $\ddot{\theta}_R = \dot{\omega}_R(t) = \alpha_R(t)$; it yields the following augmented nonlinear system:

$$
\begin{aligned}
\dot{x}(t) =&\Big(\frac{r}{2}\cos(\theta_3(t)) - \frac{r}{w}P\sin(\theta_3(t))\Big)\omega_R(t) + \Big(\frac{r}{2}\cos(\theta_3(t)) + \frac{r}{w}P\sin(\theta_3(t))\Big)\omega_L(t),\\
\dot{y}(t) =&\Big(\frac{r}{2}\sin(\theta_3(t)) + \frac{r}{w}P\cos(\theta_3(t))\Big)\omega_R(t) + \Big(\frac{r}{2}\sin(\theta_3(t)) - \frac{r}{w}P\cos(\theta_3(t))\Big)\omega_L(t),\\
\dot{\omega}_R =&\alpha_R(t),\\
\dot{\omega}_L =&\alpha_L(t).
\end{aligned} \tag{5}
$$

Equations (1)–(3) describe the kinetic and potential energy of the tires and the joint together with the kinematic model (5); all of them will be used to solve the Euler–Lagrange equations describing the robot behavior. Here, the challenge is to design a tracking controller that maintains the vertical position of the joint while the differential cart follows a trajectory. The next section proposes a method for the design of a discrete-time PDC controller for trajectory tracking while guaranteeing stability.

3. Main Results

The proposed methodology assumes that the only energy contributing to the joint motion comes from the translation of the cart, with zero rotational energy at θ_3; rotation about θ_3 will only be considered in the kinematic part of the cart. Following the Euler–Lagrange modeling procedure, the kinetic energy can be expressed as $T(t) = E_{kL}(t) + E_{kR}(t) + E_{k1}(t) + E_{k2}(t)$ and the potential energy as $U(t) = E_{pL}(t) + E_{pR}(t) + E_{p1}(t) + E_{p2}(t)$ (from Equations (1), (2), and (3), respectively). The Lagrangian is then denoted as $\mathcal{L}(q(t)) = T(t) - U(t)$, where $q(t)$ is the generalized coordinate.

Traditionally, the functional derivative is applied using a vector q together with the non-conservative forces acting on each joint; in this case, the torques applied to each wheel are considered [17]. However, a model with torques as the inputs could be more practical. This work assumes that each motor has an internal velocity controller, which is very common, as discussed in [18].

The vector of generalized coordinates is defined by $q(t) = [\theta_1(t), \theta_2(t)]^T$ and solving the Euler–Lagrange equation:

$$
\frac{d}{dt}\left(\frac{\partial\mathcal{L}(q(t))}{\partial\dot{q}(t)}\right) - \frac{\partial\mathcal{L}(q(t))}{\partial q(t)} = 0,
$$

the solution for the generalized coordinates is obtained as follows:

$$
\begin{aligned}
q_1(t) :=&\frac{r}{2}(m_1l_1 + m_2L_1)\cos(\theta_1(t))(\ddot{\theta}_L(t) + \ddot{\theta}_R(t)) + (m_1l_1^2 + m_2L_1^2 + I_1)\ddot{\theta}_1(t)\\
&+ m_2L_1l_2\cos(\theta_1(t) - \theta_2(t))\ddot{\theta}_2(t) + m_2L_1l_2\sin(\theta_1(t) - \theta_2(t))\dot{\theta}_2^2(t)\\
=&(m_1l_1 + m_2L_1)g\sin(\theta_1(t)),
\end{aligned} \tag{6}
$$

$$
\begin{aligned}
q_2(t) :=&\frac{r}{2}m_2l_2\cos(\theta_2(t))(\ddot{\theta}_L(t) + \ddot{\theta}_R(t)) + (m_2l_2^2 + I_2)\ddot{\theta}_2(t)\\
&+ m_2L_1l_2\cos(\theta_1(t) - \theta_2(t))\ddot{\theta}_1 - m_2L_1l_2\sin(\theta_1(t) - \theta_2(t))\dot{\theta}_1^2(t)\\
=&m_2gl_2\sin(\theta_2(t)).
\end{aligned} \tag{7}
$$

Considering that angles θ_1 and θ_2 variate within a small range $(-10^\circ \leq \theta_{1,2} \leq 10^\circ)$, the following approximations can be used:

$$
\cos(\theta_{1,2}) \approx 1, \quad \sin(\theta_{1,2}) \approx \theta_{1,2}, \quad \dot{\theta}_{1,2}^2 \approx 0; \tag{8}
$$

hence, (6) and (7) yield:

$$
\begin{aligned}
&\frac{r}{2}(m_1l_1 + m_2L_1)\ddot{\theta}_L(t) + \frac{r}{2}(m_1l_1 + m_2L_1)\ddot{\theta}_R(t) + (m_1l_1^2 + m_2L_1^2 + I_1)\ddot{\theta}_1(t)\\
&+ m_2L_1l_2\ddot{\theta}_2(t) - (m_1l_1 + m_2L_1)g\theta_1(t) = 0,
\end{aligned} \tag{9}
$$

$$\frac{r}{2}m_2 l_2 \ddot{\theta}_L(t) + \frac{r}{2}m_2 l_2 \ddot{\theta}_R(t) + m_2 L_1 l_2 \ddot{\theta}_1(t) + (m_2 l_2^2 + I_2)\ddot{\theta}_2(t) - m_2 g l_2 \theta_2(t) = 0, \tag{10}$$

where $\dot{\theta}_R(t)$ and $\dot{\theta}_L(t)$ represent the accelerations of each wheel that are considered as the control inputs. From (9) and (10), a system of first-order ordinary differential equations is obtained:

$$\begin{aligned} \dot{\theta}_1(t) &= \omega_1(t), \\ \dot{\theta}_2(t) &= \omega_2(t), \\ \dot{\omega}_1(t) &= \sigma_1\theta_1(t) + \sigma_2\theta_2(t) + \sigma_3\alpha_R(t) + \sigma_3\alpha_L(t), \\ \dot{\omega}_2(t) &= \sigma_4\theta_1(t) + \sigma_5\theta_2(t) + \sigma_6\alpha_R(t) + \sigma_6\alpha_L(t), \end{aligned} \tag{11}$$

with:

$$\begin{aligned} \sigma_1 &= \frac{L_1 g l_2^2 m_2^2 + g l_1 m_1 l_2^2 m_2 + I_2 L_1 g m_2 + I_2 g l_1 m_1}{I_2 m_2 L_1^2 + m_1 m_2 l_1^2 l_2^2 + I_2 m_1 l_1^2 + I_1 m_2 l_2^2 + I_1 I_2}, \\ \sigma_2 &= -\frac{L_1 g l_2^2 m_2^2}{I_2 m_2 L_1^2 + m_1 m_2 l_1^2 l_2^2 + I_2 m_1 l_1^2 + I_1 m_2 l_2^2 + I_1 I_2}, \\ \sigma_3 &= -\frac{l_1 m_1 m_2 r l_2^2 + I_2 L_1 m_2 r + I_2 l_1 m_1 r}{2(I_2 m_2 L_1^2 + m_1 m_2 l_1^2 l_2^2 + I_2 m_1 l_1^2 + I_1 m_2 l_2^2 + I_1 I_2)}, \\ \sigma_4 &= -\frac{l_2 m_2 (g m_2 L_1^2 + g l_1 m_1 L_1)}{I_2 m_2 L_1^2 + m_1 m_2 l_1^2 l_2^2 + I_2 m_1 l_1^2 + I_1 m_2 l_2^2 + I_1 I_2}, \\ \sigma_5 &= \frac{l_2 m_2 (g m_2 L_1^2 + g m_1 l_1^2 + I_1 g)}{I_2 m_2 L_1^2 + m_1 m_2 l_1^2 l_2^2 + I_2 m_1 l_1^2 + I_1 m_2 l_2^2 + I_1 I_2}, \\ \sigma_6 &= -\frac{l_2 m_2 (m_1 r l_1^2 - L_1 m_1 r l_1 + I_1 r)}{2(I_2 m_2 L_1^2 + m_1 m_2 l_1^2 l_2^2 + I_2 m_1 l_1^2 + I_1 m_2 l_2^2 + I_1 I_2)}. \end{aligned}$$

Finally, by combining (5) and (11), a nonlinear model of the inverted pendulum on a self-balancing robot yields:

$$\begin{bmatrix} \dot{x}(t) \\ \dot{y}(t) \\ \dot{\theta}_1(t) \\ \dot{\theta}_2(t) \\ \dot{\omega}_R(t) \\ \dot{\omega}_L(t) \\ \dot{\omega}_1(t) \\ \dot{\omega}_2(t) \end{bmatrix} = \begin{bmatrix} 0 & 0 & 0 & 0 & \frac{\sigma_{10}-\sigma_7}{2w} & \frac{\sigma_{10}+\sigma_7}{2w} & 0 & 0 \\ 0 & 0 & 0 & 0 & \frac{\sigma_9+\sigma_8}{2w} & \frac{\sigma_9-\sigma_8}{2w} & 0 & 0 \\ 0 & 0 & 0 & 0 & 0 & 0 & 1 & 0 \\ 0 & 0 & 0 & 0 & 0 & 0 & 0 & 1 \\ 0 & 0 & 0 & 0 & 0 & 0 & 0 & 0 \\ 0 & 0 & 0 & 0 & 0 & 0 & 0 & 0 \\ 0 & 0 & \sigma_1 & \sigma_2 & 0 & 0 & 0 & 0 \\ 0 & 0 & \sigma_4 & \sigma_5 & 0 & 0 & 0 & 0 \end{bmatrix} \begin{bmatrix} x(t) \\ y(t) \\ \theta_1(t) \\ \theta_2(t) \\ \omega_R(t) \\ \omega_L(t) \\ \omega_1(t) \\ \omega_2(t) \end{bmatrix} + \begin{bmatrix} 0 & 0 \\ 0 & 0 \\ 0 & 0 \\ 0 & 0 \\ 1 & 0 \\ 0 & 1 \\ \sigma_3 & \sigma_3 \\ \sigma_6 & \sigma_6 \end{bmatrix} \begin{bmatrix} \alpha_R(t) \\ \alpha_L(t) \end{bmatrix} \tag{12}$$

with:

$$\sigma_7 = 2Pr\sin(\theta_3(t)),\ \sigma_8 = 2Pr\cos(\theta_3(t)),\ \sigma_9 = rw\sin(\theta_3(t)),\ \text{and}\ \sigma_{10} = rw\cos(\theta_3(t)).$$

Model (12) contains information about the kinematic aspect associated with the non-holonomic constraints and the dynamic longitudinal model related to self-balancing coupled with the inverted pendulum. Despite the small angle approximations, the model contains some nonlinearities related to the angle $\theta_3(t)$ that will be handled in the next section by considering exact convex modeling.

Convex Controller Design

The nonlinear model (12) can be represented as a discrete convex Takagi–Sugeno (TS) model to simplify the controller design. This transformation is achieved using the nonlinear sector technique, as detailed in [16]. A convex representation is obtained by expressing the nonlinearities of the system as bounded functions. This rewriting of the nonlinear model allows us to design the controller by obtaining a numerical solution using Linear Matrix Inequalities (LMIs).

The resulting discrete convex model is given by the following:

$$\begin{aligned} \mathbf{x}(k+1) &= \sum_{i=1}^{r} h_i(\rho(k))G_i\mathbf{x}(k) + H\mathbf{u}(k), \\ \mathbf{y}(k) &= C\mathbf{x}(k), \end{aligned} \tag{13}$$

where $\mathbf{x} \in \mathbb{R}^8$, $\mathbf{u} \in \mathbb{R}^2$, and $\mathbf{y} \in \mathbb{R}^8$ are the state, input, and output vectors, respectively. The exogenous premise vector is $\rho(k) = [\cos(\theta_3(k)), \sin(\theta_3(k))]^T$, it is assumed to be bounded in a compact set $\mathcal{C} \in \mathbb{R}^2$, and the number of vertices is $r = 2^2 = 4$. Matrices $G_i \in \mathbb{R}^{8\times 8}$, $H \in \mathbb{R}^{8\times 2}$, and $C \in \mathbb{R}^{4\times 8}$ are the discrete equivalent matrix vertices computed utilizing the sector nonlinearity approach [16]. The membership functions $h_i(\rho(k))$, $i \in \{1,2,3,4\}$ hold the convex sum property in $\mathcal{C}$: $\sum_{i=1}^{r} h_i(\rho(k)) = 1$. Hence, as the entries of the premise vector $\rho(k)$ have natural limits, we have $-1 \leq \rho_{1,2} \leq 1$; then, the membership functions

$$h_1(k) = w_1^0(k)w_2^0(k),\ h_2(k) = w_1^0(k)w_2^1(k),\ h_3(k) = w_1^1(k)w_2^0(k),\ h_4(k) = w_1^1(k)w_2^1(k),$$

are obtained with $w_1^0(k) = \frac{1-\cos(\theta_3(k))}{2}$, $w_1^1(k) = 1 - w_1^0(k)$, $w_2^0(k) = \frac{1-\sin(\theta_3(k))}{2}$ and $w_2^1(k) = 1 - w_2^0(k)$.

For the control and stabilization of the system (13), the following PDC-type control law is employed

$$\mathbf{u}(k) = \sum_{i=1}^{r} h_i(\rho(k))K_i\mathbf{x}(k),$$

where K_i, $i \in \{1,2,3,4\}$ is the feedback gain to be designed.

As is customary in the convex community, the design of the feedback gains K_i is carried out by quadratic Lyapunov functions ending in LMI conditions. However, the polytopic representation assumes independence among all terms of the vector ρ. Because of coupling among these terms, regions within the compact set $\mathcal{C}$ do not accurately belong to the model, leading to conservatism in the polytopic representation, as discussed in [19]. Specifically, the point $\rho = [0,\ 0]^T$ within $\mathcal{C}$ results in two critically uncontrollable poles, limiting an asymptotically stable solution for the entire $\mathcal{C}$.

Therefore, the values of K_i for each vertex are computed by considering a Linear Quadratic Regulator (LQR). This approach allows for a straightforward adjustment of the cost matrices to attain the desired solution. The discrete-time LQR gains are calculated using the matrices Q and R to solve the optimization problem [20]:

$$J(k) = \sum_{k=1}^{\infty} \left(\mathbf{x}^T(k)Q\mathbf{x}(k) + \mathbf{u}^T(k)R\mathbf{u}(k)\right).$$

As a stability test, an LMI region with a radius infinitesimally greater than unity to encompass the uncontrollable poles is considered through the following [21]:

$$\begin{bmatrix} -S(1+\Delta_r) & (*) \\ S\hat{G}_i & -S(1+\Delta_r) \end{bmatrix} < 0, \quad S = S^T > 0, \tag{14}$$

where $\hat{G}_i = G_i + HK_i$, $i \in \{1,2,3,4\}$, $S \in \mathbb{R}^{8\times 8}$. These LMIs verify that all closed-loop system poles are within the circular region of radius $1 + \Delta_r$. A non-negative slack variable $\Delta_r \geq 0$ is introduced to handle numerical issues arising from critically stable uncontrollable poles.

Remark 1. *Due to the presence of uncontrollable points within the polytope defined by the nonlinear terms* $\sin\theta_3$ *and* $\cos\theta_3$*, it is not possible to find a direct solution for LMIs* (14)*. This is a computational problem that arises from the specific structure of the model and avoids those standard LMI* *solvers finding a feasible solution. However, the system is controllable, and a solution must exist, even if the solver can not solve the LMI directly. The proposed approach computes the controller gains at the vertex of the polytope by considering an LQR. Then, with the computed gains, the LMI (14) is solved to prove the asymptotic stability. We acknowledge that obtaining a direct LMI solution would be ideal. Nonetheless, the proposed method provides an effective alternative for stabilizing the inverted pendulum system from a practical point of view.*

4. Simulation Results

For the simulations and experimental tests, the parameters in Table 2 have been used; most of these parameters have been provided by the manufacturer, while others have been computed and estimated in the laboratory. Due to the characteristics of the prototype, a sampling time of $T_s = 10$ ms is considered.

Table 2. Parameters of the prototype.

Parameter	Value
g	9.8
m_1	0.9
m_2	0.1
r	0.0335
L_1	0.126
L_2	0.390
l_1	$L_1/2$
l_2	$L_2/2$
I_1	$\frac{1}{12}m_1L_1^2$
I_2	$\frac{1}{12}m_2L_2^2$
w	0.178
P	0.001

To compute the LQR at each vertex, the following cost matrices have been used:

$$Q = \text{diag}([50, 50, 70, 150, 40, 40, 150, 150]),$$
$$R = \text{diag}([0.005, 0.005]).$$

The weight matrices were adjusted empirically to balance the system performance and actuator power consumption. It should be noted that the proposed method primarily validates the stability of the convex gain configurations without ensuring optimal performance due to the presence of non-controllable points within the convex polytope. Some aspects, such as robustness, noise handling, and optimization of the weight matrices, are beyond the scope of this study. However, the stability of the system is verified by testing the LMI (14) with the computed gains, which guarantees asymptotic convergence of the desired tracking controller. The computed gains are:

$$K_1 = \begin{bmatrix} 5 & -60 & -7902 & 17945 & 55 & -3 & -201 & 3760 \\ -60 & 5 & -7902 & 17945 & -3 & 55 & -201 & 3760 \end{bmatrix},$$
$$K_2 = \begin{bmatrix} -60 & -5 & -7902 & 17945 & 55 & -3 & -201 & 3760 \\ 5 & 60 & -7902 & 17945 & -3 & 55 & -201 & 3760 \end{bmatrix},$$
$$K_3 = \begin{bmatrix} 60 & 5 & -7902 & 17945 & 55 & -3 & -201 & 3760 \\ -5 & -60 & -7902 & 17945 & -3 & 55 & -201 & 3760 \end{bmatrix},$$
$$K_4 = \begin{bmatrix} -5 & 60 & -7902 & 17945 & 55 & -3 & -201 & 3760 \\ 60 & -5 & -7902 & 17945 & -3 & 55 & -201 & 3760 \end{bmatrix}.$$

For the stability proof, using LMIs (14) with $\hat{G}_i = G_i + HK_i$, $i \in \{1, 2, \ldots, 4\}$, and $\Delta_r = 0.0005$, the following matrix S have been obtained:

$$S = \begin{bmatrix} 0.0099 & (*) & (*) & (*) & (*) & (*) & (*) & (*) \\ 0.0000 & 0.0099 & (*) & (*) & (*) & (*) & (*) & (*) \\ -0.0000 & 0.0000 & 0.6949 & (*) & (*) & (*) & (*) & (*) \\ 0.0000 & -0.0000 & -1.6826 & 6.7423 & (*) & (*) & (*) & (*) \\ 0.0000 & -0.0000 & -0.0031 & 0.0188 & 0.0002 & (*) & (*) & (*) \\ 0.0000 & -0.0000 & -0.0031 & 0.0188 & 0.0001 & 0.0002 & (*) & (*) \\ 0.0000 & -0.0000 & -0.0029 & 0.0828 & 0.0009 & 0.0009 & 0.0067 & (*) \\ 0.0000 & -0.0000 & -0.3458 & 1.3896 & 0.0077 & 0.0077 & 0.0392 & 0.5091 \end{bmatrix}.$$

Simulation results illustrate the stabilization of the pendulum from an unstable initial position, and its return to the origin are presented in Figure 3. These simulations operate within the ±10-degree range, consistent with the small-angle approximation. A Gaussian noise signal was also introduced into the control inputs to simulate possible disturbances. As can be seen in the figure, the controllers perform well, keeping the vertical position of the pendulum while maintaining the system at the origin.

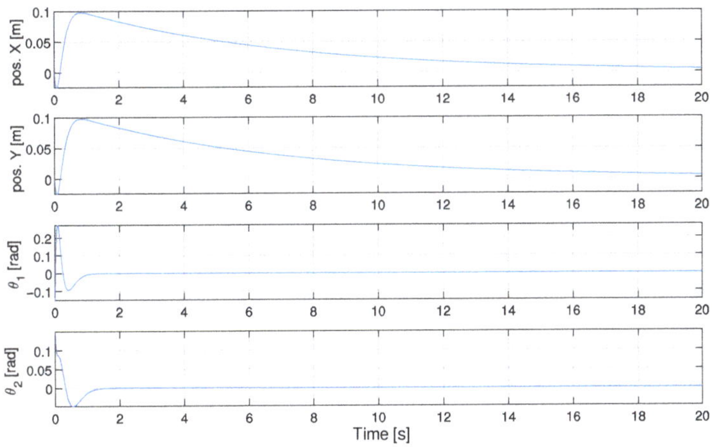

Figure 3. Transient response for initial conditions at $\theta_1 = -0.15$ rad, $\theta_2 = 0.15$ rad, $\theta_3 = \frac{\pi}{4}$ rad different from zero.

5. Experimental Results

For practical implementation, the speed in each actuator is assumed to be controlled; in our case, the manufacturer provides a tuned PI controller to control the speed in each motor. Figure 4 shows a closed-loop system diagram of the proposal. As the output of the PDC controller is in terms of acceleration (rad/s^2) at sampling instant (k), it is necessary to obtain the reference of velocity for the PI controller. To this end, a dynamic integrator is added. First, the PDC acceleration command is multiplied by the sampling time T_s to obtain the required velocity increment at time $k+1$ in rad/s. This increment is added to the current velocity to obtain the reference velocity at $k+1$. The PI controller is correctly tuned and uses a considerably smaller sampling period, inserting virtually negligible dynamics into the overall system.

The algorithm has been programmed in the embedded system provided by the manufacturer, which consists of an STM32F103—Arm Cortex-M3 microcontroller. The data have been acquired through Bluetooth serial communication, with a sampling period of 500 ms. A lemniscate trajectory was considered as a reference. Details of how to implement this reference are given in Appendix A.

Figure 5 displays the trajectory tracking response, where a slight deviation can be seen in the curves due to the variations in the linear and angular velocities of the path. These deviations show the sensitivity of the system to path accelerations. Figure 6 illustrates the behavior of angles θ_1 and θ_2 during the 60 s of testing. In this case, noise is particularly noticeable at θ_1, mainly due to the IMU's low quality and the potentiometer used to measure θ_2. However, despite the noise, the maximum deflection peaks remain within the designed values for the small angle approximations.

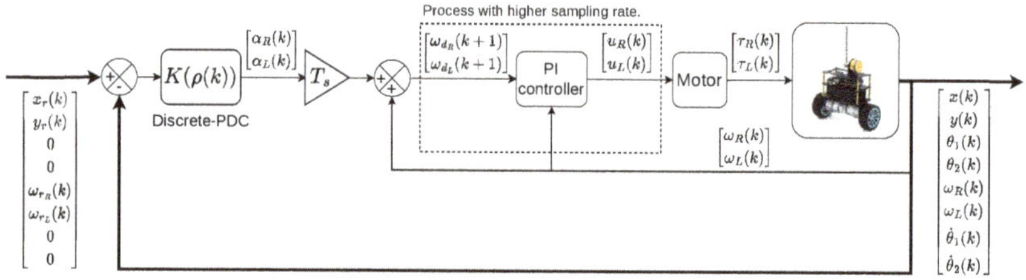

Figure 4. Overall system diagram. Symbols τ and u represent the torque and voltage applied to each DC motor.

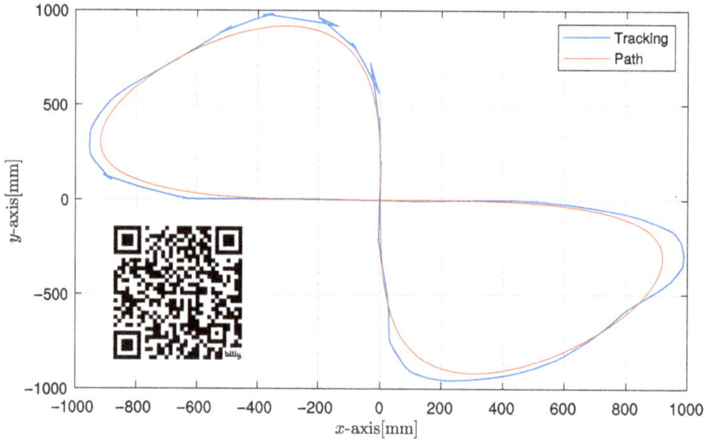

Figure 5. Experimental trajectory tracking.

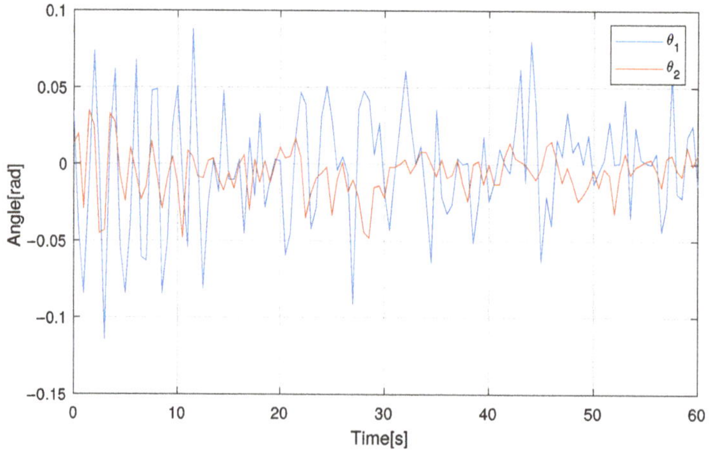

Figure 6. Transient behavior of the angles θ_1 and θ_2.

Figure 7 shows the errors concerning x and y obtained in the tracking test. Instantaneous deviations are observed due to the continuous effort to stabilize the cart, the pendulum, and the sensor noise. The different plots illustrate that the control system design implies a compromise between tracking error reduction and system stability. A video of the experimental test can be consulted at the link: https://bit.ly/gturixpendubot (accessed on 16 September 2024).

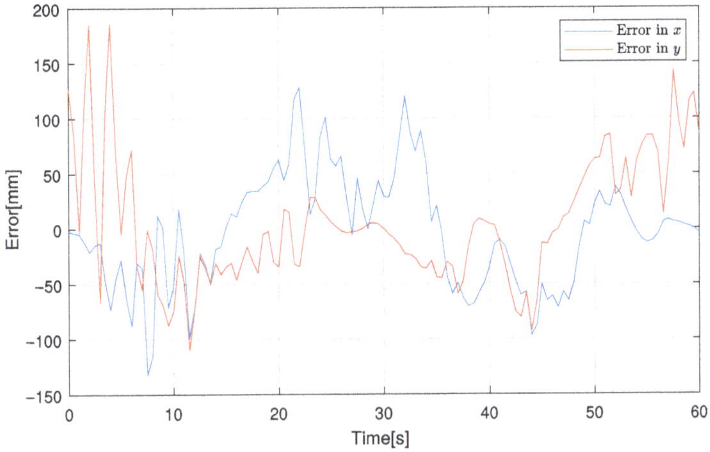

Figure 7. Transient behavior of the trajectory tracking error in the x and y axes.

6. Conclusions

A convex approach has been considered for modeling and controlling an inverted pendulum coupled with a self-balancing differential drive cart. The adopted control approach is practical for satisfactorily managing the nonholonomic constraints in the differential drive kinematic model of the cart, as stability and trajectory tracking have been achieved. Regardless of the noise and disturbances, the system presents an adequate response, which is demonstrated to be robust and practical. The main drawback encountered was the impossibility of finding a feasible solution for the LMIs due to uncontrollable points that appear due to the nonholonomic system. This led to exploring an alternative solution using LQR, followed by stability verification. Future work will focus on improving the controller's robustness and performance, optimizing its ability to handle system disturbances and uncertainties, and considering more advanced LMI techniques.

Author Contributions: conceptualization, Y.G.-C. and F.-R.L.-E.; methodology, Y.G.-C., F.-R.L.-E. and V.E.-M.; software, F.-R.L.-E.; validation, Y.G.-C. and F.-R.L.-E.; formal analysis, F.-R.L.-E., Y.G.-C. and V.E.-M.; investigation, Y.G.-C.; resources, J.D.-Z.; data curation, Y.G.-C. and M.L.-P.; writing—original draft preparation, Y.G.-C.; writing—review and editing, Y.G.-C., F.-R.L.-E. and V.E.-M.; supervision, project administration and funding acquisition, J.D.-Z. and F.-R.L.-E. All authors have read and agreed to the published version of the manuscript.

Funding: This research was funded by Tecnológico Nacional de México (TecNM) and Consejo Nacional de Humanidades, Ciencias y Tecnologías (CONAHCYT) in Mexico.

Data Availability Statement: The raw data supporting the conclusions of this article will be made available by the the authors upon request.

Acknowledgments: This research was supported by Tecnológico Nacional de México through the *Proyectos de Investigación Científica* call and by the National Council of Humanities, Science, and Technologies (CONAHCYT) under the national scholarship program. The authors are grateful for the scientific support provided by the RICCA network (*Red Internacional de Control y Cómputo Aplicados*). We also extend our appreciation to the anonymous reviewers for their valuable feedback, which has significantly contributed to improving the quality of this paper.

Conflicts of Interest: The authors declare no conflicts of interest.

Appendix A

The experiment utilized a lemniscate trajectory, mathematically described by the following equations:

$$\begin{aligned} x_r(t) &= \Theta_x \sin(\Omega t), \\ y_r(t) &= \Theta_y \sin(2\Omega t). \end{aligned} \quad \text{(A1)}$$

where Θ_x and Θ_y represent the maximum amplitude on the respective axis, $\Omega = \frac{2\pi}{T_r}$ is the angular frequency, and T_r is the period to traverse one complete cycle of the trajectory.

The linear velocity at each instant of the path, $v_{\text{path}}(t)$, is given by:

$$v_{\text{path}}(t) = \sqrt{v_x(t)^2 + v_y(t)^2},$$

where the velocity components $v_x(t)$ and $v_y(t)$ are:

$$\begin{aligned} v_x(t) &= \Theta_x \Omega \cos(\Omega t), \\ v_y(t) &= 2\Theta_y \Omega \cos(2\Omega t). \end{aligned}$$

The angle of the velocity vector $v_{\text{path}}(t) = [v_x(t), v_y(t)]$ is calculated as:

$$\theta_{\text{path}}(t) = \tan^{-1}\left(\frac{v_y(t)}{v_x(t)}\right),$$

and the angular velocity $\omega_{\text{path}}(t) = \frac{d\theta\text{path}(t)}{dt}$ is derived as:

$$\omega_{\text{path}}(t) = -\frac{4\Theta_x\Theta_y\Omega(3\sin(\Omega t) - 2\sin^3(\Omega t))}{\Theta_x^2\cos(2\Omega t) + \Theta_x^2 + 8\Theta_y^2\cos^2(2\Omega t)}.$$

The inverse kinematics for determining the reference velocity at each wheel are expressed as:

$$\begin{aligned} \omega_{r_L}(t) &= \frac{2v_{\text{path}}(t) + w\omega_{\text{path}}(t)}{2r}, \\ \omega_{r_R}(t) &= \frac{2v_{\text{path}}(t) - w\omega_{\text{path}}(t)}{2r}, \end{aligned} \quad \text{(A2)}$$

where w is the distance between the wheels and r is the radius.

The reference coordinates and velocities in their discretized Equations (A1) and (A2) form are given by:

$$\begin{aligned} x_r(k) &= \Theta_x \sin(\Omega T_s k), \\ y_r(k) &= \Theta_y \sin(2\Omega T_s k), \end{aligned}$$

$$\begin{aligned} \omega_{r_L}(k) &= \frac{2v_{\text{path}}(k) + w\omega_{\text{path}}(k)}{2r}, \\ \omega_{r_R}(k) &= \frac{2v_{\text{path}}(k) - w\omega_{\text{path}}(k)}{2r}, \end{aligned}$$

where Ω is the angular frequency, and T_s is the sampling period.

The parameters used to generate the trajectory are shown in Table A1:

Table A1. Parameters used for the experimental path.

Parameter	Value
Θ_x	1000 [mm]
Θ_y	500 [mm]
Ω	$2\pi/60$

Figure A1 illustrates the behavior of the instantaneous linear and angular velocities along the generated trajectory. In this scenario, the system encounters a trajectory with variable angular and linear velocities.

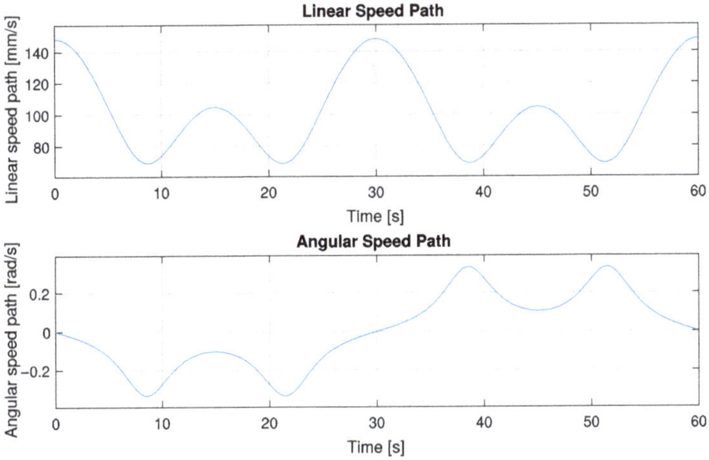

Figure A1. Instantaneous linear and angular velocity of the generated trajectory.

References

1. Brentari, M.; Zambotti, A.; Zaccarian, L.; Bosetti, P.; Biral, F. Position and speed control of a low-cost two-wheeled, self-balancing inverted pendulum vehicle. In Proceedings of the 2015 IEEE International Conference on Mechatronics (ICM), Nagoya, Japan, 6–8 March 2015; pp. 347–352.
2. Rahmani, R.; Mobayen, S.; Fekih, A.; Ro, J.S. Robust Passivity Cascade Technique-Based Control Using RBFN Approximators for the Stabilization of a Cart Inverted Pendulum. *Mathematics* **2021**, *9*, 1229. [CrossRef]
3. Valencia-Palomo, G.; Hilton, K.R.; Rossiter, J.A. Predictive control implementation in a PLC using the IEC 1131.3 programming standard. In Proceedings of the 2009 European Control Conference (ECC), Budapest, Hungary, 23–26 August 2009; IEEE: Piscataway, NJ, USA, 2009; pp. 1317–1322.
4. Valencia-Palomo, G.; Rossiter, J. Novel programmable logic controller implementation of a predictive controller based on Laguerre functions and multiparametric solutions. *IET Control Theory Appl.* **2012**, *6*, 1003–1014. [CrossRef]
5. Diaz, D.; Kelly, R. On modeling and position tracking control of the generalized differential driven wheeled mobile robot. In Proceedings of the 2016 IEEE International Conference on Automatica, ICA-ACCA 2016, Curico, Chile, 19–21 October 2016.
6. Chan, R.P.M.; Stol, K.A.; Halkyard, C.R. Review of modelling and control of two-wheeled robots. *Annu. Rev. Control* **2013**, *37*, 89–103. [CrossRef]
7. Raudys, A.; Šubonienė, A. A Review of Self-balancing Robot Reinforcement Learning Algorithms. In Proceedings of theCommunications in Computer and Information Science, Curico, Chile, 19–21 October 2020; pp. 159–170.
8. Kuntal, V.; Kumar, R.; Soni, H.; Sagarmani; Mishra, S.; Singh, S.K.; Choudhury, B. Advancements in Control Algorithms and Key Components for Self-Balancing Electric Unicycles: A Comprehensive Review. In Proceedings of the 2023 3rd International Conference on Innovative Mechanisms for Industry Applications (ICIMIA), Bengaluru, India, 21–23 December 2023; pp. 491–497.
9. Yue, M.; Wang, S.; Sun, J.Z. Simultaneous balancing and trajectory tracking control for two-wheeled inverted pendulum vehicles: A composite control approach. *Neurocomputing* **2016**, *191*, 44–54. [CrossRef]
10. Chiew, T.H.; Lee, Y.K.; Chang, K.M.; Ong, J.J.; Goh, Y.H. Design and Analysis of Super Twisting Sliding Mode-PID Controller for Two-wheeled Self-balancing Robot. *AIP Conf. Proc.* **2023**, *2680*, 020127.

11. Díaz-Téllez, J.; Gutierrez-Vicente, V.; Estevez-Carreon, J.; Ramírez-Cárdenas, O.D.; García-Ramirez, R.S. Nonlinear Control of a Two-Wheeled Self-balancing Autonomous Mobile Robot. In *Advances in Soft Computing, Proceedings of the 20th Mexican International Conference on Artificial Intelligence, MICAI 2021, Mexico City, Mexico, 25–30 October 2021*; Springer International Publishing: Cham, Switzerland, 2021; pp. 348–359.
12. Lower, M. Nonlinear Controller for an Inverted Pendulum Using the Trigonometric Function. *Appl. Sci.* **2023**, *13*, 12272. [CrossRef]
13. Unluturk, A.; Aydogdu, O. Machine Learning Based Self-Balancing and Motion Control of the Underactuated Mobile Inverted Pendulum with Variable Load. *IEEE Access* **2022**, *10*, 104706–104718. [CrossRef]
14. Nougues, S.; Guayacan, S.M.; Cuadros, D.G.; Leon-Rodriguez, H. Modelling and Design a Self-Balancing Dual-Wheeled Robot with PID Control. In Proceedings of the 2023 23rd International Conference on Control, Automation and Systems (ICCAS), Yeosu, Republic of Korea, 17–20 October 2023; pp. 489–497.
15. Sani, M.A.A.; David, J.B.A.; Jaafar, N.H.; Yusof, M.I.; Sani, N.S.; Sadikan, S.F.N. The Development of a Self-Balancing Robot Based on Complementary Filter and Arduino. In *Applied Problems Solved by Information Technology and Software*; SpringerBriefs in Applied Sciences and Technology; Springer Nature: Cham, Switzerland, 2024; Part F2025; pp. 71–78.
16. Bernal, M.; Sala, A.; Lendek, Z.; Guerra, T.M. *Analysis and Synthesis of Nonlinear Control Systems*; Studies in Systems, Decision and Control; Springer International Publishing: Cham, Switzerland, 2022; Volume 408.
17. Zhong, W.; Röck, H. Energy and passivity based control of the double inverted pendulum on a cart. In Proceedings of the 2001 IEEE International Conference on Control Applications, Mexico City, Mexico, 7 September 2001; pp. 896–901.
18. Gómez-Coronel, L.; Alvarado-Algarin, A.; Marquez-Zepeda, M.J.; Ancheyta-López, H.; López-Estrada, F.R.; Santos-Ruiz, I. Modeling and full-state feedback stabilization of a linear inverted pendulum. In Proceedings of the 24th Robotics Mexican Congress, COMRob 2022, Hidalgo, Mexico, 9–11 November 2022; pp. 42–47.
19. Atoui, H.; Sename, O.; Milanes, V.; Martinez-Molina, J.J. Toward switching/interpolating LPV control: A review. *Annu. Rev. Control* **2022**, *54*, 49–67. [CrossRef]
20. Das, S.; Pan, I.; Halder, K.; Das, S.; Gupta, A. LQR based improved discrete PID controller design via optimum selection of weighting matrices using fractional order integral performance index. *Appl. Math. Model.* **2013**, *37*, 4253–4268. [CrossRef]
21. Duan, G.-R.; Yu, H.H. *LMIs in Control Systems: Analysis, Design and Applications*; CRC Press: Boca Raton, FL, USA, 2013; pp. 99–102.

Mathematical and Computational Applications

MDPI

Article

Resolving Contrast and Detail Trade-Offs in Image Processing with Multi-Objective Optimization

Daniel Molina-Pérez [1,*] and Alam Gabriel Rojas-López [2,*]

[1] Escuela Superior de Cómputo, Instituto Politécnico Nacional, Ciudad de México 07700, Mexico
[2] Centro de Innovación y Desarrollo Tecnológico en Cómputo, Instituto Politécnico Nacional, Ciudad de México 07700, Mexico
* Correspondence: dmolinap@ipn.mx (D.M.-P.); arojasl2101@alumno.ipn.mx (A.G.R.-L.)

Abstract: This article addresses the complex challenge of simultaneously enhancing contrast and detail in an image, where improving one property often compromises the other. This trade-off is tackled using a multi-objective optimization approach. Specifically, the proposal's model integrates the sigmoid transformation function and unsharp masking highboost filtering with the NSGA-II algorithm. Additionally, a posterior preference articulation is introduced to select three key solutions from the Pareto front: the maximum contrast solution, the maximum detail solution, and the knee point solution. The proposed technique is evaluated on a range of image types, including medical and natural scenes. The final solutions demonstrated significant superiority in terms of contrast and detail compared to the original images. The three selected solutions, although all are optimal, captured distinct characteristics within the images, offering different solutions according to field preferences. This highlights the method's effectiveness across different types and enhancement requirements and emphasizes the importance of the proposed preferences in different contexts.

Keywords: multi-objective optimization; image enhancement; contrast and detail; sigmoid transformation; NSGA-II; a posterior preference articulation

1. Introduction

Image enhancement is the process of applying specific techniques to boost an image's visual quality. These techniques can imply diverse criteria, such as increasing the image contrast, noise reduction, highlighting relevant details, or adjusting brightness and color saturation. The main goal of image enhancements is to make the information within the image more easily interpretable and perceptible to human viewers, as well as to boost the automatic process of applications such as pattern recognition, medical analysis, and computer vision, among others [1,2].

Traditional image enhancement methods are divided into two well-defined categories [3,4]. The first is the spatial domain enhancement, which directly modifies pixel values to adjust aspects such as contrast and image detail. This category includes techniques such as gamma correction [5], histogram equalization [2], and sigmoid correction [6], as well as more advanced techniques such as multi-dimensional adaptive local enhancements [7] and recursive filters for color balance and shadow correction [8]. The second category is frequency domain enhancement, which transforms the image into a mathematical domain to adjust the frequency components, allowing fine detail enhancement and reduction of undesirable patterns through techniques such as high-pass or low-pass filtering. In this family, methods such as wavelet transform [9], discrete cosine transformation [10], and Fourier transform [11] are included. Both categories have specific applications and are chosen according to the type of improvement desired.

Remarkably, not all images require the same enhancement process, since the enhancement strategy relies upon the image's specific characteristics. For example, a low-contrast image can significantly benefit from spatial domain enhancement techniques, like the contrast adjustment or the histogram equalization approach, which enhance the image's sharpness [12,13]. On the other hand, in medical images such as magnetic resonances, the fine details highlighting can be critical; therefore, frequency domain enhancement techniques are more suitable for adjusting the high frequencies and spotlighting the image's internal structures [14]. In deep learning architectures that handle visual data such as point clouds [15] or drone datasets [16], image enhancement should aim to improve feature extraction by making relevant details more prominent, ensuring that critical features such as edges, shapes, or textures are effectively captured during the training process. Hence, there is no unique ideal operator for all images nor a unique quantitative metric that automatically evaluates the image quality. Automatic image enhancement is a process that produces enhanced images without human intervention and is an extremely complicated task in image processing [17].

Image enhancement methods are typically parametric, which means their effectiveness relies significantly on the fine-tuning of various parameters, leading to increased complexity in achieving optimal results. Recently, multi-objective optimization has gained prominence in image processing as it addresses the challenge of simultaneously improving multiple, conflicting quality criteria. Therefore, the current work is focused on image enhancement, considering two essential properties of image processing: contrast and details. Improving these properties through transformation functions is generally compromised, meaning that increasing contrast can lead to a significant loss of details in the image. Hence, a multi-objective optimization problem with two objective functions is established: one related to the image contrast and the other to the image details. These functions are measured employing the entropy and the standard deviation for the image contrast, while the pixels' quantity and intensity in high-frequency regions measure the image details. The contrast enhancement is performed by the sigmoid transform function, and the detail enhancement is performed by the unsharp masking and highboost filtering. To address the problem, NSGA-II [18] is used with a posterior preference articulation, meaning that specific solutions are selected once the Pareto front is finally computed. The current work offers the following contributions:

1. The trade-off between the image contrast and details is set as a multi-objective problem. Unlike the traditional mono-objective approach, which only provides an optimal solution with a predefined priority, the current proposal offers the best solutions regarding the compromises between both criteria along the Pareto front.
2. A posterior preference operator is articulated, providing three key images from the Pareto front: the image with maximum contrast, the image with maximum detail, and the image at the knee of the front, representing the image closest to the utopia point. This operator allows end-users to select the most suitable solutions for their needs.
3. An experiment is conducted with images of two categories: medical and natural scene images. Both categories represent research fields where image processing is an essential endeavor. The results of this experiment demonstrate that the NSGA-II achieves images of superior quality compared to the original instances. Furthermore, a thorough analysis is conducted regarding the suitability of the obtained images according to the established preferences. For medical images, the evaluation focuses on how the selected solutions enhance the clarity and detail of relevant structures, which is crucial for diagnostics and analysis. For natural scene images, the analysis shows how the solutions improve contrast and detail, making the images more visually appealing and impactful.

The remainder of this paper is organized as follows: Section 3 outlines the sigmoid correction and unsharp masking with highboost filtering methods used in this study. In Section 4, we present the proposed multi-objective optimization model and posterior preference articulation method. Section 5 provides the benchmark results, including experimental design, graphical analysis, and quantitative evaluations. Finally, Section 6 offers conclusions and suggestions for future research directions.

2. Related Work

A significant challenge for image enhancement methods is tuning their parameters to achieve optimal results. This task can be complex, as each parameter may affect the image characteristics in different ways. Evolutionary algorithms (EAs) stand out as highly effective tools in this context. Pioneer work in this area is that of Bhandarkar et al. [19], where a genetic algorithm (GA) was employed for an image segmentation problem. The outcomes showed that the GA outperforms traditional image segmentation methods regarding accuracy and robustness against noise. In subsequent years, numerous studies proposed diverse approaches to address image enhancement problems through EAs. An example of this is presented in [20], where an optimization-based process employs the particle swarm optimization (PSO) algorithm to enhance the contrast and details of images by adjusting the parameters of the transformation function, taking into account the relationship between local and global information. Another instance of optimization-based image enhancements is presented in [21], where an accelerated variant of PSO is used to optimize the above-mentioned transformation function, achieving a more efficient algorithm in terms of convergence.

Using EAs to solve image enhancement problems remains a common trend in recent years. This is observed in works such as [22], where the differential evolution (DE) algorithm was employed to maximize image contrast through a modified sigmoid transformation function. This function adjusts parameters that control the contrast magnitude and the balance between bright and dark areas, with optimal values determined through the evolutionary process. Similarly, Bhandari and Maurya [23] developed a novel optimized histogram equalization method, preserving average brightness while improving the contrast through a cuckoo search algorithm. The proposed method uses plateau boundaries to modify the image histogram, avoiding the extreme brightness changes often caused by traditional histogram equalization. Another interesting application integrating EAs and image histograms is presented in [24], where a GA was applied to optimize a histogram equalization through optimal subdivisions considering different delimited light exposition regions. Particularly, these optimization-based strategies have been taken to fields beyond engineering, like [25], where directed searching optimization was applied in a medical image enhancement process, improving the contrast while preserving the texture in specified zones through a threshold parameter. The popularity of EAs has led to more sophisticated approaches, as in [26], where hybridization between whales optimization and chameleon swarm algorithms was proposed specifically to find the optimal parameters of the incomplete beta function and gamma dual correction. Several other EAs have been applied to image enhancement problems, such as monarch butterfly optimization [27], chimp optimization algorithm [28], sunflower optimization [29], and slime mold algorithm [30], among others. However, despite their contributions to notable improvements in image processing, these approaches primarily focus on maximizing or minimizing a single criterion. As a result, single-objective methods may inadequately address the complex relationships among various image characteristics, thereby limiting their effectiveness in real-world applications, where multiple criteria must be optimized simultaneously.

In the last decade, multi-objective optimization in image processing has also been the subject of several investigations. The relevance of the multi-objective approach lies in the need to balance multiple quality criteria simultaneously. In many cases, improving one image characteristic may worsen another. This inherent conflict between particular objectives requires an approach that obtains a set of optimal solutions, known as the Pareto front. For example, in [31], a PSO variant was proposed to address a multi-objective problem aimed to simultaneously maximize the available information quantity (through the entropy evaluation) and minimize the resulting image distortion (measured by the structural similarity index). Similarly, in [32], a multi-objective optimization using PSO is implemented to simultaneously optimize brightness, contrast, and colorfulness in a Retinex-based method. Another instance of multi-objective image enhancement problems is presented in [33], where GA was employed to maximize the Poisson log-likelihood function (used to measure the quantitative accuracy) and the generalized scan-statistic model (measures the detection performance). In [34], a multi-objective cuckoo search algorithm was used to enhance contrast by maximizing entropy and minimizing noise variance in adaptive histogram equalization. Another trade-off problem of image enhancement was solved in [35], where through the Non-dominated Sorting Genetic Algorithm based on Reference Points (NSGA-III), the optimal parameters for anisotropic diffusion were found, aiming to produce effective filtering results while balancing the image noise and its contrast.

While the Pareto front approach provides a diverse range of solutions in image enhancement tasks, it often lacks mechanisms for preference articulation, which is essential for decision makers to select the most appropriate solution based on specific application needs [36–38]. Despite its importance, the explicit application of preference articulation in the literature on multi-objective image enhancement remains limited. This work proposes a multi-objective approach to simultaneously enhance contrast and details, as these two parameters have primarily been addressed through single-objective methods. Consequently, there is a lack of models that consider their interdependence. Additionally, this approach incorporates preference articulation to facilitate the selection of a limited set of images that effectively capture the diverse characteristics within the dataset.

3. Materials and Methods

3.1. Sigmoid Correction

Sigmoid correction is a technique used in image processing to adjust the contrast of an image in a nonlinear manner. This method is particularly useful for enhancing the contrasts in an image's dark and bright regions. The correction is achieved using a sigmoid function, which maps the input intensity values to a new range according to a sigmoid curve [22,39,40]. The sigmoid function used for image correction is defined as:

$$g(x) = \frac{1}{1 + e^{-\alpha(x-\Delta)}}, \tag{1}$$

where $g(x)$ represents the transformed pixel values; x is the original pixel intensity, scaled in the range of 0 to 1; α is the steepness of the sigmoid curve, which controls the degree of contrast adjustment; and Δ is a value that determines the midpoint of the sigmoid curve related to the normalized pixel intensity, allowing control over the balance between the bright and dark regions of the image.

The parameter α affects how rapidly the transition occurs between dark and light regions. A higher value of α results in a steeper curve, increasing contrast by making transitions more abrupt, while lower values of α produce a gentler curve with smoother transitions, as shown in Figure 1a. The Δ parameter enables fine-tuning the balance between bright and dark regions. By adjusting this value, you can control where the midpoint of the contrast adjustment occurs, thus affecting how the dark and light areas of the image are processed. A higher Δ value results in a larger range of intensities being considered dark, leading to a darker overall image, whereas a lower value results in a larger range of intensities being considered bright, making the image brighter, as depicted in Figure 1b.

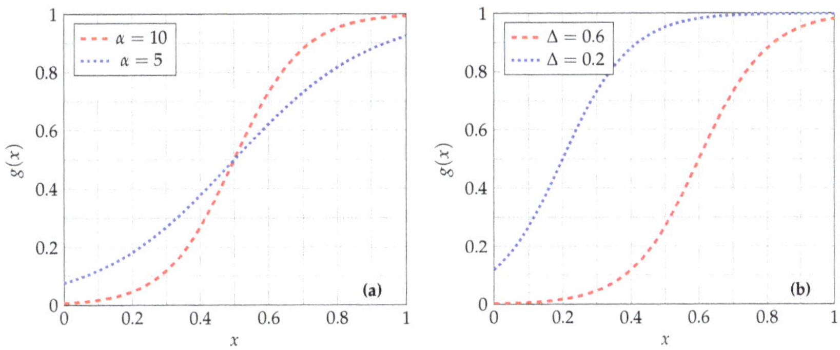

Figure 1. Comparison of sigmoid correction curves demonstrating the impact of varying parameters on image enhancement. (**a**) Effect of different α values on the gradient of the sigmoid curve, influencing the contrast adjustment. (**b**) Impact of different (Δ) values on the midpoint of the sigmoid curve, affecting the balance between light and dark regions.

3.2. Unsharp Masking and Highboost Filtering

Unsharp masking and highboost filtering (UMH) are techniques for enhancing image sharpness and detail by manipulating high-frequency components. The process begins with creating a blurred version of the original image using a smoothing filter, such as an average filter. For a filter size of S, the average filter $\mathbf{A}_f \in \mathbb{R}^{S\times S}$ is defined as an $S \times S$ squared matrix, where each $a_{ij}\ \forall\, i,j \in \{1,\ldots,S\}$ element within $\mathbf{A}_f$ is equal to $\frac{1}{S^2}$. With $S = 5$, the average filter is represented explicitly as follows:

$$\mathbf{A}_f = \frac{1}{5^2}\begin{bmatrix} 1 & 1 & 1 & 1 & 1 \\ 1 & 1 & 1 & 1 & 1 \\ 1 & 1 & 1 & 1 & 1 \\ 1 & 1 & 1 & 1 & 1 \\ 1 & 1 & 1 & 1 & 1 \end{bmatrix}. \tag{2}$$

Applying this low-pass filter to the input image yields a blurred version of the original image, denoted as $f_b(x)$. In the unsharp masking process, a mask $m(x)$ is created by subtracting the blurred image from the original image:

$$m(x) = x - f_b(x). \tag{3}$$

This mask highlights the high-frequency details that are suppressed by the blur. However, not all high-frequency components contribute meaningfully to the image's detail. To exclude insignificant details, a threshold d_{th} is applied, where any values in the mask below this threshold are set to zero:

$$m(x) = \begin{cases} 0, & \text{if } |m(x)| \le d_{th} \\ m(x), & \text{otherwise} \end{cases}. \tag{4}$$

This step ensures that only significant details are enhanced. To enhance the image, this mask is then added back to the original image with a scaling factor k to control the extent of the enhancement. The resulting enhanced image $g(x)$ is calculated as:

$$g(x) = x + k \cdot m(x). \tag{5}$$

4. The Proposed Algorithm

This section presents the novel multi-objective optimization model and a posterior preference articulation method developed specifically for this study. This method is designed to find optimal trade-offs between the objective functions while providing preferential solutions capable of capturing different characteristics of the images. Furthermore, an analysis of the complexity of the proposed algorithm is conducted to evaluate its efficiency in handling the optimization process.

4.1. Multi-Objective Optimization Problem

This work focuses on enhancing images by considering two fundamental properties of image processing: contrast and details. The enhancement of these properties through transformation functions is often compromised, meaning that an increase in contrast can result in a significant loss of details in the image. Therefore, a multi-objective problem is defined with two objective functions:

- **Contrast Function**: Defined as the product of entropy $H(I)$ and the normalized standard deviation of the pixel intensities $\sigma_{norm}(I)$ of the I image. The contrast function is expressed as:

$$f_1(\alpha, \Delta) = H(I) \times \sigma_{norm}(I). \tag{6}$$

- **Details Function**: Defined as the product of the number of pixels in high-frequency regions $N_{HF}(I)$ and the intensity of these high-frequency pixels $IN_{HF}(I)$ of the I image. The details function is denoted as:

$$f_2(\alpha, \Delta) = N_{HF}(I) \times \log_{10}(\log_{10}(IN_{HF}(I))). \tag{7}$$

Formally, the multi-objective optimization problem can be expressed as in (8), where $\mathcal{F} = -\{f_1, f_2\}$ represents the set of objective functions to be minimized. The decision vector $\boldsymbol{\phi} = [\alpha, \Delta]^T$ controls the behavior of the sigmoid correction function, with $\boldsymbol{\phi}_{min}$ as the lower and $\boldsymbol{\phi}_{max}$ as the upper boundary.

$$\begin{aligned} &\min_{\boldsymbol{\phi} \in \mathbb{R}^2} \mathcal{F}(\boldsymbol{\phi}) \\ &\text{subject to:} \quad \boldsymbol{\phi}_{min} \leq \boldsymbol{\phi} \leq \boldsymbol{\phi}_{max} \end{aligned} . \tag{8}$$

Before evaluating these objective functions, a series of image transformations are performed. Contrast enhancement is achieved using a sigmoid transformation function, where α and Δ are the decision variables that control the degree of contrast adjustment. Additionally, UMH is applied for detail enhancement. These transformations impact the objective functions by modifying the entropy, standard deviation, number of high-frequency pixels, and intensities. Figure 2 illustrates the overall procedure for solving the multi-objective optimization problem using NSGA-II.

The selection of NSGA-II in this study is based on several algorithm strengths. Firstly, its elitist mechanism ensures the preservation of the best solutions across generations, preventing the loss of optimal solutions. Additionally, the crowding operator used by NSGA-II eliminates the need for sensitive parameters associated with niche techniques, simplifying the implementation process. Furthermore, NSGA-II exhibits low time complexity relative to other multi-objective evolutionary algorithms, which is advantageous for practical applications. It has also demonstrated efficient performance in multi-objective problems involving two objective functions, making it relevant for contrast and detail enhancement tasks. Although other more recent multi-objective evolutionary algorithms have succeeded in generic benchmarks, NSGA-II remains highly competitive in solving real-world problems [41,42], making it suitable for this work.

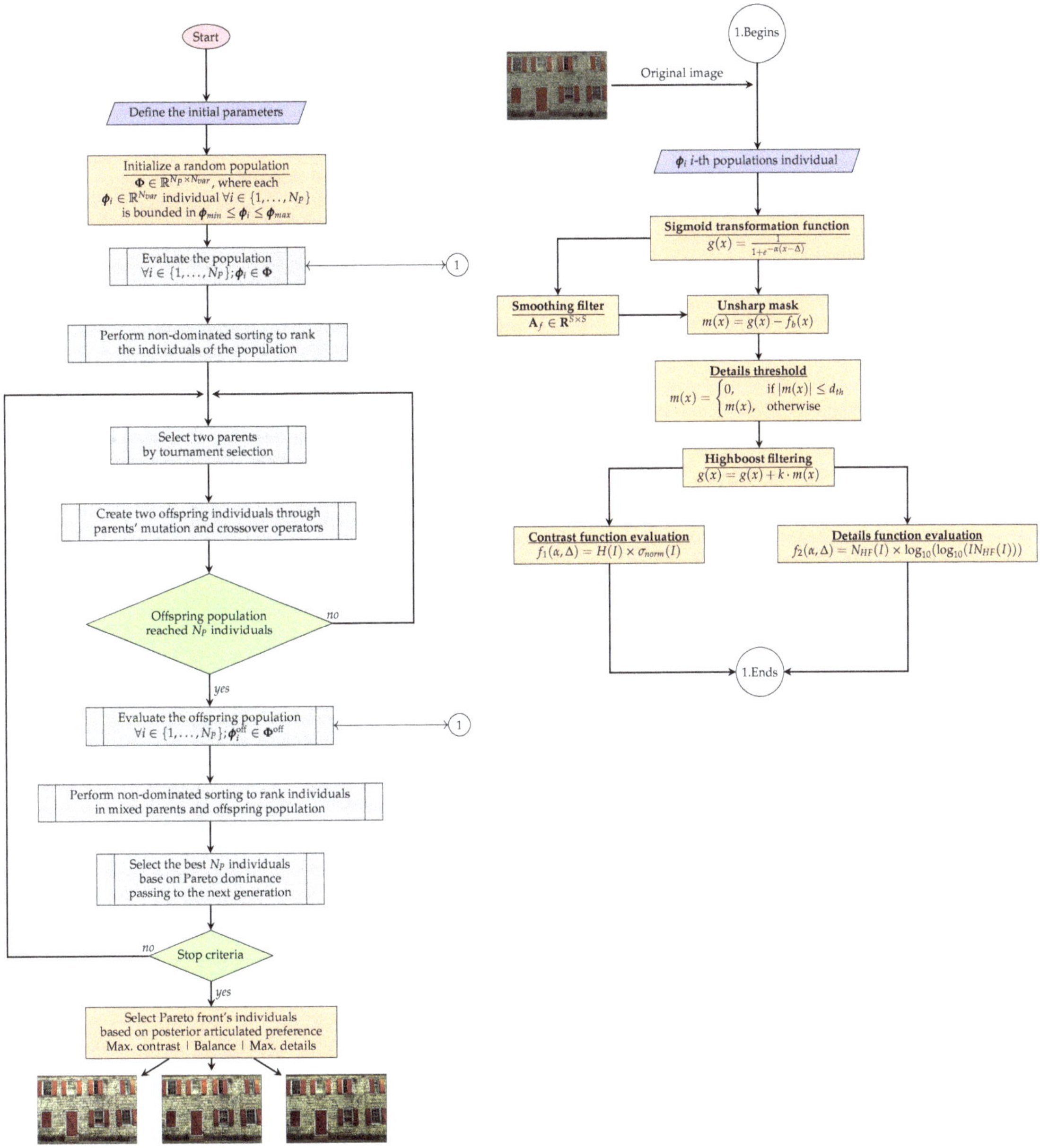

Figure 2. The flowchart depicts the multi-objective optimization process for image enhancement, starting with the initialization of a random population of decision vectors $\boldsymbol{\phi} = [\alpha, \Delta]^T$ that control the sigmoid transformation for contrast adjustment. Balloon 1 denotes the evaluation procedure, which includes image enhancement via sigmoid correction and highboost filtering, followed by evaluating individuals based on contrast and detail objectives. Non-dominated sorting ranks the individuals and tournament selection generates offspring, iterating until stop criteria are met. Finally, the algorithm returns three solutions that present opposing balances between contrast and detail.

4.2. A Posterior Preference Articulation

The main objective of preference articulation is to select a limited set of images that effectively capture diverse characteristics within the data, which may be relevant across various contexts. This approach is particularly important given that the Pareto front can encompass as many solutions as there are members in the population or even more when using an external archive. This work focuses on choosing the extreme solutions of the Pareto front and its knee point. The knee point is identified as the solution on the Pareto front that is closest to the utopia point, which typically represents the ideal but unattainable objective values. Within the optimal trade-offs of the Pareto front, these three solutions represent opposing balances of contrast and detail, helping to capture essential image characteristics that may be relevant according to the specific requirements of each application.

To determine the knee point, the Pareto front is first normalized, ensuring that the values of each objective function fall within the range [0, 1]. The utopia point in this normalized space is a vector of zeros, representing the best possible values for each objective. The Euclidean distance of each solution on the normalized Pareto front to the utopia point is calculated as follows:

$$d_i = \sqrt{\sum_{j=1}^{N_f} (f_{ij} - p_j)^2}, \tag{9}$$

where N_f is the number of objective functions; f_{ij} is the normalized value of the j-th objective for the i-th solution; and p_j is the value of the j-th objective at the utopia point (typically 0 after normalization). The knee point is identified as the solution with the minimum distance to the utopia point.

4.3. Time Complexity

The most computationally expensive process in NSGA-II is the non-dominated sorting, which is used to categorize the individuals in the population [18]. This step has a time complexity of $O(M \cdot N^2)$, where M is the number of objectives and N is the population size. Additionally, the population's objective functions evaluation has a time complexity of $O(N \cdot F)$, where F represents the time required to evaluate an individual. Since F depends on the specific application and can vary, the total complexity of NSGA-II is expressed as $O(N \cdot M^2 + N \cdot F)$.

In this approach, the evaluation of each individual includes several image processes. The sigmoid function is applied to each pixel, resulting in an $O(P)$ complexity, where P is the total number of pixels in the image. The smoothing filter has a complexity of $O(P \cdot k^2)$, where $k \times k$ is the filter size, implying that operations must be performed on k^2 neighbors. However, this term can be disregarded since k is a constant. The unsharp mask process, the application of detail thresholding, and the enhancement of details also have an $O(P)$ complexity. Finally, calculating quality metrics, such as entropy and standard deviation, requires traversing each pixel, adding $O(P)$ to the complexity. Thus, the total time complexity for evaluating the objective function is $O(P)$, meaning that the execution time grows linearly with the image size. Therefore, the overall complexity of the proposal is $O(N \cdot M^2 + N \cdot P)$.

To empirically validate this, we conducted an experiment using different image resolutions, as shown in Table 1. The results demonstrate the linear relationship between image size and execution time. This behavior is fundamental when handling high-resolution images, ensuring that processing remains feasible even as image size increases. The experiments are executed on a machine with an Apple M2 chip, an 8-core CPU, and 8 GB of unified memory.

Regarding implementation complexity, the proposed algorithm applies the correction methods before each objective function evaluation. At the end of the generations, preference articulation selects extremes and knee solutions from the Pareto front. These operations do not disrupt the conventional operators of the algorithm, enabling straightforward implementation in other evolutionary algorithms. To effectively support the preference articulation process, it is recommended that the algorithms be based on Pareto dominance.

Table 1. Execution times of the algorithm for different image resolutions. In each case, 10,000 evaluations of the objective function were performed.

Resolution	Execution Time (s)
800 × 450 pixels (WVGA)	66
1280 × 720 pixels (HD)	180
1920 × 1080 pixels (Full HD)	528
3840 × 2160 pixels (4K)	2180

5. Benchmark Results and Discussion

In this session, the fundamental aspects of the experiment are described. Subsequently, the results are presented and discussed in two parts: first, the visual analysis of the resulting images is conducted, followed by a discussion of the numerical results from the images concerning well-established indicators.

5.1. Experimental Design

A set of twenty images is selected to assess the effectiveness of the method developed in this work. This set is divided into two groups: the first includes 10 natural scene images extracted from the Kodak dataset [43], specifically from kodim01 to kodim10 (hereinafter referred to as Natural1 to Natural10, respectively). The second group consists of 10 medical images (referred to as Medical1 to Medical10, respectively) selected from various libraries, including brain images [44,45], blood composition images (white blood cells of the basophil and eosinophil types) [46,47], X-rays [48], ocular nodules [49], dental infections [50], microphotographs of pulmonary blood vessels [51], and traumatic forearm positioning [52].

The NSGA-II algorithm is executed for each image with a maximum of 30,000 function evaluations, aiming to produce a Pareto front containing the best solutions. From this front, solutions are extracted according to the defined articulation preference operator. The objective is to evaluate the quality of these solutions in terms of contrast and details, complemented by a visual analysis to determine the suitability of each image for specific purposes. Finally, the similarity of the enhanced images to the original ones is assessed using the structural similarity index (SSIM), which quantifies the degree of similarity between the processed and original images.

The parameter values used in this experiment are as follows: population size ($N_p = 50$), number of variables ($N_{var} = 2$), number of objective functions ($N_f = 2$), number of evaluations (N_{eval} = 30,000), mutation probability ($P_m = 0.5$), crossover probability ($P_c = 0.7$), simulated binary crossover parameter ($N_c = 5$), polynomial mutation parameter ($N_m = 5$), details threshold ($d_{th} = 0.1$), lower bound ($\boldsymbol{\phi}_{min} = [0, 0]^T$), and upper bound ($\boldsymbol{\phi}_{max} = [10, 10]^T$).

5.2. Graphical Results

Tables 2–5 present the results obtained through the multi-objective optimization image enhancement approach. Specifically, Tables 2 and 3 show the results for natural images, while Tables 4 and 5 display medical images. The tables are organized as follows: the first and second columns list the image names and their corresponding original, unenhanced versions. The third to fifth columns showcase the selected points from the Pareto front, representing the maximum contrast, knee point, and maximum detail, in that order. The final column illustrates the obtained Pareto front through the optimization process, with red, green, and orange points indicating the images that achieved maximum contrast, knee point, and maximum detail, respectively.

Table 2. Natural image result—1.

	Original	Max. Contrast •	Knee Point •	Max. Detail •	Pareto Front
Natural1					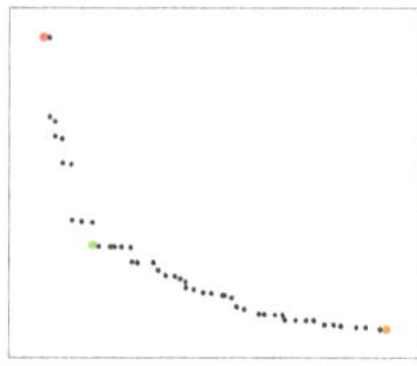
Natural2					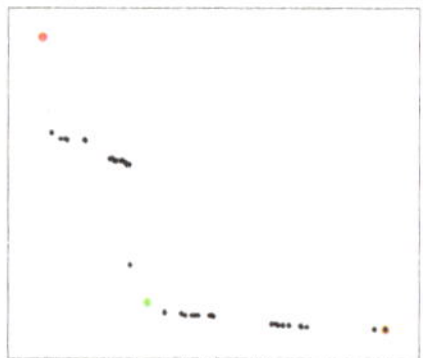
Natural3					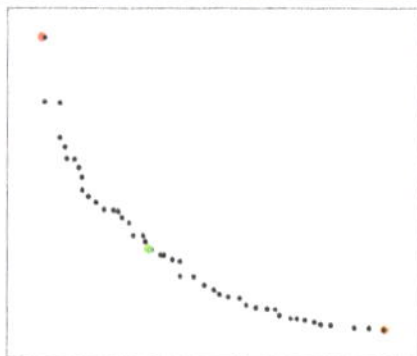
Natural4					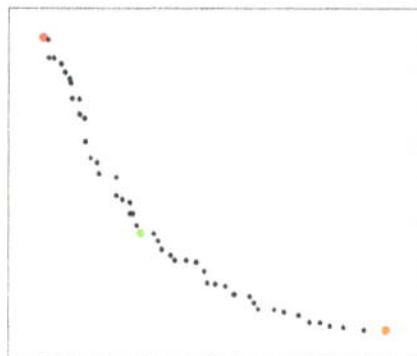
Natural5					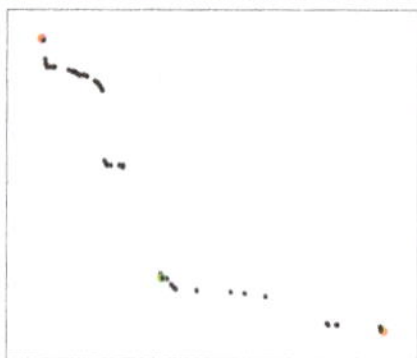

Table 3. Natural image results—2.

	Original	Max. Contrast ●	Knee Point ●	Max. Detail ●	Pareto Front
Natural6					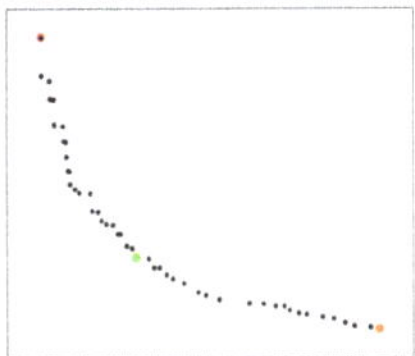
Natural7					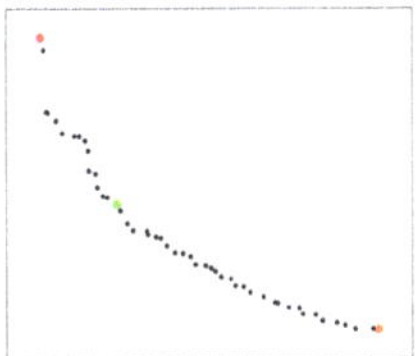
Natural8					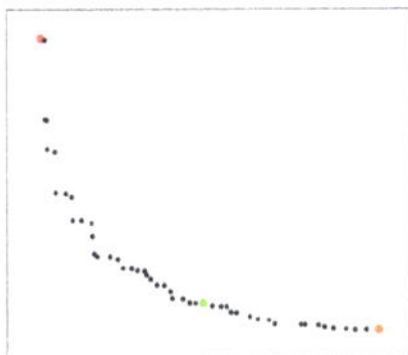
Natural9					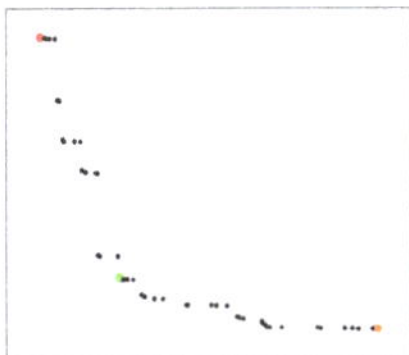
Natural10					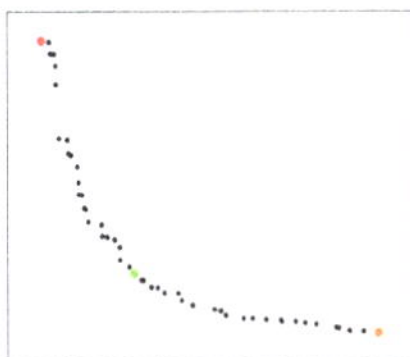

Table 4. Medical image results—1.

	Original	Max. Contrast ●	Knee Point ●	Max. Detail ●	Pareto Front
Medical1	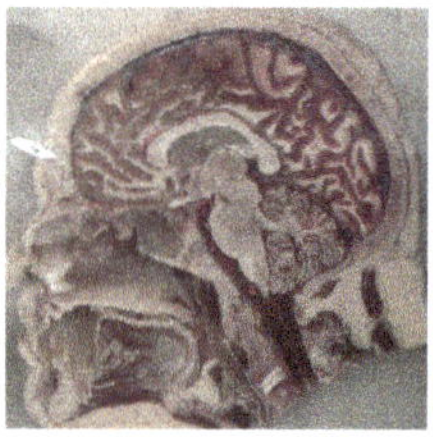	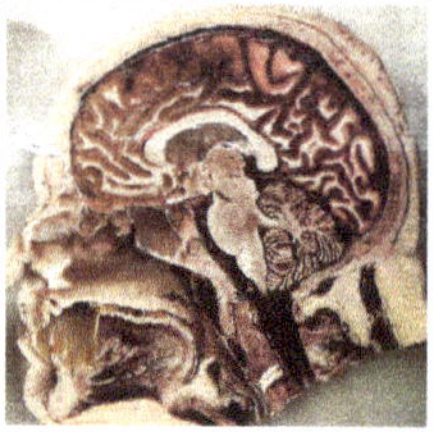	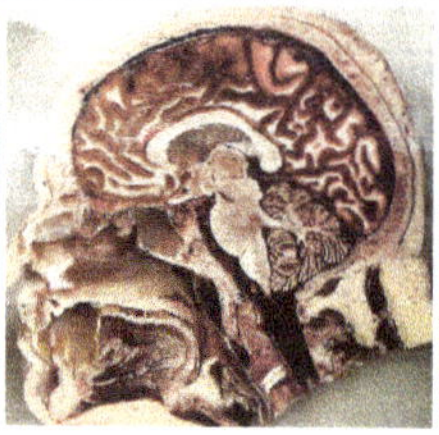	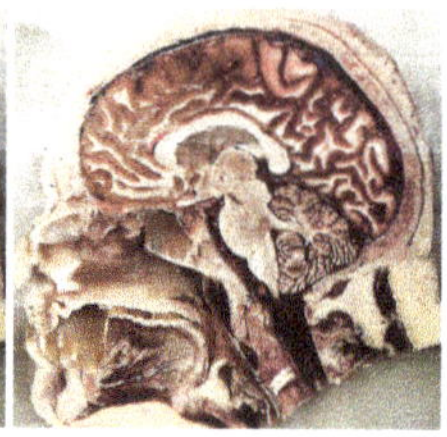	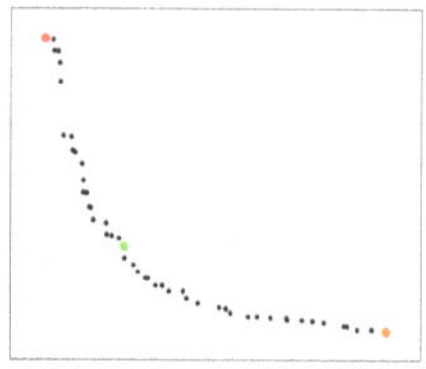
Medical2	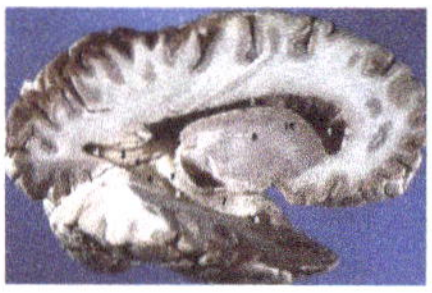				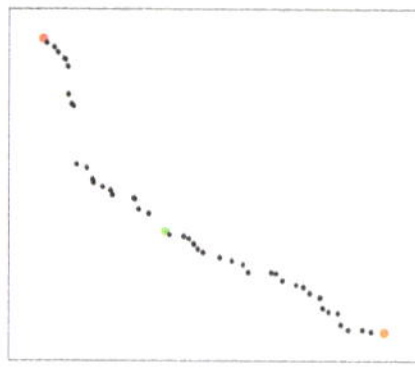
Medical3	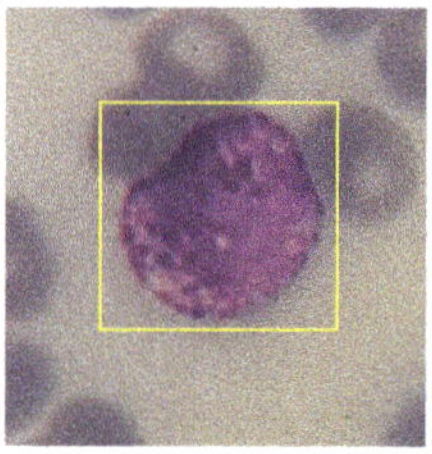	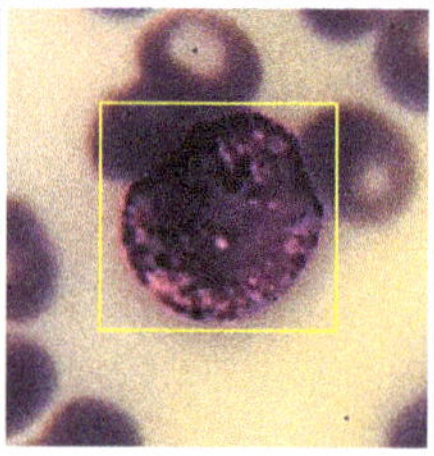	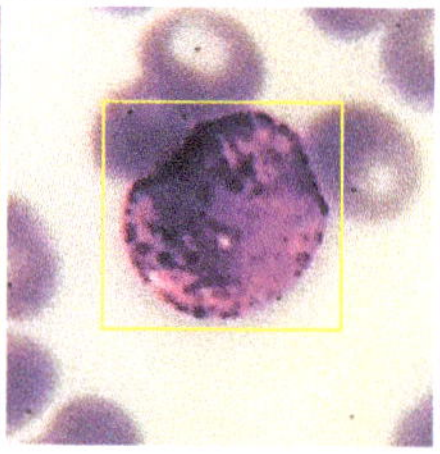	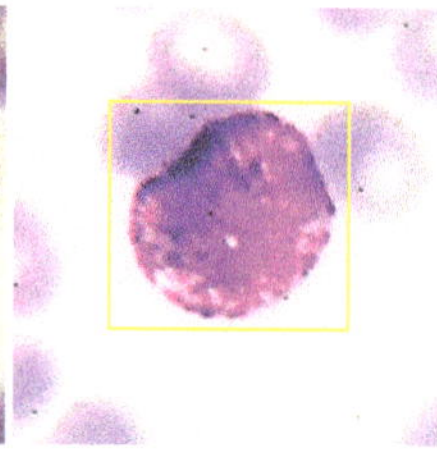	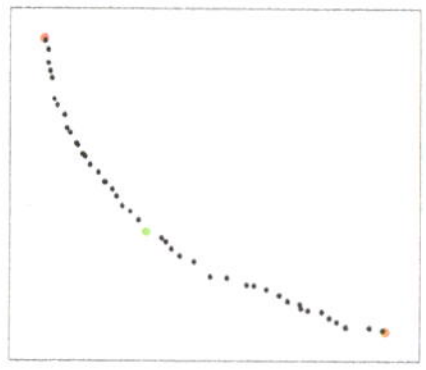
Medical4	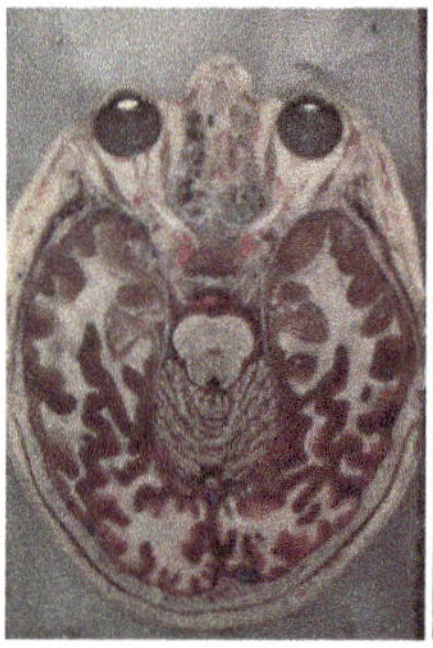	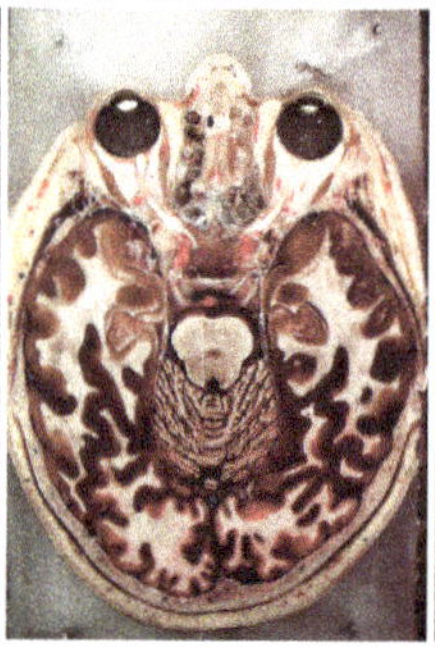	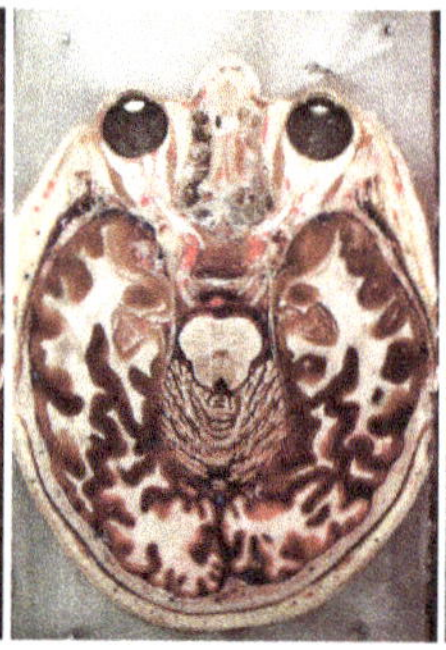	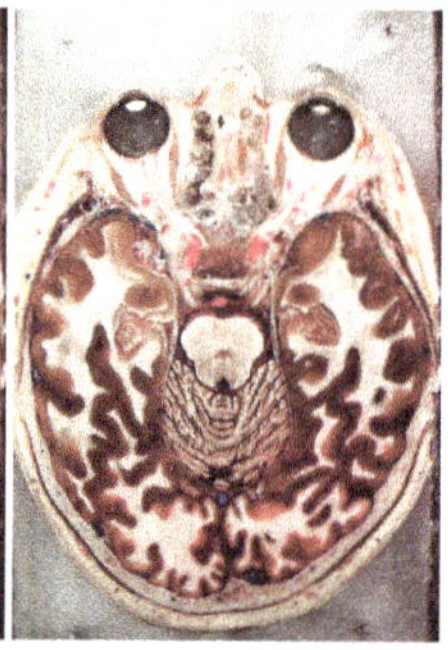	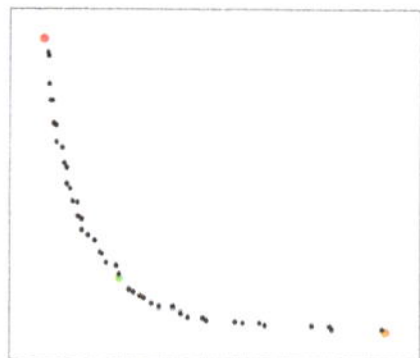
Medical5		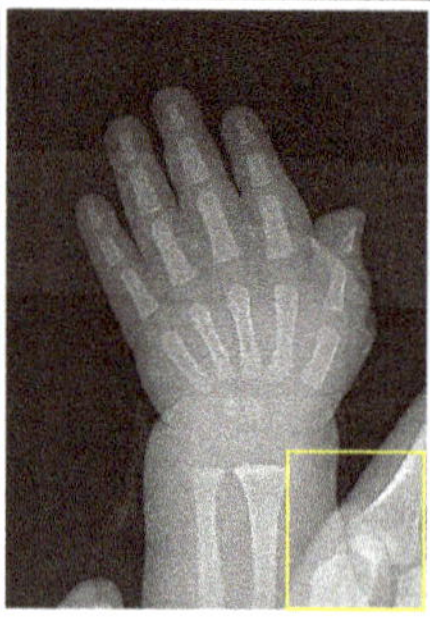	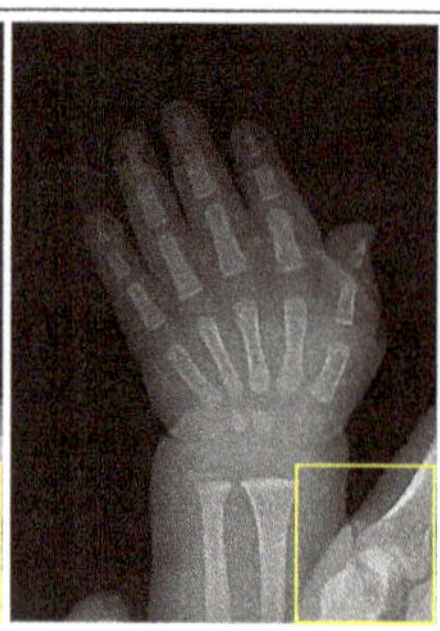	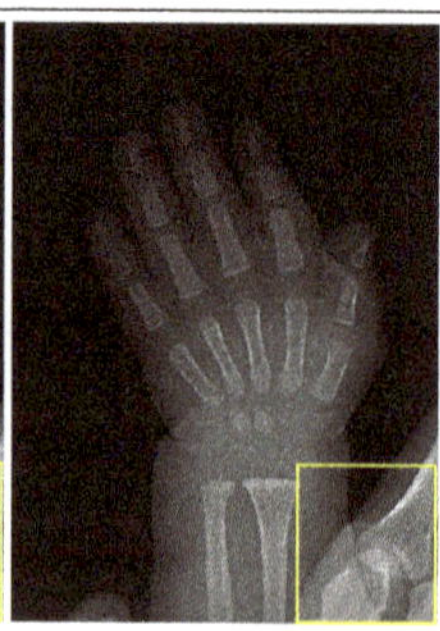	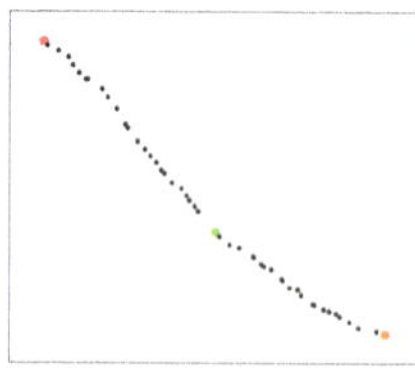

Table 5. Medical image results—2.

	Original	Max. Contrast •	Knee Point •	Max. Detail •	Pareto Front
Medical6	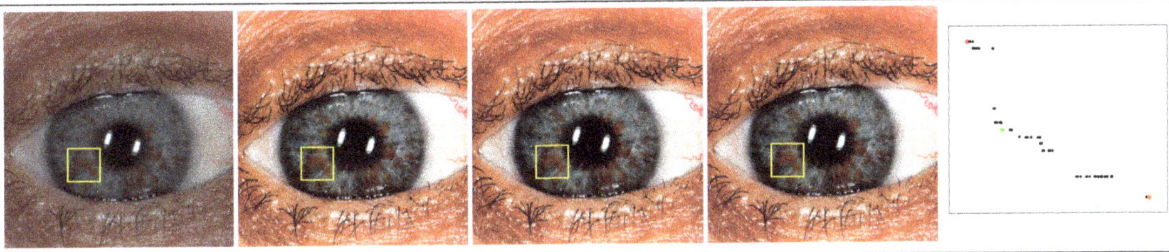				
Medical7	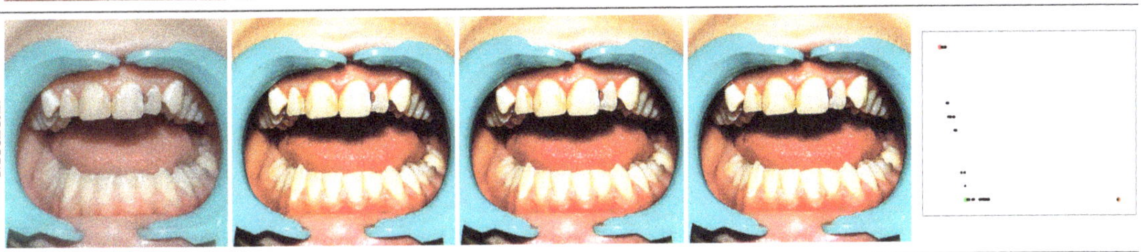				
Medical8	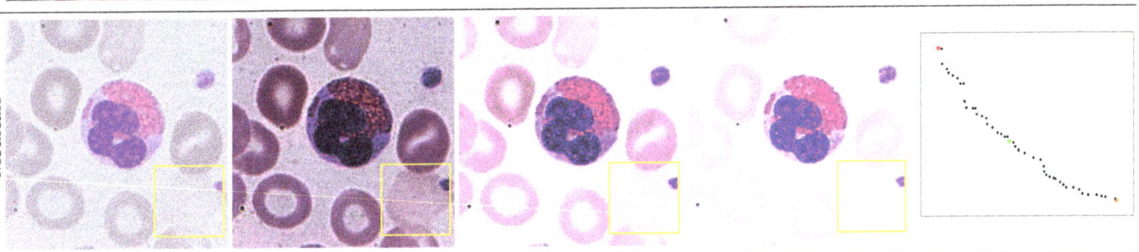				
Medical9	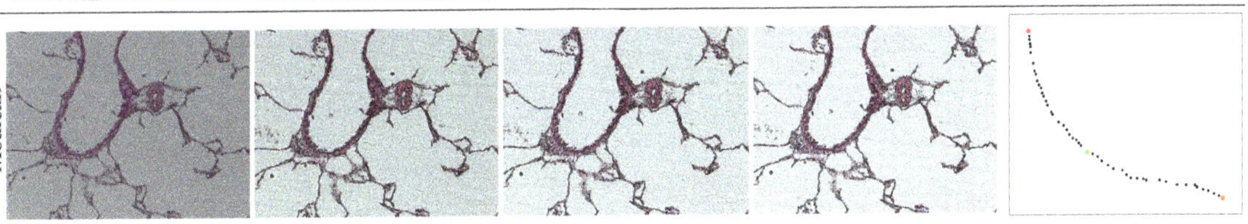				
Medical10	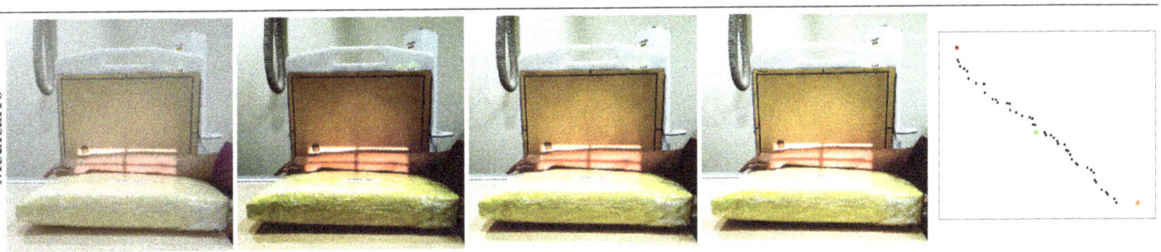				

As observed in the results, the images extracted from the Pareto front significantly maximize contrast and detail compared to the original images. In all study cases, the original image is dominated by the solutions extracted from the fronts, demonstrating the approach's effectiveness in improving visual quality. However, the differences among the three enhanced images for each problem require a more detailed analysis.

In the natural images, the differences among the three preferred images are more subtle, given that these are high-quality images with inherently low contrast, specifically selected for contrast enhancement exercises. More pronounced differences are observed in Natural1, Natural6, Natural7, and Natural8 images. For the Natural7 image, there is a general improvement in overall details. However, specific regions, such as the highlighted flower in the yellow box, may lose details compared to the original image, which retains more information. This suggests that for future work, it may be advisable to apply local and/or adaptive image enhancement techniques to preserve details in specific regions while maintaining overall image quality.

For medical images, there are instances where differences are more perceptible. For example, in the Medical3 image, the maximum contrast solution makes it difficult to visualize the internal details of the basophil (a white blood cell highlighted in the box), which could result in a less accurate interpretation. In contrast, the knee and maximum detail solutions provide a clearer view of the interior of the white blood cell. Similarly, in the Medical5 image, the maximum contrast solution highlights the hand and arm bone structures. However, the maximum detail image offers a more precise view of the internal structures within the bones (see the highlighted region), which is crucial for a more detailed evaluation. Another notable example is the Medical8 image, where the maximum detail solution offers a more detailed view of the internal structure of the eosinophil (another type of white blood cell). However, the maximum contrast image improves the visibility of red blood cells. As shown in the yellow box, this solution reveals a red blood cell that is nearly imperceptible in the other solutions. An interesting case is the Medical6 image, where only a few non-dominated solutions are present on the Pareto front. Despite the similarities among the preferred solutions, the nodules are much more perceptible in the enhanced images than in the original image, as observed in the highlighted region.

The solutions extracted from the Pareto front represent optimal trade-offs between contrast and detail. For natural images, these three alternatives can be considered useful based on aesthetic criteria or in subsequent automatic processes that require prioritizing one property over another. In the case of medical images, these alternatives allow for a more precise evaluation suited to different diagnostic needs, providing a flexible approach to enhance the visualization of critical details according to the clinical context.

5.3. *Quantitative Results*

Tables 6 and 7 present noteworthy information regarding several criteria following the next structure. For each image whose name is presented in the first column, a set of three rows displays the outcomes of the Pareto front's maximum contrast, knee point, and maximum details solutions. The results per individual regarding their entropy, normalized standard deviation, number of pixels, and pixel intensity are displayed from the third to the sixth column. The seventh and eighth columns display the objective function values of the multi-objective optimization problem. The last column displays the SSIM with respect to the original images, where all those images archived values above 0.7, i.e., SSIM > 0.7, are in boldface, implying that these images accomplished the enhancement with an acceptable similarity to the original image.

As can be seen, the maximum contrast solutions generally yield higher entropy and normalized standard deviation, indicating a broader range of pixel intensities and greater variability in the enhanced images. In contrast, maximum detail solutions focus on enhancing the finer details within the images. This often results in lower entropy and normalized standard deviation compared to maximum contrast solutions but may increase the number and the intensity of high-frequency pixels (indicating more detailed textures). These results highlight differences in contrast and detail between the solutions extracted from the Pareto front that may not be perceptible in the previous visual discussion. If we examine some cases, we can find images such as Natural2, Medical1, and Medical4, where the extreme points on the Pareto front do not show a visual difference. However, their associated values for contrast and detail exhibit numerical differences.

The reported values of the objective functions show that all exposed solutions are non-dominated, indicating that they represent an optimal trade-off between the two evaluated criteria. Regarding the SSIM index, 65% of the solutions exhibit values above 0.7, indicating a generally high level of structural similarity. Furthermore, by only analyzing the medical images, 85% of the solutions reached SSIM outcomes beyond 0.7, indicating that the proposal is a trustworthy tool for dealing with this kind of information. Nonetheless, if only the natural scene images are evaluated, the number of solutions that archived this SSIM outcome decreases to 50%. This may influence the artificially imposed low contrast in this set of images. Consequently, future work should consider incorporating SSIM as an additional objective function, especially for problems where image fidelity is crucial.

Table 6. Optimization results for natural images.

Image	Solution	$H(I)$	$\sigma_{norm}(I)$	$N_{HF}(I)$	Intensity$_{HF}(I)$	f_1	f_2	SSIM
Natural1	Max. Contrast	7.5531	0.5957	136,328	28,907.9607	4.4997	100,511.1934	0.6658
	Knee	7.5593	0.5919	140,554	29,866.6697	4.4741	103,785.0868	0.6434
	Max. Detail	7.4855	0.5771	142,301	30,043.5271	4.3198	105,104.0062	0.6153
Natural2	Max. Contrast	6.3958	0.2890	42,665	7256.3766	1.8482	29,298.3087	0.6388
	Knee	6.3867	0.2889	43,005	7323.3415	1.8449	29,547.1093	0.6484
	Max. Detail	6.3633	0.2888	43,039	7331.7722	1.8374	29,572.3889	0.6509
Natural3	Max. Contrast	7.5080	0.5562	27,263	4478.1528	4.1760	18,199.8145	**0.8264**
	Knee	7.4174	0.5437	31,497	5253.1676	4.0331	21,228.6115	**0.7635**
	Max. Detail	7.2310	0.5140	33,082	5554.9158	3.7168	22,370.5067	0.6750
Natural4	Max. Contrast	7.6028	0.5582	38,563	6506.9944	4.2437	26,317.5588	**0.7603**
	Knee	7.5705	0.5498	41,095	6840.4895	4.1620	28,125.8294	**0.7430**
	Max. Detail	7.4589	0.5305	42,358	6941.6262	3.9571	29,014.4705	**0.7154**
Natural5	Max. Contrast	7.6397	0.6084	123,121	25,158.0587	4.6481	90,179.9275	0.6053
	Knee	7.6277	0.6070	123,275	25,177.7133	4.6297	90,296.0866	0.5938
	Max. Detail	7.5816	0.6062	123,311	25,173.0702	4.5959	90,321.6616	0.5894
Natural6	Max. Contrast	7.2965	0.7151	92,537	18,795.8519	5.2174	66,825.3294	0.6586
	Knee	7.3302	0.7005	96,356	19,321.6818	5.1345	69,678.1741	0.6931
	Max. Detail	7.2851	0.6758	97,675	19,209.1750	4.9234	70,611.6297	**0.7175**
Natural7	Max. Contrast	7.4601	0.5546	45,798	8129.0078	4.1374	31,650.5187	**0.7839**
	Knee	7.3807	0.5442	48,599	8487.5329	4.0163	33,666.6691	**0.7652**
	Max. Detail	7.1648	0.5160	50,088	8529.1422	3.6970	34,707.5355	**0.7388**
Natural8	Max. Contrast	7.2975	0.7166	135,043	30,609.3702	5.2293	99,829.9165	0.6525
	Knee	7.3574	0.7065	142,322	31,648.0378	5.1979	105,373.9500	0.6738
	Max. Detail	7.3067	0.6903	144,811	31,899.5143	5.0441	107,256.0717	0.6770
Natural9	Max. Contrast	7.4927	0.5278	41,161	7399.7872	3.9550	28,296.7289	**0.7953**
	Knee	7.4455	0.5261	42,537	7682.5907	3.9174	29,304.3657	**0.8037**
	Max. Detail	7.3050	0.5196	42,823	7772.1429	3.7954	29,520.5423	**0.8067**
Natural10	Max. Contrast	7.5721	0.5280	32,834	5719.8726	3.9983	22,240.9423	**0.7855**
	Knee	7.5501	0.5198	35,058	6002.2180	3.9246	23,814.2353	**0.7937**
	Max. Detail	7.4591	0.5000	35,621	6062.6948	3.7295	24,210.7577	**0.7868**

Table 7. Optimization results for medical images.

Image	Solution	$H(I)$	$\sigma_{norm}(I)$	$N_{HF}(I)$	Intensity$_{HF}(I)$	f_1	f_2	SSIM
Medical1	Max. Contrast	7.8448	0.5823	28,921	4292.0292	4.5682	19,256.7633	**0.7469**
	Knee	7.8279	0.5820	29,564	4414.5851	4.5558	19,718.7333	**0.7608**
	Max. Detail	7.7870	0.5767	29,861	4481.5297	4.4908	19,935.0599	**0.7732**
Medical2	Max. Contrast	7.1626	0.7809	20,764	3818.1387	5.5935	13,726.0797	0.6879
	Knee	7.1339	0.7809	21,025	3876.9084	5.5710	13,911.8213	0.6584
	Max. Detail	7.1054	0.7784	21,162	3905.2624	5.5309	14,008.8062	0.6366
Medical3	Max. Contrast	7.4796	0.6403	501	68.4979	4.7890	227.2635	**0.7746**
	Knee	7.1434	0.5984	875	118.2316	4.2747	427.0364	**0.8652**
	Max. Detail	6.4043	0.4811	1058	145.9017	3.0808	529.6895	**0.8658**
Medical4	Max. Contrast	7.8136	0.5744	15,156	1923.4562	4.4882	9576.7837	**0.7955**
	Knee	7.7803	0.5708	16,008	2042.4497	4.4411	10,157.2992	**0.8073**
	Max. Detail	7.6851	0.5563	16,190	2081.7596	4.2751	10,286.2666	**0.8092**
Medical5	Max. Contrast	4.1156	0.4317	1017	117.5567	1.7767	495.9842	0.5690
	Knee	4.1156	0.3829	1486	182.3434	1.5757	763.2875	0.6173
	Max. Detail	4.1156	0.3346	1729	217.4919	1.3772	905.4307	0.6017
Medical6	Max. Contrast	7.6461	0.4482	16,721	2343.1399	3.4268	10,709.4627	**0.7703**
	Knee	7.6442	0.4482	16,732	2347.1401	3.4259	10,717.7392	**0.7703**
	Max. Detail	7.6363	0.4482	16,742	2346.8756	3.4224	10,724.0633	**0.7702**
Medical7	Max. Contrast	7.7687	0.6045	6378	970.3478	4.6965	3831.1820	**0.7633**
	Knee	7.7658	0.6047	6389	972.1292	4.6958	3838.3433	**0.7612**
	Max. Detail	7.7591	0.6047	6389	972.2033	4.6916	3838.3663	**0.7607**
Medical8	Max. Contrast	6.7448	0.4860	354	45.3329	3.2779	150.7481	**0.7133**
	Knee	6.1648	0.4642	496	64.1421	2.8618	222.8542	**0.8958**
	Max. Detail	5.3591	0.4163	585	74.7138	2.2309	268.6718	**0.9375**
Medical9	Max. Contrast	5.6986	0.4041	29,117	7292.7190	2.3027	20,000.4591	**0.8336**
	Knee	5.5245	0.4116	29,587	7479.4364	2.2737	20,352.2764	**0.8403**
	Max. Detail	5.3411	0.4130	29,762	7561.0648	2.2061	20,485.1491	**0.8429**
Medical10	Max. Contrast	7.7175	0.5087	16,293	2996.5663	3.9260	10,606.1673	**0.8023**
	Knee	7.5585	0.5002	16,562	3094.9172	3.7810	10,803.7683	**0.8674**
	Max. Detail	7.3872	0.4864	16,812	3100.2399	3.5931	10,968.0618	**0.8806**

5.4. Image Quality Metrics

Besides the analysis related to the optimization problem, an evaluation of well-recognized image enhancement metrics is conducted. Table 8 displays the metrics evaluations related to the natural set (left side) and medical image set (right side). The particular points of interest (Max. contrast, Knee, and Max. Detail) are divided by rows on each set. The study encompasses the Contrast-to-Noise Ratio (CNR), Blind/Referenceless Image Spatial Quality Evaluator (BRISQUE), and Natural Image Quality Evaluator (NIQE), which are tagged in that order in the column headers. Next, a brief explanation of how these metrics are addressed is offered.

- The Contrast-to-Noise Ratio (CNR): This metric quantifies the contrast between a signal and the background. The higher the CNR outcome, the better the image enhancement. Values from zero to one indicate a poor contrast, while ranges from one to three indicate a moderated contrast, and greater values indicate a good contrast, implying that it is easy to identify features within the image [53]. This metric is reference-based, i.e., it compares the enhanced image w.r.t. the original one. This work uses Otsu's method to compute the threshold to differentiate noise from signal [54].
- The Blind/Referenceless Image Spatial Quality Evaluator (BRISQUE): This is a no-reference-based metric computed directly from the enhanced image. It measures the degradation of the image quality after an image processing task. Based on statistical approaches, it compares the image against a Gaussian mixture model that scores the

image, where the lower the score, the better the image quality. Values lower than twenty mean a high-quality image, values from twenty to forty reflect good image quality, values from forty to sixty indicate fair image quality, and greater values reflect poor-quality images [55].

- The Natural Image Quality Evaluator (NIQE): This is also a no-reference-based metric that directly assesses the quality of images by comparing them to a statistical model built from natural images instead of a Gaussian mixture model. The lower the NIQE score, the better the image quality. Scores below five indicate high-quality images, a range between five and ten reflects good image quality, and greater values indicate poor quality [56].

Table 8. Results for key points of interest (Max. contrast, Knee, and Max. Detail) using three metrics: Contrast-to-Noise Ratio (CNR), Blind/Referenceless Image Spatial Quality Evaluator (BRISQUE), and Natural Image Quality Evaluator (NIQE).

		CNR	BRISQUE	NIQE			CNR	BRISQUE	NIQE
Natural1	Max. Contrast	**3.6344**	40.5179	5.6368	Medical1	Max. Contrast	**3.6809**	24.507	**4.2045**
	Knee	**3.7547**	39.6382	5.6135		Knee	**3.7298**	24.2549	**4.2751**
	Max. Detail	**3.8288**	37.4743	5.4585		Max. Detail	**3.7766**	22.7663	**4.0709**
Natural2	Max. Contrast	**6.9888**	23.0218	5.2016	Medical2	Max. Contrast	**4.969**	**19.1406**	**3.6733**
	Knee	**7.2094**	23.9762	5.2328		Knee	**4.9564**	**17.8147**	**3.6678**
	Max. Detail	**7.3308**	24.2154	5.2739		Max. Detail	**4.9139**	**17.1296**	**3.621**
Natural3	Max. Contrast	**3.346**	27.0687	**3.9379**	Medical3	Max. Contrast	**3.5856**	26.0907	**4.7967**
	Knee	**3.6162**	20.4811	**3.8339**		Knee	**3.388**	**17.4607**	**4.4942**
	Max. Detail	**4.1426**	**9.3008**	**3.7158**		Max. Detail	**4.2904**	24.1414	**3.6347**
Natural4	Max. Contrast	**3.9906**	20.2948	**3.8824**	Medical4	Max. Contrast	**3.5482**	36.0147	**4.798**
	Knee	**4.1183**	**18.7291**	**3.7746**		Knee	**3.7009**	36.7916	**4.4748**
	Max. Detail	**4.3539**	**16.0336**	**3.5537**		Max. Detail	**3.8495**	36.3804	**4.0389**
Natural5	Max. Contrast	**4.0943**	27.3883	**3.8463**	Medical5	Max. Contrast	**4.5134**	42.16	5.5734
	Knee	**4.0401**	27.2286	**3.8365**		Knee	**3.1547**	42.0232	5.5161
	Max. Detail	**4.0405**	26.7598	**3.8073**		Max. Detail	2.9729	38.0933	5.1749
Natural6	Max. Contrast	**5.5369**	32.0397	**4.0916**	Medical6	Max. Contrast	**4.0207**	23.052	**4.6726**
	Knee	**5.2404**	35.2761	**4.0959**		Knee	**4.0261**	22.7929	**4.6967**
	Max. Detail	**5.041**	37.5524	**4.1249**		Max. Detail	**3.9962**	22.9169	**4.7093**
Natural7	Max. Contrast	**3.4593**	32.261	**4.1593**	Medical7	Max. Contrast	**4.0663**	**17.5317**	**4.1831**
	Knee	**3.757**	27.3056	**3.8866**		Knee	**4.0174**	**16.9241**	**4.2208**
	Max. Detail	**3.901**	32.4678	**3.3908**		Max. Detail	**4.0262**	**17.1**	**4.2797**
Natural8	Max. Contrast	**5.3661**	40.1719	**4.4922**	Medical8	Max. Contrast	2.578	38.0246	5.5089
	Knee	**4.8581**	36.9644	**4.4318**		Knee	**4.5542**	38.5449	5.0139
	Max. Detail	**4.5455**	29.4885	**4.2977**		Max. Detail	**5.2398**	46.1957	**4.9405**
Natural9	Max. Contrast	2.9738	23.1393	**4.6043**	Medical9	Max. Contrast	**4.3413**	32.277	5.731
	Knee	**3.228**	21.1312	**4.525**		Knee	**4.5623**	25.0689	5.7618
	Max. Detail	**3.5316**	**17.2083**	**4.4686**		Max. Detail	**4.656**	25.8006	5.7614
Natural10	Max. Contrast	**3.3885**	**10.7718**	**4.4499**	Medical10	Max. Contrast	2.8411	23.3222	**3.2685**
	Knee	**3.2393**	**9.9866**	**4.3771**		Knee	**3.8337**	24.7217	**3.1833**
	Max. Detail	2.9976	**7.067**	**4.1238**		Max. Detail	**4.3031**	24.559	**3.1317**

According to this information, on each column (evaluated metric) of Table 8, the results that reflect high-quality images are in boldface. The following highlights are offered to provide a straightforward understanding of the outcomes.

- Regarding the CNR metric, 93.33% (28/30) of the enhanced natural images achieved good noise contrast, indicating clarity in the obtained images. Similarly, 90% of the enhanced medical images (27/30) reached good noise contrast.

- Regarding the BRISQUE criterion, 23.33% (7/30) of the enhanced natural images can be considered high-quality images, while the rest accomplish good quality standards. Interestingly, the enhanced medical images achieved the same number of high-quality images.
- In terms of the NIQUE metric, 80% (24/30) of the enhanced natural images reached high-quality standards. On the other hand, only 73.33% (22/30) of the enhanced medical images reach this standard. Nonetheless, the rest of the images of both sets possess good-quality features.

These results demonstrate that the current proposal not only improves image contrast and details while providing a range of trade-off solutions but also preserves image quality and naturalness without introducing noise. This is particularly significant, as these criteria were not explicitly chosen for the optimization problem.

5.5. Comparison with Baseline Methods

In this section, a comparison is made between the results obtained by the proposed multi-objective approach and two well-known contrast enhancement methods: contrast stretching (CS) and adaptive histogram equalization (CLAHE), as illustrated in Table 9. The first advantage of the multi-objective approach is that it produces a set of optimal solutions, each offering a different balance between contrast and detail. On the other hand, traditional methods only provide a single solution. For comparison, we selected the knee points from the Pareto front generated by the proposed method.

The comparison is performed by assessing the dominance of the obtained solutions in terms of contrast (f_1) and details (f_2). If the solution from the multi-objective approach outperforms the traditional methods in both criteria, it is considered dominant (denoted by status "+"). If it improves only one of the two criteria, it is considered non-dominated (denoted by status "||"). If the traditional method outperforms both criteria, the multi-objective solution is dominated (denoted by status "-").

The results show that the multi-objective approach dominated CS in 19 images, being dominated only in the case of Medical5. Although CS is a classical method that adjusts the histogram to improve the overall contrast of an image, the multi-objective approach demonstrates superior performance as it focuses on enhancing the contrast–detail balance. Compared with CLAHE, 13 non-dominated solutions were obtained, 6 where the proposal dominates, and 1 where CLAHE outperforms our approach. Adaptive equalization performs better in detail enhancement (f_2), which explains the lack of clear dominance in several cases compared to the multi-objective approach. Adaptive correction methods generally have advantages in highlighting contrast and details over general approaches. However, the optimization-based proposal can outperform the adaptive approach in several images while maintaining a competitive performance in most images, demonstrating the proposal's robustness and flexibility in various scenarios.

To complement the analysis, a Wilcoxon rank-sum test was conducted to assess the statistical significance of the differences between the methods. The p-values indicate the following:

- CS vs. Multi-objective approach (Contrast): $p = 0.01058$, showing a significant difference in contrast performance, favoring the multi-objective approach.
- CS vs. Multi-objective approach (Details): $p = 0.12643$, suggesting no significant difference in detail enhancement.
- CLAHE vs. Multi-objective approach (Contrast): $p = 0.00771$, demonstrating a statistically significant improvement in contrast with the multi-objective approach.
- CLAHE vs. Multi-objective approach (Details): $p = 0.59786$, indicating no significant difference in detail enhancement between CLAHE and the multi-objective approach.

These statistical results reinforce the superiority of the multi-objective approach in improving contrast, particularly when compared with CS and CLAHE, while its performance in detail enhancement remains competitive.

Table 9. Comparison of dominance between the proposed multi-objective approach, contrast stretching (CS), and adaptive histogram equalization (CLAHE). The symbols represent dominance status: "+" indicates the multi-objective solution dominates, "||" indicates non-dominance, and "-" indicates the multi-objective solution is dominated.

Problem	CE			CLAHE			Multi-Objective (Knee Point)	
	*f*1	*f*2	Status	*f*1	*f*2	Status	*f*1	*f*2
Natural1	3.6024	80,366.9308	+	4.0675	119,318.1151	||	4.4741	103,785.0868
Natural2	1.5731	17,045.7257	+	2.0435	39,844.8025	-	1.8449	29,547.1093
Natural3	2.7972	9570.6512	+	2.8002	26,295.5886	||	4.0331	21,228.6115
Natural4	3.2772	13,866.4487	+	3.4698	41,296.5049	||	4.1620	28,125.8294
Natural5	2.9780	45,449.1717	+	4.1502	91,934.6577	||	4.6297	90,296.0866
Natural6	4.1140	37,133.3034	+	3.9282	98,788.5865	||	5.1345	69,678.1741
Natural7	3.1986	21,209.0719	+	3.3195	33,578.2905	+	4.0163	33,666.6691
Natural8	4.0220	70,655.2488	+	4.2470	108,329.5632	||	5.1979	105,373.9500
Natural9	3.0300	18,327.6153	+	2.8849	34,909.6415	||	3.9174	29,304.3657
Natural10	2.4836	10,523.4740	+	2.9333	32,775.9795	||	3.9246	23,814.2353
Medical1	3.7188	9277.9447	+	3.8531	23,295.9010	||	4.5558	19,718.7333
Medical2	4.7342	7518.9552	+	3.9691	12,665.8688	+	5.5710	13,911.8213
Medical3	4.2625	232.5334	+	2.6556	543.0372	||	4.2747	427.0364
Medical4	3.9715	3712.3943	+	3.8346	11,225.1916	||	4.4411	10,157.2992
Medical5	2.2046	5069.1086	-	0.5166	0.0000	+	1.5757	763.2875
Medical6	2.5440	3904.2629	+	3.1897	20,304.1639	||	3.4259	10,717.7392
Medical7	3.6151	1660.6467	+	3.5522	4639.7335	||	4.6958	3838.3433
Medical8	2.7290	73.8758	+	2.4340	117.5608	+	2.8618	222.8542
Medical9	2.1427	18,930.2122	+	1.8009	16,273.9174	+	2.2737	20,352.2764
Medical10	2.8411	6341.1905	+	2.8874	10,778.6971	+	3.7810	10,803.7683

6. Conclusions

The conflict between contrast and detail in image processing is presented as a multi-objective problem. This approach obtains a set of optimal solutions, forming a Pareto front in all cases, highlighting the trade-off between these two properties. Therefore, it is demonstrated that a single-objective approach to this problem will only lead to a particular solution among all the optimal solutions obtained through the multi-objective approach.

The proposed model integrates the sigmoid transformation function and UMH into the NSGA-II. Additionally, a posterior preference articulation is added, which selects three key solutions from the Pareto front: the maximum contrast solution, the maximum detail solution, and the knee point solution. These three solutions showed significant superiority in terms of contrast and detail compared to the original images. Furthermore, the outcomes visually and numerically demonstrated how these three image solutions, though all optimal solutions, differ in terms of entropy, standard deviation, number of detail pixels, and detail intensity. This variability allows fundamental characteristics to emerge across the images, underscoring the relevance of the proposed preferences across various contexts. Moreover, the proposed method demonstrated its effectiveness against traditional contrast enhancement techniques, such as contrast stretching and adaptive histogram equalization, by achieving a superior balance between contrast and detail, as evidenced by the dominance of its solutions in several cases.

A post hoc analysis regarding popular image quality metrics, such as CNR, BRISQUE, and NIQE, showed that part of the images created through the current proposal achieved high-quality image standards, while none of the generated enhanced images decreased below good-quality standards. These results demonstrate that the method is a trustworthy tool for image enhancement, offering a range of solutions that not only meet diverse user needs but also consistently maintain high-quality outcomes

Despite the overall improvement in image details, specific regions may lose details compared to the original image. This suggests that for future work, it may be advisable to apply local and/or adaptive image enhancement techniques to preserve details in specific regions while maintaining overall image quality. Future reserach should develop adaptive preference articulations that can identify all solutions from the Pareto front that reveal singular information within the image. Moreover, a broader investigation into the method's versatility is planned, including experiments with alternative evolutionary algorithms. Furthermore, multi-objective micro-evolutionary algorithms are a promising and challenging research path in image enhancement tasks. Their potential lies in reducing computational overhead, though they also face challenges related to limited solution diversity and the risk of premature convergence.

Author Contributions: Conceptualization, D.M.-P.; methodology, D.M.-P. and A.G.R.-L.; validation, D.M.-P. and A.G.R.-L.; investigation, D.M.-P.; writing, D.M.-P. and A.G.R.-L.; visualization, A.G.R.-L. All authors have read and agreed to the published version of the manuscript.

Funding: This research received no external funding.

Data Availability Statement: The source codes for the multi-objective optimization algorithm used in this study are available at https://github.com/dani90molinaperez/Multi-Objective-Optimization-for-Image-Processing.

Acknowledgments: The authors acknowledge the support from the Consejo Nacional de Humanidades, Ciencia y Tecnología (CONAHCYT) and its support in Mexico through the institutions ESCOM-IPN and CIDETEC-IPN.

Conflicts of Interest: The authors declare no conflict of interest.

References

1. Russ, J.C.; Russ, J.C. *Introduction to Image Processing and Analysis*; CRC Press: Boca Raton, FL, USA, 2017.
2. Gonzalez, R.C.; Woods, R.E. *Digital Image Processing*, 4th ed.; Global, Ed.; Pearson: New York, NY, USA, 2018.
3. Aşuroğlu, T.; Sümer, E. Performance analysis of spatial and frequency domain filtering in high resolution images. In Proceedings of the 2015 23nd Signal Processing and Communications Applications Conference (SIU), Malatya, Turkey, 16–19 May 2015; pp. 935–938.
4. Lepcha, D.C.; Goyal, B.; Dogra, A.; Sharma, K.P.; Gupta, D.N. A deep journey into image enhancement: A survey of current and emerging trends. *Inf. Fusion* **2023**, *93*, 36–76. [CrossRef]
5. Kubinger, W.; Vincze, M.; Ayromlou, M. The role of gamma correction in colour image processing. In Proceedings of the 9th European Signal Processing Conference (EUSIPCO 1998), Island of Rhodes, Greece, 8–11 September 1998; pp. 1–4.
6. Braun, G.J.; Fairchild, M.D. Image lightness rescaling using sigmoidal contrast enhancement functions [3648-13]. In *Proceedings of the Proceedings-SPIE the International Society for Optical Engineering*; SPIE International Society for Optical: Bellingham, WA, USA, 1999; pp. 96–107.
7. Stimper, V.; Bauer, S.; Ernstorfer, R.; Schölkopf, B.; Xian, R.P. Multidimensional contrast limited adaptive histogram equalization. *IEEE Access* **2019**, *7*, 165437–165447. [CrossRef]
8. Albu, F.; Vertan, C.; Florea, C.; Drimbarean, A. One scan shadow compensation and visual enhancement of color images. In Proceedings of the 2009 16th IEEE International Conference on Image Processing (ICIP), Cairo, Egypt, 7–10 November 2009; pp. 3133–3136.
9. Zhang, D.; Zhang, D. Wavelet transform. In *Fundamentals of Image Data Mining: Analysis, Features, Classification and Retrieval*; Springer: New York, NY, USA, 2019; pp. 35–44.
10. Strang, G. The discrete cosine transform. *SIAM Rev.* **1999**, *41*, 135–147. [CrossRef]
11. Chittora, N.; Babel, D. A brief study on Fourier transform and its applications. *Int. Res. J. Eng. Technol.* **2018**, *5*, 1127–1130.
12. Guo, X.; Li, Y.; Ling, H. LIME: Low-light image enhancement via illumination map estimation. *IEEE Trans. Image Process.* **2016**, *26*, 982–993. [CrossRef]
13. Wang, W.; Wu, X.; Yuan, X.; Gao, Z. An experiment-based review of low-light image enhancement methods. *IEEE Access* **2020**, *8*, 87884–87917. [CrossRef]

14. Ullah, Z.; Farooq, M.U.; Lee, S.H.; An, D. A hybrid image enhancement based brain MRI images classification technique. *Med. Hypotheses* **2020**, *143*, 109922. [CrossRef]
15. Wang, C.; Wu, M.; Lam, S.K.; Ning, X.; Yu, S.; Wang, R.; Li, W.; Srikanthan, T. GPSFormer: A Global Perception and Local Structure Fitting-based Transformer for Point Cloud Understanding. *arXiv* **2024**, arXiv:2407.13519.
16. Wang, R.; Lam, S.K.; Wu, M.; Hu, Z.; Wang, C.; Wang, J. Destination intention estimation-based convolutional encoder-decoder for pedestrian trajectory multimodality forecast. *Measurement* **2025**, *239*, 115470. [CrossRef]
17. Dhal, K.G.; Ray, S.; Das, A.; Das, S. A survey on nature-inspired optimization algorithms and their application in image enhancement domain. *Arch. Comput. Methods Eng.* **2019**, *26*, 1607–1638. [CrossRef]
18. Deb, K.; Pratap, A.; Agarwal, S.; Meyarivan, T. A fast and elitist multiobjective genetic algorithm: NSGA-II. *IEEE Trans. Evol. Comput.* **2002**, *6*, 182–197. [CrossRef]
19. Bhandarkar, S.M.; Zhang, Y.; Potter, W.D. An edge detection technique using genetic algorithm-based optimization. *Pattern Recognit.* **1994**, *27*, 1159–1180. [CrossRef]
20. Braik, M.; Sheta, A.F.; Ayesh, A. Image Enhancement Using Particle Swarm Optimization. *World Congr. Eng.* **2007**, *1*, 978–988.
21. Behera, S.K.; Mishra, S.; Rana, D. Image enhancement using accelerated particle swarm optimization. *Int. J. Eng. Res. Technol.* **2015**, *4*, 1049–1054.
22. Nguyen-Thi, K.N.; Che-Ngoc, H.; Pham-Chau, A.T. An efficient image contrast enhancement method using sigmoid function and differential evolution. *J. Adv. Eng. Comput.* **2020**, *4*, 162–172. [CrossRef]
23. Bhandari, A.K.; Maurya, S. Cuckoo search algorithm-based brightness preserving histogram scheme for low-contrast image enhancement. *Soft Comput.* **2020**, *24*, 1619–1645. [CrossRef]
24. Acharya, U.K.; Kumar, S. Genetic algorithm based adaptive histogram equalization (GAAHE) technique for medical image enhancement. *Optik* **2021**, *230*, 166273. [CrossRef]
25. Acharya, U.K.; Kumar, S. Directed searching optimized texture based adaptive gamma correction (DSOTAGC) technique for medical image enhancement. *Multimed. Tools Appl.* **2024**, *83*, 6943–6962. [CrossRef]
26. Braik, M. Hybrid enhanced whale optimization algorithm for contrast and detail enhancement of color images. *Clust. Comput.* **2024**, *27*, 231–267. [CrossRef]
27. Rani, S.S. Colour image enhancement using weighted histogram equalization with improved monarch butterfly optimization. *Int. J. Image Data Fusion* **2024**, *15*, 510–536. [CrossRef]
28. Du, N.; Luo, Q.; Du, Y.; Zhou, Y. Color image enhancement: A metaheuristic chimp optimization algorithm. *Neural Process. Lett.* **2022**, *54*, 4769–4808. [CrossRef]
29. Krishnan, S.N.; Yuvaraj, D.; Banerjee, K.; Josephson, P.J.; Kumar, T.C.A.; Ayoobkhan, M.U.A. Medical image enhancement in health care applications using modified sun flower optimization. *Optik* **2022**, *271*, 170051. [CrossRef]
30. Ma, G.; Yue, X.; Zhu, J.; Liu, Z.; Zhang, Z.; Zhou, Y.; Li, C. A novel slime mold algorithm for grayscale and color image contrast enhancement. *Comput. Vis. Image Underst.* **2024**, *240*, 103933. [CrossRef]
31. More, L.G.; Brizuela, M.A.; Ayala, H.L.; Pinto-Roa, D.P.; Noguera, J.L.V. Parameter tuning of CLAHE based on multi-objective optimization to achieve different contrast levels in medical images. In Proceedings of the 2015 IEEE International Conference on Image Processing (ICIP), Quebec City, QC, Canada, 27–30 September 2015; pp. 4644–4648.
32. Matin, F.; Jeong, Y.; Park, H. Retinex-based image enhancement with particle swarm optimization and multi-objective function. *IEICE Trans. Inf. Syst.* **2020**, *103*, 2721–2724. [CrossRef]
33. Abouhawwash, M.; Alessio, A.M. Multi-objective evolutionary algorithm for PET image reconstruction: Concept. *IEEE Trans. Med. Imaging* **2021**, *40*, 2142–2151. [CrossRef]
34. Kuran, U.; Kuran, E.C. Parameter selection for CLAHE using multi-objective cuckoo search algorithm for image contrast enhancement. *Intell. Syst. Appl.* **2021**, *12*, 200051. [CrossRef]
35. Cuevas, E.; Zaldívar, D.; Pérez-Cisneros, M. Multi-objective Optimization of Anisotropic Diffusion Parameters for Enhanced Image Denoising. In *New Metaheuristic Schemes: Mechanisms and Applications*; Springer: New York, NY, USA, 2023; pp. 241–268.
36. Jaimes, A.L.; Martınez, S.Z.; Coello, C.A.C. An introduction to multiobjective optimization techniques. *Optim. Polym. Process.* **2009**, *1*, 29.
37. Lee, D.H.; Kim, K.J.; Köksalan, M. A posterior preference articulation approach to multiresponse surface optimization. *Eur. J. Oper. Res.* **2011**, *210*, 301–309. [CrossRef]
38. Wang, H.; Olhofer, M.; Jin, Y. A mini-review on preference modeling and articulation in multi-objective optimization: Current status and challenges. *Complex Intell. Syst.* **2017**, *3*, 233–245. [CrossRef]
39. Imtiaz, M.S.; Wahid, K.A. Image enhancement and space-variant color reproduction method for endoscopic images using adaptive sigmoid function. In Proceedings of the 2014 36th Annual International Conference of the IEEE Engineering in Medicine and Biology Society, Chicago, IL, USA, 26–30 August 2014; pp. 3905–3908.
40. Srinivas, K.; Bhandari, A.K. Low light image enhancement with adaptive sigmoid transfer function. *IET Image Process.* **2020**, *14*, 668–678. [CrossRef]
41. Folkersma, L. The Impact of Problem Features on NSGA-II and MOEA/D Performance. Master's Thesis, Utrecht University, Utrecht, The Netherlands, 2020.
42. Sağlican, E.; Afacan, E. MOEA/D vs. NSGA-II: A Comprehensive Comparison for Multi/Many Objective Analog/RF Circuit Optimization through a Generic Benchmark. *ACM Trans. Des. Autom. Electron. Syst.* **2023**, *29*, 1–23. [CrossRef]

43. Eastman Kodak Company. Kodak Lossless True Color Image Suite. 2013. Available online: https://r0k.us/graphics/kodak/ (accessed on 18 August 2024).
44. Hartung, M. Visible Human Project—Brain Images. Case Study. 2024. Available online: https://radiopaedia.org/cases/visible-human-project-brain-images-1 (accessed on 10 August 2024). [CrossRef]
45. Al Kabbani, A. Human Brain—Lateral View. Case Study. 2024. Available online: https://radiopaedia.org/cases/human-brain-lateral-view (accessed on 10 August 2024). [CrossRef]
46. Nickparvar, M. White Blood Cells Dataset: A Large Dataset of Five White Blood Cells Types. 2022. Available online: https://www.kaggle.com/datasets/masoudnickparvar/white-blood-cells-dataset/data (accessed on 10 August 2024).
47. Mooney, P. Blood Cell Image Dataset. 2018. Available online: https://www.kaggle.com/datasets/paultimothymooney/blood-cells/discussion/437393, (accessed on 10 August 2024).
48. Radiological Society of North America. Pediatric Bone Age Machine Learning Challenge Dataset. 2017. Available online: https://www.kaggle.com/code/plarmuseau/image-contrast-enhancement-techniques/input (accessed on 10 August 2024).
49. Thurston, M. Lisch Nodules (Photo). 2024. Available online: https://radiopaedia.org/cases/lisch-nodules-photo (accessed on 10 August 2024).
50. Dhairya, S. Dental Condition Dataset. 2024. Available online: https://www.kaggle.com/datasets/sizlingdhairya1/oral-infection (accessed on 10 August 2024).
51. Sinitca, A.M.; Lyanova, A.I.; Kaplun, D.I.; Hassan, H.; Krasichkov, A.S.; Sanarova, K.E.; Shilenko, L.A.; Sidorova, E.E.; Akhmetova, A.A.; Vaulina, D.D.; et al. Microscopy Image Dataset for Deep Learning-Based Quantitative Assessment of Pulmonary Vascular Changes. *Sci. Data* **2024**, *11*, 635. [CrossRef] [PubMed]
52. Er, A. Trauma Forearm Positioning (Photo). 2020. Available online: https://radiopaedia.org/cases/trauma-forearm-positioning-photo (accessed on 10 August 2024).
53. Baker, M.E.; Dong, F.; Primak, A.; Obuchowski, N.A.; Einstein, D.; Gandhi, N.; Herts, B.R.; Purysko, A.; Remer, E.; Vachani, N. Contrast-to-noise ratio and low-contrast object resolution on full-and low-dose MDCT: SAFIRE versus filtered back projection in a low-contrast object phantom and in the liver. *Am. J. Roentgenol.* **2012**, *199*, 8–18. [CrossRef] [PubMed]
54. Otsu, N. A threshold selection method from gray-level histograms. *Automatica* **1975**, *11*, 23–27. [CrossRef]
55. Saad, M.A.; Bovik, A.C.; Charrier, C. Blind image quality assessment: A natural scene statistics approach in the DCT domain. *IEEE Trans. Image Process.* **2012**, *21*, 3339–3352. [CrossRef]
56. Zvezdakova, A.; Kulikov, D.; Kondranin, D.; Vatolin, D. Barriers towards no-reference metrics application to compressed video quality analysis: On the example of no-reference metric NIQE. *arXiv* **2019**, arXiv:1907.03842.

Mathematical and Computational Applications

Article

New Metaheuristics to Solve the Internet Shopping Optimization Problem with Sensitive Prices

Miguel A. García-Morales [1], José Alfredo Brambila-Hernández [1,*], Héctor J. Fraire-Huacuja [1], Juan Frausto [1], Laura Cruz [1], Claudia Gómez [1] and Alfredo Peña-Ramos [2]

1 National Technology of Mexico/Madero City Technological Institute, Madero City 89460, Tamaulipas, Mexico; soporteclusterlanti@gmail.com (M.A.G.-M.); hector.fh@cdmadero.tecnm.mx (H.J.F.-H.); juan.frausto@gmail.com (J.F.); lauracruzreyes@itcm.edu.mx (L.C.); claudia.gs@cdmadero.tecnm.mx (C.G.)

2 Faculty of Engineering, Autonomous University of Tamaulipas, Tampico 89109, Tamaulipas, Mexico; apramos@docentes.uat.edu.mx

* Correspondence: jabrambila@gmail.com

Abstract: In this research, two new methods for solving the Internet shopping optimization problem with sensitive prices are proposed, incorporating adaptive adjustment of control parameters. This problem is classified as NP-hard and is relevant to current electronic commerce. The first proposed solution method corresponds to a Memetic Algorithm incorporating improved local search and adaptive adjustment of control parameters. The second proposed solution method is a particle swarm optimization algorithm that adds a technique for diversification and adaptive adjustment of control parameters. We assess the effectiveness of the proposed algorithms by comparing them with the Branch and Bound algorithm, which presents the most favorable outcomes of the state-of-the-art method. Nine instances of three different sizes are used: small, medium, and large. For performance validation, the Wilcoxon and Friedman non-parametric tests are applied. The results show that the proposed algorithms exhibit comparable performance and outperform the Branch and Bound algorithm.

Keywords: memetic algorithm; particle swarm; technique diversification; improved local search; internet shopping problem; adaptive adjustment; branch and bound

1. Introduction

Currently, most Internet transactions are primarily due to online purchases, which have become increasingly important because of the COVID-19 pandemic [1]. Customers seek to minimize the cost of their purchases by considering the discounts offered by online stores, since prices change daily depending on the advertised offers. The Internet shopping problem was formally defined by Błażewicz [2]; furthermore, this author shows that this problem corresponds to the NP-hard type by constructing a pseudo-polynomial transformation of the Exact Cover by 3-Sets problem P1.

Currently, metaheuristic algorithms are widely used to solve optimization problems. They are generally used to find optimal solutions or to work towards the optimal value in NP-hard optimization problems [3]. A study by Li [4] identifies the main trends and areas of application of metaheuristic algorithms such as logistics, manufacturing, artificial intelligence, and computer science; the main reason for its popularity is its flexibility in addressing NP-hard problems. Valencia-Rivera [5] presents a complete analysis of trends and the impact of the metaheuristic algorithms that are most frequently used to solve optimization problems today; the review includes particle swarm optimization algorithms, genetic algorithms, and Ant Colony algorithms.

Therefore, developing new solution methods and metaheuristics is essential for identifying which stores offer the best discount to satisfy a shopping list. The described issue is known as the Internet shopping problem with sensitive prices (IShOPwD); related works are described below:

Błażewicz [6] designed the IShOPwD model for the first time, considering only the customer's perspective. This model explicitly addresses a scenario where a client aims to buy several products from multiple online stores, focusing on minimizing the total cost of all products and incorporating applicable discounts into the overall cost. The solution method developed is a branch-and-bound algorithm (BB).

Musial [7] proposes a set of algorithms (Greedy Algorithm (GA), Forecasting Algorithm (FA), Cellular Processing Algorithm (CPA), MinMin Algorithm, and Branch and Bound algorithm) to solve the problem of IShOPwD optimization. These algorithms are crafted to generate various solutions at different computation times, aiming to achieve results that are close to the optimal solution. The results of these proposed algorithms are compared with the optimal solutions and calculated using a BB.

In the work of Błażewicz et al. [8], the authors define and study some price-sensitive extensions of the IShOP. The study includes the IShOP with price-sensitive discounts and a newly defined optimization problem: the IShOP that includes two discount functions (a shipping cost discount function and a product price discount function). They formulate mathematical programming problems and develop some algorithms (new heuristic: new forecasting), and exhaustive feasibility tests are carried out.

Another related work is that of Chung and Choi [9]. This work aims to introduce an optimal bundle search problem called "OBSP" that allows integration with an online recommendation system (Bundle Search Method) to provide an optimized service considering pairwise discounting and the delivery cost. The results are integrated into an online recommendation system to support user decision-making.

Mahrundinda [10] and Morales [11] analyze the trends and solution methods applied to the IShOP and determine that no solution methods have used non-deterministic metaheuristics to solve the IShOPwD until now. That is why it was decided to adopt the solution methods proposed by Huacuja [12] and García-Morales [13], which are metaheuristic algorithms that provide the best results for the IShOP variant with shipping costs.

In this research, we develop two metaheuristic algorithms (a Memetic Algorithm (MAIShOPwD) and a particle swarm optimization algorithm (PSOIShOPwD)) that allow us to solve the Internet shopping problem with sensitive prices.

The main contributions of the proposed MAIShOPwD algorithm are as follows:

- A novel method for calculating changes in the objective values of the current solution during local search, with a time complexity of $O(1)$.
- Adaptive adjustment of control parameters.

On the other hand, the contributions of the proposed PSOIShOPwD algorithm are as follows:

- A neighborhood diversification technique and a local search at the end of each iteration.
- Adaptive adjustment of control parameters.

A revised approach derived from the bandit-based adaptive operator selection technique, Fitness-Rate-Rank-Based Multiarmed Bandit (FRRMAB) [14], was used to adjust the control parameters in both proposed algorithms adaptively.

A series of feasibility experiments, including the instances used in [12], were carried out to identify the advantages of the proposed algorithms. Subsequently, the non-parametric tests were applied with a significance level of 5% to establish certain conclusions.

This research comprises the following sections: Section 2 formally defines the problem addressed. Section 3 describes the main components of the proposed algorithms. Sections 4 and 5 describe the general structure of the MAIShOPwD and PSOIShOPwD algorithms. Section 6 describes the computational experiments carried out to validate the feasibility of the proposed algorithms. Section 7 presents the results obtained through the Wilcoxon and Friedman non-parametric tests. Moreover, Section 8 defines the conclusions obtained and potential areas for future work.

2. Definition of the Problem

The data set N_j includes the available products in each store j. Each product i is associated with a cost c_{ij} and a shipping cost d_i. The shipping cost is added if the client purchases one or more items in the store j.

Formally, the problem considers that a client aims to purchase items from a specified set $N = (1, \ldots, n)$, in a set of online stores $M = (1, \ldots, m)$ with the lowest total cost, including a discount f_k, which is applicable in the calculation of the objective function. We calculate the objective value of the solution I using Equation (1):

$$min\ f_k\left(\sum_{j=1}^{m} \sum_{i=1\&\&I(i)=j}^{n} (d_i + c_{ij})\right) \tag{1}$$

A candidate solution is represented by a vector I of length N, which specifies the store from which each product should be purchased. A more detailed description can be found in the work of Błażewicz et al. [6].

Equation (2) below shows the criteria used to select the discount for the total cost of the shopping list for the function f_k.

$$f_k(x)\begin{cases} x\ if\ x \leq 25, \\ 0.95x\ if\ 25 < x \leq 50, \\ 0.90x\ if\ 50 < x \leq 100, \\ 0.85x\ if\ 100 < x \leq 200, \\ 0.80x\ if\ 200 < x \end{cases} \tag{2}$$

3. The General Structure of the Proposed Algorithms

This section describes all the main elements of the proposed solution methods, which are part of the following algorithms: the Memetic Algorithm (MAIShOPwD) and the particle swarm optimization algorithm (PSOIShOPwD).

3.1. *Representation of a Candidate Solution*

Related state-of-the-art works use a matrix representation of the candidate solutions such that the time complexity of evaluating the objective function for each candidate solution is $O(n^2)$. This work proposes using a vector representation to reference the solution proposed in [12], in which store j is established and each product i must be purchased. Figure 1 shows the vector representation of a candidate solution.

Products

1	2	...	n	
j_1	j_2	...	j_n	Stores

Figure 1. Candidate solution [12].

3.2. *Adaptive Parameter Adjustment*

The two algorithms proposed in this paper use initial values in their parameters. Later, these values are adjusted based on specific actions taken during the search process. The work in [12] expands the description of the modified FRRMAB [14] method for adaptive tuning of control parameters.

3.2.1. Assignment of Credits

Fialho [15] recommends not using the reward value directly to prove fitness. Because this affects the algorithm's robustness, the adaptive adjustment method of control uses Fitness Improvement Rates (FIRs) to avoid the problem, as indicated in Equation (3).

$$FIR_{i,t} = \frac{pf_{i,t} - cf_{i,t}}{pf_{i,t}} \tag{3}$$

where

$pf_{i,t}$ represents the fitness value of the parent.

$cf_{i,t}$ represents the fitness value of the children.

Each FIR value of the actions used is stored in a FIFO (first in, first out) structure of fixed size W. This setup enables the latest actions to be retained while the earliest ones are discarded within the sliding window, which stores the stock index and FIR value.

Calculating the reward ($Reward_i$) for each action, the cumulative sum of FIR values for each action within the sliding window is calculated. They are arranged in descending order based on the rank ($Rank_i$) of each action i. Finally, a decay factor D within the range of 0 to 1 is applied to modify the $Reward_i$ and this allows the best actions to be selected using Equation (4):

$$Decay_i = D^{Rank_i} \times Reward_i \tag{4}$$

Next, each credit is allocated to action i according to Equation (5).

$$FRR_{i,t} = \frac{Decay_i}{\sum_{j=1}^{K} Decay_j} \tag{5}$$

As the decay factor D decreases, the likelihood of selecting the best action increases. Algorithm 1 shows how credits are assigned to an action. The algorithm used is a modified version of the FRRMAB algorithm proposed by Li [14], which was initially used to perform adaptive selection of genetic operators (AOS) but has been adapted to perform adaptive selection of actions. First, the rewards for each operator ($Reward_i$) are initialized to 0, and the application counter of each action (n_i) is also initialized to 0 (lines 1, 2). It is iterated through the sliding window (lines 3–8), which contains a fixed number of recent applications of the actions. For each entry in the window, the index of the action is obtained (line 4). Its corresponding fitness improvement rate (line 5), the improvement rate (FIR), is added to the total reward of the action (line 6), and the application counter of the action is incremented (line 7). Once the improvement rates have been added, all rewards are sorted in descending order (line 9). Based on this ranking, each action is given a rank value ($Rank_i$). A decay factor (D) is applied to its reward for each action to adjust for the influence of better-performing actions (lines 10–12). The new adjusted reward ($Decay_{action}$) is calculated by multiplying the trader's rank by its total reward (line 11). Finally, the adjusted reward for each action is normalized (lines 14–16) to obtain the final credit value (FRR_{action}), which is subsequently used in the trader selection process.

Algorithm 1. Credit allocation.

1: Initialize each reward $Reward_i = 0$
2: Initialize$n_i = 0$;
3: ***for*** $i \leftarrow 1$ ***to*** SlidingWindow.length ***do***
4: $\quad action =$ SlidingWindow.GetIndexAction(i)
5: $\quad FIR =$ SlidingWindow.GetFIR(i)
6: $\quad Reward_{action} = Reward_{action} + FIR$
7: $\quad n_{action} + +$
8: ***end for***
9: Rank $Reward_i$ in descending order and set $Rank_i$ to be the rank value of action i
10: ***for*** *action* $\leftarrow$ ***to*** *K* ***do***
11: $\quad Decay_{action} = D^{Rank_{action}} \times Reward_{action}$
12: ***end for***
13: $\quad DecaySum = \sum_{action=1}^{K} Decay_{action}$
14: ***for*** *action* $\leftarrow$ ***to*** *K* ***do***
15: $\quad FRR_{action} = Decay_{action} / DecaySum$
16: ***end for***

3.2.2. Bandit-Based Action Selection

The bandit-based approach utilizes Fitness Rate Rank (FRR) values to assess quality for action selection, as depicted in Algorithm 2. Algorithm 2 is used for the action selection based on the multiarmed bandit approach [16] used in the FRRMAB algorithm [14]. The algorithm starts by checking if any actions still need to be selected. If so, one of these operators is chosen uniformly at random (lines 1–3). This mechanism ensures that the algorithm tries to use all possible actions at least once. Suppose all operators have already been selected at least once. In that case, the algorithm selects the operator that maximizes a function, combining the estimated quality of the operator (FRR) and further exploration based on the number of times that operator has been selected (line 4); C is a factor that controls the balance between exploitation and exploration.

Algorithm 2. Bandit-based action selection.

1: ***if*** There are actions that have not been selected ***then***
2: $action_t$ = randomly select a security from the action pool
3: ***else***
4: $action_t = \underset{i=\{1,\ldots,K\}}{\operatorname{argmax}} \left(FRR_{i,t} + C \times \sqrt{\frac{2 \times \ln\left(\sum_{j=1}^{K} n_{j,t}\right)}{n_{i,t}}} \right)$
5: ***end if***

3.3. *Actions Pool for the Memetic Algorithm*

Below are the six adjustment actions for parameters crossover probability (pc) and mutation probability (pm) of the MAIShOPwD algorithm: Action 1 increases the values of the crossover and mutation probabilities by one ten-thousandth. Action 2 decreases the values of the crossover and mutation probabilities by one ten-thousandth. Action 3 increases the value of crossover probability by one ten-thousandth. Action 4 increases the value of the mutation probability by one ten-thousandth. Action 5 decreases the value of crossover probability by one ten-thousandth. Finally, action 6 decreases the value of the mutation probability by one ten-thousandth.

(1) Action 1

$pc = pc + 0.0001;$
$pm = pm + 0.0001;$

(2) Action 2

$pc = pc - 0.0001;$
$pm = pm - 0.0001;$

(3) Action 3

$pc = pc + 0.0001;$

(4) Action 4

$pm = pm + 0.0001;$

(5) Action 5

$pc = pc - 0.0001;$

(6) Action 6

$pm = pm - 0.0001;$

3.4. *Actions Pool for the Particle Swarm Optimization Algorithm*

Below are the six adjustment actions for the personal learning quotient (c_1) and global learning quotient (c_2) of the PSOIShOPwD algorithm: Action 1 increases the values of c_1 and c_2 by one ten-thousandth. Action 2 decreases the values of c_1 and c_2 by one ten-thousandth. Action 3 increases the value of c_1 by one ten-thousandth. Action 4 increases the value of c_2 by one ten-thousandth. Action 5 decreases the value of c_1 by one ten-thousandth. Finally, action 6 decreases the value of c_2 by one ten-thousandth.

(1) Action 1

$c_1 = c_1 + 0.0001;$
$c_2 = c_2 + 0.0001;$

(2) Action 2

$c_1 = c_1 - 0.0001;$
$c_2 = c_2 - 0.0001;$

(3) Action 3

$c_1 = c_1 + 0.0001;$

(4) Action 4

$c_2 = c_2 + 0.0001;$

(5) Action 5

$c_1 = c_1 - 0.0001;$

(6) Action 6

$c_2 = c_2 - 0.0001;$

4. The General Structure of the Memetic Algorithm

Huacuja [12] introduced a Memetic Algorithm to tackle the IShOP with shipping costs. The results obtained by the Memetic Algorithm demonstrate that it could feasibly be used to solve other variants related to IShOP. Considering this contribution as a feasible solution method, the decision was made to improve the local search to allow a significant reduction in the quality and efficiency of the results for the IShOPwD. Additionally, a mechanism that allows adaptive adjustment of some control parameters was added, considering the contribution of García-Morales [17].

4.1. Used Heuristics

The heuristics used to generate candidate solutions are described below. They are called best first and best [12]. These heuristics identify the store where a product can be purchased at the lowest cost, considering the discount associated with the total cost of the shipping list. If more detailed information on the operation of these heuristics is required, please consult the work conducted in [12].

4.1.1. Heuristic: Best First

The best first heuristic changes the stores assigned to a solution vector until the first cost improvement associated with the current product is found. This process is repeated for all products in the shopping list. This heuristic has a time complexity of $O(nm)$ [12].

4.1.2. Heuristic: The Best

The best heuristic works the same as the best first heuristic. The difference is mainly that this heuristic continues when it finds the first improvement in the cost of each product but continues to search all stores for the lowest cost for each product. The process concludes when all stores have been checked for all products in the solution vector. This process is applied to all products. The best heuristic has a complexity of $O(nm)$ [12].

4.2. Binary Tournament

A certain number of individuals are randomly selected from the population, and the fittest advance to the next generation. Each individual participates in two comparisons, and the winning individuals form a new population; this ensures that any individual in the population has a maximum of two copies in the new population. A more extensive description of the binary tournament can be found in [12].

4.3. Crossover Mechanism

The crossover mechanism is applied to a certain percentage of the individuals sequentially. This operator selects two solution vectors called parent one and parent two. Both parent vectors are divided in half to generate two children. The first offspring is created by combining the first half of parent 1 with the second half of parent 2. The second offspring is formed by merging the first half of parent 2 with the second half of parent 1.

4.4. Mutation Operator

This operator selects a percentage of elite solutions from the population. Subsequently, it goes through each elite solution and generates a random number. If the value of the mutation probability parameter is less than the heuristic mutation process, the heuristic mutation process will be applied to each selected elite solution. Detailed information on the mutation heuristic can be found in [12].

4.5. Improved Local Search

The local search algorithm improves a given vector solution, reducing the total cost. To achieve this goal, we change the assigned store for each product, reducing the solution's total cost. This procedure improves a given solution, and speeding up the local search algorithm is costly. This procedure was modified to reduce the computational costs and determine the change in the objective value of the current solution. Let $I = (j_1, j_2, j_3, \ldots, j_n)$ be the current solution and $I' = j'_1, j_2, j_3, \ldots, j_n)$ be the solution we obtain when the store one should buy the product from changes from j_1 to j'_1. The change in the objective values of the current solution is given by $\delta = F(I') - F(I)$. If this is directly realized using Equation 1, the computational complexity required to determine δ is $O(nm)$. The following equation shows how to calculate δ in $O(1)$.

Theorem 1. $\delta = \left(d_{j'_1} + c_{1j'_1}\right) - \left(d_{j_1} + c_{1j_1}\right)$.

Proof. Let $I = (j_1, j_2, j_3, \ldots, j_n)$ the current solution and $I' = (j'_1, j_2, j_3, \ldots, j_n)$ the solution that we obtain when the store one should buy the product from changes from j_1 to j'_1.

Then, using Equation (6):

$$F(I) = \sum_{i=1}^{n} \sum_{\forall j \mid I(j)=i} (d_j + c_{ij}) = (d_{j_1} + c_{1j_1}) + \sum_{i=2}^{n} \sum_{\forall j \mid I(j)=i} (d_j + c_{ij}) \tag{6}$$

$$F(I\prime) = \sum_{i=1}^{n} \sum_{\forall j \mid I'(j)=i} (d_j + c_{ij}) = (d_{j'_1} + c_{1j'_1}) + \sum_{i=2}^{n} \sum_{\forall j \mid I'(j)=i} (d_j + c_{ij}) \tag{7}$$

Therefore, $\delta = F(I') - F(I) = (d_{j'_1} + c_{1j'_1}) - (d_{j_1} + c_{1j_1})$. □

The local search tries to improve the current solution by changing the stores assigned to the products. For each product, the store that produces the maximal reduction in the objective value is identified. To determine if a change in the store assigned to the product produces a reduction or not, Theorem 1 is applied again in the search process. For this reason, these results significantly impact the local search performance.

4.6. Proposed Memetic Algorithm

Algorithm 3 shows the structure of the proposed MAISHOPwD algorithm. In steps 1 and 2, the values of the initial parameters are defined, and the instance to be used is loaded. From steps 3 to 7, an initial population is generated, the objective value and discount for the entire population are calculated, and the best local and global solution can be identified. In step 9, the binary tournament is applied to the population, and a new population will be generated based on the results obtained. In step 10, a percentage of elite solutions are moved from the new population to an intermediate population. In step 11, the crossover operator affects the new population, and the missing elements of the intermediate population are generated. In steps 12 and 13, the mutation operator and the improved local search are applied to the intermediate population to obtain a population of children. In step 15, the local solution is checked to determine if it is better than the global solution; if so, the global solution is updated. In steps 16 and 17, the entire population is reset, the best global solution is inserted, and all other individuals in the population will be ruled out. In steps 18,19 and 20, the sliding window is updated, and the rewards of the adaptive control parameter adjustment method are ranked. In step 21, the process of each iteration ends, and finally, in step 22, the best solution and the global cost are obtained.

Algorithm 3. MAIShOPwD Algorithm.

Input: *MaxIter*: Maximum number of iterations
Instance: m (stores), n (products), c_{ij} (product cost), d_j (shipping cost)
Parameters/Variables: *ps*: initial population size
pc: crossover probability
pm: mutation probability
er: rate elitism
pop: Initial population
BestGlobalSolution, *BestGlobalCost*: Best overall solution and cost
BestLocalSolution, *BestLocalCost*: Best solution and cost in each generation
Functions:
BestGlobal(pop, BestGlobalSolution, BestGlobalCost): from *pop* obtain the *BestGlobalSolution* and the *BestGlobalCost*
CrossoverOperator(NewPop, IntermediatePop): among non − elite individuals in *NewPop*, the crossover operator is applied to randomly selected $ps *$ pc solutions and the remaining solutions are copied as − is to *IntermediatePop*
MutationOperator(IntermediatePop, ChildPop): with probability *pm*, apply the mutation operator to all the solutions in the intermediate population and move the offspring to the population of children
ImprovedLocalSearch(IntermediatePop, ChildPop): apply improved *LocalSearch* to all solutions in *IntermediatePop* and move the offspring to *ChildPop*
BestLocal(ChildPop, BestLocalSolution, BestLocalCost): obtain the *BestLocalSolution* and the *BestLocalCost* in *ChildPop*
ObjectiveFunction( ): calculate the objective value
Discount(ObjectiveFunction( )): apply the discount to the objective value
FRRMAB( ): obtains the index for executing an action
ExecuteAction(indexaction): performs an action based on an index
SlidingWindow(indexaction, improvement[i]): a sliding window that stores both the frecuency of action execution and the decrease in individual costs
UpgradeRewards(SlidingWindow): enhance the rewards held within the sliding window
Output: *BestGlobalSolution*: Best overall solution
BestGlobalCost : Best overall cost
1: Set initial parameters : population size (*ps*), percentage of cross (*pc*), probability of mutation (*pm*), Rate elitism (*er*)
2: Load Instance
3: Randomly generate initial population of size *ps*
4: *ObjectiveFunction(pop)*
5: *Discount(ObjectiveFunction(pop))*
6: *BestGlobal(pop, BestGlobalSolution, BestGlobalCost)*
7: *BestLocalSolution = BestGlobalSolution; BestLocalCost = BestGlobalCost;*

Algorithm 3. *Cont.*

while (*not MaxIter*) *do*
    Apply binary tournament in *pop* to get *NewPop*
    Move from *NewPop* to *IntermediatePop* the elite solutions
    $CrossoverOperator(NewPop, IntermediatePop)$
    $MutationOperator(IntermediatePop, ChildPop)$
    $ImprovedLocalSearch\ (IntermediatePop,\ ChildPop)$
    $BestLocal(ChildPop,\ BestLocalSolution, BestLocalCost$
    if $BestLocalCost < BestGlobalCost$ **then** {*BestGlobalCost* = *BestLocalCost*; *BestGloblalSolution* = *BestLocalSolution*}
    $pop = \phi;\ pop = pop \cup \{BestGlobalSolution\}$
    Randomly generate $ps - 1$ solutions and add each solution to *pop*
    Add to $SlidingWindow(indexaction, improvement[i])$
    $UpgradeRewards(SlindingWindow)$
    Rank Rewards
end while
return $(BestGlobalSolution, BestGlobalCost)$

5. Overview of the Particle Swarm Optimization Algorithm Structure

This section describes the second proposed solution method for the IShOPwD. This method utilizes the PSOIShOPwD algorithm. In this algorithm, two strategies are considered to avoid premature convergence [18]: the first corresponds to the diversification of neighborhoods [13], and the second is an improved local search; both strategies are executed only once per iteration to prevent excessive execution time consumption.

In addition, this proposed algorithm incorporates the adaptive adjustment of control parameters using an adaptation of the FRRMAB method [14].

5.1. Neighborhood Diversification Technique

The diversification technique runs through each particle position in the population [18]. For each position in a particle, it generates a random value r from the interval $[1,\ M]$. Then, it adjusts the value of the current position based on the following criteria: if the particle's position is even, the operation subtracts the sum of the best position of the current particle and a random value r from the total number of stores, M. If the previous condition is not satisfied, then M is reduced by the difference between the best position of the current particle and the random r. Subsequently, the algorithm checks for compliance with minimum and maximum position constraints. Following this, it computes the objective value of the modified particle and assesses if there has been an improvement in its best cost; if so, the best overall particle is modified. The described process concludes once the entire population has been modified and evaluated. The process is detailed in Algorithm 4.

Algorithm 4. Neighborhood diversification technique.

Inputs: Population of solutions
Outputs: Updated Population.
Variables:
particles: Population
particles.size: Population size
N: Number of products
M: Number of stores
Functions:
$random\_number\ [1, M]$: Generates a random number in the range [1, M].
$ObjectiveFunction(particles_i)$: Calculate objective value of the particle.
$Max(particles_i.position[j], 1)$: Calculates the maximum value between the j position of a particle and 1.

Algorithm 4. *Cont.*

$Min(particles_i.position[j],\ M)$: Calculates the minimum value between the position j of a particle and M.
$Discount(ObjectiveFunction(\,))$: Apply the discount to the objective value

```
for i = 1 to Particles.size do
  r = random_number [1, M]
    for j = 1 to N do
      if i % 2 == 0 then
          particles_i.Position[j] = M − (particles_i.BestPosition[j] + r)
          particles_i.position[j] = Max(particles_i.position[j], 1)
          particles_i.position[j] = Min(particles_i.position[j], M)
      else
          particles_i.Position[j] = M − (particles_i.BestPosition[j] − r)
          particles_i.position[j] = Max(particles_i.position[j], 1)
          particles_i.position[j] = Min(particles_i.position[j], M)
      end if
    end for
    Discount(ObjectiveFunction(particles_i))
    if particles_i.Cost < particles_i.bestCost then
        particles_i.bestPosition = particles_i.position
        particles_i.bestCost = particles_i.cost
        if particles_i.bestCost < GlobalBest.cost then
          GlobalBest = particles_i
        end if
    end if
  end for
return Particles
```

5.2. Local Search

The local search aims to improve the best global particle [18]. The process consists of modifying the particle's positions and evaluating whether the modifications reduce the particle's overall cost. This process is repeated in all positions of the particle. Finally, the total cost of the updated particle is calculated and compared with the best cost of the global particle. The newly updated particle will replace the best global particle if the cost is lower than the global particle. The process is detailed in Algorithm 5.

Algorithm 5. Local Search.

Inputs: Best Global solution
Outputs: Best Global Solution Updated.
Variables:
N: Number of products
M: Number of stores
$Cost$: Objective value cost
Functions:
$ObjectiveFunction(GlobalBest)$: Calculate objective value of the Global Best
$Discount(ObjectiveFunction(\,))$: apply the discount to the objective value

```
Cost = Discount(ObjectiveFunction(GlobalBest))
for i = 1 to N do
        for j = 1 to M do
          GlobalBest.Position[i] = j
          if Discount(ObjectiveFunction(GlobalBest)) < Cost then
             Cost = Discount(ObjectiveFunction(GlobalBest))
             j' = j
```

Algorithm 5. *Cont.*

```
8.             end if
9.         end for
10.    GlobalBest.Position[i] = j'
11.  end for
12. Discount(Objectivefunction(GlobalBest))
13. if (GlobalBest.Cost < GlobalBest.bestCost) then
14.        GlobalBest.bestPosition = GlobalBest.Position
15.        GlobalBest.bestCost = GlobalBest.Cost
16. end if
17. return GlobalBest
```

5.3. Proposed Particle Swarm Optimization Algorithm (PSOIShOPwD)

Algorithm 6 outlines the fundamental framework of the proposed PSOIShOPwD algorithm. In steps 1 to 4, the initial values of the parameters are defined, a random population of particles is generated, the overall best particle is obtained, and the minimum and maximum particle velocities are calculated. In steps 7 and 8, the algorithm retrieves the index linked to an action and executes that action to adjust the values c_1 and c_2. From steps 10 to 15, the algorithm adjusts the position and velocity of each particle, ensuring they remain within the permissible range; otherwise, a corrective method is applied. In steps 17, 18, and 19, the algorithm calculates the objective value, applies the discount to each particle, and determines the cost improvement of each particle relative to its best cost. From steps 20 to 27, the algorithm assigns the previously computed objective value to an improvement vector. It then compares the best position of the current particle with the global particle; if it is superior, the global particle's objective value is replaced with it. In step 28, the sliding window is updated with the action index and the particle cost improvement. In steps 29 and 30, the sliding window rewards are updated, and descending order is applied. In steps 32 and 33, diversification and a local search are applied. Finally, in steps 34 and 35, the inertial weight (w) is updated, and the best global solution is obtained.

Algorithm 6. PSOIShOPwD Algorithm.

Inputs: Population.
Outputs: Global Best Solution.
Variables/Parameters:
$particles$: Population
$particles.size$: Population size
N: Number of products
M: Number of stores
$MaxIt$: Maximum number of iterations
w: Inertial weight
$wdamp$: Inertial weight damping radius
c_1: Personal learning quotient
c_2: Global learning quotient
$GlobalBest$: Global Best Solution
Functions:
$random\_number[0, M]$: Generates a Random number in the range $[1, M]$.
$ObjectiveFunction(particles_i)$: Calculate the objective value of a particle
$Max\left(particles_i.position_j, 1\right)$: Calculates the maximum value between the j position of a particle and 1.
$Min\left(particles_i.position_j, M\right)$: Calculates the minimum value between the j position of a particle and M.

Algorithm 6. *Cont.*

$NDT(particles)$: Applies the neighborhood diversification technique to all particles.
$LS(GlobalBest)$: Apply local search to the best global solution.
$FRRMAB()$: Gets an index of an action to perform.
$executeAction(actionIndex)$: Execute an action according to an index.
$SlidingWindow(actionIndex, improvement[i])$: Sliding window that stores the index of action to be performed and the improvement in the cost of the particle.
$UpgradeRewards(SlidingWindow)$: Update the sliding window rewards.
$Discount(ObjectiveFunction(particles_i))$: Apply the assigned discount to the total cost of the purchase.

```
Initialize parameters MaxIt, Particles.size, w, wdamp, c1, c2, N, M.
particles = GeneratePopulation(Particles.size)
GlobalBest = getBestGlobal(particles)
velMax = 0.2 * M, velMin = -velMax
for it = 1 to MaxIt do
   for i = 1 to nPop do
      actionIndex = FRRMAB()
      executeAction(actionIndex)
     for j = 1 to d do
     particles_i = w × particles_i.velocity_j + c1 × random_number[0,1] × (particles_i.bestPosition_j − particles_i.position_j) + c2 × random_number[0,1] × (GlobalBest.position_j − particles_i.position_j);
         particles_i.velocity_j = Max(particles_i.velocity_j, velMin);
         particles_i.velocity_j = Min(particles_i.velocity_j, velMax);
         particles_i.position_j = particles_i.posicion + particles_i.velocity_j;
         particles_i.position_j = Max(particles_i.position_j, 1);
         particles_i.position_j = Min(particles_i.position_j, M);
      end for
      Objectivefunction(particles_i)
      Discount(Objectivefunction(particles_i))
      fitnessimprovement = (particles_i.bestCost − particles_i.cost) / particles_i.bestCost
      if particles_i.cost < particles_i.bestCost then
         improvement[i] = fitnessimprovement
         particles_i.bestPosition = particless_i.position
         particles_i.bestCost = particles_i.cost
         if particles_i.bestCost < GlobalBest.cost then
            GlobalBest = particles_i
         end if
      end if
      Add to SlidingWindow(actionIndex, improvement[i])
      Upgrade_Rewards(SlidingWindow)
      Rank Rewards
   end for
  NDT(particles)
  LS(GlobalBest)
  w = w * wdamp
end for
return GlobalBest
```

6. Computational Experiments

The proposed algorithms are validated using the same instances created in [12], which are presented in Table 1. This table displays the subgroup sizes of instances used in the computational experiments. Each instance name indicates the number of products denoted by n and the number of stores denoted by m. Each algorithm was executed independently

30 times, and the averages of their 30 medians were calculated for the objective value (*OV*) and time execution (*Time*). The time used by the algorithms is measured in milliseconds.

Table 1. Configuration of instance size.

Small Instances	Medium Instances	Large Instances
$3n20m$	$5n240m$	$50n400m$
$4n20m$	$5n400m$	$100n240m$
$5n20m$	$50n240m$	$100n400m$

Table 2 shows the assigned values of variables and parameters utilized in the computational experiments for each analyzed algorithm. These assigned initial values were taken from the state-of-the-art works of Huacuja [12] and García-Morales [18], in which different values for the corresponding parameters were analyzed.

Table 2. Variables and parameters of the proposed algorithms.

Variables/Parameters	MAIShOPwD	PSOIShOPwD
$population.size$	100	100
pc	0.6 *	0.6
pm	0.01 *	0.01
er	0.05	
$wdamp$		0.99
$MaxIt$	100	100
w		1
c_1, c_2		1.5, 2 *

* Start values.

7. Results

Tables 3–5 show the outcomes obtained from the computational experiments carried out, and subsequently, the non-parametric Wilcoxon test is applied with a significance level of 5%. For each table, the first column corresponds to the evaluated instance. The second and fourth columns show the *OV* and the *Time* achieved by the reference algorithm and, as a subindex, the standard deviation. Correspondingly, the third and fifth columns show these values for the comparison algorithm. Columns six and seven, however, represent the $p - value$ obtained by the Wilcoxon test for the objective value and shortest time, respectively. The cells shaded in gray represent the lowest *OV* or the shortest time when the best solution is found according to each column. The symbol ↑ indicates a significant difference in favor of the reference algorithm. The symbol ↓ indicates a significant difference in favor of the comparison algorithm, and the symbol = indicates no significant differences.

Table 3. Comparative performance of BB vs. MAIShOPwD algorithms applying the Wilcoxon test.

Instances	BB	MAIShOPwD	BB	MAIShOPwD	*p*-Value (*OV*)	*p*-Value (*Time*)
	OV	*OV*	*Time*	*Time*		
$3n20m$	$56.56_{0.04}$=	56.56_{0}	$0.084_{6.2\times10^{-5}}$↑	$1.2222 \times 10^{-5}{}_{0.05}$	0.625	0.00001
$4n20m$	$70.68_{0.21}$=	70.68_{0}	$0.048_{4.3\times10^{-5}}$↑	$6.6667 \times 10^{-6}{}_{0.007}$	0.625	0.00001
$5n20m$	$89.16_{0.62}$=	$89.04_{0.004}$	$0.040_{0.0002}$↑	$0.0002_{0.029}$	0.625	0.00001
$5n240m$	$75.70_{1.94}$↑	$67.68_{1.67}$	$0.067_{1.76}$↓	$0.274_{0.1}$	0.00001	0.00001
$5n400m$	$70.56_{2.44}$↑	$62.66_{1.68}$	$0.243_{0.288}$↓	$0.790_{0.2}$	0.00001	0.00001
$50n240m$	$503.49_{12.83}$↑	$434.29_{17.60}$	$22.50_{7.50}$↑	$16.444_{8.03}$	0.00001	0.00001
$50n400m$	$433.61_{9.78}$↑	$389.14_{24.08}$	$33.11_{40.10}$↓	$48.661_{25.30}$	0.00148	0.00001
$100n240m$	$992.72_{15.32}$↑	$757.29_{38.25}$	$50.49_{39.76}$↓	$73.857_{25.64}$	0.00001	0.00001
$100n400m$	$796.87_{14.15}$↑	$655.72_{35.04}$	$176.11_{92.54}$=	$181.952_{95.30}$	0.00001	0.3843
Total average	343.261	287.007	31.41	35.775		

Table 4. Comparative performance of BB vs. PSOIShOPwD algorithms applying the Wilcoxon test.

Instances	BB	PSOIShOPwD	BB	PSOIShOPwD	p-Value (OV)	p-Value (Time)
	OV	OV	Time	Time		
$3n20m$	$56.56_{0.04} =$	56.64_{0}	$0.084_{6.2\times10^{-5}}\uparrow$	$0.01040_{6.1\times10^{-5}}$	0.625	0.00001
$4n20m$	$70.68_{0.21} =$	70.76_{0}	$0.048_{4.3\times10^{-5}}\uparrow$	0.0103_{0}	0.625	0.00001
$5n20m$	$89.16_{0.62} =$	$89.21_{3.3726\times10^{-14}}$	$0.040_{0.0002}\uparrow$	$0.008_{6.6944\times10^{-5}}$	0.625	0.00001
$5n240m$	$75.70_{1.94}\uparrow$	$68.21_{2.674\times10^{-14}}$	$0.067_{1.76}\uparrow$	$0.0104_{5.85\times10^{-18}}$	0.00001	0.00001
$5n400m$	$70.56_{2.44}\uparrow$	$64.19_{2.529\times10^{-14}}$	$0.243_{0.288}\uparrow$	$0.0110_{5.403\times10^{-18}}$	0.00001	0.00001
$50n240m$	$503.49_{12.83}\uparrow$	$463.52_{12.83}$	$22.50_{7.50} =$	$24.5541_{10.18}$	0.00001	0.24604
$50n400m$	$433.61_{9.78}\uparrow$	$355.97_{1.3876\times10^{-13}}$	$33.11_{40.10}\uparrow$	$0.0894_{2.21\times10^{-7}}$	0.00001	0.00001
$100n240m$	$992.72_{15.32}\uparrow$	$712.79_{2.3897\times10^{-13}}$	$50.49_{39.76}\uparrow$	$0.092_{3.1759\times10^{-17}}$	0.00001	0.00001
$100n400m$	$796.87_{14.15}\uparrow$	$605.63_{2.0428\times10^{-13}}$	$176.11_{92.54}\uparrow$	$0.12396_{4.1404\times10^{-17}}$	0.00001	0.00001
Total average	343.26	276.32	31.41	2.768		

Table 5. Comparative performance of MAIShOPwD vs. PSOIShOPwD algorithms applying the Wilcoxon test.

Instances	MAIShOPwD	PSOIShOPwD	MAIShOPwD	PSOIShOPwD	p-Value (OV)	p-Value (Time)
	OV	OV	Time	Time		
$3n20m$	$56.56_{0} =$	56.64_{0}	$1.2222\times10^{-5}{}_{0.05}$	$0.0104_{6.1\times10^{-5}}$	0.625	0.00001
$4n20m$	$70.68_{0} =$	70.76_{0}	$6.6667\times10^{-6}{}_{0.007}$	0.0103_{0}	0.625	0.00001
$5n20m$	$89.04_{0.004} =$	$89.21_{3.3726\times10^{-14}}$	$0.0002_{0.029}$	$0.008_{6.6944\times10^{-5}}$	0.625	0.00001
$5n240m$	$67.68_{1.67}\downarrow$	$68.21_{2.674\times10^{-14}}$	$0.274_{0.1}$	$0.0104_{5.85\times10^{-18}}$	0.00001	0.00001
$5n400m$	$62.66_{1.68}\downarrow$	$64.19_{2.529\times10^{-14}}$	$0.790_{0.2}$	$0.0110_{5.403\times10^{-18}}$	0.00001	0.00001
$50n240m$	$434.29_{17.60}\downarrow$	$463.52_{12.83}$	$16.444_{8.03}$	$24.5541_{10.18}$	0.00001	0.00016
$50n400m$	$389.14_{24.08}\uparrow$	$355.97_{1.3876\times10^{-13}}$	$48.661_{25.30}$	$0.0894_{2.21\times10^{-7}}$	0.00001	0.00001
$100n240m$	$757.29_{38.25}\uparrow$	$712.79_{2.3897\times10^{-13}}$	$73.857_{25.64}$	$0.092_{3.1759\times10^{-17}}$	0.00001	0.00001
$100n400m$	$655.72_{35.04}\uparrow$	$605.63_{2.0428\times10^{-13}}$	$181.952_{95.30}$	$0.12396_{4.1404\times10^{-17}}$	0.00001	0.00001
Total average	287.007	276.32	37.775	2.7678		

Based on the Wilcoxon test results, Table 3 illustrates that the proposed MAIShOPwD algorithm outperforms the state-of-the-art algorithm regarding the objective value in six out of nine instances. Regarding the shortest time metric, the MAIShOP algorithm performs better in four out of nine instances evaluated.

The results obtained by the Wilcoxon test shown in Table 4 indicate that the proposed PSOIShOPwD algorithm outperforms the state-of-the-art algorithm in six out of nine instances when evaluating the objective value. Regarding the shortest time metric, this algorithm outperforms the others in eight out of nine instances.

The results obtained by the Wilcoxon test shown in Table 5 indicate that the two proposed algorithms perform similarly to the objective value, since the MAIShOPwD algorithm performs better in medium instances. In contrast, the PSOIShOPwD algorithm performs better in large instances.

Table 6 contains the results obtained from the evaluation of the Friedman test at a 5% significance level; the first column lists the instances, while the second and third columns show the results for the *OV* and the *Time* achieved by the state-of-the-art algorithm. The fourth and fifth columns present equivalent data but for the MAIShOP algorithm. The sixth and seventh columns display results similar to those in the second and third columns, but specifically for the PSOIShOPwD algorithm. The last two columns contain the results of the *p*-value for each instance evaluated by the Friedman test.

The results obtained by the Friedman test indicate that in small instances concerning the objective value, the performances of the three algorithms are the same. For medium-sized instances, the algorithm that performs the best is MAIShOPwD. In the case of large instances, the best algorithm is PSOIShOPwD. The time analysis reveals that the two proposed algorithms outperform the state-of-the-art algorithm.

Table 6. Comparative performance of the BB vs. MAIShOPwD vs. PSOIShOPwD algorithms applying the Friedman test.

Instances	BB		MAIShOPwD		PSOIShOPwD		*p*-Value (*OV*)	*p*-Value (*Time*)
	OV	*Time*	*OV*	*Time*	*OV*	*Time*		
$3n20m$	$56.56_{0.04}=$	$0.084_{6.2\times10^{-5}}\downarrow$	56.56_{0}	$1.2222\times10^{-5}{}_{0.05}$	56.64_{0}	$0.0104_{6.1\times10^{-5}}$	0.97531	0.00001
$4n20m$	$70.68_{0.21}=$	$0.048_{4.3\times10^{-5}}\downarrow$	70.68_{0}	$6.6667\times10^{-6}{}_{0.007}$	70.76_{0}	0.0103_{0}	0.97531	0.00001
$5n20m$	$89.16_{0.62}=$	$0.040_{0.0002}\downarrow$	$89.04_{0.004}$	$0.0002_{0.029}$	$89.21_{3.3726\times10^{-14}}$	$0.008_{6.6944\times10^{-5}}$	0.97531	0.00001
$5n240m$	$75.70_{1.94}\downarrow$	$0.067_{1.76}\downarrow$	$67.68_{1.67}$	$0.274_{0.1}$	$68.21_{2.674\times10^{-14}}$	$0.0104_{5.85\times10^{-18}}$	0.00001	0.00001
$5n400m$	$70.56_{2.44}\downarrow$	$0.243_{0.288}\downarrow$	$62.66_{1.68}$	$0.790_{0.2}$	$64.19_{2.529\times10^{-14}}$	$0.0110_{5.403\times10^{-18}}$	0.00001	0.00001
$50n240m$	$503.49_{12.83}\downarrow$	$22.50_{7.50}\downarrow$	$434.29_{17.60}$	$16.444_{8.03}$	$463.52_{12.83}$	$24.5541_{10.18}$	0.00001	0.00001
$50n400m$	$433.61_{9.78}\downarrow$	$33.11_{40.10}\downarrow$	$389.14_{24.08}$	$48.661_{25.30}$	$355.97_{1.3876\times10^{-13}}$	$0.0894_{2.21\times10^{-7}}$	0.00001	0.00001
$100n240m$	$992.72_{15.32}\downarrow$	$50.49_{39.76}\downarrow$	$757.29_{38.25}$	$73.857_{25.64}$	$712.79_{2.3897\times10^{-13}}$	$0.092_{3.1759\times10^{-17}}$	0.00001	0.00001
$100n400m$	$796.87_{14.15}\downarrow$	$176.11_{92.54}\downarrow$	$655.72_{35.04}$	$181.952_{95.30}$	$605.63_{2.0428\times10^{-13}}$	$0.12396_{4.1404\times10^{-17}}$	0.00001	0.00001
Total average	343.261	31.41	287.007	35.775	276.32	2.768		

8. Conclusions and Future Work

This research addresses the IShOPwD, a variant of the IShOP that has become very relevant for buyers in the current electronic commerce scenario because online stores offer endless benefits for customers acquiring their products. This variant allows us to identify the most economical cost of a shopping list, considering a discount associated with the total cost. In this article, two main metaheuristic algorithms that have not yet been considered are proposed to solve IShOPwD. The first is a MAIShOPwD algorithm that incorporates an improved local search that contains a method used to calculate the change in the objective value of the current solution with a time complexity of $O(1)$, thus avoiding excessive expenditure of computing time and adaptive adjustment of control parameters that, during the execution of the algorithm, adjust the best values of the crossover and mutation probability. The second is a PSOIShOPwD algorithm that, unlike the state-of-the-art algorithms, also uses a vector representation of the candidate solutions. It includes neighborhood diversification to avoid local stagnation of the algorithm and incorporates the adaptive adjustment of control parameters that directly benefit the personal and global learning parameters. These parameters allow better positioning of the particles in the search space. The proposed algorithms are validated using the non-parametric Wilcoxon and Friedman tests, which were utilized to assess the results obtained from both the proposed algorithms and the state-of-the-art BB. According to the results obtained in the Wilcoxon test, the MAIShOPwD algorithm achieves a better performance compared to the BB; however, in inefficiency, the BB is better, and in the case of the PSOIShOPwD algorithm, it is better in terms of both quality and efficiency compared to the BB. This same test concludes that the two proposed algorithms have similar performance. The Friedman test indicates that the two proposed algorithms exhibit superior performances in both quality and efficiency when compared to the state-of-the-art algorithm.

Finally, in future work related to the IShOPwD variant, all the control parameters used by the algorithms could be incorporated into the adaptive adjustment, adding the restriction of being able to purchase more than one product of the same type and being able to use other types of discount such as coupons and lightning offers.

Author Contributions: Conceptualization: H.J.F.-H.; methodology: L.C.; research: J.F.; software: M.A.G.-M. and J.A.B.-H.; formal analysis: C.G.; writing, proofreading, and editing: H.J.F.-H., M.A.G.-M., J.A.B.-H. and A.P.-R. All authors have read and agreed to the published version of the manuscript.

Funding: This research received no external funding.

Data Availability Statement: The source code of the proposed algorithms, as well as the instances used, can be downloaded from: https://github.com/JAlfredoBrambila/ISHOPwD (accessed on 3 December 2024).

Acknowledgments: The authors want to thank Laboratorio Nacional de Tecnologías de la Información and the support of (a) the TecNM project 21336.24-P and (b) the support granted through the Scholarship for Graduate Studies to the students Miguel Ángel García Morales and José Alfredo Brambila Hernández with CVU 658787 and 1011850, respectively.

Conflicts of Interest: The authors declare that they have no known competing financial interests or personal relationships that could have appeared to influence the work reported in this paper.

References

1. Martínez Valdez, R.I.; Catache Mendoza, M.d.C.; Pedroza Cantú, G.; Huerta Cerda, Z.M. El Impacto del COVID-19 en la incidencia de compras en Línea de los Millenials. *Rev. Ing. Gestión Ind.* **2022**, *1*, 17–27. [CrossRef]
2. Błażewicz, J.; Kovalyov, M.Y.; Musiał, J.; Urbański, A.P.; Wojciechowski, A. Internet shopping optimization problem. *Int. J. Appl. Math. Comput. Sci.* **2010**, *20*, 385–390.
3. Hussain, K.; Mohd Salleh, M.N.; Cheng, S.; Shi, Y. Metaheuristic research: A comprehensive survey. *Artif. Intell. Rev.* **2019**, *52*, 2191–2233. [CrossRef]
4. Li, G.; Zhang, T.; Tsai, C.Y.; Yao, L.; Lu, Y.; Tang, J. Review of the metaheuristic algorithms in applications: Visual analysis based on bibliometrics (1994–2023). *Expert Syst. Appl.* **2024**, *255*, 124857. [CrossRef]
5. Valencia-Rivera, G.H.; Benavides-Robles, M.T.; Morales, A.V.; Amaya, I.; Cruz-Duarte, J.M.; Ortiz-Bayliss, J.C.; Avina-Cervantes, J.G. A systematic review of metaheuristic algorithms in electric power systems optimization. *Appl. Soft Comput.* **2024**, *150*, 111047. [CrossRef]
6. Błażewicz, J.; Bouvry, P.; Kovalyov, M.Y.; Musial, J. Erratum to: Internet shopping with price-sensitive discounts. *4OR* **2014**, *12*, 403–406. [CrossRef]
7. Musial, J.; Pecero, J.E.; Lopez, M.C.; Fraire, H.J.; Bouvry, P.; Błażewicz, J. How to efficiently solve Internet Shopping Optimization Problem with price sensitive discounts? In Proceedings of the 2014 11th International Conference on e-Business (ICE-B) IEEE, Vienna, Austria, 28–30 August 2014; pp. 209–215.
8. Błażewicz, J.; Cheriere, N.; Dutot, P.F.; Musial, J.; Trystram, D. Novel dual discounting functions for the Internet shopping optimization problem: New algorithms. *J. Sched.* **2016**, *19*, 245–255. [CrossRef]
9. Chung, J.; Choi, B. Complexity and algorithms for optimal bundle search problem with pairwise discount. *J. Distrib. Sci.* **2017**, *15*, 35–41. [CrossRef]
10. Mahrudinda, M.; Chaerani, D.; Rusyaman, E. Systematic literature review on adjustable robust counterpart for internet shopping optimization problem. *Int. J. Data Netw. Sci.* **2022**, *6*, 581–594. [CrossRef]
11. Morales, M.Á.G.; Huacuja, H.J.F.; Solís, J.F.; Reyes, L.C.; Santillán, C.G.G. A Survey of Models and Solution Methods for the Internet Shopping Optimization Proble. In *Fuzzy Logic and Neural Networks for Hybrid Intelligent System Design*; Studies in Computational Intelligence; Castillo, O., Melin, P., Eds.; Springer: Cham, Switzerland, 2023; Volume 1061. [CrossRef]
12. Huacuja, H.J.F.; Morales, M.Á.G.; Locés, M.C.L.; Santillán, C.G.G.; Reyes, L.C.; Rodríguez, M.L.M. Optimization of the Internet Shopping Problem with Shipping Costs. In *Fuzzy Logic Hybrid Extensions of Neural and Optimization Algorithms: Theory and Applications*; Springer: Cham, Switzerland, 2021; pp. 249–255.
13. García-Morales, M.Á.; Fraire-Huacuja, H.J.; Brambila-Hernández, J.A.; Frausto-Solís, J.; Cruz-Reyes, L.; Gómez-Santillán, C.G.; Carpio-Valadez, J.M. Particle Swarm Optimization Algorithm with Improved Opposition-Based Learning (IOBL-PSO) to Solve Continuous Problems. In *Hybrid Intelligent Systems Based on Extensions of Fuzzy Logic, Neural Networks and Metaheuristics*; Springer: Cham, Switzerland, 2023; pp. 115–126.
14. Li, K.; Fialho, A.; Kwong, S.; Zhang, Q. Adaptive Operator Selection with Bandits for a Multiobjective Evolutionary Algorithm Based on Decomposition. *IEEE Trans. Evol. Comput.* **2014**, *18*, 114–130. [CrossRef]
15. Fialho, A.; da Costa, L.; Schoenauer MSebag, M. Analyzing bandit-based adaptive operator selection mechanisms. *Ann. Math. Artif. Intell.* **2010**, *60*, 25–64. [CrossRef]
16. Auer, P.; Cesa-Bianchi, N.; Fischer, P. Finite-time analysis of the multiarmed bandit problem. *Mach. Learn.* **2002**, *47*, 235–256. [CrossRef]
17. García-Morales, M.A.; Brambila-Hernández, J.A.; Fraire-Huacuja, H.J.; Frausto-Solis, J.; Cruz-Reyes, L.; Gómez-Santillan, C.G.; Carpio-Valadez, J.M. Multi-objective Evolutionary Algorithm Based on Decomposition with Adaptive Adjustment of Control Parameters to Solve the Bi-objective Internet Shopping Optimization Problem (MOEA/D-AACPBIShOP). *Comput. Sist.* **2024**, *28*, 727–738. [CrossRef]
18. García-Morales, M.; Brambila, A.; Fraire-Huacuja, H.; Cruz-Reyes, L.; Frausto-Solís, J.; Gómez-Santillán, C.; Rangel-Valdez, N. A Novel Particle Swarm Optimization Algorithm to Solve the Internet Shopping Optimization Problem with Shipping Costs. *Int. J. Comb. Optim. Probl. Inform.* **2024**, *15*, 101–114. [CrossRef]

Mathematical and Computational Applications

MDPI

Article

Data-Driven Fault Diagnosis in Water Pipelines Based on Neuro-Fuzzy Zonotopic Kalman Filters

Esvan-Jesús Pérez-Pérez [1], Yair González-Baldizón [2,3], José-Armando Fragoso-Mandujano [2], Julio-Alberto Guzmán-Rabasa [4] and Ildeberto Santos-Ruiz [2,*]

1 Research Group of Advanced Control Systems, Universitat Politècnica de Catalunya, Rambla Sant Nebridi 22, 08222 Terrassa, Spain; esvan.de.jesus.perez@upc.edu
2 Turix-Dynamics Diagnosis and Control Group, I.T. Tuxtla Gutiérrez, Tecnológico Nacional de México, Carretera Panamericana S/N, Tuxtla Gutiérrez 29050, Mexico; yair.gonzalez@unach.mx (Y.G.-B.); jose.fm@tuxtla.tecnm.mx (J.-A.F.-M.)
3 Faculty of Accounting and Administration, Universidad Autónoma de Chiapas, Blvd. Belisario Domínguez Km. 1081, Tuxtla Gutiérrez 29050, Mexico
4 Turix-Dynamics Diagnosis and Control Group, I.T. Hermosillo, Tecnológico Nacional de México, Av. Tecnológico 115, Hermosillo 83170, Mexico; jaguzmanrabasa@gmail.com
* Correspondence: ildeberto.dr@tuxtla.tecnm.mx

Abstract: This work presents a data-driven approach for diagnosing sensor faults and leaks in hydraulic pipelines using neuro-fuzzy Zonotopic Kalman Filters (ZKF). The approach involves two key steps: first, identifying the nonlinear pipeline system using an adaptive neuro-fuzzy inference system (ANFIS), resulting in a set of Takagi–Sugeno fuzzy models derived from pressure and flow data, and second, implementing a neuro-fuzzy ZKF bench to detect pipeline leaks and sensor faults with adaptive thresholds. The learning phase of the neuro-fuzzy systems considers only fault-free data. Fault isolation is achieved by comparing zonotopic sets and evaluating a fault signature matrix. The method accounts for parametric uncertainty and measurement noise, ensuring robustness. Experimental validation on a hydraulic pipeline demonstrated high precision (up to 99.24%), recall (up to 99.20%), and low false positive rates (as low as 0.76%) across various fault scenarios and operational points.

Keywords: pipeline diagnosis; leak detection; water monitoring; neuro-fuzzy system; zonotopic Kalman filter; sensor faults

1. Introduction

Pipelines are essential for the distribution of liquids, gases, and hydrocarbons, facilitating the efficient and safe transport of these resources. However, sensor faults and leaks can cause infrastructure damage, economic losses, environmental contamination, and health risks. The lack of timely attention to these faults exacerbates their repercussions, highlighting the need to improve maintenance strategies [1]. Although significant efforts are made in preventive and corrective maintenance, these are not always sufficient to anticipate and mitigate all possible faults. Factors such as corrosion, aging, installation errors, natural events, and human activities present constant challenges [2]. Despite these hurdles, crafting a robust FD methodology remains an essential undertaking, albeit an intricate and costly one. This is particularly true in the context of multi-component systems where complexity scales exponentially, such as pressurized water pipelines.

Various fault detection methods in pipelines can be performed, including vibration analysis, leak detection based on interferometric fiber sensors, acoustic techniques, and the filter diagonalization method, among others [3]. In [4], hybrid algorithms were used to identify faults using data fusion in two pharmaceutical manufacturing cases. The authors investigated the use of Raman spectra with collected data for multivariate statistical process control with a data-driven approach in Industry 4.0 contexts. Similarly, the methodology proposed in [5] reduced the amount of data needed for the learning process and achieved higher accuracy in the detection and classification of faults in photovoltaic systems. In [6], the authors exploited cooperation between cloud and edge servers. The cloud server employed a new approach that included feature selection based on a genetic algorithm to identify correlations and redundancies among features. A notable strategy was presented in [7], where an early time series classification algorithm was applied to support fault diagnosis in a hydraulic system with multiple interconnected components. The early classification model successfully balanced accuracy and timeliness.

In the field of hydraulic systems, ref. [8] presented a fault detection method based on machine learning to locate leak faults using data variation rates from flow meters and pressure sensors. The proposed method includes a delayed alert activation algorithm that periodically checks the time series of replacement flow to identify network leaks. In [9], the authors introduced evaluation metrics for classification using machine learning and deep learning to enhance the diagnosis of anomalies and defects in each component of the hydraulic system. Data were collected from IoT sensor data in the time domain. Similarly, the approach proposed by [10] combined a fault isolation system with an inference algorithm. The beneficial synergy effect was demonstrated and confirmed for the example of diagnosing a two-tank system.

In the realm of state observers, a fault diagnosis method using a discrete-time approach based on interval observers was proposed in [11] to enhance the integration of fault detection and isolation tasks. The results were applied to the urban sewer system of Barcelona. In [12], the authors addressed the problem of robust fault diagnosis in a four-tank hydraulic system in the presence of unknown input disturbances and sensor noise. Fault detection was investigated by generating residuals. Similarly, ref. [13] presented a hydraulic system as a case study. The methodology estimated faults in both sensors and actuators. The proposed approach included an observer that estimated induced faults, and the observer gains were calculated by solving linear matrix inequality constraints. An example using an extended Kalman filter was shown by [14], where an extended Kalman filter was used to detect and locate blockages in a pipeline. Additionally, observability and controllability in hydraulic networks were analyzed. Recently, ref. [15] proposed a methodology for fault diagnosis in hydraulic systems. Two features of the proposed technique were highlighted, namely, the reduction of redundant information and the improvement of diagnostic performance, with the reduced number of sensors simplifying the model and its complexity.

On the other hand, one of the challenges in pipeline systems is related to the uncertainty of parameters such as roughness, hydraulic friction, fluid turbulence, and measurement noise. Robust fault detection and identification schemes have been implemented to address these issues; for example, complex computational schemes using CNN [16], transformer neural network [17], and LSTM approaches have been explored [18]. In [19], the use of multilabel classification techniques was proposed to simultaneously predict pipe failures in water supply systems. Three models (discriminant analysis, logistic regression, and random forest) were analyzed, and different prediction periods were used. In [20], the authors proposed an adaptive constrained clustering approach for real-time fault detection in industrial systems. Their method employs a cyclical Adaptive Constrained Clustering algorithm that combines micro-clustering and macro-clustering steps. This dynamic process incorporates must-link and cannot-link constraints to classify nominal and non-nominal working conditions, leveraging a mix of unlabeled data and limited labeled information. In [21], the authors developed a condition-based maintenance (CBM) scheme for detecting air leaks in pneumatic cylinders by monitoring piston speed profiles using Hall effect sensors. In addition, they employed machine learning techniques, including a majority voting ensemble of three models. The method achieved an accuracy of 88.4% in experimental tests. However, machine learning methods require prior knowledge of faults for detection and classification, which entails high computational costs. In the case of model-based methods, a refined mathematical model of the system is required, making it crucial to design methods that are robust to parametric uncertainty or mathematical model mismatches.

The present work introduces a data-driven method for detecting faults and leaks in hydraulic pipeline systems using neuro-fuzzy ZKF. Our approach begins with identification of the nonlinear pipeline system using an ANFIS, which generates Takagi–Sugeno fuzzy models from measured pressure and flow data. Note that this method does not require prior knowledge of the faults, as only fault-free data are used for training. Subsequently, a neuro-fuzzy ZKF bench is implemented to detect sensor faults and pipeline leaks through adaptive thresholds. The method accounts for parametric uncertainty arising from hydraulic friction and fluid turbulence as well as measurement noise, ensuring system robustness. Neuro-fuzzy ZKF observers are employed for fault detection, while fault isolation is achieved by comparing zonotopic sets and evaluating a fault signature matrix. Experimental validation in a hydraulic pipeline system demonstrates the method's high precision and reliability in fault and leak detection.

The rest of this document is organized as follows: Section 2 presents the preliminaries and notations used throughout the paper; Section 3 outlines the methodology for data-based fault diagnosis, describes the pipeline setup for industrial applications and system requirements, and details the ANFIS architecture as well as the design of the neuro-fuzzy ZKF observers; Section 4 presents the results obtained from the methodology under different operating conditions, using various metrics to evaluate the method's effectiveness; finally, Section 5 discusses the conclusions and outlines future work.

2. Preliminaries

Throughout this paper, $\mathbb{N}$ and $\mathbb{R}$ denote the set of natural and real numbers, respectively; $\mathbb{R}^m$ denotes the set of m-dimensional real vectors; the superscript T denotes the matrix transpose, whereas $\det(.)$ denotes the matrix determinant; I_n is an $n \times n$ identity matrix; $\mathrm{Tr}(.)$ and $\mathrm{Cov}(.)$ denote the trace and covariance, respectively; and $||.||_p$ indicates the p-norm, whereas $||.||_F$ and $||.||_\infty$ indicate the Frobenius and infinity norm, respectively.

Definition 1 (Zonotope [22]). *A zonotope $\langle c, R\rangle \subset \mathbb{R}^m$ with center $c \in \mathbb{R}^m$ and generator matrix $R \in \mathbb{R}^{m\times p}$ is a particular form of polytope defined as the affine transformation of a unitary hypercube $[-1,1]^p \subset \mathbb{R}^p$, as follows:*

$$\langle c, R\rangle = \{c + R\xi : ||\xi||_\infty \leq 1\}. \tag{1}$$

Definition 2 (Centered zonotope [23]). *A centered zonotope is defined as $\langle R\rangle = \langle 0, R\rangle \subset \mathbb{R}^m$. For any permutation of the columns of R, the zonotope remains invariant.*

Definition 3 (Interval hull [24]). *The interval hull $b(R) \subset \mathbb{R}^{m\times m}$ of a given zonotope $\langle c, R\rangle \subset \mathbb{R}^m$ with $R \in \mathbb{R}^{m\times p}$ is the smallest aligned box in which $\langle c, R\rangle$ can be enclosed, i.e., $\langle c, R\rangle \subset \langle c, b(R)\rangle$. The term $b(R)$ is a diagonal matrix determined by the sum of the rows of R:*

$$b(R)_{ii} = \sum_{j=1}^{p} |R_{ij}|. \quad (2)$$

Definition 4 (Reduction operator [24]). *Given the zonotope $\langle c, R \rangle \subset \mathbb{R}^m$ with $c \in \mathbb{R}^m$ and $R \in \mathbb{R}^{m\times p}$, the reduction operator $\downarrow_q$, where $q \in \mathbb{N}$ specifies the maximum number of columns of R such that $\langle c, R \rangle \subset \langle c, \downarrow_q (R) \rangle$ consists of the following:*

- *Sort p columns of $R = [r_1, r_2, \ldots, r_p]$ on the decreasing Euclidean norm, i.e., $\|r_j\| \geq \|r_{j+1}\|$.*
- *$\downarrow_q (R) = [R_\downarrow, b(R_\uparrow)] \in \mathbb{R}^{m\times q}$, where $R_\downarrow = [r_1, \ldots, r_{q-m}]$ and $R_\uparrow = [r_{q-m+1}, \ldots, r_p]$*

Definition 5 (F_W-radius [25]). *Given the zonotope $\langle c, R \rangle \subset \mathbb{R}^m$ and the weighting matrix $W \in \mathbb{R}^{m\times m}$, s.t., $W = W^T \succ 0$. The F_W-radius is determined by computing the weighted Frobenius norm of R as follows:*

$$\|\langle c, R \rangle\|_{F,W} = \sqrt{Tr(R^T W R)}, W = I_n. \quad (3)$$

Definition 6 (Covariation [25]). *The covariation of a given zonotope $\langle c, R \rangle \subset \mathbb{R}^m$ is defined as $Cov(\langle c, R \rangle) = RR^T$.*

Property 1. *Representing the Minkowski sum and the linear image as $\oplus$ and $\odot$, respectively, s.t., M is a matrix of appropriate dimension; then, $\langle c_1, R_1 \rangle \oplus \langle c_2, R_2 \rangle = \langle c_1 + c_2, [R_1, R_2] \rangle$ and $M \odot \langle c, R \rangle = \langle Mc, MR \rangle$.*

3. Data-Driven Fault Diagnosis Approach for Pipeline System

This section presents a data-driven method for pipeline system fault diagnosis utilizing only the measurement data available during the operation process, as illustrated in Figure 1. The proposed approach employs ANFIS to model the inherent nonlinear dynamics of the pipeline. ANFIS combines the learning capability of neural networks with the interpretability of fuzzy logic, making it effective for identifying complex nonlinear behaviors in dynamic systems [26]. By training ANFIS with data from fault-free operation, the proposed method captures the pipeline's dynamics under normal conditions and generates Takagi–Sugeno fuzzy systems, allowing for accurate modeling of the pipeline's behavior across varying operating scenarios.

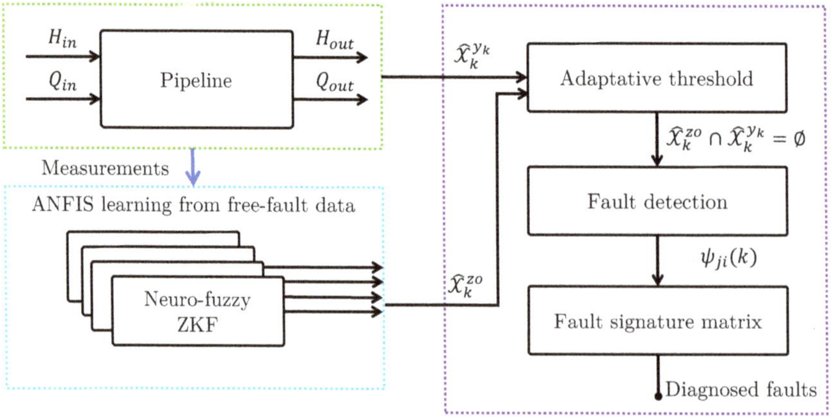

Figure 1. Diagram of the proposed method for fault diagnosis.

To improve fault detection, neuro-fuzzy ZKF with adaptive thresholds is applied. ZKF operates using deterministic set-based estimation, representing uncertainties and disturbances within bounded convex sets known as zonotopes. This approach ensures robust performance under uncertain conditions and provides a reliable mechanism for detecting faults in noisy environments [27]. The adaptive thresholds adjust dynamically to changes in operating conditions, enabling early detection of a variety of fault types with high sensitivity and precision.

Finally, fault isolation is achieved through a fault signature matrix analysis strategy that systematically correlates observed fault signatures with potential fault types. This efficient strategy facilitates the identification and classification of faults within the pipeline system, ensuring rapid and accurate fault diagnosis and enabling timely corrective actions.

3.1. Pipeline Setup for Industry Applications and System Requirements

The experimental setup consists of a hydraulic pipeline system designed to simulate and study various operational scenarios, including fault and leakage detection. The P&ID diagram of the system is shown in Figure 2. It features a single pipeline configuration that can be reconfigured into a branched network. The pipeline is constructed with 2″ Schedule 80 PVC, providing an internal diameter of 0.052 m and sufficient durability for high-pressure applications.

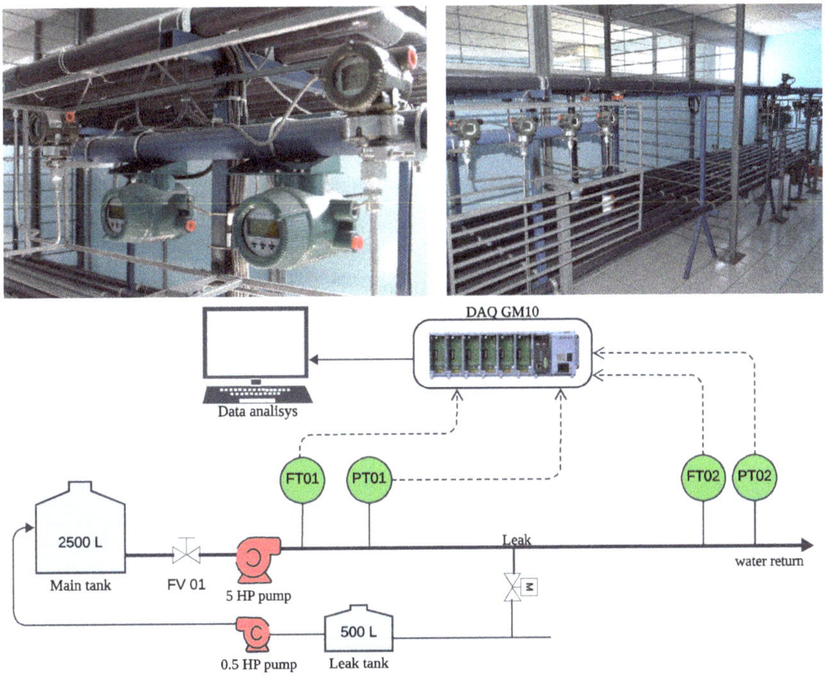

Figure 2. Physical hydraulic system (**top**) and P&ID Diagram (**bottom**).

The pipeline spans a total length of 115 m, and is arranged in a serpentine layout to optimize space and flow dynamics. Pressure (PT01, PT02) and flow (FT01, FT02) sensors are strategically installed at the inlet and outlet of the system to capture key operational parameters. These measurements are collected and stored using a Yokogawa GM10 data acquisition (DAQ) system, which communicates via the Modbus TCP/IP protocol over Ethernet.

To simulate different operating conditions, a Siemens Micromaster 420 variable frequency drive is employed to control the pumping power, with a capacity of up to 5 HP; additionally, a manually operated valve is included to simulate leak scenarios along the pipeline. The system components, including the sensors and their respective models, are summarized in Table 1.

This experimental setup provides a robust platform for evaluating various fault detection and diagnosis methodologies under controlled yet realistic conditions. Implementing the proposed method requires placement of sensors at the inlet and outlet of the pressurized pipeline. Additionally, a DAQ system and a computer are necessary to process the data collected from the sensors.

Table 1. Components and sensors in the experimental hydraulic pipeline setup.

Component	Sensor/Element and Model	Key Features
Pipeline	N/A	Internal diameter: 0.052 m, Material: PVC, High pressure resistance
Flow Sensors	Coriolis Mass Flow Meter (Yokogawa ROTAMASS)	Mass flow measurement, ±0.1% accuracy, wide range
Pressure Sensors	Gauge Pressure Sensor (Yokogawa EJA530E)	±0.055% accuracy, 4 a 20 mA range, HART communication, DPharp sensor, IP67 rated
SCADA System	Data Acquisition Recorder (Yokogawa GM10)	Modbus TCP/IP communication, reliable data storage
Pump Control	Variable Frequency Drive (Siemens Micromaster 420)	Up to 5 HP, precise speed control
Valves	Manual	Flow control, configuration adjustments
Leak Valve	Manual	Simulates leaks, no integrated sensor

3.2. Flow Dynamics in a Pipeline

A model for fluid flowing through a horizontal pipeline is derived from the physical principles of momentum conservation and mass conservation (continuity equation) in terms of time t and a spatial variable z conveniently defined in the axial direction of the pipeline. The spatiotemporal dynamics of the flow are modeled by a set of hyperbolic partial differential equations [28], commonly referred to as the water hammer equations:

$$\frac{\partial H(z,t)}{\partial t} + \frac{b^2}{gA_r}\frac{\partial Q(z,t)}{\partial z} = 0, \tag{4}$$

$$\frac{1}{A_r}\frac{\partial Q(z,t)}{\partial t} + g\frac{\partial H(z,t)}{\partial z} + \frac{\lambda\, Q(z,t)|Q(z,t)|}{2dA_r^2} = 0, \tag{5}$$

where H is the pressure head [m], Q is the flow rate [m^3/s], b is the wave speed [m/s], g is the gravity acceleration [m/s^2], A_r is the cross-sectional area [m^2] of the pipeline, and d is its diameter [m]. To obtain accurate results with the above model, it is necessary to precisely know the friction factor λ and wave speed b, although the latter only influences transient calculations and not the steady-state case. A challenge in estimating these parameters is their dependence on temperature, and in the case of λ its nonlinear variation with flow rate.

For constant flow (steady-state) conditions, the friction factor can be determined from the Darcy–Weisbach equation:

$$\rho g \Delta H = \lambda \frac{\rho q^2}{2A_r^2 d}\Delta z \tag{6}$$

where Δz is the length traveled by the fluid and ΔH is the pressure drop over that length.

In this work, turbulent flow conditions were considered throughout the experiments, as determined by the Reynolds number (Re) calculations for the pipeline system. Under turbulent flow, the friction factor λ is a nonlinear implicit function of the Reynolds number and the internal roughness ε of the pipeline. The Swamee–Jain approximation was used to estimate λ, providing a practical and accurate approach for turbulent regimes. This dependence is provided by the Colebrook–Moody formula [29], although in calculations the explicit Swamee–Jain approximation is often used [30]:

$$\lambda = 0.25\left[\log_{10}\left(\frac{\varepsilon}{3.7d} + \frac{5.74}{\mathrm{Re}^{0.9}}\right)\right]^{-2}. \tag{7}$$

It should be noted that the partial differential Equations (4) and (5) model the spatial and temporal variations of the pressure and flow rate in the pipeline subject to certain constraints, namely, the uniform velocity distribution, one-dimensional flow, constant liquid density, and constant cross-sectional area. Continuous flow and no branches are assumed without considering the existence of lateral connections.

An explicit analytical solution for the partial differential Equations (4) and (5) is not known. The solution is often obtained numerically using finite differences or the method of characteristics [31], and two boundary conditions are required; these can be known pressures or flow rates at the pipeline ends.

For a pipeline of length L, considering a spatial partition with N sections and $N+1$ points, $\{z_i\} := \{z_1 = 0, z_2 = z_1 + \Delta z_1, \ldots, z_{N+1} = z_N + \Delta z_N = L\}$, and approximating the partial derivatives by first-order finite differences, a discretized version of (4) and (5) in the spatial variable z is obtained:

$$H(z_i,t) = H_i, \quad Q(z_i,t) = Q_i, \tag{8}$$

$$\frac{\partial H(z,t)}{\partial t} = \dot{H}_i, \quad \frac{\partial Q(z,t)}{\partial t} = \dot{Q}_i, \tag{9}$$

$$\frac{\partial H(z,t)}{\partial z} \approx \frac{\Delta H_i}{\Delta z_i} = \frac{H_i(t) - H_{i+1}(t)}{\Delta z_i}, \tag{10}$$

$$\frac{\partial Q(z,t)}{\partial z} \approx \frac{\Delta Q_i}{\Delta z_i} = \frac{Q_i(t) - Q_{i+1}(t)}{\Delta z_i}, \tag{11}$$

where i represents the considered spatial section.

Incorporating the approximations in (8)–(10) into (4) and (5), the dynamic model of the flow is expressed by

$$\dot{H}_{i+1} = a_2\frac{Q_i(t) - Q_{i+1}(t)}{\Delta z_i}, \tag{12}$$

$$\dot{Q}_i = a_1\frac{H_i(t) - H_{i+1}(t)}{\Delta z_i} + \mu\, Q_i(t)|Q_i(t)|, \tag{13}$$

where

$$a_1 = gA_r, \quad a_2 = \frac{b^2}{gA_r}, \quad \mu = \frac{-\lambda}{2dA_r}. \tag{14}$$

In Table 2, the most relevant parameters of the experimental pipeline are presented for nominal operating conditions.

These equations and parameterizations can be used to test methods for detecting sensor faults and leaks. The experimental plant provides four measured variables: the inlet pressure variable H_{in} is measured by sensor PT01, and the inlet flow variable Q_{in} is measured by FT01; at the outlet, sensors PT02 and FT02 measure the pressure H_{out} and flow Q_{out}, respectively. Additionally, faults can be induced in the sensors and leaks in the pipeline. The faults are introduced as deviations in the four available measurements. Specifically, four types of synthetic faults are induced in the sensors, including a step-type bias of 10% with respect to the nominal value. The leak is created by actuating one of the leak valves in the experimental plant.

Table 2. Pipeline parameters for nominal operating conditions.

Parameter	Value
Pipe Length, L	115 m
Pipe Diameter, d	0.052 m
Cross-sectional area, A_r	0.0021 m^2
Relative Roughness, ε	2.47×10^{-4}
Fluid Density, ρ	995.736 kg/m^3
Kinematic Viscosity, ν	8.03×10^{-7} m^2/s
Wave Speed, b	422.754 m/s
Gravity Acceleration, g	9.81 m/s^2

3.3. Preparation of the Dataset for Training the Neuro-Fuzzy System

Experiments were carried out under various fault-free operating conditions, each lasting 600 s per experiment and having a sampling rate of 5 samples per second. It is important to note that each experiment differed due to the turbulent flow of pressurized water. The measurements contain noise, and no data preprocessing is required. To capture the nonlinearity of the system, the four variables H_{in}, H_{out}, Q_{in}, and Q_{out} were structured in the form of a regressor considering two instances of k. For each measured variable, an identified variable was considered in its regressor form, as shown in Table 3:

Table 3. System variables to be identified in regressive form.

Variable	Regressive Form
$\hat{H}_{in}(k)$	$(H_{in}(k), H_{in}(k-1), H_{in}(k-2), Q_{in}(k), Q_{out}(k))$
$\hat{H}_{out}(k)$	$(H_{out}(k), H_{out}(k-1), H_{out}(k-2), Q_{in}(k), Q_{out}(k))$
$\hat{Q}_{in}(k)$	$(Q_{in}(k), Q_{in}(k-1), Q_{in}(k-2), H_{in}(k), H_{out}(k))$
$\hat{Q}_{out}(k)$	$(Q_{out}(k), Q_{out}(k-1), Q_{out}(k-2), H_{in}(k), H_{out}(k))$

These regressive expressions function as inputs for the ANFIS networks; through the learning process, they can estimate the variables and develop Takagi–Sugeno fuzzy models to design the neuro-fuzzy ZKF observers.

3.4. System Identification-Based ANFIS Learning

The system identification process uses the regressive form of the variables presented in Table 3. The inputs to the ANFIS are structured for system identification as illustrated in Figure 3. It is important to mention that the training process uses fault-free data to ensure the accuracy and reliability of the model. Figure 3 shows the regressive variables $Q_{in}(k)$, $Q_{out}(k)$, $H_{in}(k-2)$, $H_{in}(k-1)$, and $H_{in}(k)$ used as inputs for the fuzzy inference system. The ANFIS utilizes these inputs to identify the system's behavior and estimate the output variable $\hat{H}_{in}(k)$.

The regressive expressions serve as inputs to ANFIS. Through training with fault-free data, these can accurately identify the estimated variables and develop Takagi–Sugeno models. These models are essential for designing zonotopic Kalman filters and enhancing the system's fault detection capabilities.

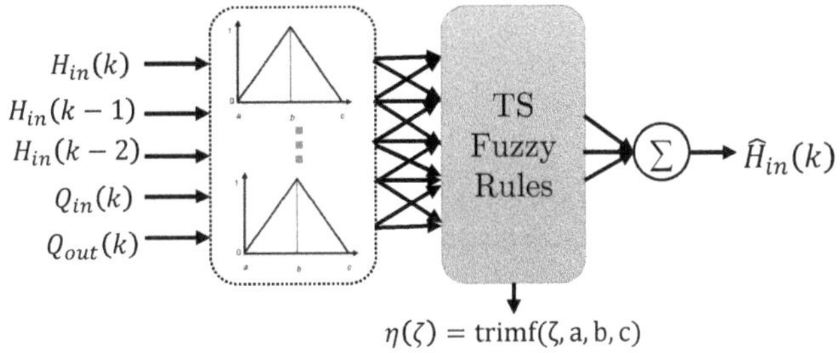

Figure 3. Structure of the ANFIS model for system identification using regressive inputs.

ANFIS is employed to approximate the nonlinear behavior of each variable of the pipeline. The input data for ANFIS are made up of each variable, as shown in Table 3, and are structured as follows:

$$\zeta = \begin{bmatrix} H_{in}(k) & H_{in}(k-1) & H_{in}(k-2) & Q_{in}(k) & Q_{out}(k) \end{bmatrix}^T. \tag{15}$$

Layer 1. *This is the fuzzification layer, where membership functions (MF) are used for fuzzification. Each MF, denoted by $\eta(\cdot)$, has three parameters (a_{nf}, b_{nf}, c_{nf}), and is defined as follows:*

$$\eta_{nf}(\zeta_o) = \begin{cases} 0 & \text{if } \zeta_o \leq a_{nf} \\ \dfrac{\zeta_o - a_{nf}}{b_{nf} - a_{nf}} & \text{if } a_{nf} < \zeta_o \leq b_{nf} \\ \dfrac{c_{nf} - \zeta_o}{c_{nf} - b_{nf}} & \text{if } b_{nf} < \zeta_o \leq c_{nf} \\ 0 & \text{if } \zeta_o > c_{nf} \end{cases}, \quad \begin{matrix} \forall nf = 1, \ldots, N_{MF}, \\ \forall o = 1, \ldots, N_\zeta, \end{matrix} \tag{16}$$

where ζ stands for the vector of the ANFIS input variables (from now on referred to as the antecedent parameter) and N_{MF} signifies the number of membership functions for each antecedent parameter.

Layer 2. *This layer formulates the rules using the preceding membership functions; in each of the $N_v = (N_{MF})^{N_\zeta} = 32$ nodes, there is a fixed node that multiplies the incoming signals and transmits the product:*

$$\mu_i(\zeta) = \prod_{o=1}^{N_v} \eta_{nf}(\zeta_o), \quad \forall i = 1, \ldots, N_v. \tag{17}$$

Layer 3. *This is the normalization layer:*

$$\bar{\mu}_i(\zeta) = \frac{\mu_i(\zeta)}{\sum_{i=1}^{N_v} \mu_i(\zeta)}, \quad \forall i = 1, \ldots, N_v. \tag{18}$$

Layer 4. *Known as the defuzzification or consequent layer, the if–then fuzzy rules of Takagi and Sugeno are used [32]:*

$$\begin{aligned} \mathcal{R}_i : IF \quad \zeta_1 \quad is \quad \eta_{m1} \quad AND, \ldots, AND \quad \zeta_{N_\zeta} \quad is \quad \eta_{mN_\zeta} \\ THEN \quad \bar{\mu}_i(\zeta_i \alpha_i + \gamma_i), \quad \forall i = 1, \ldots, N_v. \end{aligned} \tag{19}$$

Output 1. *This layer calculates the final output by aggregating all incoming signals from the defuzzification layer:*

$$\sum_{i=1}^{N_v} \bar{\mu}_i(\zeta_i \alpha_i + \gamma_i). \tag{20}$$

After training the ANFIS and calculating the normalized weights (18) and consequent parameters (19), the fuzzy TS representation is constructed. As an example, the variable $\hat{H}_{in}$ can be used, which is formulated as follows:

$$\hat{H}_{in}(k) = \sum_{i=1}^{N_v} \bar{\mu}_i(\zeta(k))\Big(\alpha_{1i} H_{in}(k) + \alpha_{2i} H_{in}(k-1) + \alpha_{3i} H_{in}(k-2) + \alpha_{4i} Q_{in}(k) + \alpha_{5i} Q_{out}(k) + \gamma_i\Big). \tag{21}$$

The terms in (21) can be rearranged as

$$\hat{H}_{in}(k) = \sum_{i=1}^{N_v} \bar{\mu}_i(\zeta(k))\left(\begin{bmatrix} \alpha_{1i}^1 & \alpha_{2i}^1 & \alpha_{3i}^1 \\ \alpha_{1i}^2 & \alpha_{2i}^2 & \alpha_{3i}^2 \\ \alpha_{1i}^3 & \alpha_{2i}^3 & \alpha_{3i}^3 \end{bmatrix} x + \begin{bmatrix} \alpha_{4i}^1 & \alpha_{5i}^1 \\ \alpha_{4i}^2 & \alpha_{5i}^2 \\ \alpha_{4i}^3 & \alpha_{5i}^3 \end{bmatrix} u + \begin{bmatrix} \gamma_i^1 \\ \gamma_i^2 \\ \gamma_i^3 \end{bmatrix}\right), \tag{22}$$

where $x = \begin{bmatrix} H_{in}(k) & H_{in}(k-1) & H_{in}(k-2) \end{bmatrix}^T$ represents the state vector and $u = \begin{bmatrix} Q_{in} & Q_{out} \end{bmatrix}^T$ represents the input vector. The superscript $1, 2, 3$ denotes the output number for ANFIS. The polytopic structure is reformulated as a discrete-time state space representation:

$$\begin{aligned} x(k+1) &= \sum_{i=1}^{N_v} \bar{\mu}_i(\zeta(k))(A_i x(k) + B_i u(k) + \gamma_i) \\ y(k) &= Cx(k), \end{aligned} \tag{23}$$

where N_v is determined by $N_v = (N_{\mathrm{MF}})^{N_\zeta}$, $\bar{\mu}_i(\zeta(k))$ represents the premise functions, $A_i \in \mathbb{R}^{n_x \times n_x}$, $B_i \in \mathbb{R}^{n_x \times n_u}$, $\gamma_i \in \mathbb{R}^{n_x}$, and $C \in \mathbb{R}^{n_y \times n_x}$ denote the system matrices, and $y(k) \in \mathbb{R}^{n_y}$ denotes the computed output vector. It is important to note that the system is subject to uncertainties due to model mismatches caused by friction and turbulent flow. The matrices Ψ_i and Ω_i represent the uncertainties, whereas F_v accounts for noise.

$$\begin{aligned} x(k+1) &= \sum_{i=1}^{N_v} \bar{\mu}_i(\zeta(k))((A_i + \Psi_i)x(k) + (B_i + \Omega_i)u(k) + \gamma_i), \\ y(k) &= Cx(k) + F_\sigma \sigma(k). \end{aligned} \tag{24}$$

The values of the uncertain matrices are obtained from the error covariance matrix of all consequent neuro-fuzzy parameters, which is produced by the hybrid learning algorithm. To build the Ψ_i and Ω_i matrices, standard deviations are derived from the variances present in the covariance matrix; F_σ denotes the noise matrices with fixed dimensions and $\sigma(k) \in \mathbb{R}^{n_y}$ represents the measurement noise based on the sensor's additive noise in the pipeline system. The uncertain parameters can be approximated in a single term according to [33]. Thus, (24) is reformulated as follows:

$$\begin{aligned} x(k+1) &= \sum_{i=1}^{N_v} \bar{\mu}_i(\zeta(k))(A_i x(k) + B_i u(k) + \gamma_i + E_i \delta(k)), \\ y(k) &= Cx(k) + F_\sigma \sigma(k) \end{aligned} \tag{25}$$

with

$$E_i\delta(k) = \Psi_i x(k) + \Omega_i u(k), \tag{26}$$

where E_i is the uncertainty distribution matrix of suitable dimensions and $\delta(k) \in \mathbb{R}^{n_x}$ is a vector that captures the impact of uncertainty. The next section outlines the procedure for designing the set-based neuro-fuzzy ZKF observer.

3.5. Design of a Neuro-Fuzzy ZKF Observer

The design of a neuro-fuzzy ZKF observer using a Takagi–Sugeno fuzzy system is obtained through ANFIS learning. The observer is based on the ZKF, which offers several advantages over the traditional Kalman filter (KF). This design is called the neuro-fuzzy ZKF observer due to its combination of neuro-fuzzy logic and robust zonotopic estimation. Unlike a KF, the proposed observer uses bounded uncertainties instead of probabilistic distributions, allowing it to handle large deterministic disturbances more effectively. Both techniques share a similar gain structure and use prediction–update cycles to refine state estimates; however, the neuro-fuzzy ZKF provides zonotopic sets that enclose all possible admissible system states, whereas a KF offers confidence ellipses based on state covariance.

The neuro-fuzzy ZKF does not require assumptions about the distribution of uncertainties, making it more flexible and robust against large deterministic disturbances. Additionally, it provides zonotopic sets that enclose all possible state values, offering explicit guarantees of state inclusion. This design leverages these advantages by combining the adaptability and flexibility of fuzzy systems with the robustness and precision of the ZKF, enhancing state estimation in complex dynamic systems.

To provide a simplified notation, the TS system (24) is reformulated as follows:

$$\begin{aligned} x(k+1) &= A_\omega x(k) + B_\omega u(k) + \gamma_\omega + E_\omega \delta(k), \\ y(k) &= Cx(k) + F_\sigma \sigma(k), \end{aligned} \tag{27}$$

where $A_\omega = \sum_{i=1}^{N_v} \bar{\mu}_i(\zeta(k))A_i$, $B_\omega = \sum_{i=1}^{N_v} \bar{\mu}_i(\zeta(k))B_i$, $\gamma_\omega = \sum_{i=1}^{N_v} \bar{\mu}_i(\zeta(k))\gamma_i$, and $E_\omega = \sum_{i=1}^{N_v} \bar{\mu}_i(\zeta(k))E_i$.

The uncertainties and noise are considered as a zonotopic representation, as follows:

$$\begin{aligned} \delta &\in \langle c_\delta, R_\delta \rangle, \\ \sigma &\in \langle c_\sigma, R_\sigma \rangle, \end{aligned} \tag{28}$$

where c_δ and c_σ denote the centers of the zonotopes that bound the uncertainty and noise, respectively, with their associated generator matrices $R_\delta \in \mathbb{R}^{n_x \times n_x}$ and $R_\sigma \in \mathbb{R}^{n_y \times n_y}$. The following assumption is considered when designing the observer.

Assumption 1. *The uncertainties and noise in* (28) *are assumed to be bounded by a unit hypercube zonotope centered at the origin. Specifically, for all* $k \geq 0$, *we have* $\delta \in [-1,1]^{n_\delta} = \langle 0, I_{n\delta} \rangle$ *and* $\sigma \in [-1,1]^{n_\sigma} = \langle 0, I_{n_\sigma} \rangle$, *where* $I_{n\delta} \in \mathbb{R}^{n_\delta \times n_\delta}$ *and* $I_{n_\sigma} \in \mathbb{R}^{n_\sigma \times n_\sigma}$ *are identity matrices.*

Considering the assumption that uncertainties and noise are bounded by the unit hypercube zonotope centered at the origin and that the initial state x_0 belongs to the set $\mathcal{X}_0^{zo} = \langle c_{k,0}^{zo}, R_{k,0}^{zo} \rangle$, where $c_{k,0}^{zo} \in \mathbb{R}^{n_x}$ represents the center and $R_{k,0}^{zo} \in \mathbb{R}^{n_x \times n_{R_{k,0}^{zo}}}$ is a non-empty matrix containing the generator matrices of the initial zonotope $\mathcal{X}_0^{zo}$, the following neuro-fuzzy ZKF observer is structured.:

$$\hat{x}(k+1) = A_\omega \hat{x}(k) + B_\omega u(k) + \gamma_\omega(k) + E_\omega \delta(k) + L_\omega(y - Cx(k) - F_\sigma \sigma(k)) \tag{29}$$

where the vector $\hat{x}(k+1) \in \mathbb{R}^{n_x}$ denotes the estimated states and $L_\omega \in \mathbb{R}^{n_x \times n_y}$ represents the observer gains that need to be determined. The following proposition is essential for calculating the observer gains.

Proposition 1. *Given the system* (27) *and observer structure* (29), *the zonotope* $\hat{\mathcal{X}}_k^{zo} = \langle c_{k+1}^{zo}, R_{k+1}^{zo} \rangle$ *is recursively forward-predicted as follows:*

$$\begin{aligned} c_{k+1}^{zo} &= (A_\omega - L_\omega C)c_k^{zo} + B_\omega u_k + \gamma_k + L_\omega y_k \\ R_{k+1}^{zo} &= [(A_\omega - L_\omega C)\bar{R}_k^{zo}, E_\omega, -L_\omega F_\sigma] \\ \bar{R}_k^{zo} &= \downarrow_q (R_k^{zo}) \end{aligned} \tag{30}$$

such that the reduction operator $\downarrow_q$ *satisfies* $\bar{R}_k = \downarrow_q (R)$ *and* $\hat{x}(k) \in \langle c_k, R_k \rangle \subset \langle c_k, \bar{R}_k \rangle$.

Proof. Assuming $\hat{\mathcal{X}}^{zo} = \langle c_{k+1}^{zo}, R_{k+1}^{zo} \rangle$ at time k (true at $k = 0$) and $\downarrow_{q,W}$ preserving inclusion, $\hat{\mathcal{X}}^{zo} = \langle c_k, \bar{R}_k \rangle$; because $\delta = \langle 0, I_{n\delta} \rangle$ and $\sigma \in \langle 0, I_{n_\sigma} \rangle$ (27), from (29) we have

$$\begin{aligned} \hat{x}(k+1) = {} & ((A_\omega - L_\omega C) \odot \langle c_k, R_k \rangle) \oplus (B_\omega \odot \langle u_k, 0 \rangle) \oplus (\langle \gamma_\omega, 0 \rangle) \\ & \oplus (E_\omega \odot \langle 0, I_{n_\delta} \rangle) \oplus (L_\omega \odot \langle y_k, 0 \rangle) \oplus (-L_\omega F_\sigma \odot \langle 0, I_{n_\sigma} \rangle). \end{aligned} \tag{31}$$

Then, applying the zonotope property 1 yields (30). Hence, the proof is completed. □

As highlighted in Proposition 1, the zonotopic state-bounding observer (30) is characterized by the zonotopic observer gain L_z at each time step k. According to [25,34], the size of the state-bounding zonotope $\hat{\mathcal{X}}^{zo} = \langle c_{k+1}^{zo}, R_{k+1}^{zo} \rangle$ can be minimized using its F-radius. The following theorem offers a method to calculate L_z for this purpose.

Theorem 1. *Consider the nonlinear TS fuzzy system* (27) *and its associated zonotopic observer* (29). *The size of the zonotope defined in* (30) *can be optimized by using the following observer gain:*

$$L_\omega = \Gamma_\omega \Phi_k^{-1} \tag{32}$$

with

$$\Gamma_\omega = A_\omega P_k C^T, \quad P_k = R_{x_k} R_{x_k}^T, \quad \Phi_k = C P_k C^T + F_\sigma F_\sigma^T.$$

Proof. According to [25], minimizing the F-radius and F_W-radius of a zonotope is equivalent to minimizing the trace of its covariance; therefore, minimizing the F-radius of the zonotope $\hat{\mathcal{X}}^{zo} = \langle c_{k+1}^{zo}, R_{k+1}^{zo} \rangle$ is equivalent to minimizing the trace of its covariance $P_{k+1} = R_{k+1}^{zo} R_{k+1}^{zo\ T}$, i.e.,

$$\mathcal{F} = \|R_{k+1}^{zo}\|_F^2 = \text{Tr}(R_{k+1}^{zo} R_{k+1}^{zo\ T}) = \text{Tr}(P_{k+1}), \tag{33}$$

where $\mathcal{F}$ denotes the Frobenius radius and P is the covariance of the zonotope matrix R_x. Similarly, minimizing the F_W-radius of the zonotope $\mathcal{X}_{k+1}$ is equivalent to minimizing the criterion, i.e.,

$$\mathcal{F}_W = \|R_{k+1}^{zo}\|_{F,W}^2 = \text{Tr}(W P_{k+1}), \tag{34}$$

where $\mathcal{F}_W$ denotes the weighted Frobenius radius and WP is the weighted function covariance of the zonotope matrix R_{x_k}. Note that W is any positive definite weighting matrix. It follows from (30) that

$$P_{k+1} = (A_\omega - L_\omega C)P_k(A_\omega - L_\omega C)^T + D_\omega + L_\omega D_\sigma L_\omega^T, \tag{35}$$

where $D_\omega = E_\omega E_\omega^T$ and $D_\sigma = F_\sigma F_\sigma^T$. From (35), the criterion $\mathcal{F}_W$ in (34) can be rewritten as

$$\mathcal{F}_W = \mathrm{Tr}(W(A_\omega - L_\omega C)P_k(A_\omega - L_\omega C)^T + WD_\omega + WL_\omega D_\sigma L_\omega^T). \tag{36}$$

Then, the optimal value of the observer gain L_ω is determined such that

$$\frac{\partial \mathcal{F}_W}{\partial L_\omega} = 0. \tag{37}$$

Considering (36), this latter yields

$$\frac{\partial \mathcal{F}_W}{\partial L_\omega} = \frac{\partial}{\partial L_\omega}\mathrm{Tr}(WL_\omega(CP_kC^T + D_\sigma)L_\omega^T) - 2\frac{\partial}{\partial L_\omega}\mathrm{Tr}(WA_\omega P_k C^T L_\omega^T) = 0. \tag{38}$$

Developing Equation (38), we obtain

$$\frac{\partial}{\partial L_\omega}\mathrm{Tr}(WL_\omega(CP_kC^T + D_\sigma)L_\omega^T) = 2WL_\omega(CP_kC^T + D_\sigma), \tag{39}$$

$$\frac{\partial}{\partial L_\omega}\mathrm{Tr}(WA_\omega P_k C^T L_\omega^T) = WA_\omega P_k C^T. \tag{40}$$

Then, Equation (38) becomes

$$2WL_\omega(CP_kC^T + D_\sigma) - 2WA_\omega P_k C^T = 0. \tag{41}$$

Simplifying, we obtain

$$WL_\omega(CP_kC^T + D_\sigma) = WA_\omega P_k C^T. \tag{42}$$

Because W is a positive definite weighting matrix, we can multiply by W^{-1} and transpose

$$L_\omega(CP_kC^T + D_\sigma) = A_\omega P_k C^T. \tag{43}$$

Finally, we arrive at the desired equation:

$$L_\omega \Phi_k = A_\omega P_k C^T \tag{44}$$

which in turn leads to the observer gain expression in (32). □

3.6. Neuro-Fuzzy ZKF Fault Detection Scheme

The fault detection and isolation process using the neuro-fuzzy ZKF involves estimating the pipeline variables with ANFIS and propagating disturbances and noise through a zonotopic observer. This process identifies faults by checking the intersection between the estimated zonotope and the measured strip at each time instant. If the intersection is empty, a fault is indicated. The results are stored in a Fault Signature Matrix (FSM), which aids in diagnosing the faults. The detailed steps of this fault detection scheme are outlined in Algorithm 1.

Algorithm 1 Zonotopic Fault Detection Scheme

Input: Pipeline variables estimated by ANFIS, as in Table 3
Output: Fault Signature Matrix (FSM)
Initialize:
  Number of estimated variables of system s
  Zonotopic observer parameters and disturbance bounds
for each time instant k **do**
  Estimate the pipeline variables using ANFIS
  Propagate uncertainty and noise through the zonotopic observer
  Calculate the strip $\mathcal{X}_k^{y_k}$ for each measured variable:

$$\mathcal{X}_k^{y_k} = \{x(k) \in \mathbb{R}^{n_x} : |Cx(k) - y_s(k)| \leq F_\sigma\} \tag{45}$$

  Check for the existence of a fault by verifying:

$$\hat{\mathcal{X}}_k^{zo} \cap \mathcal{X}_k^{y_k} = \emptyset \tag{46}$$

  if $\hat{\mathcal{X}}_k^{zo} \cap \mathcal{X}_k^{y_k} = \emptyset$ **then**
    A fault is indicated
    Generate residuals $r_s(k)$
    Update Fault Signature Matrix (FSM) as follows:

$$\psi_{s,j}(k) = \begin{cases} 0 & \text{if } r_s(k) \text{ is consistent (No fault)} \\ 1 & \text{if } r_s(k) \text{ is not consistent (Fault)} \end{cases} \tag{47}$$

  else
    No fault is indicated
  end if
end for
Return: Fault Signature Matrix (FSM)

4. Results

This section presents the results of the proposed method for faults and leak detection in hydraulic pipeline systems using neuro-fuzzy ZKF. The method was evaluated in the experimental pipeline specifically configured for this purpose.

4.1. Experimental Data Setup

The experimental conditions for the hydraulic pipeline system are summarized in Table 4. Each operating condition was tested multiple times to ensure data consistency and reliability.

Table 4. Summary of experimental conditions.

Parameter	Details
Pump power frequency	60 Hz, 50 Hz, 40 Hz, 30 Hz
Number of repetitions per condition	5
Duration of each experiment	600 s
Sampling frequency	5 samples per second
Total samples per experiment	3000
Fault condition	free

All experiments were conducted under fault-free conditions to establish a baseline for model training and validation. The collected dataset was partitioned into two subsets to facilitate the development of the Adaptive Neuro-Fuzzy Inference System (ANFIS) models. Specifically:

- Training Set (80% of the dataset): Used to optimize the model's parameters during the training process.
- Validation Set (20% of the dataset): Employed to evaluate model performance and fine-tune the neuro-fuzzy parameters.

4.2. Identification Results Using ANFIS and ZKF Observers

The system variables $H_{in}, H_{out}, Q_{in}, Q_{out}$ were identified using ANFIS, the structure of which is described in Section 3.4.

Subsequently, the neuro-fuzzy ZKF observers were implemented as described in Section 3.5. The results of the observers are graphically presented below, corresponding to an operating point of 60 Hz for the pump. Figure 4 shows the inlet pressure H_{in} under fault-free conditions, where the zonotopic observer (red and green lines) envelops the measured signal shown by the blue line. Similarly, Figure 5 shows the outlet pressure H_{out} under fault-free conditions, Figure 6 shows the inlet flow Q_{in}, and Figure 7 shows the outlet flow Q_{out}, all under fault-free conditions, with the zonotopic observer enveloping the measured signals.

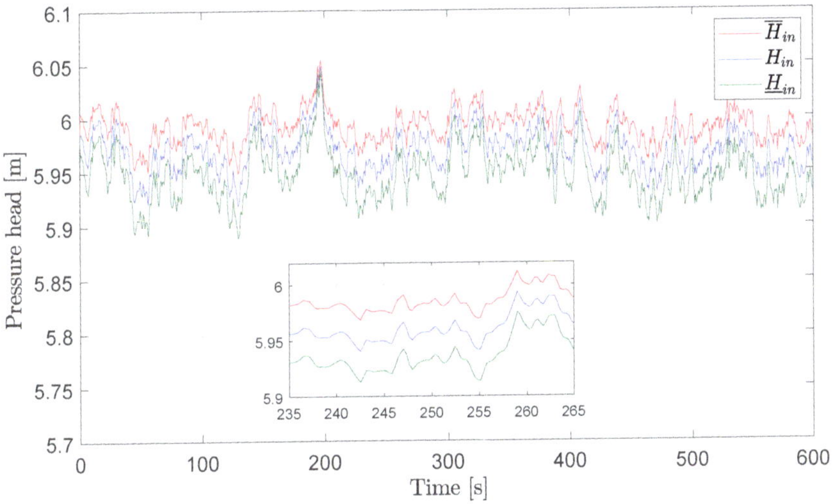

Figure 4. Inlet pressure H_{in} under fault-free conditions.

4.3. Evaluation Under Fault Conditions

To assess the effectiveness of the proposed method, various tests were conducted by intentionally inducing faults in the sensors and creating a leak in the pipeline. The specific faults we induced were as follows:

- Fault 1: A step-type bias of 10% with respect to the nominal value in the inlet pressure sensor (H_{in}).
- Fault 2: A step-type bias of 10% with respect to the nominal value in the outlet pressure sensor (H_{out}).
- Fault 3: A step-type bias of 10% with respect to the nominal value in the inlet flow sensor (Q_{in}).
- Fault 4: A step-type bias of 10% with respect to the nominal value in the outlet flow sensor (Q_{out}).
- Fault 5: A leak induced by actuating one of the leak valves in the pipeline.

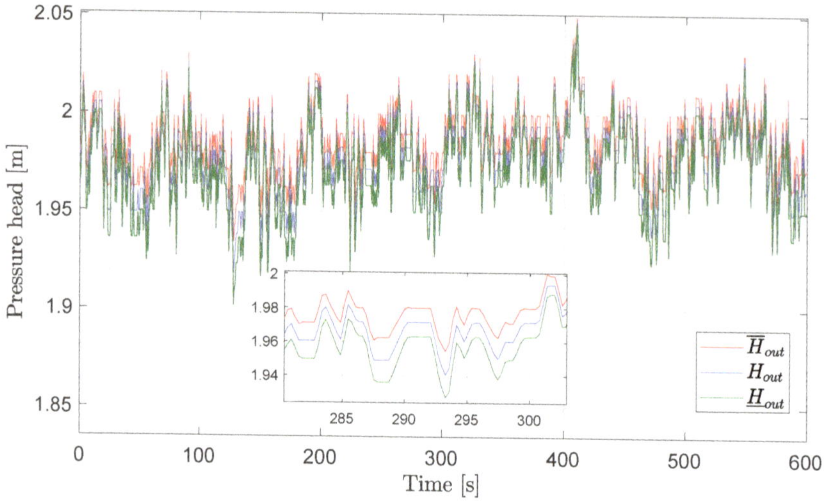

Figure 5. Outlet pressure H_{out} under fault-free conditions.

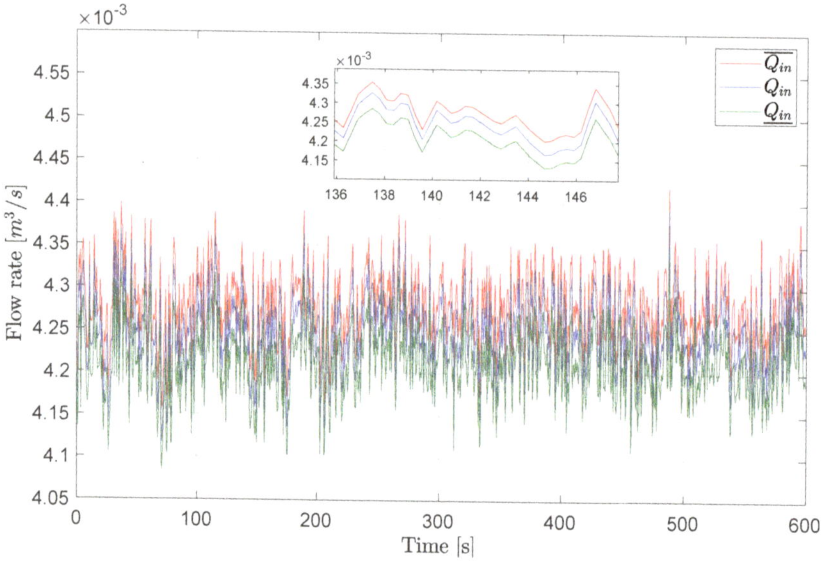

Figure 6. Inlet flow Q_{in} under fault-free conditions.

The faults can be visualized graphically in the following figures under pump conditions at 60 Hz. Due to space limitations in the article, only two graphs are shown for the types of sensor faults. In Figure 8, a fault in the inlet pressure sensor (H_{in}) is activated at time $t = 150$ s, exceeding the observer's upper threshold. In Figure 9, a fault in the inlet flow sensor (Q_{in}) is shown at time $t = 230$ s, which also exceeds the upper threshold. Finally, Figure 10 illustrates an induced leak in the pipeline at time $t = 310$ s. The upper subplot shows the increased Q_{in} due to the leak flow, while the lower subplot shows the decreased Q_{out} due to the leak. In the figures, red and green lines indicate the lower and upper thresholds, respectively.

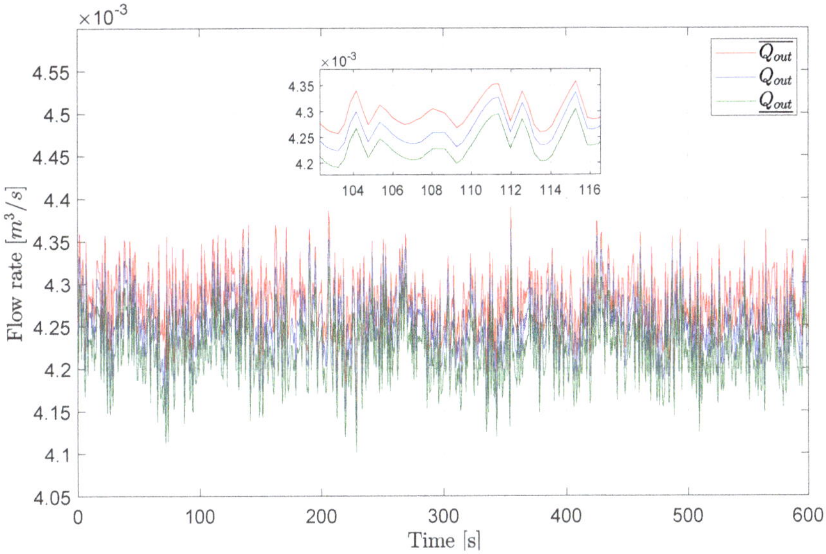

Figure 7. Outlet flow Q_{out} under fault-free conditions.

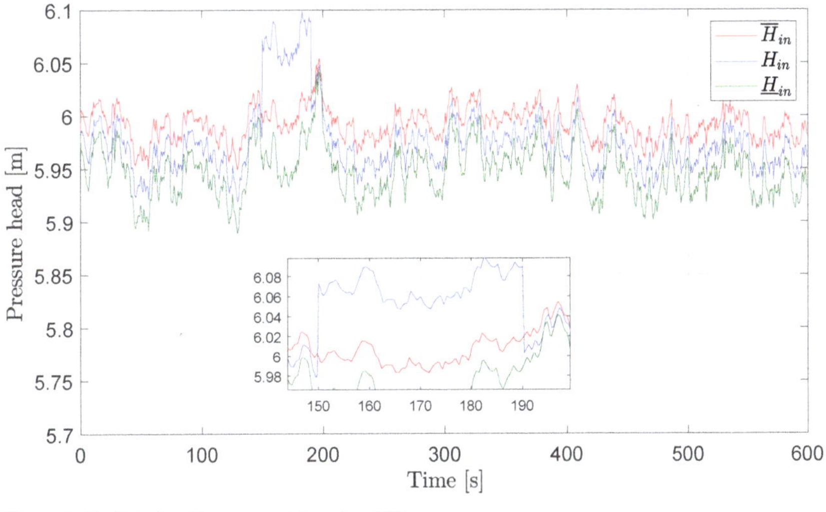

Figure 8. Fault in the H_{in} sensor at time $t = 150$ s.

Using Algorithm 1, the fault detection results are summarized in Table 5, which presents the Fault Signature Matrix (FSM). The FSM summarizes the detection results for each fault scenario. Each row corresponds to a residual r_s generated by the algorithm, while each column corresponds to a specific fault. A value of 1 indicates that the fault was detected in the respective scenario, whereas an empty cell indicates no detection. The matrix clearly shows which residuals are associated with each fault. When a leak occurs, all residuals are activated. Each fault creates a unique signature, allowing for accurate fault diagnosis.

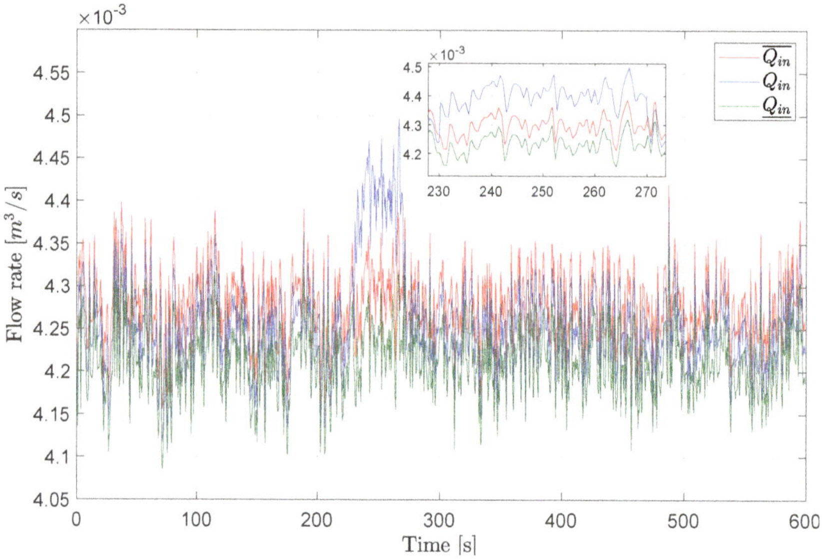

Figure 9. Fault in the Q_{in} sensor at time $t = 230$ s.

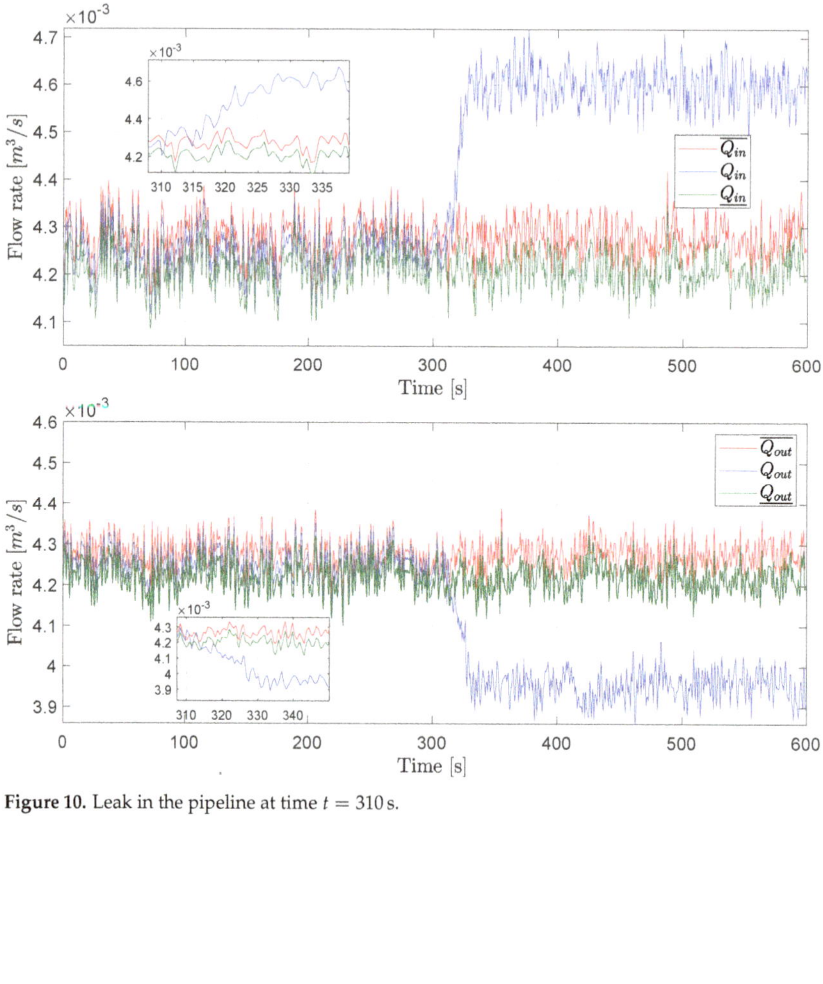

Figure 10. Leak in the pipeline at time $t = 310$ s.

Table 5. Fault signature matrix (FSM).

Residual	Fault 1	Fault 2	Fault 3	Fault 4	Fault 5
r_1	1				1
r_2		1			1
r_3			1		1
r_4				1	1

To evaluate the effectiveness and robustness of the proposed method across all fault cases and at four different pump operating points, the following metrics were used [35].

- Precision: The percentage of true positives over the total number of positive predictions made by the model, calculated as follows:

$$\text{Precision} = \frac{\text{TP}}{\text{TP} + \text{FP}}. \tag{48}$$

- Recall (Detection Rate): The percentage of true positives over the total number of actual positives (the model's ability to correctly identify faults), calculated as follows:

$$\text{Recall} = \frac{\text{TP}}{\text{TP} + \text{FN}}. \tag{49}$$

- False Positive Rate: The percentage of incorrect positive predictions (false positives) over the total number of actual negatives, calculated as follows:

$$\text{False Positive Rate} = \frac{\text{FP}}{\text{FP} + \text{TN}}. \tag{50}$$

These metrics were evaluated across all fault cases and at four different pump operating points, with TP representing the true positives, FP representing the false positives, FN representing the false negatives, and TN representing the true negatives.

The performance metrics obtained at 60 Hz indicate high precision and recall (detection rate) and low false positive rates across five fault types (see Table 6). For Fault 1, the system demonstrated a precision of 98.44% and a recall of 98.72%, with a false positive rate of 1.56%; this reflects a high identification capability, although there is room to reduce false alerts. In the case of Fault 2, the results were outstanding, with a precision of 99.24% and a detection rate of 99.12% along with a very low false positive rate of 0.76%. This indicates extreme accuracy and reliability. Fault 3 showed a precision of 97.51% and a detection rate of 97.04%, but with a false positive rate of 2.48%. Despite the high precision and recall, the higher false positive rate indicates a greater number of incorrect alerts. For Fault 4, the results showed a precision of 98.82% and a detection rate of 98.68%, with a false positive rate of 1.18%. This reflects excellent performance, with a good balance between precision and low false alarms. Fault 5 demonstrated outstanding performance, with a precision of 99.18%, a detection rate of 99.20%, and a false positive rate of only 0.82%.

Table 6. Results obtained with the pump at 60 Hz using the proposed fault diagnosis method.

Fault	Precision	Recall (Detection Rate)	False Positive Rate
1	98.44%	98.72%	1.56%
2	99.24%	99.12%	0.76%
3	97.51%	97.04%	2.48%
4	98.82%	98.68%	1.18%
5	99.18%	99.20%	0.82%

At 50 Hz, the outcomes were highly favorable, showcasing high precision and detection rates across the five faults, as shown in Table 7. For Fault 1, a precision of 97.60% and a detection rate of 97.40% were achieved, with a false positive rate of 2.40%. Despite the high effectiveness, there is potential to minimize incorrect alerts. Fault 2 recorded a precision of 98.06% and a detection rate of 97.98%, with a false positive rate of 1.94%, indicating robust system reliability. For Fault 3, the precision was 97.70%, and the detection rate was 97.58%, with a false positive rate of 2.30%. Although these figures are impressive, the false positive rate could still be improved. Fault 4 showed a precision of 98.48% and a detection rate of 98.78%, with a false positive rate of 1.52%, reflecting a good balance between precision and low false alarms. Fault 5 demonstrated a precision of 98.86% and a detection rate of 99.00%, with a false positive rate of only 1.14%.

Table 7. Results obtained with the pump at 50 Hz using the proposed fault diagnosis method.

Fault	Precision	Recall (Detection Rate)	False Positive Rate
1	97.60%	97.40%	2.40%
2	98.06%	97.98%	1.94%
3	97.70%	97.58%	2.30%
4	98.48%	98.78%	1.52%
5	98.86%	99.00%	1.14%

At 40 Hz, the results exhibited exceptional performance, displaying high precision and detection rates, as shown in Table 8. For Fault 1, a precision of 98.29% and a detection rate of 98.70% were attained, with a false positive rate of 1.72%. This signifies a high level of effectiveness in identifying faults. For Fault 2, the system achieved an impressive precision of 99.02% and a detection rate of 99.16%, with a low false positive rate of 0.98%. This confirms the system's high reliability. Fault 3 achieved a precision of 97.39% and a detection rate of 97.70%, with a false positive rate of 2.62%. Despite the high values, there is potential to reduce the false positive rate. Fault 4 exhibited a precision of 98.48% and a detection rate of 98.46%, with a false positive rate of 1.52%, demonstrating a good balance between precision and minimizing false alarms. Fault 5 showed excellent performance, with a precision of 98.92%, a detection rate of 99.02%, and a false positive rate of just 1.08%.

Table 8. Results obtained with the pump at 40 Hz using the proposed fault diagnosis method.

Fault	Precision	Recall (Detection Rate)	False Positive Rate
1	98.29%	98.70%	1.72%
2	99.02%	99.16%	0.98%
3	97.39%	97.70%	2.62%
4	98.48%	98.46%	1.52%
5	98.92%	99.02%	1.08%

At 30 Hz, the results consistently showed high precision and detection rates for the five faults (see Table 9). For Fault 1, a precision of 97.61% and a detection rate of 97.96% were achieved, with a false positive rate of 2.40%. This indicates high effectiveness, although there is room to reduce false alarms. In Fault 2, the results showed a precision of 98.24% and a detection rate of 98.24%, with a false positive rate of 1.76%. These results suggest high system reliability. Fault 3 presented a precision of 97.79% and a detection rate of 97.48%, with a false positive rate of 2.20%. Although the figures are high, the false positive rate could be improved. For Fault 4, the results showed a precision of 98.82% and a detection rate of 98.68%, with a false positive rate of 1.18%. This reflects an excellent balance between

precision and low false alarms. Fault 5 showed a precision of 98.68% and a detection rate of 98.64%, with a false positive rate of only 1.32%.

Table 9. Results obtained with the pump at 30 Hz using the proposed fault diagnosis method.

Fault	Precision	Recall (Detection Rate)	False Positive Rate
1	97.61%	97.96%	2.40%
2	98.24%	98.24%	1.76%
3	97.79%	97.48%	2.20%
4	98.82%	98.68%	1.18%
5	98.68%	98.64%	1.32%

To provide a comprehensive overview of the fault diagnosis system's performance at different frequencies, the results obtained at four operational frequencies (60 Hz, 50 Hz, 40 Hz, and 30 Hz) were averaged. The averaged results are presented in Table 10:

Table 10. Averaged results obtained at different operational frequencies.

Fault	Averaged Precision	Averaged Recall	Averaged False Positive Rate
1	98.00%	98.20%	2.02%
2	98.89%	98.63%	1.36%
3	97.60%	97.45%	2.40%
4	98.65%	98.65%	1.35%
5	98.91%	98.97%	1.09%

The averaged results reflect very consistent and robust performance of the fault diagnosis system under various operational conditions.

For Fault 1, an averaged precision of 98.00% and a detection rate of 98.20% were observed, with a false positive rate of 2.02%. Although the values are high, there is room to improve the false positive rate. Fault 2 showed an averaged precision of 98.89% and a detection rate of 98.63%, with a low false positive rate of 1.36%, suggesting high system reliability for this fault.

In the case of Fault 3, the averaged precision achieved is 97.60%, with a detection rate of 97.45% and a false positive rate of 2.40%. Although the figures are high, the false positive rate is the highest, indicating areas for improvement. For Fault 4, both the averaged precision and detection rate are 98.65%, and the false positive rate is 1.35%, indicating an excellent balance and system performance.

Finally, Fault 5 shows an average precision of 98.91% and a detection rate of 98.97%, with the lowest false positive rate of 1.09%, underscoring the high effectiveness and reliability of the diagnosis method. The fault diagnosis system has demonstrated high effectiveness and reliability across different operational frequencies, with consistently high precision and detection rates and low false positive rates. These results validate the robustness of the proposed method and its applicability under various operating conditions.

Table 11 compares various leak diagnosis methods and their respective accuracies. In [36], the authors utilized support vector machines (SVM) and achieved an accuracy of 90%. The approach by [37] combined SVM, decision trees, and Naive-Bayes, achieving 98.25%. In [38], the authors employed a deep neural network classifier, yielding an accuracy of 97.89%. In [39] deep learning with convolutional autoencoder networks was applied, achieving 96.67%. The proposed method integrating ANFIS and ZKF outperformed these methods with an accuracy of 98.41%. These results demonstrate that hybrid approaches offer superior performance compared to traditional machine learning techniques such as SVM.

Table 11. Comparative analysis of leak diagnosis methods.

Reference	Method Used	Accuracy of Leak Diagnosis (%)
Sun et al., 2016 [36]	Support Vector Machines	90.00
El-Zahab et al., 2018 [37]	SVM, Decision Trees, and Naive-Bayes	98.25
Zadkarami et al., 2020 [38]	Deep Neural Network Classifier	97.89
Ahmad et al., 2022 [39]	Deep Learning Based on Convolutional Autoencoder Networks	96.67
Proposed Method	ANFIS and ZKF	98.41

5. Conclusions

Results obtained from the proposed data-driven method for fault and leak detection in hydraulic pipeline systems using neuro-fuzzy ZKF demonstrate high effectiveness and robustness. The proposed method combines the strengths of data-driven techniques with neuro-fuzzy observers, providing a robust solution for detecting anomalies in complex hydraulic systems.

Unlike other machine learning techniques, this method offers the notable advantage of relying exclusively on fault-free data. This makes the data collection process easier by eliminating the requirement for prior knowledge of faults. Additionally, it enhances the system's adaptability to various operating conditions. The proposed approach began with identifying the nonlinear pipeline system using an adaptive ANFIS. This system generates Takagi–Sugeno fuzzy models based on measured pressure and flow data, effectively capturing the nonlinear dynamics of the pipeline. These models are then integrated into a neuro-fuzzy ZKF framework, allowing for adaptive thresholding and enhanced fault detection capabilities. The proposed method takes into account parametric uncertainty due to hydraulic friction and fluid turbulence as well as measurement noise, ensuring robustness in real-world applications. Fault isolation is achieved by comparing zonotopic sets and evaluating a matrix of fault signatures, enabling precise identification of various fault types.

The experimental validation of the proposed method demonstrated high precision (up to 99.24%) and recall (up to 99.20%) across various fault scenarios and operational points, indicating its capability to accurately identify sensor faults and pipeline leaks. Additionally, with false positive rates as low as 0.76%, the method minimized incorrect alerts, enhancing its reliability for practical applications.

Moreover, the method's successful application across different operational frequencies (60 Hz, 50 Hz, 40 Hz, and 30 Hz) demonstrated its scalability and applicability under various operating conditions. This versatility is crucial for practical deployment, ensuring that the method can be adapted to different scenarios without significant modifications. Our method's ability to scale across different operational conditions further enhances its practical utility, making it a valuable contribution to the field of fault detection and diagnosis in hydraulic systems.

Although the proposed method shows significant promise, several avenues for future research have been identified. Extending the method to detect and isolate other types of faults in hydraulic systems, such as valve malfunctions or pump failures, would broaden its applicability. Further research could also focus on optimizing the neuro-fuzzy models to improve the accuracy and speed of fault detection and isolation. Exploring advanced data fusion techniques might enhance the method's ability to handle complex and noisy datasets, potentially improving detection performance.

By pursuing these future directions, the capabilities and applicability of the proposed fault detection method could be further enhanced, ensuring its relevance and effectiveness in a wide range of practical scenarios.

Author Contributions: Conceptualization, E.-J.P.-P. and I.S.-R.; methodology, E.-J.P.-P.; software, Y.G.-B.; validation, E.-J.P.-P., I.S.-R. and Y.G.-B.; formal analysis, E.-J.P.-P.; investigation, J.-A.F.-M.; data curation, J.-A.G.-R.; writing—original draft preparation, E.-J.P.-P.; writing—review and editing, I.S.-R.; visualization, Y.G.-B. and J.-A.F.-M.; funding acquisition, I.S.-R. All authors have read and agreed to the published version of the manuscript.

Funding: This research was funded by Tecnológico Nacional de México (TecNM) grant number 20212.24-P and Consejo Nacional de Humanidades, Ciencias y Tecnologías (CONAHCYT) in Mexico.

Data Availability Statement: The raw data supporting the conclusions of this article will be made available by the authors on request.

Acknowledgments: The authors thank the RICCA (Red Internacional de Control y Cómputo Aplicados) research network for their invaluable scientific support.

Conflicts of Interest: The authors declare no conflict of interest.

References

1. Qian, J.; Song, Z.; Yao, Y.; Zhu, Z.; Zhang, X. A review on autoencoder based representation learning for fault detection and diagnosis in industrial processes. *Chemom. Intell. Lab. Syst.* **2022**, *231*, 104711. [CrossRef]
2. Cheded, L.; Doraiswami, R. A novel integrated framework for fault diagnosis with application to process safety. *Process Saf. Environ. Prot.* **2021**, *154*, 168–188. [CrossRef]
3. Datta, S.; Sarkar, S. A review on different pipeline fault detection methods. *J. Loss Prev. Process Ind.* **2016**, *41*, 97–106. [CrossRef]
4. Jul-Jørgensen, I.; Facco, P.; Gernaey, K.; Barolo, M.; Hundahl, C. Data fusion of Raman spectra in MSPC for fault detection and diagnosis in pharmaceutical manufacturing. *Comput. Chem. Eng.* **2024**, *184*, 108647. [CrossRef]
5. Eskandari, A.; Milimonfared, J.; Aghaei, M. Fault detection and classification for photovoltaic systems based on hierarchical classification and machine learning technique. *IEEE Trans. Ind. Electron.* **2020**, *68*, 12750–12759. [CrossRef]
6. Fawwaz, D.Z.; Chung, S.H. Real-time and robust hydraulic system fault detection via edge computing. *Appl. Sci.* **2020**, *10*, 5933. [CrossRef]
7. Askari, B.; Carli, R.; Cavone, G.; Dotoli, M. Data-driven fault diagnosis in a complex hydraulic system based on early classification. *IFAC-PapersOnLine* **2022**, *55*, 187–192. [CrossRef]
8. Xue, P.; Jiang, Y.; Zhou, Z.; Chen, X.; Fang, X.; Liu, J. Machine learning-based leakage fault detection for district heating networks. *Energy Build.* **2020**, *223*, 110161. [CrossRef]
9. Kim, D.; Heo, T.Y. Anomaly detection with feature extraction based on machine learning using hydraulic system IoT sensor data. *Sensors* **2022**, *22*, 2479. [CrossRef]
10. Kościelny, J.M.; Bartyś, M. A new method of diagnostic row reasoning based on trivalent residuals. *Expert Syst. Appl.* **2023**, *214*, 119116. [CrossRef]
11. Meseguer, J.; Puig, V.; Escobet, T. Fault diagnosis using a timed discrete-event approach based on interval observers: Application to sewer networks. *IEEE Trans. Syst. Man Cybern. A Syst. Hum.* **2010**, *40*, 900–916. [CrossRef]
12. Tahraoui, S.; Meghebbar, A.; Boubekeur, D.; Boumediene, A. Fault Detection in a Five Tank Hydraulic System. *Electroteh. Electron. Autom.* **2015**, *63*, 51.
13. Youssef, T.; Chadli, M.; Karimi, H.R.; Wang, R. Actuator and sensor faults estimation based on proportional integral observer for TS fuzzy model. *J. Frankl. Inst.* **2017**, *354*, 2524–2542. [CrossRef]
14. Jafari, R.; Razvarz, S.; Vargas-Jarillo, C.; Gegov, A. Blockage detection in pipeline based on the extended Kalman filter observer. *Electronics* **2020**, *9*, 91. [CrossRef]
15. Tao, H.; Jia, P.; Wang, X.; Wang, L. Real-Time Fault Diagnosis for Hydraulic System Based on Multi-Sensor Convolutional Neural Network. *Sensors* **2024**, *24*, 353. [CrossRef]
16. Rousseau, P.; Laubscher, R. A Condition-Monitoring Methodology Using Deep Learning-Based Surrogate Models and Parameter Identification Applied to Heat Pumps. *Math. Comput. Appl.* **2024**, *29*, 52. [CrossRef]
17. Qiu, S.; Cui, X.; Ping, Z.; Shan, N.; Li, Z.; Bao, X.; Xu, X. Deep learning techniques in intelligent fault diagnosis and prognosis for industrial systems: A review. *Sensors* **2023**, *23*, 1305. [CrossRef] [PubMed]

18. Frausto-Solís, J.; Galicia-González, J.C.d.J.; González-Barbosa, J.J.; Castilla-Valdez, G.; Sánchez-Hernández, J.P. SSA-Deep Learning Forecasting Methodology with SMA and KF Filters and Residual Analysis. *Math. Comput. Appl.* **2024**, *29*, 19. [CrossRef]
19. Robles-Velasco, A.; Cortés, P.; Muñuzuri, J.; De Baets, B. Prediction of pipe failures in water supply networks for longer time periods through multi-label classification. *Expert Syst. Appl.* **2023**, *213*, 119050. [CrossRef]
20. Askari, B.; Bozza, A.; Cavone, G.; Carli, R.; Dotoli, M. An adaptive constrained clustering approach for real-time fault detection of industrial systems. *Eur. J. Control* **2023**, *74*, 100858. [CrossRef]
21. Barakat, N.; Hajeir, L.; Alattal, S.; Hussein, Z.; Awad, M. Air leaks fault detection in maintenance using machine learning. *J. Qual. Maint. Eng.* **2024**, *30*, 391–408. [CrossRef]
22. Le, V.T.H.; Stoica, C.; Alamo, T.; Camacho, E.F.; Dumur, D. *Zonotopes: From Guaranteed State-Estimation to Control*; John Wiley & Sons: Hoboken, NJ, USA, 2013.
23. Combastel, C. A state bounding observer based on zonotopes. In Proceedings of the 2003 European Control Conference (ECC), Cambridge, UK, 1–4 September 2003; IEEE: Cambridge, UK, 2003; pp. 2589–2594.
24. Combastel, C. A state bounding observer for uncertain non-linear continuous-time systems based on zonotopes. In Proceedings of the 44th IEEE Conference on Decision and Control, Seville, Spain, 12–15 December 2005; IEEE: Seville, Spain, 2005; pp. 7228–7234.
25. Combastel, C. Zonotopes and Kalman observers: Gain optimality under distinct uncertainty paradigms and robust convergence. *Automatica* **2015**, *55*, 265–273. [CrossRef]
26. Haznedar, B.; Kalinli, A. Training ANFIS structure using simulated annealing algorithm for dynamic systems identification. *Neurocomputing* **2018**, *302*, 66–74. [CrossRef]
27. Zhang, W.; Wang, Z.; Guo, S.; Shen, Y. Interval estimation of sensor fault based on zonotopic Kalman filter. *Int. J. Control* **2021**, *94*, 1641–1650. [CrossRef]
28. Chaudhry, M.H. *Applied Hydraulic Transients*; Springer: Berlin/Heidelberg, Germany, 2014; Volume 415.
29. White, F.M.; Ng, C.; Saimek, S. *Fluid Mechanics*; McGraw-Hill, Cop.: New York, NY, USA, 2011.
30. Zeghadnia, L.; Robert, J.L.; Achour, B. Explicit solutions for turbulent flow friction factor: A review, assessment and approaches classification. *Ain Shams Eng. J.* **2019**, *10*, 243–252. [CrossRef]
31. Santos-Ruiz, I.; Bermúdez, J.R.; López-Estrada, F.R.; Puig, V.; Torres, L.; Delgado-Aguiñaga, J. Online leak diagnosis in pipelines using an EKF-based and steady-state mixed approach. *Control Eng. Pract.* **2018**, *81*, 55–64. [CrossRef]
32. Takagi, T.; Sugeno, M. Fuzzy identification of systems and its applications to modeling and control. *IEEE Trans. Syst. Man Cybern.* **1985**, *SMC-15*, 116–132. [CrossRef]
33. Chen, J.; Patton, R.J. *Robust Model-Based Fault Diagnosis for Dynamic Systems*; Springer Science & Business Media: Berlin/Heidelberg, Germany, 2012; Volume 3.
34. Alamo, T.; Bravo, J.M.; Camacho, E.F. Guaranteed state estimation by zonotopes. *Automatica* **2005**, *41*, 1035–1043. [CrossRef]
35. Savva, M.; Ioannou, I.; Vassiliou, V. Fuzzy-logic based IDS for detecting jamming attacks in wireless mesh IoT networks. In Proceedings of the 2022 20th Mediterranean Communication and Computer Networking Conference (MedComNet), Pafos, Cyprus, 1–3 June 2022; IEEE: Pafos, Cyprus, 2022; pp. 54–63.
36. Sun, J.; Xiao, Q.; Wen, J.; Zhang, Y. Natural gas pipeline leak aperture identification and location based on local mean decomposition analysis. *Measurement* **2016**, *79*, 147–157. [CrossRef]
37. El-Zahab, S.; Abdelkader, E.M.; Zayed, T. An accelerometer-based leak detection system. *Mech. Syst. Signal Process.* **2018**, *108*, 276–291. [CrossRef]
38. Zadkarami, M.; Safavi, A.A.; Taheri, M.; Salimi, F.F. Data driven leakage diagnosis for oil pipelines: An integrated approach of factor analysis and deep neural network classifier. *Trans. Inst. Meas. Control* **2020**, *42*, 2708–2718. [CrossRef]
39. Ahmad, S.; Ahmad, Z.; Kim, C.H.; Kim, J.M. A method for pipeline leak detection based on acoustic imaging and deep learning. *Sensors* **2022**, *22*, 1562. [CrossRef] [PubMed]

Mathematical and Computational Applications

MDPI

Article

Cyber–Physical System Attack Detection and Isolation: A Takagi–Sugeno Approach

Angel R. Guadarrama-Estrada, Gloria L. Osorio-Gordillo *, Rodolfo A. Vargas-Méndez, Juan Reyes-Reyes and Carlos M. Astorga-Zaragoza

Tecnológico Nacional de México/CENIDET, Interior Internado Palmira S/N, Col. Palmira, Cuernavaca 62490, Morelos, Mexico; d19ce018@cenidet.tecnm.mx (A.R.G.-E.); rodolfo.vm@cenidet.tecnm.mx (R.A.V.-M.); juan.rr@cenidet.tecnm.mx (J.R.-R.); carlos.az@cenidet.tecnm.mx (C.M.A.-Z.)

* Correspondence: gloria.og@cenidet.tecnm.mx

Abstract: This paper presents an approach for designing a generalized dynamic observer (GDO) aimed at detecting and isolating attack patterns that compromise the functionality of cyber–physical systems. The considered attack patterns include denial-of-service (DoS), false data injection (FDI), and random data injection (RDI) attacks. To model an attacker's behavior and enhance the effectiveness of the attack patterns, Markovian logic is employed. The design of the generalized dynamic observer is grounded in the mathematical model of a system, incorporating its dynamics and potential attack scenarios. An attack-to-residual transfer function is utilized to establish the relationship between attack signals and the residuals generated by the observer, enabling effective detection and isolation of various attack schemes. A three-tank interconnected system, modeled under the discrete Takagi–Sugeno representation, is used as a case study to validate the proposed approach.

Keywords: generalized dynamic observer; denial-of-service attack (DoS); false data injection attack (FDI); random data injection attack (RDI); cyber–physical system; Takagi–Sugeno system (T-S); markovian logic

1. Introduction

Cyber–physical systems (CPSs) have applications in various engineering fields, including smart grids, water distribution systems, intelligent transportation, and industrial plants. The connectivity and integration of these systems offer numerous benefits by enabling remote monitoring and control of sensor measurements and control signals. However, the use of different communication architectures can also increase the vulnerability of the systems to cyber-attacks, which can cause human losses, economic and environmental damage, and disruptions in essential activities [1–3].

CPSs are composed of two primary layers: a physical layer and a cyber layer. The physical layer comprises sensors and actuators responsible for gathering information about a physical system's behavior. In contrast, the cyber layer consists of a control network that determines the system's behavior by making decisions and a communication infrastructure that transmits the acquired data and control signals to the actuators [4–6].

A cyber-attack is defined as the intervention of an external agent that adversely affects the behavior of the physical system. The literature identifies various types of attacks, including denial-of-service (DoS) attacks, replay attacks, zero-dynamics attacks, and false data injection (FDI) [7–10]. Addressing cyber-attacks in CPSs from a control perspective offers the advantage of modeling attack signals as disturbances to system operation. This approach facilitates the understanding of their behavior and enables strategies for predicting and mitigating their effects [11–13].

The Takagi–Sugeno (T-S) representation is an approach that is widely used in the field of automatic control to represent nonlinear systems using a set of fuzzy rules and local linear models associated with each rule. This allows for the design of robust and adaptive control strategies that can address uncertainties, disturbances, failures, or attacks directed at CPSs [14–18]. Several recent studies have addressed significant challenges in the field of CPSs. For instance, the authors of [9] investigate the H_∞ performance of discrete-time networked systems affected by network-induced delays (NIDs) and malicious packet dropouts (MPDs). This study highlights the critical impact of communication impairments on system stability and performance. Similarly, the authors of [6] propose an innovative approach to synchronizing master–slave neural networks using event-triggered mechanisms. These mechanisms effectively reduce data transmission over communication channels subject to stochastic attacks modeled by Markov processes. The proposed method leverages static output feedback and conditions formulated through linear matrix inequalities (LMIs), ensuring synchronization while minimizing resource utilization. These contributions underscore the importance of advanced modeling and control techniques to mitigate the vulnerabilities of CPSs in various disruptive scenarios.

Generalized dynamic observers (GDOs) are highly effective tools for detecting and isolating attack schemes in cyber–physical systems [19,20]. These structures offer significant advantages due to their additional degrees of freedom compared to other observers, such as proportional observers (POs) and proportional–integral observers (PIOs). This flexibility enables GDOs to achieve higher accuracy, especially under steady-state conditions. Unlike simpler observer designs, GDOs are particularly suited for systems with high nonlinearity or dynamic and complex attacks that require greater adaptability and robustness. This approach ensures that the observer can adapt to the specific characteristics of the system, enhancing its precision and robustness in attack detection and isolation.

One of the key strengths of GDOs lies in their adaptability. The flexibility offered by the design based on LMIs allows the incorporation of various performance criteria, such as robustness against noise, resilience to slowly varying attack signals, and the ability to handle complex system dynamics, including nonlinearities represented by T-S models.

In this work, a system of three interconnected tanks will be considered as a cyber–physical system due to its integration of physical and logical components that interact with each other. It should be noted that the three-tank system is susceptible to various attack schemes. Such systems are fundamental tools in chemical process engineering.

The main contribution of this paper is the design of a GDO to detect and isolate different attack schemes targeting the input of a CPS under the T-S representation. To ensure observer stability, various techniques (Lyapunov, elimination lemma, and Schur complement) will be used to solve LMIs. Additionally, a residual attack transfer function is used to isolate different attack scenarios.

To provide clarity and guide the reader, a brief description of the structure of the paper is included. Section 2 presents the problem formulation, detailing the mathematical modeling of a CPS within the T-S approach and incorporating potential attack scenarios. Section 3 describes the formulation of various attack schemes, including DoS, FDI, and RDI. Section 4 outlines the design of the GDO, emphasizing the role of residual attack transfer functions in detection and isolation. Section 5 provides an in-depth analysis of the stability of the proposed observer using the Lyapunov theory, the elimination lemma, and the Schur complement. Section 6 introduces a case study featuring a three-tank interconnected system modeled under the T-S framework, followed by simulation results to validate the proposed approach. Finally, Section 7 summarizes the conclusions and potential future research directions.

2. Problem Formulation

This section introduces a discrete-time nonlinear cyber–physical system under the Takagi–Sugeno (T-S) representation, an actuator attack model, a GDO structure, and a residual equation used for attack detection. The CPS is described as follows:

$$\begin{aligned} x(k+1) &= \sum_{i=1}^{\kappa} \mu_i(x(k))(A_i x(k)) + Bu(k) + F_a a_a(k), \\ y(k) &= Cx(k), \end{aligned} \tag{1}$$

where $x(k) \in \mathbb{R}^n$ is the system state vector, $u(k) \in \mathbb{R}^m$ is the input, and $y(k) \in \mathbb{R}^p$ represents the measured output variables.

The system's nonlinearity is represented through the matrices A_i, which vary with the nonlinear operating points of the system. These points are characterized by the weighting functions $\mu_i(x(k))$, which satisfy the following convex sum condition:

$$\sum_{i=1}^{\kappa} \mu_i(x(k)) = 1, \quad \mu_i(x(k)) \geq 0, \forall i \in \{1, \ldots, \kappa\}. \tag{2}$$

Here, $\kappa = 2^s$, where s denotes the number of nonlinearities in the system, and each A_i corresponds to a local linear model associated with a specific operating region. The matrices $A_i \in \mathbb{R}^{n \times n}$, $B \in \mathbb{R}^{n \times m}$, $C \in \mathbb{R}^{p \times n}$, and $F_a \in \mathbb{R}^{n \times r}$ are known.

The actuator attack $a_a(k) \in \mathbb{R}^r$ is defined as

$$a_a(k) = \alpha(k) a(k), \tag{3}$$

where $\alpha(k)$ refers to stochastic Markovian processes whose values lie between 0 and 1, and $a(k)$ denotes the actuator attack. To enable the design of the generalized dynamic observer, it is assumed that the T-S cyber–physical system satisfies the following observability condition:

$$rank \begin{bmatrix} C \\ CA_i \\ \vdots \\ CA_i^{n-1} \end{bmatrix} = n. \tag{4}$$

Figure 1 shows the GDO structure used to detect attacks directed at the actuator.

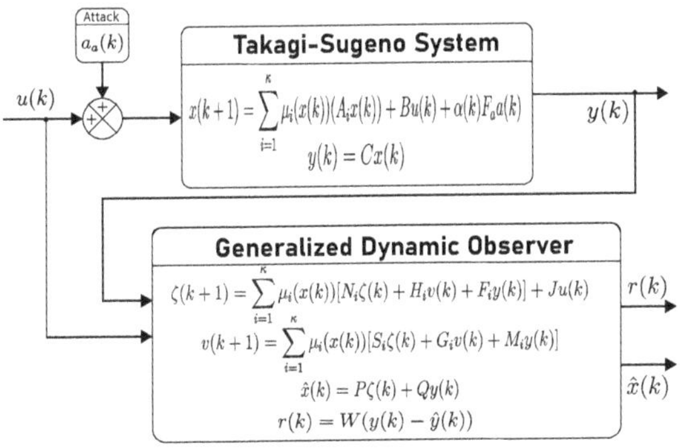

Figure 1. Generalized dynamic observer scheme.

This observer allows one to obtain the estimation of the residuals generated by different attack schemes applied to the cyber–physical system under the T-S representation.

Consider a GDO structure for System (1).

$$\zeta(k+1) = \sum_{i=1}^{\kappa} \mu_i(x(k))[N_i\zeta(k) + H_i v(k) + F_i y(k)] + Ju(k), \tag{5}$$

$$v(k+1) = \sum_{i=1}^{\kappa} \mu_i(x(k))[S_i\zeta(k) + G_i v(k) + M_i y(k)], \tag{6}$$

$$\hat{x}(k) = P\zeta(k) + Qy(k), \tag{7}$$

where $\zeta(k) \in \mathbb{R}^{q_0}$ represents the functional observer state vector, $v(k) \in \mathbb{R}^{q_1}$ is an auxiliary vector, and $\hat{x}(k) \in \mathbb{R}^n$ is the observer estimation vector. N_i, H_i, F_i, J, S_i, M_i, P, and Q are unknown matrices of appropriate dimensions.

The following is the residual equation used for the purpose of detecting and isolating attacks:

$$r(k) = W(y(k) - \hat{y}(k)), \tag{8}$$

Lemma 1. *There exists an observer of the form* (5)–(7) *for System* (1) *if and only if the following two statements are valid.*

1. *There exists a matrix T with appropriate dimensions such that the following conditions are satisfied:*

$$N_i T + F_i C - TA_i = 0, \tag{9}$$

$$J = TB, \tag{10}$$

$$S_i T + M_i C = 0, \tag{11}$$

$$PT + QC = I_n. \tag{12}$$

2. *The matrix* $\begin{bmatrix} N_i & H_i \\ S_i & L_i \end{bmatrix}$ *is stable* $\forall i = 1, \dots, \kappa$.

Consider a matrix $T \in \mathbb{R}^{q_0 \times (n+r)}$ to define the transformed error vector $\varepsilon(k) = \zeta(k) - T(k)$, whose dynamics can be expressed as follows:

$$\varepsilon(k+1) = \zeta(k+1) - Tx(k+1). \tag{13}$$

Solving for $\zeta(k)$, the expression $\zeta(k) = \varepsilon(k) + Tx(k)$ is acquired. Substituting $\zeta(k)$ into (13), the following is obtained:

$$\begin{aligned} \varepsilon(k+1) = \sum_{1=i}^{\kappa} \mu_i(x(k))[N_i\varepsilon(k) + \underbrace{(N_i T + F_i C - TA_i)}_{=0}x(k) \\ + H_i v(k)] + \underbrace{(J - TB)}_{=0}u(k) - TF_a a_a(k). \end{aligned} \tag{14}$$

Subsequently, Equations (6) and (7) can be rewritten as follows:

$$v(k+1) = \sum_{i=1}^{\kappa} \mu_i(x(k))[\underbrace{(S_i T + M_i C)}_{=0}x(k) + S_i\varepsilon(k) + G_i v(t)], \tag{15}$$

$$\hat{x}(k) = P\varepsilon(k) + \underbrace{(PT + QC)}_{=I}x(k). \tag{16}$$

Thus, the residual equation takes the following form:

$$\begin{aligned} r(k) &= W(\hat{y}(k) - y(k)), \\ &= WC\underbrace{(\hat{x}(k) - x(k))}_{e(k)}. \end{aligned} \tag{17}$$

Then, the dynamics of the observation error formed by (15) and (15) can be written as follows:

$$\begin{bmatrix} \varepsilon(k+1) \\ v(k+1) \end{bmatrix} = \begin{bmatrix} N_i & H_i \\ S_i & G_i \end{bmatrix} \begin{bmatrix} \varepsilon(k) \\ v(k) \end{bmatrix} + \begin{bmatrix} -TF_a \\ 0 \end{bmatrix} a_a(k). \tag{18}$$

Now, the problem of attack isolation is reduced to determining the matrices N_i, H_i, F_i, S_i, G_i, M_i, P, Q, W, and T.

3. Formulation of Attack Schemes

In this section, the different attack schemes are defined.

3.1. Denial-of-Service (DoS) Attacks

DoS attacks are aimed at disrupting the transmission of information in control and monitoring systems. These attacks can be carried out through interference in communication channels, the overload of data packets on the network, or other similar means.
The cyber-attack schemes on actuators have the following form:

$$a(k) = -u(k). \tag{19}$$

Replacing the actuator attack of Equation (19) in (3) and using the system in Equation (1), the following system is obtained:

$$\begin{aligned} x(k+1) &= \sum_{1=i}^{\kappa} \mu_i(x(k))[A_i x(k)] + Bu(k) - \alpha(k)F_a u(k), \\ y(k) &= Cx(k), \end{aligned} \tag{20}$$

where F_a is a known matrix by the attacker with the dimensions of the matrix B. $\alpha(k)$ refers to Markovian stochastic processes that take values of 0 and 1.

3.2. False Data Injection (FDI) Attack

The purpose of these attacks is to affect the system's operation through the manipulation of actuators. An actuator attack aims to alter the control signal that the system receives, generating instability in the system.

When System (1) is affected by a false data injection attack, the equation described in (3) should consider the following structure:

$$a(k) = -u(k) + b_a(k), \tag{21}$$

where $b_a(k)$ are deceptive data that the adversary tries to inject into the actuator. Substituting Equation (21) into (3) and, in turn, into (20), the following system is derived:

$$\begin{aligned} x(k+1) &= \sum_{1=i}^{\kappa} \mu_i(x(k))[A_i x(k)] + Bu(k) - \alpha(k)F_a u(k) + \alpha(k)F_a b_a(k), \\ y(k) &= Cx(k). \end{aligned} \tag{22}$$

3.3. Random Data Injection (RDI) Attack

This type of attack aims to make the sensor reading or controller signal different from the real value. Recall that a random data injection attack assumes incomplete knowledge of the system being attacked, unlike an FDI attack. An RDI attack has the following form:

$$a(k) = \pm b_a(k). \tag{23}$$

The positive and negative signs in Equation (23) are included because the effect of the random data injection can either add or subtract dynamics to the system, depending on the nature of the attack. This is due to the random nature of the attack scheme, which does not follow a fixed pattern and can, therefore, unpredictably alter the system's input matrix. Substituting Equation (23) into (3) yields the following system:

$$\begin{aligned} x(k+1) &= \sum_{1=i}^{\kappa} \mu_i(x(k))[A_i x(k)] + Bu(k) \pm \alpha(k)F_a b_a(k), \\ y(k) &= Cx(k), \end{aligned} \tag{24}$$

Finally, the classification shown in Figure 2 is presented for the different attack schemes to be used. These schemes have been divided by type of attack, followed by a sub-classification based on the logic of the attack. Likewise, a more detailed classification has been made based on the location of the application of the attack scheme.

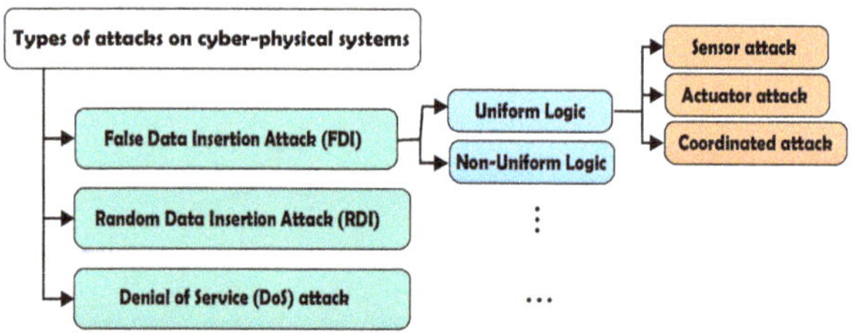

Figure 2. Types of attacks and structure.

4. Design of the Generalized Dynamic Observer

This section undertakes the parameterization of the observer matrices alongside the formulation of a design targeting the detection and isolation of attacks.

4.1. Parameterization of the Observer

It is considered that the following conditions hold for Equations (15) and (16):

$$\begin{aligned} &a)\ N_iT + F_iC - TA_i = 0, && b)\ J = TB, \\ &c)\ S_iT + M_iC = 0, && d)\ PT + QC = I_n. \end{aligned}$$

We define a matrix $E \in \mathbb{R}^{q_0 \times n}$ of full row rank, $\Sigma = \begin{bmatrix} E \\ C \end{bmatrix}$ and $\Omega = \begin{bmatrix} I_n \\ C \end{bmatrix}$. Conditions (c–d) can be written as follows:

$$\begin{bmatrix} S_i & M_i \\ P & Q \end{bmatrix} \begin{bmatrix} T \\ C \end{bmatrix} = \begin{bmatrix} 0 \\ I_n \end{bmatrix}. \tag{25}$$

The necessary and sufficient conditions for (25) to have a solution are

$$rank \begin{bmatrix} T \\ C \end{bmatrix} = rank \begin{bmatrix} T \\ C \\ I_n \end{bmatrix} = n. \tag{26}$$

From [21], since $rank \begin{bmatrix} \Omega \\ \Sigma \end{bmatrix} = rank(\Omega) = n$, condition (26) is equivalent to the existence of two matrices $T \in \mathbb{R}^{q_0 \times n}$ and $K \in \mathbb{R}^{q_0 \times p}$ such that

$$T + KC = E, \tag{27}$$

where $E \in \mathbb{R}^{q_0 \times n}$ is an arbitrary matrix such that $rank \begin{bmatrix} E \\ C \end{bmatrix} = n$.

Equation (27) can also be written as

$$\begin{bmatrix} T & K \end{bmatrix} \Omega = E. \tag{28}$$

Using the particular solution to solve for the matrices T and K, the ensuing results are obtained:

$$T = E\Omega^{+} \begin{bmatrix} I_n \\ 0 \end{bmatrix}, \qquad K = E\Omega^{+} \begin{bmatrix} 0 \\ I_p \end{bmatrix}. \tag{29}$$

Considering matrix T from Equation (27) into condition a), it can be rewritten as

$$N_i E + \underbrace{(F_i - N_i K)}_{\tilde{K}} = TA_i, \tag{30}$$

where $\tilde{K}_i = F_i - N_i K$, and Equation (30) can be written as

$$\begin{bmatrix} N_i & \tilde{K}_i \end{bmatrix} \underbrace{\begin{bmatrix} E \\ C \end{bmatrix}}_{\Sigma} = TA_i. \tag{31}$$

Using the general solution, the subsequent formulation is derived for matrices N_i and $\tilde{K}_i$.

$$N_i = \underbrace{TA_i \Sigma^{+} \begin{bmatrix} I_{q_0} \\ 0 \end{bmatrix}}_{N_{i1}} - \underbrace{Y_{i1}(I_{q_0+p} - \Sigma\Sigma^{+}) \begin{bmatrix} I_{q_0} \\ 0 \end{bmatrix}}_{N_2}, \tag{32}$$

$$\tilde{K}_i = \underbrace{TA_i \Sigma^{+} \begin{bmatrix} 0 \\ I_p \end{bmatrix}}_{\tilde{K}_{i1}} - \underbrace{Y_{i1}(I_{q_0+p} - \Sigma\Sigma^{+}) \begin{bmatrix} 0 \\ I_p \end{bmatrix}}_{\tilde{K}_2}, \tag{33}$$

where Y_{i1} is a matrix of appropriate dimensions with arbitrary elements.

From Equation (30), we can deduce the value of matrix F_i as follows:

$$\begin{aligned} F_i &= \tilde{K}_{i1} + N_i K, \\ &= \tilde{K}_{i1} - Y_{i1}\tilde{K}_2 + N_{i1}K - Y_{i1}N_2 K, \\ &= F_{i1} - Y_{i1}F_2, \end{aligned} \tag{34}$$

where $F_{i1} = TA_i\Sigma^+ \begin{bmatrix} K \\ I_p \end{bmatrix}$ and $F_2 = (I_{q_0+p} - \Sigma\Sigma^+)\begin{bmatrix} K \\ I_p \end{bmatrix}$. To obtain the matrices S_i, M_i, P, and Q, we consider the following equation and incorporate it into (25):

$$\begin{bmatrix} T \\ C \end{bmatrix} = \begin{bmatrix} I_{q_0} & -K \\ 0 & I_p \end{bmatrix} \underbrace{\begin{bmatrix} E \\ C \end{bmatrix}}_{\Sigma}, \tag{35}$$

resulting in the following expression:

$$\begin{bmatrix} S_i & M_i \\ P & Q \end{bmatrix} \begin{bmatrix} I_{q_0} & -K \\ 0 & I_p \end{bmatrix} \Sigma = \begin{bmatrix} 0 \\ I_n \end{bmatrix}, \tag{36}$$

which has the following general solution:

$$\begin{bmatrix} S_i & M_i \\ P & Q \end{bmatrix} \begin{bmatrix} I_{q_0} & -K \\ 0 & I_p \end{bmatrix} = \begin{bmatrix} 0 \\ I_n \end{bmatrix} \Sigma^+ - \begin{bmatrix} W_{i1} \\ W_2 \end{bmatrix} (I_{q_0+p} - \Sigma\Sigma^+), \tag{37}$$

where W_{i1} and W_2 are matrices of appropriate dimensions with arbitrary elements. Then, the matrices S_i, M_i, P, and Q can be written as

$$S_i = -W_{i1}N_2, \tag{38}$$

$$M_i = -W_{i1}F_2, \tag{39}$$

$$P = \Sigma^+ \begin{bmatrix} I_{q_0} \\ 0 \end{bmatrix} - W_2 N_2, \tag{40}$$

$$Q = \Sigma^+ \begin{bmatrix} K \\ I_p \end{bmatrix} - W_2 F_2. \tag{41}$$

By employing (32) and (38), the error dynamics (18) are expressed as

$$\varphi(k+1) = (\mathbb{A}_{i1} - \mathbb{Y}_i\mathbb{A}_2)\varphi(k) + \mathbb{B}a_a(k), \tag{42}$$

where $\mathbb{A}_{i1} = \begin{bmatrix} N_1 & 0 \\ 0 & 0 \end{bmatrix}$, $\mathbb{A}_2 = \begin{bmatrix} N_2 & 0 \\ 0 & -I_{q_1} \end{bmatrix}$, $\mathbb{B} = \begin{bmatrix} -TF_a \\ 0 \end{bmatrix}$, $\mathbb{Y}_i = \begin{bmatrix} Y_{i1} & Hi \\ W_{i1} & G_i \end{bmatrix}$ and $\varphi(k) = \begin{bmatrix} \varepsilon(k) \\ v(k) \end{bmatrix}$.

Finally, the system residue is as follows:

$$r(k) = W \underbrace{\begin{bmatrix} CP & 0 \end{bmatrix}}_{\mathbb{C}} \begin{bmatrix} \varepsilon(k) \\ v(k) \end{bmatrix}, \tag{43}$$

where, without loss of generality, $W_2 = 0$ is taken for simplicity.

4.2. Design of the Attack Detection and Isolation Observer

The objective of attack isolation in this article is to obtain a residual attack transfer function equal to a diagonal in order to deal with attacks that may occur simultaneously.

To carry out attack detection and isolation, it is necessary to represent Equations (42) and (43) in transfer function form, where G_z is a transfer function from the attack $a_a(k)$ to the residual $r(k)$:

$$G_z = \left(\begin{array}{c|c} \mathbb{A}_{i1} - \mathbb{Y}_i\mathbb{A}_2 & \mathbb{B} \\ \hline W\mathbb{C} & 0 \end{array}\right). \tag{44}$$

The next step involves proposing a transfer function that has an input–output dependence, where the desired behavior is that when there is a change in an input, only one output reacts.

Proposition 1. *The transfer function* (44) *can be diagonalized if and only if* $(\mathbb{C}\mathbb{B})$ *has a full column, i.e.,* $p \leq r$.

Proposition 1 verifies the output separability condition, which means that to isolate r attacks, it is required to measure p outputs.

The following theorem shows how to design an observer of the form (5)–(7) to isolate attacks.

Theorem 1. *Consider that* $p \leq r$ *and let*

$$\Lambda = diag(\lambda_1, \ldots, \lambda_r) \in \mathbb{R}^{r \times r}, \ -1 < \lambda_i < 1, \tag{45}$$

$$\Gamma = diag(\gamma_1, \ldots, \gamma_r) \in \mathbb{R}^{r \times r}, \ |\gamma_i| > 0, \ \forall i = 1, \ldots r, \tag{46}$$

be given. Then, there exist matrices $\mathbb{Y}_i$ *and* W *such that*

$$(\mathbb{A}_{i1} - \mathbb{Y}_i\mathbb{A}_2)\mathbb{B} = \mathbb{B}\Lambda, \tag{47}$$

$$W\mathbb{C}\mathbb{B} = \Gamma, \tag{48}$$

Then, the solutions of matrices $\mathbb{Y}_i$ *and* W *are given by*

$$\mathbb{Y}_i = (\mathbb{A}_{i1}\mathbb{B} - \mathbb{B}\Lambda)(\mathbb{A}_2\mathbb{B})^+ - Z_i(I - (\mathbb{A}_2\mathbb{B})(\mathbb{A}_2\mathbb{B})^+), \tag{49}$$

$$W = \Gamma(\mathbb{C}\mathbb{B})^+, \tag{50}$$

where Z_i *is an arbitrary matrix of appropriate dimensions.*

If there exist matrices $\mathbb{Y}_i$ *and* W *satisfying* (47) *and* (48)*, then*

$$G_z = \left(\begin{array}{c|c} \Lambda & I \\ \hline \Gamma & 0 \end{array}\right), = diag\left(\frac{\gamma_1}{z - \lambda_1}, \ldots, \frac{\gamma_r}{z - \lambda_r}\right), \tag{51}$$

Proof. Consider that $\mathbb{C}\mathbb{B}$ has full column rank, and hence, for $\mathbb{B}$, there exists a matrix completion $\mathbb{B}^{\perp}$ such that

$\tilde{\mathbb{B}} = \begin{bmatrix} \mathbb{B} & \mathbb{B}^{\perp} \end{bmatrix}$ is a nonsingular matrix that allows

$\tilde{\mathbb{B}}^{-1} = \begin{bmatrix} \tilde{\mathbb{B}}_1 & \tilde{\mathbb{B}}_2 \end{bmatrix}^T$. Thus, the following expression is obtained:

$$G_z = \left(\begin{array}{c|c} \tilde{\mathbb{B}}^{-1}(\mathbb{A}_{i1} - \mathbb{Y}_i\mathbb{A}_2)\tilde{\mathbb{B}} & \tilde{\mathbb{B}}^{-1}\mathbb{B} \\ \hline W\mathbb{C}\tilde{\mathbb{B}} & 0 \end{array} \right), \tag{52}$$

$$= \left(\begin{array}{c|c} \begin{bmatrix} \tilde{\mathbb{B}}_1^T \\ \tilde{\mathbb{B}}_2^T \end{bmatrix} (\mathbb{A}_{i1} - \mathbb{Y}_i\mathbb{A}_2) \begin{bmatrix} \mathbb{B} & \mathbb{B}^\perp \end{bmatrix} & \begin{bmatrix} \tilde{\mathbb{B}}_1^T \\ \tilde{\mathbb{B}}_2^T \end{bmatrix} \mathbb{B} \\ \hline W\mathbb{C} \begin{bmatrix} \mathbb{B} & \mathbb{B}^\perp \end{bmatrix} & 0 \end{array} \right), \tag{53}$$

$$= \left(\begin{array}{cc|c} \tilde{\mathbb{B}}_1^T(\mathbb{A}_{i1} - \mathbb{Y}_i\mathbb{A}_2)\mathbb{B} & \tilde{\mathbb{B}}_1^T(\mathbb{A}_{i1} - \mathbb{Y}_i\mathbb{A}_2)\mathbb{B}^\perp & I_{q_0+q_1} \\ \tilde{\mathbb{B}}_2^T(\mathbb{A}_{i1} - \mathbb{Y}_i\mathbb{A}_2)\mathbb{B} & \tilde{\mathbb{B}}_2^T(\mathbb{A}_{i1} - \mathbb{Y}_i\mathbb{A}_2)\mathbb{B} & 0 \\ \hline W\mathbb{C}\mathbb{B} & W\mathbb{C}\mathbb{B}^\perp & 0 \end{array} \right). \tag{54}$$

We consider $\begin{bmatrix} \tilde{\mathbb{B}}_1 & \tilde{\mathbb{B}}_2 \end{bmatrix}^T \begin{bmatrix} \mathbb{B} & \mathbb{B}^\perp \end{bmatrix} = I_{2(q_0+q_1)}$ to obtain

$$G_z = \left(\begin{array}{cc|c} \tilde{\mathbb{B}}_1^T(\mathbb{A}_{i1} - \mathbb{Y}_i\mathbb{A}_2)\mathbb{B} & \tilde{\mathbb{B}}_1^T(\mathbb{A}_{i1} - \mathbb{Y}_i\mathbb{A}_2)\mathbb{B}^\perp & I_{q_0+q_1} \\ 0 & \tilde{\mathbb{B}}_2^T(\mathbb{A}_{i1} - \mathbb{Y}_i\mathbb{A}_2)\mathbb{B}^\perp & 0 \\ \hline W\mathbb{C}\mathbb{B} & W\mathbb{C}\mathbb{B}^\perp & 0 \end{array} \right). \tag{55}$$

Removing an uncontrollable subspace, the following result is obtained:

$$G_z = \left(\begin{array}{c|c} \Lambda & I_r \\ \hline \Gamma & 0 \end{array} \right). \tag{56}$$

From (56), we find that

$$(\mathbb{A}_{i1} - \mathbb{Y}_i\mathbb{A}_2)\mathbb{B} = \mathbb{B}\Lambda, \tag{57}$$

$$W\mathbb{C}\mathbb{B} = \Gamma, \tag{58}$$

□

Then, from (57), the general solution of matrix $\mathbb{Y}_i$ is given by

$$\mathbb{Y}_i = (\mathbb{A}_{i1}\mathbb{B} - \mathbb{B}\Lambda)(\mathbb{A}_2\mathbb{B})^+ - Z_i(I - (\mathbb{A}_2\mathbb{B})(\mathbb{A}_2\mathbb{B})^+), \tag{59}$$

where Z_i is an arbitrary matrix of appropriate dimensions. From (58), the particular solution of W is

$$W = \Gamma(\mathbb{C}\mathbb{B})^+, \tag{60}$$

Substituting Equations (59) and (60) into Equation (42) yields

$$\varphi(k+1) = [\underbrace{\mathbb{A}_{i1} - (\mathbb{A}_{i1}\mathbb{B} - \mathbb{B}\Lambda)(\mathbb{A}_2\mathbb{B})^+\mathbb{A}_2}_{\bar{\mathbb{A}}_{i1}} +$$

$$Z_i \underbrace{(I - (\mathbb{A}_2\mathbb{B})(\mathbb{A}_2\mathbb{B})^+)\mathbb{A}_2}_{\bar{\mathbb{A}}_2}]\varphi(k) + \mathbb{B}a_a(k), \tag{61}$$

$$r(k) = \Gamma(\mathbb{C}\mathbb{B}^+)\mathbb{C}\varphi(k). \tag{62}$$

Equations (61) and (62) can be rewritten as

$$\varphi(k+1) = (\bar{\mathbb{A}}_{i1} + Z_i\bar{\mathbb{A}}_2)\varphi(k) + \mathbb{B}a_a(k), \tag{63}$$

$$r(k) = \Gamma(\mathbb{C}\mathbb{B})^+\mathbb{C}\varphi(k), \tag{64}$$

The problem is now to find the matrix Z_i such that (42) is stable.

5. Observer Stability Analysis

In this section, an observer stability analysis is carried out. Consider the following Lyapunov function:

$$V(\varphi(k)) = \varphi^T(k)X\varphi(k) > 0, \tag{65}$$

where $X = \begin{bmatrix} X_1 & 0 \\ 0 & X_2 \end{bmatrix} > 0$, with $X_1 \in \mathbb{R}^{q_0 \times q_0}$ and $X_2 \in \mathbb{R}^{q_1 \times q_1}$. The difference in $V(\varphi(k))$ is

$$\Delta V(\varphi(k)) = \varphi^T(k+1)X\varphi(k+1) - \varphi^T(k)X\varphi(k) < 0, \tag{66}$$

$$= \varphi^T(k)\Big((\bar{\mathbb{A}}_{i1} - Z_i\bar{\mathbb{A}}_2)^T X(\bar{\mathbb{A}}_{i1} - Z_i\bar{\mathbb{A}}_2) - X\Big)\varphi(k) < 0. \tag{67}$$

The inequality $\Delta V(\varphi(k)) < 0$ holds for all $\varphi(k) \neq 0$ if and only if

$$(\bar{\mathbb{A}}_{i1} - Z_i\bar{\mathbb{A}}_2)^T X(\bar{\mathbb{A}}_{i1} - Z_i\bar{\mathbb{A}}_2) - X < 0. \tag{68}$$

Considering the Schur complement [22], the inequality (68) is equivalent to

$$\begin{bmatrix} -X & (\bar{\mathbb{A}}_{i1} - Z_i\bar{\mathbb{A}}_2)^T X \\ X(\bar{\mathbb{A}}_{i1} - Z_i\bar{\mathbb{A}}_2) & -X \end{bmatrix} < 0, \tag{69}$$

which can be rewritten as

$$\mathcal{C}\mathcal{X}_i\mathcal{D} + (\mathcal{C}\mathcal{X}_i\mathcal{D})^T + \mathcal{E}_i < 0, \tag{70}$$

where $\mathcal{E}_i = \begin{bmatrix} -X & \bar{\mathbb{A}}_{i1}^T X \\ X\bar{\mathbb{A}}_{i1} & -X \end{bmatrix}$, $\mathcal{C} = \begin{bmatrix} 0 \\ -I \end{bmatrix}$, $\mathcal{D} = \begin{bmatrix} \bar{\mathbb{A}}_2 & 0 \end{bmatrix}$, and $\mathcal{X}_i = XZ_i$.

According to the elimination lemma, inequality (70) is equivalent to the following two conditions [23]:

$$\mathcal{C}^{\perp}\mathcal{E}_i\mathcal{C}^{\perp T} < 0, \tag{71}$$

$$\mathcal{D}^{T\perp}\mathcal{E}_i\mathcal{D}^{T\perp T} < 0. \tag{72}$$

Replacing matrices $\mathcal{C}^{\perp} = [I \;\; 0]$, $\mathcal{E}_i = \begin{bmatrix} X_1 & 0 & \Pi_a^T & 0 \\ 0 & X_2 & 0 & 0 \\ \Pi_a & 0 & -X_1 & 0 \\ 0 & 0 & 0 & -X_2 \end{bmatrix}$, where $\Pi_a = N_1 + (TF_a\Lambda - N_1TF_a)(N_2TF_a)^+N_2$, and the inequality $X > 0$ is obtained.

From (72), $\mathcal{D}^{T\perp} = \begin{bmatrix} \Pi_b^{T\perp} & 0 & 0 & 0 \\ 0 & 0 & I & 0 \\ 0 & 0 & 0 & I \end{bmatrix}$ where $\Pi_b = N_2 - (N_2TF_a)(N_2TF_a)^+N_2$, and matrix $\mathcal{E}_i$, the following result is obtained:

$$\begin{bmatrix} \Pi_b^{T\perp}X_1\Pi_b^{T\perp T} & \Pi_b^{T\perp}\Pi_a^T & 0 \\ \Pi_a\Pi_b^{T\perp T} & -X_1 & 0 \\ 0 & 0 & -X_2 \end{bmatrix} < 0. \tag{73}$$

Then, the matrix $\mathcal{X}_i$ in (70) can be obtained as follows:

$$\mathcal{X}_i = \mathcal{C}_r{}^+\mathcal{K}_i\mathcal{D}_l{}^+ + \mathcal{Z} - \mathcal{C}_r{}^+\mathcal{C}_r\mathcal{Z}\mathcal{D}_l\mathcal{D}_l{}^+, \tag{74}$$

where $\mathcal{Z}$ is an arbitrary matrix of appropriate dimensions, and $(\mathcal{D}_l, \mathcal{D}_r)$ and $(\mathcal{C}_l, \mathcal{C}_r)$ represent complete rank constituents of $\mathcal{C}$ and $\mathcal{D}$, respectively. This arrangement holds such that $\mathcal{C} = \mathcal{C}_l\mathcal{C}_r$ and $\mathcal{D} = \mathcal{D}_l\mathcal{D}_r$.

The matrices $\mathcal{K}_i$, $\mathcal{S}_i$, and $\mathcal{V}_i$ are obtained as follows:

$$\mathcal{K}_i = -\mathcal{R}^{-1}\mathcal{C}_l{}^T\mathcal{V}_i\mathcal{D}_r{}^T(\mathcal{D}_r\mathcal{V}_i\mathcal{D}_r{}^T)^{-1} + \mathcal{S}_i^{\frac{1}{2}}\mathcal{L}(\mathcal{D}_r\mathcal{V}_i\mathcal{D}_r{}^T)^{-\frac{1}{2}}, \tag{75}$$

$$\mathcal{S}_i = \mathcal{R}^{-1} - \mathcal{R}^{-1}\mathcal{C}_l^T[\mathcal{V}_i - \mathcal{V}_i\mathcal{D}_r^T(\mathcal{D}_r\mathcal{V}_i\mathcal{D}_r{}^T)^{-1}\mathcal{D}_r\mathcal{V}_i]\mathcal{C}_l\mathcal{R}^{-1}, \tag{76}$$

$$\mathcal{V}_i = (\mathcal{C}_l\mathcal{R}^{-1}\mathcal{C}_l{}^T - \mathcal{E}_i)^{-1} > 0, \tag{77}$$

where $\mathcal{L}$ and $\mathcal{R}$ are arbitrary matrices satisfying $||\mathcal{L}|| < 1$ and $\mathcal{R} > 0$ such that $\mathcal{V}_i$ is positive definite.

Matrix Z_i is parameterized as follows:

$$Z_i = X^{-1}\mathcal{X}_i. \tag{78}$$

To summarize the stability analysis, it is necessary to ensure the feasibility of the solution of inequalities (71) and (72) to obtain a matrix $X > 0$; then, using (74)–(77), matrix $\mathcal{X}_i$ must be determined from (70) to finally obtain matrix Z_i from (78).

6. Case Study

Model of an Three-Tank Interconnected System

Interconnected tank systems are a fundamental tool in chemical process engineering. These systems consist of several tanks that are interconnected through pipes and valves. The aim of these systems is to control the liquid level in the tanks, as shown in Figure 3.

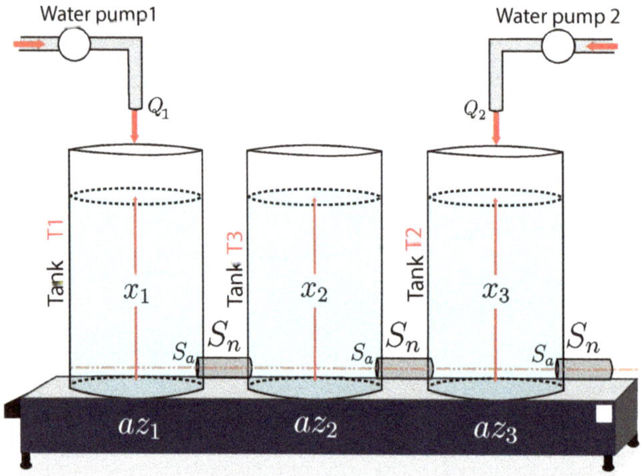

Figure 3. The case study diagram (three interconnected tanks).

The interconnected three-tank system [24] is characterized by the following set of three differential equations:

$$\begin{aligned}
x_1(k+1) &= x_1(k) - \frac{T_d}{S_a}[a_{z1}S_n\sqrt{2g(x_1(k) - x_3(k))}] + \frac{T_d}{S_a}Q_1(k), \\
x_2(k+1) &= x_2(k) + \frac{T_d}{S_a}[a_{z3}S_n\sqrt{2g(x_1(k) - x_3(k)}] - \frac{T_d}{S_a}a_{z2}S_n\sqrt{2gx_2(k)} + \frac{T_d}{S_a}Q_2(k), \\
x_3(k+1) &= x_3(k) + \frac{T_d}{S_a}[a_{z1}S_n\sqrt{2g(x_1(k) - x_3(k))}] - \frac{T_d}{S_a}a_{z3}S_n\sqrt{2g(x_3(k) - x_2(k))},
\end{aligned} \tag{79}$$

where $T_d = 0.01$ is the discretization time, and $Q_1(k)$ and $Q_2(k)$ represent the flow rates of pump 1 and pump 2, respectively. $x_1(k)$, $x_2(k)$, and $x_3(k)$ denote the heights of the three interconnected tanks. Table 1 presents the parameters of the model.

Table 1. Parameters of the system.

Parameter	Value	Units	Definition
S_a	0.0154	m^2	Tank cross-section
S_n	5×10^{-5}	m^2	Pipe cross-section
g	9.82	m^2/s	Gravity
az_1	0.46	-	Outlet coefficient of tank 1
az_2	0.58	-	Outlet coefficient of tank 2
az_3	0.48	-	Outlet coefficient of tank 3

By writing model (79) in a nonlinear state space representation, it is found that

$$\begin{aligned} x(k+1) &= A(x(k))x(k) + Bu(k), \\ y(k) &= Cx(k), \end{aligned} \tag{80}$$

where $x(k) = \begin{bmatrix} x_1(k) \\ x_2(k) \\ x_3(k) \end{bmatrix}, y(k) = \begin{bmatrix} x_1(k) \\ x_2(k) \end{bmatrix}, u(k) = \begin{bmatrix} Q_1(k) \\ Q_2(k) \end{bmatrix}, A(x(k)) = \begin{bmatrix} A_{11} & 0 & A_{13} \\ 0 & A_{22} & A_{23} \\ A_{31} & A_{32} & A_{33} \end{bmatrix}$,

$B = \begin{bmatrix} \frac{Td}{S_a} & 0 \\ 0 & \frac{Td}{S_a} \\ 0 & 0 \end{bmatrix}, C = \begin{bmatrix} 1 & 0 & 0 \\ 0 & 1 & 0 \end{bmatrix}$,

$$\begin{aligned}
A_{11} =& 1 - \frac{T_d az_1 S_n \sqrt{2g}}{S_a}\rho_1(k), \quad A_{13} = -\frac{T_d az_1 S_n \sqrt{2g}}{S_a}\rho_1(k), \\
A_{22} =& 1 - \frac{T_d az_3 S_n \sqrt{2g}}{S_a}\rho_2(k) - \frac{T_d az_2 S_n \sqrt{2g}}{S_a}\rho_3(k), \\
A_{23} =& \frac{T_d az_3 S_n \sqrt{2g}}{S_a}\rho_2(k), \quad A_{31} = \frac{az_1 S_n T_d \sqrt{2g}}{S_a}\rho_2(k), \quad A_{23} = \frac{T_d az_3 S_n \sqrt{2g}}{S_a}\rho_2(k), \\
A_{33} =& 1 - \frac{T_d az_1 S_n \sqrt{2g}}{S_a}\rho_1(k) - \frac{T_d az_3 S_n}{S_a}\rho_2(k),
\end{aligned}$$

where az_1, az_2 and az_3 are the outlet coefficients taking values from 0 to 1; $Q_1(k)$ and $Q_2(k)$ are input flow rates 1 and 2, respectively, and S_a and S_n are the cross-sections of the tank and pipe.

Examining the nonlinear model of System (80), three premise variables can be identified, and they are

$$\begin{aligned}
\rho_1(k) &= \frac{1}{\sqrt{x_1(k) - x_3(k)}}, \quad \rho_2(k) = \frac{1}{\sqrt{x_3(k) - x_2(k)}}, \\
\rho_3(k) &= \frac{1}{\sqrt{x_2(k)}},
\end{aligned}$$

To determine the maximum and minimum variations in each nonlinearity, the input behavior in Figure 4 and the parameters in Table 1 are considered to obtain the following values.

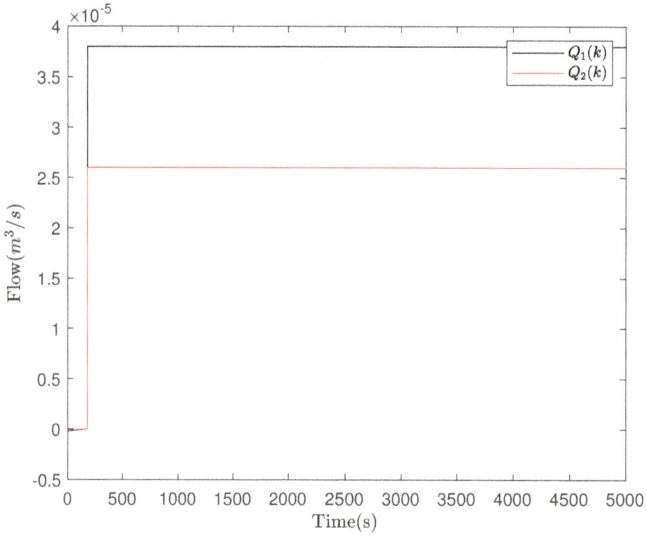

Figure 4. Variable flow inputs for the water pumps.

$$\rho_1(k):\begin{cases}\underline{\rho_1}=\min\left\{\frac{1}{\sqrt{x_1(k)-x_3(k)}}\right\}=2.2305\\ \overline{\rho_1}=\max\left\{\frac{1}{\sqrt{x_1(k)-x_3(k)}}\right\}=4.7567\end{cases}\tag{81}$$

$$\rho_2(k):\begin{cases}\underline{\rho_2}=\min\left\{\frac{1}{\sqrt{x_3(k)-x_2(k)}}\right\}=3.1623\\ \overline{\rho_2}=\max\left\{\frac{1}{\sqrt{x_3(k)-x_2(k)}}\right\}=7.4649\end{cases}\tag{82}$$

$$\rho_3(k):\begin{cases}\underline{\rho_3}=\min\left\{\frac{1}{\sqrt{x_2(k)}}\right\}=1.5342\\ \overline{\rho_3}=\max\left\{\frac{1}{\sqrt{x_2(k)}}\right\}=2.0930\end{cases}\tag{83}$$

The weighting functions are defined as follows:

$$n_0^1(\rho_1(k))=\frac{\overline{\rho}_1-\rho_1(k)}{\overline{\rho_1}-\underline{\rho_1}},\qquad n_1^1(\rho_1(k))=1-n_0^1,\tag{84}$$

$$n_0^2(\rho_2(k))=\frac{\overline{\rho}_2-\rho_2(k)}{\overline{\rho_2}-\underline{\rho_2}},\qquad n_2^1(\rho_2(k))=1-n_0^2,\tag{85}$$

$$n_0^3(\rho_3(k))=\frac{\overline{\rho}_3-\rho_3(k)}{\overline{\rho_3}-\underline{\rho_3}},\qquad n_3^1(\rho_3(k))=1-n_0^3.\tag{86}$$

In this case, the number of membership functions is $\kappa=2^3=8$, and they are defined by

$$\begin{aligned}\mu_1(\rho)&=n_0^1n_0^2n_0^3, & \mu_2(\rho)&=n_0^1n_0^2n_1^3,\\ \mu_3(\rho)&=n_0^1n_1^2n_0^3, & \mu_4(\rho)&=n_0^1n_1^2n_1^3,\\ \mu_5(\rho)&=n_1^1n_0^2n_0^3, & \mu_6(\rho)&=n_1^1n_0^2n_1^3,\\ \mu_7(\rho)&=n_1^1n_1^2n_0^3, & \mu_8(\rho)&=n_1^1n_1^2n_1^3.\end{aligned}$$

Then, the variables $\rho_1(k)$, $\rho_2(k)$, and $\rho_3(k)$ vary within a bounded region $\rho(k)=[\underline{\rho}(k),\ \overline{\rho}(k)]$, where $\underline{\rho}(k)$ and $\overline{\rho}(k)$ are the lower and upper bounds, respectively. Consequently, the following system matrices are obtained:

$$A(\rho(t)) = \begin{bmatrix} 1 - \frac{T_d C_p \sqrt{pg}}{S}\rho_1(k) & \frac{T_d C_p \sqrt{pg}}{S}\rho_1(k) & 0 \\ \frac{T_d C_p \sqrt{pg}}{S}\rho_1(k) & 1 - \frac{T_d C_p \sqrt{pg}}{S}(\rho_1(k) + \rho_2(k)) & \frac{T_d C_p \sqrt{pg}}{S}\rho_2(k) \\ 0 & \frac{T_d C_p \sqrt{pg}}{S}\rho_2(k) & 1 - \frac{T_d C_p \sqrt{pg}}{S}(\rho_2(k) + \rho_3(k)) \end{bmatrix}. \quad (87)$$

Then, the T-S model that represents the dynamics of model (1) is

$$x(k+1) = [\mu_1(\rho(k))A_1 + \cdots + \mu_8(\rho(k))A_8]x(k) + Bu(k), \quad (88)$$

where the matrices that correspond to each local model are

$$\begin{aligned} &A_1 = A(\underline{\rho_1},\ \underline{\rho_2},\ \underline{\rho_3}),\ \ A_2 = A(\underline{\rho_1},\ \underline{\rho_2},\ \overline{\rho_3}),\ \ A_3 = A(\underline{\rho_1},\ \overline{\rho_2},;\ \underline{\rho_3}), \\ &A_4 = A(\underline{\rho_1},\ \overline{\rho_2},\ \overline{\rho_3}),\ \ A_5 = A(\overline{\rho_1},\ \underline{\rho_2},\ \underline{\rho_3}),\ \ A_6 = A(\overline{\rho_1},\ \underline{\rho_2},\ \overline{\rho_3}), \\ &A_7 = A(\overline{\rho_1},\ \overline{\rho_2},\ \underline{\rho_3}),\ \ A_8 = A(\overline{\rho_1},\ \overline{\rho_2},\ \overline{\rho_3}), \end{aligned}$$

where matrix $A(\rho)$ is defined in (87).

Finally, the mathematical T-S model of (88) can be rewritten as

$$\begin{aligned} x(k+1) &= \sum_{i=1}^{8} \mu_i(\rho(k))(A_i x(k)) + Bu(k), \\ y(k) &= Cx(k). \end{aligned} \quad (89)$$

7. Simulation Results

The performance of the proposed GDO for detecting and isolating attacks is evaluated using the discrete T-S system model defined in (89), where $F_a = B$. The input signals, corresponding to the operation of two water pumps, are illustrated in Figure 4. These signals represent the variable flow rates that govern the system dynamics both under normal conditions and during various attack scenarios. The analysis of these signals serves as a foundation for understanding the effectiveness of the GDO in maintaining system stability and detecting deviations caused by external disruptions such as DoS, FDI, and RDI attacks.

The input signals $u(k)$ control the flow of water in the three-tank system and serve as the basis for system behavior. These signals are crucial for the simulation, as they establish a baseline for comparison when different attack scenarios are introduced. The following subsections present the system's response in the presence of various attack types, focusing on state estimation accuracy, residual generation, and attack detection effectiveness.

Figure 5 shows the Markovian distribution logic signals used in the simulation, comparing the behavior of a **uniform signal (black line)** and a **non-uniform signal (red line)**, which influences the attack activation patterns.

Three simulations will be presented below considering the different attack schemes $a(k)$ shown in Section 3 with uniform or non-uniform Markovian logic $\alpha(k)$, affecting one or both actuators of the CPS.

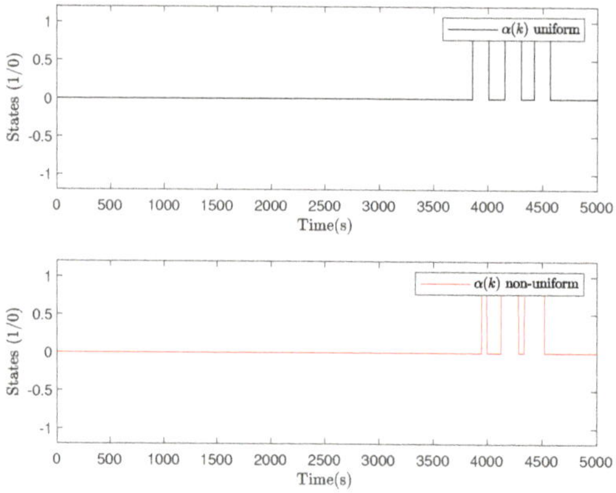

Figure 5. Markovian distribution logic.

7.1. Simulated DoS Attack

The first simulation considers a DoS attack affecting the input $u_1(k)$ with uniform logic distribution. Figure 6 shows the system states and their estimations under these conditions. The observer effectively tracks the states despite the disruption caused by the attack.

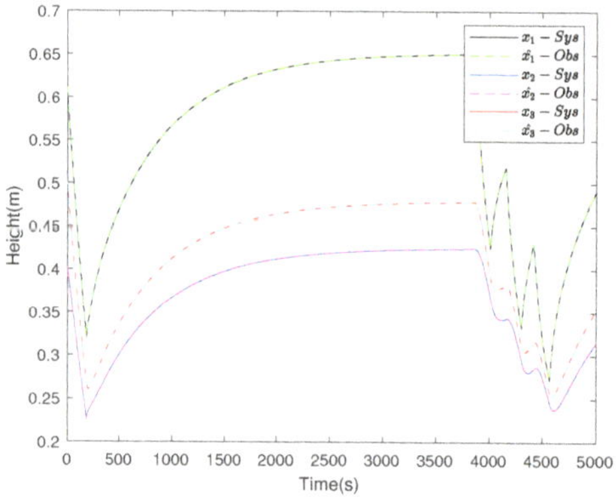

Figure 6. Behavior in the presence of a uniform DoS attack on u_1.

Figure 7 presents the residuals generated by the observer. Residual 1 detects the attack on $u_1(k)$, demonstrating the observer's ability to isolate the disruption effectively.

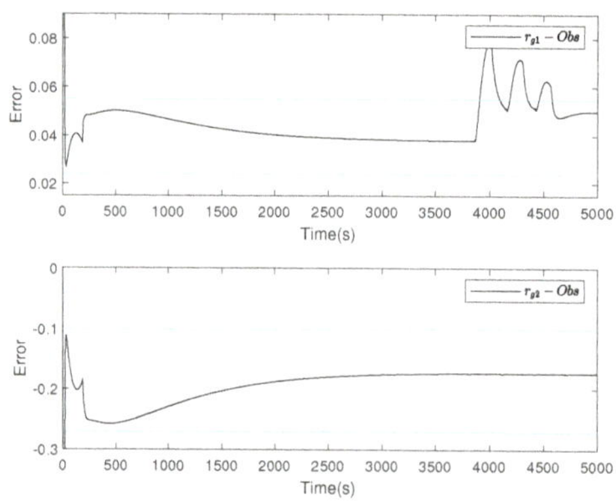

Figure 7. Residues generated in the presence of the DoS attack type.

7.2. Simulated FDI Attack

The second simulation explores the impact of an FDI attack on $u_2(k)$ using a non-uniform logic distribution. Figure 8 illustrates the false data $b_a(k)$ injected into the system, which replace the original data to alter the system's behavior.

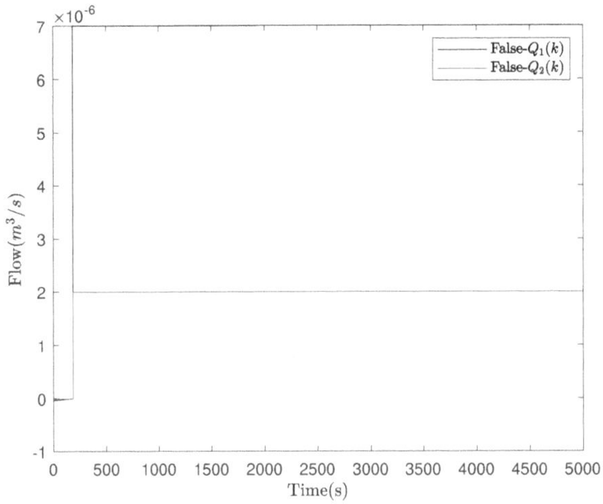

Figure 8. False data $b_a(k)$.

Figure 9 shows the system states and their estimations. The observer handles the FDI attack's challenges caused by its variable intensity and activation frequency. Residual 2, as shown in Figure 10, successfully detects the attack on $u_2(k)$.

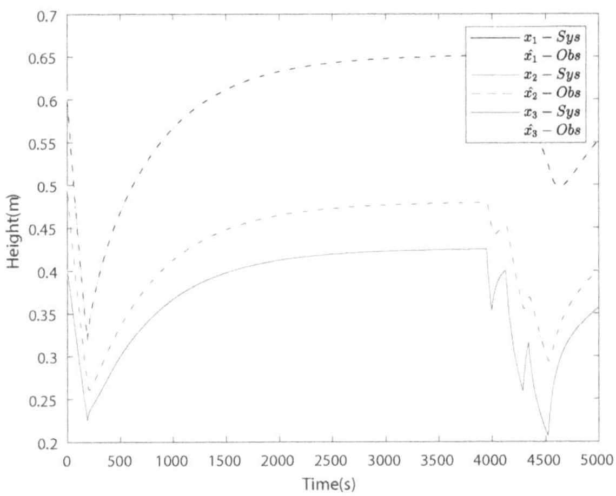

Figure 9. Behavior in the presence of an FDI attack on input u_2.

Figure 10 shows the residuals in the scenario of the FDI attack, where only residual 2 detects the attack in $u_2(k)$.

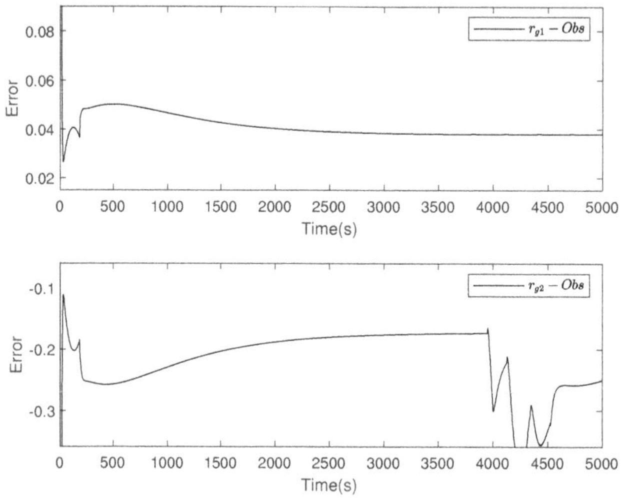

Figure 10. Residues generated in the presence of an FDI attack.

7.3. Simulated RDI Attack

The final simulation investigates a mixed attack scenario where $u_1(k)$ is subjected to an FDI attack and $u_2(k)$ is subjected to an RDI attack, with both under a uniform logic distribution. Figure 11 illustrates the system states and their estimations, while Figure 12 shows the residuals. Both residuals successfully identify the attacks, confirming the robustness of the observer.

Figure 11 shows the state estimation obtained from the observer.

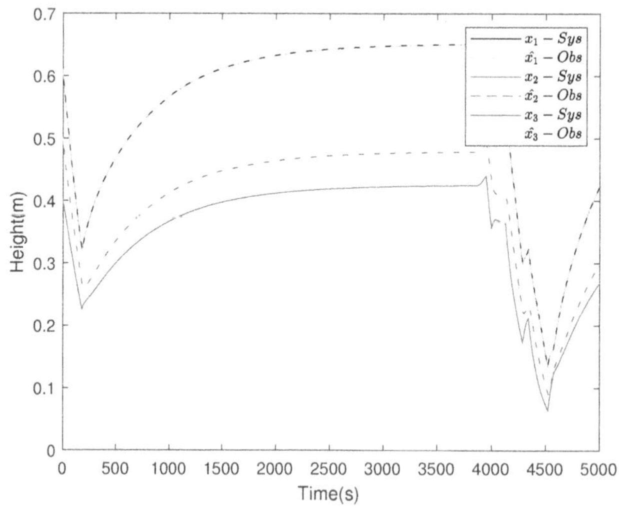

Figure 11. Behavior in the presence of an RDI attack on inputs u_1 and u_2.

In Figure 12, the residuals generated by the observer are shown, where both residues detect the scenario of simultaneous attacks.

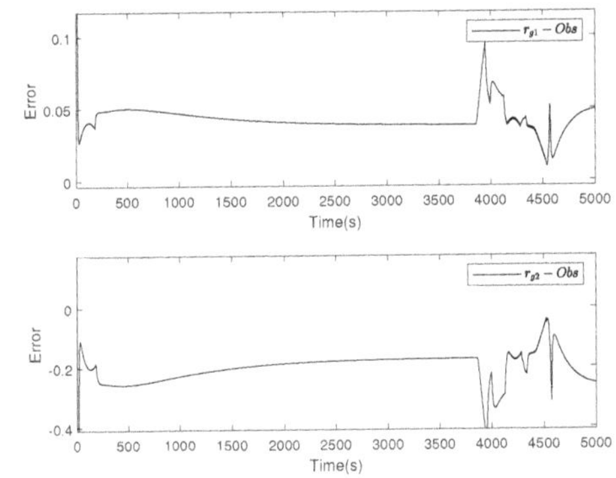

Figure 12. Residues generated in the presence of an RDI attack.

7.4. Different Attacks

In this simulation, a different scheme of attacks in both inputs is considered. The input $u_1(k)$ is affected by an FDI attack, while the input $u_2(k)$ is affected by an RDI attack, both with a uniform logic distribution.

Figure 13 shows the system states and their estimations. In Figure 14, the residuals generated by the observer are presented, and both residues detect the attacks.

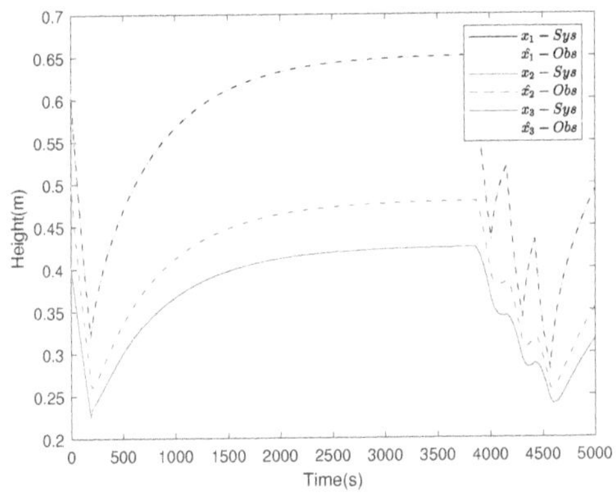

Figure 13. Behavior in the presence of a uniform simultaneous attack on u_1 and u_2.

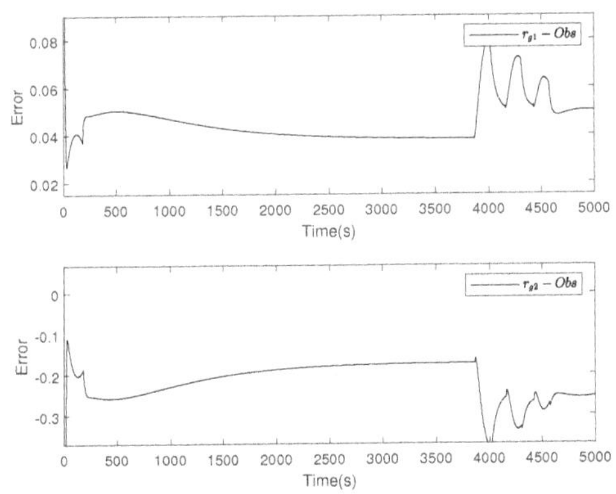

Figure 14. Residues generated in the presence of a simultaneous attack.

8. Conclusions

This article presents an approach for designing a GDO capable of detecting and isolating various attack schemes with Markovian logic in a discrete T-S cyber–physical system. The considered attack schemes include DoS attacks, FDI attacks, and RDI attacks. To realistically simulate attacker behavior, Markovian logic was employed, enabling the modeling of complex attack dynamics and enhancing the detection scheme's effectiveness.

A discrete-time interconnected three-tank system was used as a case study due to its cyber–physical nature, which integrates physical components and digital communication. The stability conditions for the observer were verified using LMIs and leveraging the Schur complement and the elimination lemma to ensure feasible and robust solutions.

This approach demonstrates the effectiveness of a GDO in detecting and isolating attack schemes in CPSs. Moreover, the proposed scheme lays the groundwork for establishing an integrated diagnostic and detection system that can support the implementation of resilient or adaptive control schemes. These features aim to ensure both the security and functionality of the system in various attack scenarios.

However, while GDOs provide significant advantages, their application is not without limitations. A key limitation of the detection and isolation framework is that it addresses only part of the broader diagnostic and control process. Although the GDO effectively identifies and isolates attacks, these capabilities alone are insufficient to fully counteract the effects of the identified attack schemes. Achieving comprehensive system protection requires integrating additional control strategies based on the estimated attack signals, enabling the system inputs to be dynamically adjusted to mitigate adverse effects. Future research should focus on extending this approach to include attack compensation or robust control mechanisms that complement the observer's current capabilities.

Author Contributions: Conceptualization, A.R.G.-E., G.L.O.-G.; methodology, A.R.G.-E., G.L.O.-G. and R.A.V.-M.; software, A.R.G.-E., G.L.O.-G. and J.R.-R.; validation, G.L.O.-G. and C.M.A.-Z.; formal analysis, A.R.G.-E., G.L.O.-G. and C.M.A.-Z.; investigation, A.R.G.-E., G.L.O.-G. and R.A.V.-M.; writing—original draft preparation, A.R.G.-E., G.L.O.-G. and J.R.-R.; writing—review and editing, A.R.G.-E., G.L.O.-G. and R.A.V.-M.; supervision, G.L.O.-G., C.M.A.-Z. All authors have read and agreed to the published version of the manuscript.

Funding: This research received no external funding.

Data Availability Statement: Data are contained within the article.

Acknowledgments: The authors acknowledge CONAHCYT for supporting Angel Rodrigo Guadarrama Estrada through a Ph.D. Scholarship.

Conflicts of Interest: The authors declare no conflicts of interest.

Abbreviations

The following abbreviations are used in this manuscript:

RDI	Random Data Injection
DoS	Denial of Service
FDI	False Data Injection
T-S	Takagi–Sugeno Systems
CPS	Cyber–Physical System
LMI	Linear Matrix Inequalities
GDO	Generalized Dynamic Observer

References

1. Sharma, A.K.; Galav, R.K.; Sharma, B. A Comprehensive Survey of various Cyber Attacks. In Proceedings of the 2023 6th International Conference on Information Systems and Computer Networks (ISCON), Mathura, India, 3–4 March 2023; pp. 1–4. [CrossRef]
2. Geetha, G.; Rajagopal, M.; Purna Chand, K. A Survey on Cyber Attack Detection: Techniques, Datasets, and Challenge. In Proceedings of the 2023 International Conference on Research Methodologies in Knowledge Management, Artificial Intelligence and Telecommunication Engineering (RMKMATE), Chennai, India, 1–2 November 2023; pp. 1–7. [CrossRef]
3. Balarengadurai, C.; C D, D.; C, H. Survey on Cyber Crime Problems and Prevention. In Proceedings of the 2022 International Conference on Smart and Sustainable Technologies in Energy and Power Sectors (SSTEPS), Mahendragarh, India, 7–11 November 2022; pp. 117–121. [CrossRef]
4. Karnik, N.; Bora, U.; Bhadri, K.; Kadambi, P.; Dhatrak, P. A comprehensive study on current and future trends towards the characteristics and enablers of industry 4.0. *J. Ind. Inf. Integr.* **2022**, *27*, 100294. [CrossRef]
5. Arghandeh, R.; von Meier, A.; Mehrmanesh, L.; Mili, L. On the definition of cyber-physical resilience in power systems. *Renew. Sustain. Energy Rev.* **2016**, *58*, 1060–1069. [CrossRef]
6. Somers, R.J.; Douthwaite, J.A.; Wagg, D.J.; Walkinshaw, N.; Hierons, R.M. Digital-twin-based testing for cyber–physical systems: A systematic literature review. *Inf. Softw. Technol.* **2023**, *156*, 107145. [CrossRef]
7. Zhao, X.; Yin, Y.; Bu, X. Resilient iterative learning control for a class of discrete-time nonlinear systems under hybrid attacks. *Asian J. Control* **2023**, *25*, 1167–1179. [CrossRef]
8. Huang, X.; Han, X. Adaptive output-feedback resilient tracking control using virtual closed-loop reference model for cyber-physical systems with false data injection attacks. *Asian J. Control* **2023**, *25*, 1380–1391. [CrossRef]
9. Zhang, L.; Chen, Y.; Li, M. Resilient Predictive Control for Cyber—Physical Systems Under Denial-of-Service Attacks. *IEEE Trans. Circuits Syst. II Express Briefs* **2022**, *69*, 144–148. [CrossRef]
10. Deng, C.; Wen, C. MAS-Based Distributed Resilient Control for a Class of Cyber-Physical Systems With Communication Delays Under DoS Attacks. *IEEE Trans. Cybern.* **2021**, *51*, 2347–2358. [CrossRef] [PubMed]
11. Nateghi, S.; Shtessel, Y.; Edwards, C.; Barbot, J.P. Resilient control of cyber-physical systems using adaptive super-twisting observer. *Asian J. Control* **2023**, *25*, 1775–1790. [CrossRef]
12. Li, Y.; Voos, H.; Darouach, M.; Hua, C. An application of linear algebra theory in networked control systems: Stochastic cyber-attacks detection approach. *Ima J. Math. Control. Inf.* **2016**, *33*, 1081–1102. [CrossRef]
13. Taheri, M.; Khorasani, K.; ImanShames; Meskin, N. Cyberattack and Machine-Induced Fault Detection and Isolation Methodologies for Cyber-Physical Systems. *IEEE Trans. Control. Syst. Technol.* **2023**, *32*, 502–517. [CrossRef]
14. Peixoto, M.L.C.; Coutinho, P.H.S.; Bessa, I.; Pessim, P.S.P.; Palhares, R.M. Event-triggered control of Takagi-Sugeno fuzzy systems under deception attacks. *Int. J. Robust Nonlinear Control.* **2023**, *33*, 7471–7487. [CrossRef]
15. Asai, Y.; Itami, T.; Yoneyama, J. Static Output Feedback Stabilizing Control for Takagi-Sugeno Fuzzy Systems. In Proceedings of the 2021 Joint 10th International Conference on Informatics, Electronics & Vision (ICIEV) and 2021 5th International Conference on Imaging, Vision & Pattern Recognition (icIVPR), Kitakyushu, Japan, 16–20 August 2021; pp. 1–6. [CrossRef]
16. Huang, X.; Chang, C.; Li, J.; Xiao, S.; Su, Q. Cooperative Interaction Observer-Based Security Control for T-S Fuzzy Cyber-Physical Systems Against Sensor and Actuator Attacks. *IEEE Trans. Reliab.* **2024**, *73*, 1982–1992. [CrossRef]
17. Hmaiddouch, I.; Essabre, M.; El Assoudi, A.; El Yaagoubi, E.H. Discrete-time Takagi-Sugeno Fuzzy Systems with Unmeasurable Premise Variables: Application to an Electric Vehicle. In Proceedings of the 2021 Fifth International Conference On Intelligent Computing in Data Sciences (ICDS), Fez, Morocco, 20–22 October 2021; pp. 1–6. [CrossRef]
18. Kavikumar, R.; Kaviarasan, B.; Lee, Y.G.; Kwon, O.M.; Sakthivel, R.; Choi, S.G. Robust dynamic sliding mode control design for interval type-2 fuzzy systems. *Discret. Contin. Dyn. Syst. -S* **2022**, *15*, 1839–1858. [CrossRef]
19. Bezzaoucha Rebaï, S.; Voos, H.; Darouach, M. Attack-tolerant Control and Observer-based Trajectory Tracking for Cyber-Physical Systems. *Eur. J. Control.* **2018**, *47*, 30–36. [CrossRef]
20. Osorio-Gordillo, G.L.; Darouach, M.; Boutat-Baddas, L.; Astorga-Zaragoza, C.M. Dynamical observer-based fault detection and isolation for linear singular systems. *Syst. Sci. Control. Eng.* **2015**, *3*, 189–197. [CrossRef]
21. Bernstein, D.S. *Matrix Mathematics: Theory, Facts, and Formulas—Second Edition*, 2nd ed.; Princeton University Press: Princeton, NJ, USA, 2011; English Edition.
22. Boyd, S.; Ghaoui, L.; Feron, E.; Balakrishnan, V. *Linear Matrix Inequalities in Systems and Control Theory*; SIAM: Delhi, India, 1994.

23. Skelton, R.; Iwasaki, T.; Grigoriadis, D. *A Unified Algebraic Approach To Control Design*; Taylor & Francis: Abingdon, UK, 1997.
24. Li, H.; He, X.; Zhang, Y.; Guan, W. Attack Detection in Cyber-Physical Systems Using Particle Filter: An Illustration on Three-Tank System. In Proceedings of the 2018 IEEE 8th Annual International Conference on CYBER Technology in Automation, Control, and Intelligent Systems (CYBER), Tianjin, China, 19–23 July 2018; pp. 504–509. [CrossRef]

Mathematical and Computational Applications

MDPI

Article

Multiclass Evaluation of Vision Transformers for Industrial Welding Defect Detection

Antonio Contreras Ortiz [1], Ricardo Rioda Santiago [1], Daniel E. Hernandez [2] and Miguel Lopez-Montiel [1,*]

[1] ITJ Labs, Blvd. Salinas 10485-Interior 1403, Aviacion, Tijuana 22014, BC, Mexico; jose.contreras@itj.com (A.C.O.); ricardo.santiago@itj.com (R.R.S.)
[2] Departamento de Ingeniería Industrial, TecNM/Instituto Tecnológico de Tijuana, Calzada Tecnológico S/N, Fracc, Tomás Aquino, Tijuana 22300, BC, Mexico; daniel.hernandezm@tectijuana.edu.mx
* Correspondence: miguel.lopez@itj.com

Abstract: Automating industrial processes, particularly quality inspection, is a key objective in manufacturing. While welding tasks are frequently automated, inspection processes remain largely manual. Advances in computer vision and AI, especially ViTs, now enable more effective defect detection and classification, offering opportunities to automate these workflows. This study evaluates ViTs for identifying defects in aluminum welding using the Aluminum 5083 TIG dataset. The analysis spans binary classification (detecting defects) and multiclass categorization (Good Weld, Burn Through, Contamination, Lack of Fusion, Misalignment, and Lack of Penetration). ViTs achieved 98% to 99% accuracy across both tasks, significantly outperforming prior models such as dense and CNNs, which struggled to surpass 80% accuracy in binary and 70% in multiclass tasks. These results, achieved with datasets of 2400 to 8000 images, highlight ViTs' efficiency even with limited data. The findings underline the potential of ViTs to enhance manufacturing inspection processes by enabling faster, more reliable, and cost-effective automated solutions, reducing reliance on manual inspection methods.

Keywords: Vision Transformer (ViT); weld defect detection; computer vision in manufacturing; automated welding inspection; multiclass classification

1. Introduction

The manufacturing industry has experienced a significant transformation in recent years, particularly since 2017, as numerous authors have underscored [1–4] the growing impact of Artificial Intelligence (AI) in the sector. This shift is largely due to the release and public availability of several powerful AI development frameworks, such as TensorFlow, Keras, and PyTorch. These frameworks have accelerated advancements not only in manufacturing, but also across various scientific fields by simplifying AI development and enabling broader experimentation and application. Previous studies [1] have emphasized the transformative potential of AI in manufacturing, particularly in areas such as predictive maintenance, production optimization, and defect detection within industrial applications. Recent discussions in [4] highlighted predictive maintenance, which enables companies to schedule maintenance before problems arise, reducing unexpected downtime and improving machine reliability. Another area, degradation estimation, helps predict when equipment may fail, allowing companies to plan replacements during scheduled downtime or periods of lower production demand. Furthermore, AI-driven product inspection, the primary focus of this paper, aims to improve quality control by automating the detection of defects and inconsistencies in products, addressing the inefficiencies and limitations of manual inspection processes. A common goal among many companies is to achieve fully automated production pipelines. Although production processes are increasingly automated, inspection workflows remain largely manual, with studies indicating that manual inspection accounts for more than 50% of quality control efforts in some industries. For example, the PHMSA Office of Pipeline Safety (OPS) has reported that OPS workers spend approximately 55% of their time ensuring the safety and consistency of the production pipelines [5]. This inspection process is time-intensive and could benefit significantly from AI-driven automation, potentially reducing manual inspection processes to close to 0% in an ideal scenario. Such automation would allow workers to focus on more complex and high-impact tasks, enhancing both productivity and operational efficiency. Manufacturing companies have long sought to automate both production and inspection processes [6–8]. For example, automation can enhance productivity, quality control, and safety measures in smart factories, such as fine-grained activity classification in assembly based on multi-visual modalities, as presented by Chen in [9]. Although automation of product manufacturing is outside the scope of this study, inspection processes [10,11]—particularly for welding defect detection—are highly relevant. Early examples of automation in this area include neurofuzzy architectures introduced in 1997 [12] and more physics-based models in 2012, such as those used for specular weld pool surface characterization [13]. However, most of these approaches provided only partial solutions and struggled to generalize. With advancements in technology, such as increasingly powerful CPUs and GPUs, and the availability of frameworks like TensorFlow, Keras, and PyTorch, a paradigm shift toward deep learning became inevitable. For example, in 2019, a group led by Daniel Bacioiu explored various neural network architectures, ranging from simple dense networks to more complex combinations of dense and convolutional layers, achieving promising results. Although most approaches could detect the presence of a defect, the advantage of deep learning models was their ability to accurately classify specific defect types. This approach resulted in the detection of weld defects with an accuracy of approximately 70% [14]. This research opened new possibilities for testing diverse architectures, including the use of pre-trained models and transformers. Introduced in 2017, the transformer architecture revolutionized deep learning by enabling the modeling of complex relationships in data. Initially applied to Natural Language Processing (NLP), transformers have since been adapted for computer vision tasks, offering a promising avenue for defect detection. Its standout feature, the attention mechanism, enables the transformer to understand complex relationships within data effectively. Initially designed for sequential data processing—specifically text—the attention mechanism facilitates the creation of contextual embeddings. These embeddings encode the relationships between words in a phrase, helping the model differentiate meanings based on the surrounding context. For instance, the word "apple" may refer to a fruit in some contexts, while in others, it could denote the technology company. This ability to capture nuanced meanings is one of the main innovations of transformers. As highlighted in the original paper [15], this architecture builds upon the encoder-decoder framework of recurrent neural networks. Although recurrent networks could handle basic NLP tasks, such as machine translation, they often struggled with more complex challenges. The emergence of the transformer generated significant interest among researchers and companies alike, leading to its application in a variety of sophisticated NLP tasks, including text classification, sentiment analysis, Natural Language Understanding (NLU), text summarization, text generation, and question answering [16]. This discovery also led to the formation of communities, with Hugging Face [17] emerging as one of the most notable examples. Initially focused on NLP, Hugging Face has expanded its reach into other AI subfields, such as computer vision. However, since transformers were originally designed for sequential data, it was recognized that images, which are inherently non-sequential, posed a challenge. In 2020, researchers at Google tackled this issue by introducing a method called image patching, which converts image data into embeddings in a manner similar to text processing [18]. This approach involves dividing images into smaller segments, or "patches", which are processed independently by a neural network composed of dense layers, convolutional layers, or a combination of both. Each patch is transformed into an embedding, and the embeddings are then concatenated in sequence, creating a structure that allows the use of the attention mechanism to generate contextual embeddings. Unlike text contextual embeddings, image contextual embeddings help the model understand spatial relationships within the image, identifying which areas are most relevant for accurate predictions. Vision Transformer (ViT) architectures have shown significant potential across various fields, often surpassing traditional state-of-the-art models. For instance, in real-time object detection, where models like YOLO previously dominated [19], new models based on transformer architectures, such as Detection Transformer (DETR) [20], have demonstrated remarkable performance. More recently, Real-Time Detection Transformer (RT-DETR) [21] was introduced, claiming to outperform YOLO and other object detection models. These examples illustrate only a fraction of the applications and variations that have evolved from the ViT architecture, which excels at identifying the most relevant regions within an image for accurate predictions. Building on these advancements, our study seeks to surpass the current state-of-the-art in welding defect detection by leveraging ViTs for multiclass classification and exploring their robustness with reduced dataset sizes. We will focus on enhancing accuracy and reliability by leveraging the pre-trained nature of these models to achieve better results. In addition, we plan to assess how many images are necessary to obtain optimal outcomes, highlighting that many existing models can benefit from this approach. This paper presents the following contributions:

- Adapting and implementing ViTs for welding defect detection, including both binary classification to determine defect presence and multiclass classification to identify specific defect types, addressing challenges unique to industrial applications.
- The proposed model represents a substantial improvement over previous implementations, achieving over 90% accuracy in a multiclass classification setting.
- Investigating the impact of dataset size reduction and balancing techniques, identifying minimal data requirements for reliable performance.

The structure of this paper is organized as follows: Section 1 outlines the fundamental concepts and recent advancements in ViTs and their relevance in the detection of industrial defects. It also provides a review of the literature on defect detection techniques, with a focus on machine learning, deep learning, and transformer-based approaches, particularly ViTs. Section 2 describes the methodology, including data collection, multiclass labeling, preprocessing steps, and model training processes. Section 3 presents the experiments conducted, detailing the evaluation metrics and performance results of the models in multiclass defect classification. A discussion compares these results with established benchmarks and previous studies in the field. Finally, Section 4 summarizes the findings, highlights the contributions, and suggests directions for future work in industrial defect detection.

2. Materials and Methods

This section provides a comprehensive overview of the dataset, methodology, models, and evaluation metrics used in this study.

2.1. Dataset

This study uses the public Al5083 Tungsten Inert Gas (TIG) weld defects dataset, created by Bacioiu et al. in 2019 [14]. This dataset was selected for its relevance to industrial defect detection, particularly for Al5083, an alloy widely used in the aerospace and automotive sectors due to its strength and corrosion resistance. It includes 33,254 weld images spanning six defect categories, making it well-suited for evaluating classification models. Although various studies have used different datasets for the detection of welding defects [22–29], the Al5083 dataset has emerged as a prominent choice due to its relevance in the advancement of detection techniques. Recent research [30–33], including the study by Wang et al. (2024) [34], highlights its significance in improving the accuracy of defect detection methods. The Al5083 dataset includes TIG weld images on 5083 aluminum, captured with an HDR camera focused on the weld pool area during the welding process. The dataset has an imbalanced class distribution, which is evident in the default distribution shown in Table 1.

Table 1. Number of categories in the training and test sets.

Label	Train	Test
Good weld	8758	2189
Burn through	1783	351
Contamination	6325	2078
Lack of fusion	4028	1007
Misalignment	2953	729
Lack of penetration	2819	234
Total	26,666	6588

Distribution of TIG 5083 aluminum dataset. The table shows the distribution of training and test samples across different welding defect categories.

Figure 1 presents image samples from the dataset, along with their respective classes: Good Weld, Burn Through, Contamination, Lack of Fusion, Misalignment, and Lack of Penetration.

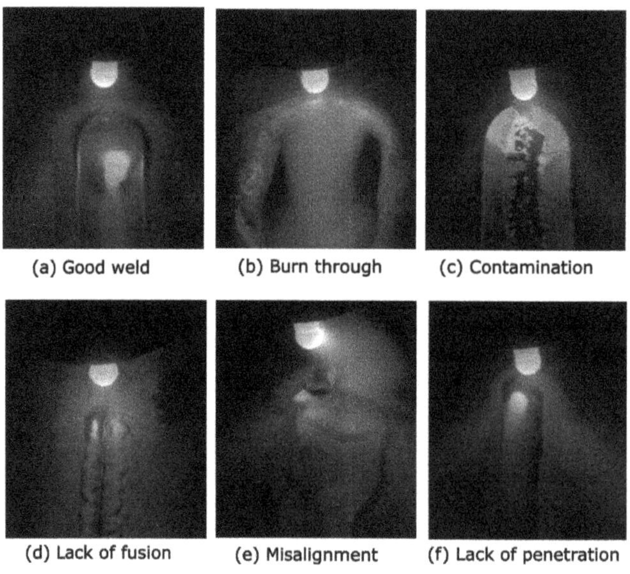

Figure 1. Sample images of six classes.

2.2. Hugging Face

The Hugging Face platform was selected to store the dataset and enable collaborative access due to its robust tools to manage datasets and pre-trained models. Hugging Face's

Transformers library simplified the fine-tuning process for ViTs, streamlining model development and experimentation [35]. Hugging Face is a data science platform that aims to promote collaboration in machine learning research and development [17]. It provides a space for sharing datasets, AI models, and tools, allowing users to upload, share, and access a vast collection of pre-trained models. These models can be fine-tuned for specific tasks or serve as a basis for training new ones. We utilize the `Transformers` and `Datasets` libraries from Hugging Face, which simplify the process of deploying models and managing datasets in multiple domains.

2.3. ViT

In this study, we use ViTs for defect detection and classification due to their ability to process images as patch sequences, capturing long-range dependencies more effectively than traditional Convolutional Neural Networks (CNNs) [36–38]. This global context-awareness makes ViTs particularly suited for identifying subtle patterns in welding defects.

Previous studies have used CNNs, achieving favorable results in the TIG Aluminum 5083 dataset [14,34]. However, we proposed using ViTs to explore whether this approach could yield improvements in metrics. For our work, we have fine-tuned a pre-trained ViT model, named BEiT, for our classification tasks. We will provide a detailed explanation of the BEiT model in Section 2.4. ViT is an architecture that is used for image recognition tasks such as image classification. It employs a Transformer-like architecture that processes images by dividing them into smaller patches. These patches are treated as sequences, allowing the model to learn spatial relationships across the image, similar to how Transformers process sequences of words in NLP tasks [18].

An example of the ViT architecture is shown in Figure 2.

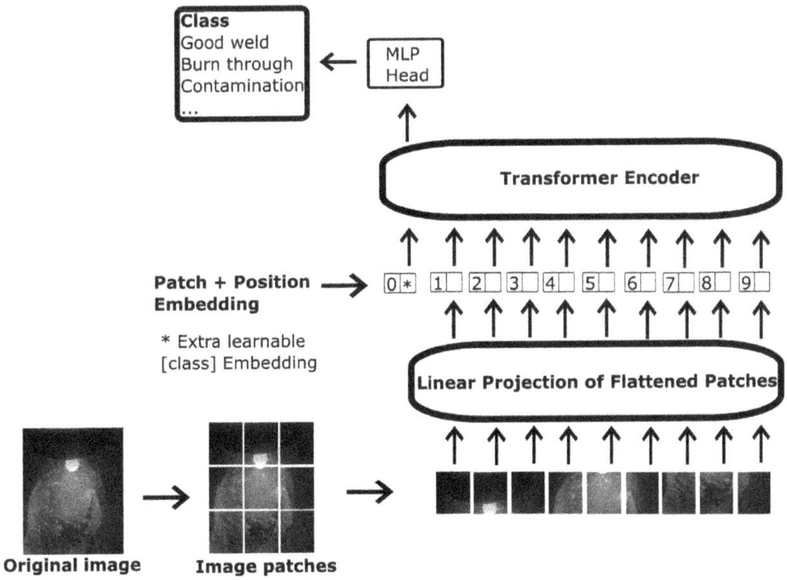

Figure 2. ViT Architecture. The image is divided into fixed-size patches, linearly embedded with position embeddings, and passed to a Transformer encoder. An extra learnable "classification token" is added for classification.

2.4. BEiT

In this work, we propose using a ViT model as an alternative to CNNs. We selected the BEiT model, a popular pre-trained transformer-based model commonly used for image classification. BEiT is a self-supervised ViT model pre-trained on ImageNet-21k at 224 × 224 resolution and can be fine-tuned on specific image classification tasks [39]. We chose BEiT for this study because of its outstanding performance in image classification tasks [38]. Unlike traditional CNNs, BEiT captures global dependencies and contextual information more effectively, which makes it particularly well suited for image defect detection [36,37]. Additionally, its pre-trained architecture significantly enhances its ability to generalize across various image-related tasks.

2.5. Classification Tasks

In this work, we approach the problem of detecting welding defects by defining two classification tasks, defining both binary and multiclass classification tasks for detecting welding defects. The binary classification task distinguishes between good and bad welding, while the multiclass classification task categorizes different types of welding defects. Previous studies have used both approaches, highlighting their relevance in real-world quality control applications. The classification tasks are defined as follows:

1. **Binary Classification Task:** This task simplifies the original six classes into two categories: good or defective welding.
2. **Multiclass Classification Task:** This task involves classifying the weld into its original six categories, preserving the specific types of defects.

Figure 3 illustrates the workflow for each classification task, with different classes assigned to each. In the binary task, the classes are Good Weld and Defective, while in the multiclass task, there are six classes: Good Weld, Burn Through, Contamination, Lack of Fusion, Misalignment, and Lack of Penetration. The figure shows that for each classification task, multiple images may be involved. Each image is processed through its corresponding model, ultimately producing a predicted class.

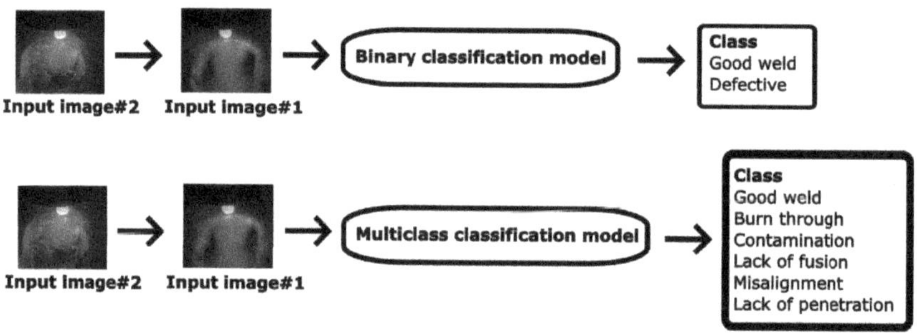

Figure 3. Classification tasks.

2.6. Design of Experiments

We defined five experiments to evaluate the performance of ViTs across varying dataset sizes and task complexities, from binary to multiclass classification. This setup aims to assess the model's robustness, scalability, and ability to generalize across diverse scenarios. For each experiment, modified versions of the original dataset (shown in Table 1) were created to evaluate the performance of the model with varying data volumes. The goal of the experiments was to evaluate the impact of reducing the number of images to 3000 or less on model performance, comparing binary classification and multiclass classification to assess how effectively the model could be trained with this reduced dataset. Each experiment maintained balanced class proportions to prevent bias in performance metrics due to data imbalance. The purpose of the first two experiments was to tackle the simpler, binary version of the problem first, followed by the more complex, multiclass version, allowing us to compare our results with the state-of-the-art models for each version of the problem.

Table 2 presents the data distribution in all experiments. Experiments 1 and 2 focused on binary classification, while Experiments 3 to 5 addressed multiclass classification. Experiment 2 used 25% of Experiment 1's data, Experiment 4 used approximately 57% of Experiment 3's data, and Experiment 5 used around 28% of Experiment 3's data.

Table 2. Dataset distribution across different experiments.

Experiment	Training Samples	Validation Samples	Testing Samples	Classification Task
1	12,000	2000	6000	Binary
2	3000	11,000	6000	Binary
3	8400	1200	2400	Multiclass
4	4800	4800	2400	Multiclass
5	2400	7200	2400	Multiclass

The table presents the number of samples in the training, validation, and testing sets for each experiment, along with the classification task type.

2.7. Experimental Setup and Hardware Configuration

The hardware configuration for training, evaluation, and testing comprised an Intel(R) Xeon(R) CPU running at 2.00 GHz, identified under the GenuineIntel vendor. The processor belongs to CPU family 6, model 85, and features 2 cores per socket, with 2 threads per core, totaling 4 logical cores. The system was equipped with 30 GB of RAM and operated within a KVM hypervisor environment. The computer architecture was x86_64 with a 64-bit CPU mode, running Ubuntu 22.04.4 LTS Linux operating system. The experiments were run using Kaggle notebooks, `Python 3.10`, `PyTorch 2.4.0`, and the transformers library version 4.45.1.

2.8. Training Pipeline

Once the list of experiments was defined, the process of running each experiment involved training five classifiers per experiment. Each classifier was trained using a stratified cross-validation approach, where datasets were randomly partitioned into training and validation subsets to ensure balanced representation across classes. This method reduced the variance in performance metrics and improved reliability. At the end of the training, all five classifiers in the experiment were evaluated and their metrics were averaged. Each classifier was trained for five epochs, based on preliminary experiments showing that model performance stabilized after this number of iterations. This choice balanced computational efficiency and performance consistency. This approach, implemented to perform cross-validation, involved training five classifiers per experiment to reduce variance and improve the reliability of the model. The steps remained consistent for all experiments and proceeded as follows:

1. **Select experiment:** Define the parameters and settings for each experiment, such as the task, the number of training samples and the number of folds F, which corresponds to the number of runs per experiment.
2. **Create a dataset for each fold:** For each fold, create a dataset by randomly selecting N images for training, with the remaining images allocated for validation.
3. **Model training:** Train a model for each fold and validate it using the validation dataset.
4. **Model testing:** Evaluate each model in the test dataset, ensuring that the same evaluation process is applied in all experiments and folds.

As shown in Figure 4, the training pipeline consists of defining experiments, creating balanced class datasets, training classifiers, and evaluating their performance. The pipeline ensures consistency between experiments and facilitates comparison between models.

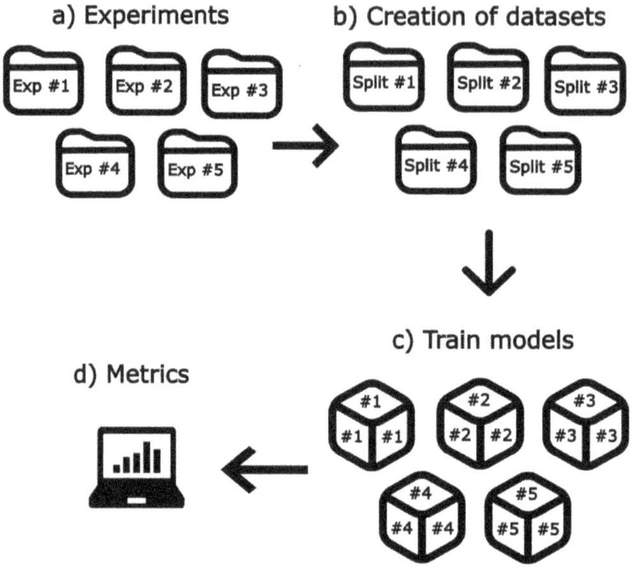

Figure 4. Training pipeline. (**a**) The pipeline consists of five experiments. (**b**) For each experiment, we create five datasets based on the specific experiment setup. (**c**) Then, we train a classifier on each generated dataset. (**d**) Finally, each model is evaluated and the metrics of all models are averaged to summarize the results.

2.9. Model Evaluation Metrics

These metrics were selected because they comprehensively capture different aspects of classifier performance, including general precision, sensitivity to false positives (precision), ability to detect true positives (recall), and balanced evaluation (F1 score), as they provide a comprehensive assessment of their effectiveness [14,29]. Metrics such as Receiver Operating Characteristic Area Under the Curve (ROC-UAC) were not used due to their limited applicability in imbalanced multiclass scenarios.

In the context of defect detection, particularly for welding, each evaluation metric provides unique insights into the model's performance:

- **Accuracy:** Measures overall correctness but can be misleading with imbalanced data, such as rare welding defects.
- **Precision:** Important to minimize false positives, avoiding unnecessary repairs or actions.
- **Recall:** Ensures most defects are identified, reducing the risk of missing critical issues.
- **F1-score:** Balances precision and recall, making it crucial for detecting welding defects, where both false positives and false negatives are costly.

Before defining the metrics, we should define the common terms. The key terms are defined as follows:

- **True Positives (*TP*):** These represent the number of instances where the model correctly predicts the positive class.
- **True Negatives (*TN*):** These refer to instances where the model accurately predicts the negative class.

- **False Positives (FP):** These indicate instances where the model incorrectly predicts the positive class for an actual negative instance.
- **False Negatives (FN):** These occur when the model predicts the negative class for an instance that is actually positive.

2.9.1. Accuracy

Accuracy measures the proportion of correct predictions made by a model of all predictions [14]. It is calculated as:

$$\text{Accuracy} = \frac{TP + TN}{TP + TN + FP + FN} \tag{1}$$

2.9.2. Precision

Precision indicates the model's ability to predict positive outcomes [29,40]. It is calculated as the ratio of true positives to the sum of true and false positives:

$$\text{Precision} = \frac{TP}{TP + FP} \tag{2}$$

2.9.3. Recall

Recall indicates the model's ability to correctly identify all relevant instances [29,40]. It is the ratio of true positives to the total of true positives and false negatives:

$$\text{Recall} = \frac{TP}{TP + FN} \tag{3}$$

2.9.4. F1-Score

The F1-score balances precision and recall as their harmonic mean, ranging from 0 to 1, where 1 indicates perfect performance [29,40]. It is especially useful when both precision and recall are important:

$$\text{F1-Score} = 2 \times \frac{\text{Precision} \times \text{Recall}}{\text{Precision} + \text{Recall}} \tag{4}$$

3. Results & Discussion

After defining the selected model, detailing the training process, outlining the experimental configurations, and specifying the evaluation metrics, we will conduct a thorough analysis of the results, organized by experiment and fold. This section aims to explore how the model performs after applying proper balancing and data undersampling techniques, both for binary and multiclass tasks. In addition, it will highlight the minimal amount of data required to achieve strong performance, demonstrating the efficiency of the ViTs.

3.1. Training Results

Before analyzing the individual evaluation results, it is essential to review the model's performance during training. Table 3 presents the accuracy values obtained on the validation set for each experiment. As previously mentioned, each experiment was repeated five times to ensure consistency, totaling 25 runs across all experiments. Overall, the models showed strong performance, with an average accuracy of 97% across all runs. For the binary classification experiments (Exp. 1 and Exp. 2), the models achieved an average accuracy of 99% and 98%, respectively. Notably, despite a significant reduction in dataset size (from 12,000 to 3000 samples), accuracy dropped by only 1%, highlighting the model's robustness. Similarly, in the multiclass classification experiments (Exp. 3 to Exp. 5), the models achieved average accuracies of 98%, 97%, and 96%, respectively. Even with only 2400 images (400 per class), Experiment 5 achieved state-of-the-art results, maintaining an accuracy above 95%. Additionally, across the 25 runs, the epoch loss in binary classification experiments ranged from 0.015 to 0.030, while in multiclass experiments, it ranged from 0.05 to 0.09. These results present the model's stability across different dataset sizes, showing that high performance can still be achieved with reduced data availability.

Table 3. Final result of each run.

Experiment	Run 1	Run 2	Run 3	Run 4	Run 5	Avg. Accuracy
Exp. #1	0.9915	0.9910	0.9905	0.9910	0.9910	0.9910
Exp. #2	0.9827	0.9827	0.9824	0.9824	0.9824	0.9825
Exp. #3	0.9725	0.9833	0.9833	0.9783	0.9833	0.9801
Exp. #4	0.9742	0.9721	0.9708	0.9708	0.9725	0.9721
Exp. #5	0.9553	0.9550	0.9640	0.9640	0.9640	0.9605

The table summarizes the accuracy results for each experiment taking the last result of the last epoch.

3.2. Binary Models

The first two experiments focused on a binary classification task aimed at determining whether an image depicted a well-executed weld or a defective one. The primary difference between these experiments was the number of images used for training. As mentioned before in the design of experiments section, Experiment 1 used 12,000 images for training. In Experiment 2, however, we aimed to assess how much we could reduce the dataset size while maintaining model performance, so in this case, only 3000 images were used for training.

3.2.1. Experiment 1

In this first experiment focused on binary classification, we used a balanced dataset of 12,000 images, 6000 showing good welds and 6000 showing defective ones. This setup was chosen to establish a baseline for performance comparison, reflecting the importance of balanced datasets in binary classification tasks, as remarked in previous studies. As described in the design of experiments section, each of the five runs generated a new dataset by automatically selecting 6000 good weld images and 6000 defective weld images for training, along with 2000 images for validation, while maintaining a constant set of 6000 images for evaluation. As seen in Figure 5, we observed minimal variability in metrics across runs, consistently ranging between 98% and 99%. This highlights the robustness of ViTs when applied to a well-curated, high-quality dataset. The balanced distribution of classes ensures impartial learning, allowing the model to detect both good and defective welds equally well. Furthermore, effective model tuning enhances feature generalization across runs, further contributing to the stability and reliability of the results.

In Figure 6, which presents the confusion matrix for each run of the experiment on the evaluation dataset, we observe that most models generalize well, exhibiting a similar number of errors, at approximately 60 across runs. This consistency indicates that, with appropriate balancing techniques, favorable results can be achieved even when the dataset varies, as the error count remains relatively constant.

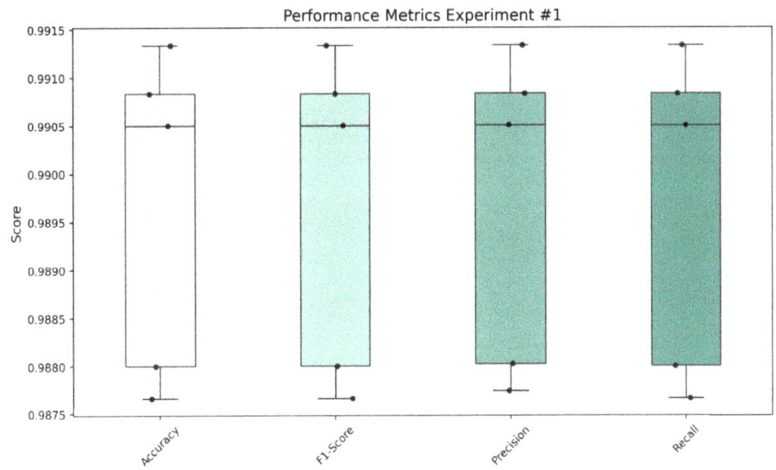

Figure 5. Box plot of different performance metrics from Experiment 1, showing the distribution of results across multiple runs. The model demonstrates consistent stability, with metrics consistently around 99%.

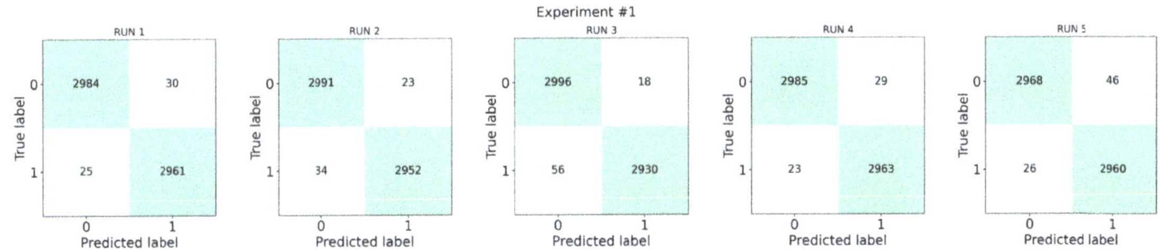

Figure 6. Confusion matrices for the 5 different runs of the dataset used in Experiment 1.

3.2.2. Experiment 2

In this second experiment, we explore the feasibility of reducing the size of a dataset for practical applications, such as early-stage development or resource-constrained environments. Focusing on binary classification, we reduced the training set to 25% of its original size, using only 3000 images for training in each run. As in the previous experiment, each dataset was unique, while the remaining 9000 images were reserved for validation and evaluation. Despite the significant reduction in training data, performance decreased by only 1 to 2% between metrics, as shown in the Figure 7. This result underscores the efficiency of pre-trained transformers, which require less data than traditional models to achieve high performance. These findings suggest that industries could reduce the time and effort spent on dataset creation, enabling earlier deployment of models for production purposes. This experiment highlights the feasibility of using smaller datasets for practical applications, such as early-stage development or resource-constrained environments, while maintaining reliable model performance. The ability of pre-trained transformers to generalize effectively with limited data opens pathways for faster, cost-effective integration into manufacturing workflows.

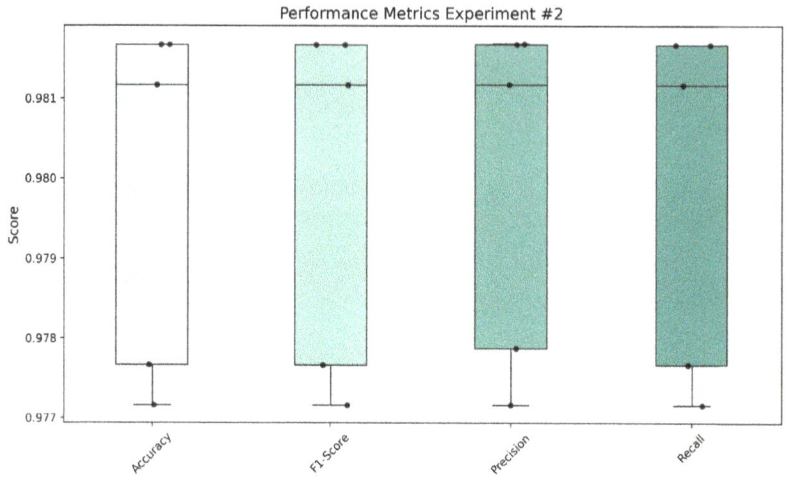

Figure 7. Box plot of different performance metrics from Experiment 2, showing the distribution of results across multiple runs. Minimal reduction in metrics in comparison with Experiment 1.

As can be seen Figure 8, compared to the first experiment, we observed approximately twice the number of errors per run, with around 120 misclassifications in all classes. This increase in misclassifications highlights the trade-off between dataset size and accuracy. Although the models in Experiment 1 outperformed those in Experiment 2, the fact that the error rate only doubled on the test dataset despite using only a quarter of the original data suggests that the models still performed relatively well. This result demonstrates that, while larger datasets can improve accuracy, smaller datasets can still yield good performance, offering a practical advantage in scenarios where reducing data volume or speeding up the training process is necessary, such as in early-stage production or resource-constrained environments.

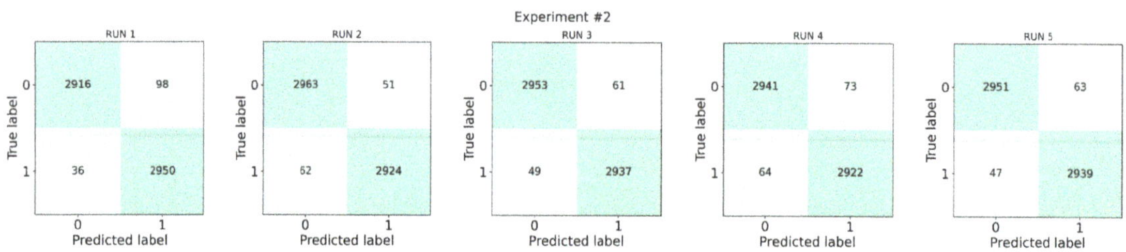

Figure 8. Confusion matrices for the 5 different runs of the dataset used in Experiment 2.

3.3. Multiclass Models

In the second section, we address the multiclass classification challenge, evaluating whether the ViT can accurately distinguish among six different classes. Our experiments aimed to minimize the amount of data used for training, reducing the dataset from 8400 images (1400 images per class) to just 400 images per class in the final experiment.

3.3.1. Experiment 3

Experiment 3 serves as the baseline for the multiclass problem, using a total of 8400 images (1400 per class). This dataset size was selected to provide a robust baseline for multiclass classification, enabling a balanced evaluation across all six classes and serving as a reference for subsequent experiments with reduced datasets. As shown in the following Figure 9, all metrics consistently fall within the 97 to 98% range, significantly surpassing the 70% state-of-the-art results reported by other authors. Notably, this level of performance was achieved while using only around a third of the original dataset size (33,000 images). These results highlight the effectiveness of ViTs in leveraging reduced data volumes when combined with proper data balancing and subsampling. This outstanding accuracy demonstrates that transformers may offer more reliable and efficient defect detection than traditional models, potentially setting a new standard for industrial applications.

As with the previous experiments, Figure 10 indicates that approximately 60 images were misclassified in each run, similar to Experiment 1. Upon reviewing the dataset, including sample images from earlier sections (e.g., Figure 1), we identified several images that appeared to be incorrectly labeled. This suggests that the misclassifications may result from labeling errors or from challenging features that complicate classification, particularly for visually similar defect types. Correcting these labeling errors in future studies could further improve the model's performance and reliability.

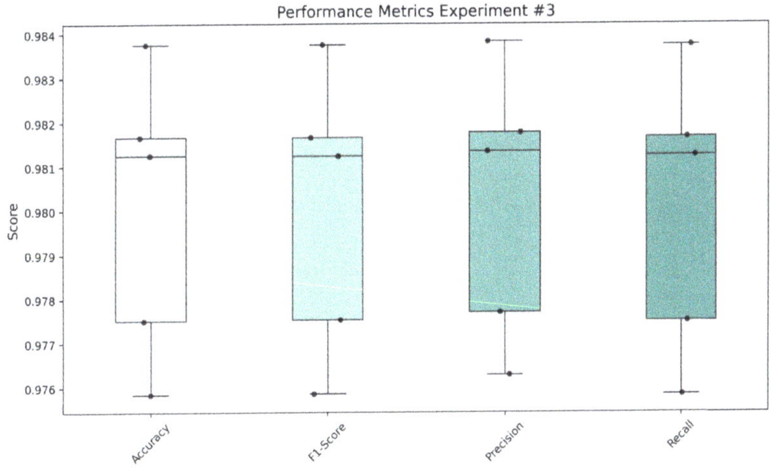

Figure 9. Box plot of different performance metrics from Experiment 3, showing the distribution of results across multiple runs. Baseline for multiclass welding detection.

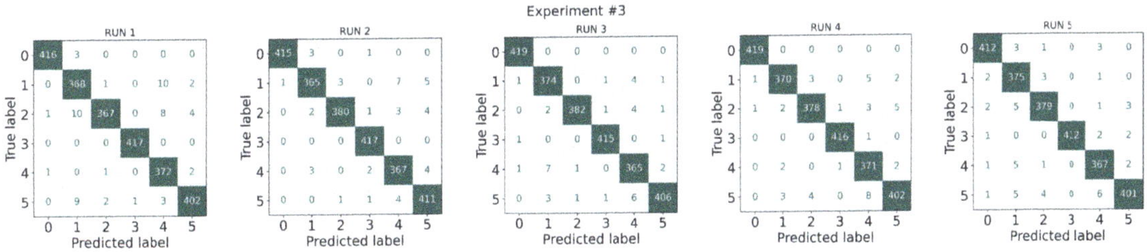

Figure 10. Confusion matrices for the 5 different runs of the dataset used in Experiment 3.

3.3.2. Experiment 4

In the fourth experiment, we reduced the dataset to 4800 images, with each class balanced, representing about 60% of the baseline dataset presented in Experiment 3. As shown in the following Figure 11, even with just 60% of the data, each metric decreased by only 1%, still achieving strong performance at around 97%. Similar to Experiment 2 in the binary case, this result suggests that even with reduced training data, high accuracy can still be achieved, indicating the potential to develop and deploy models faster and reduced the time consuming task of data labeling.

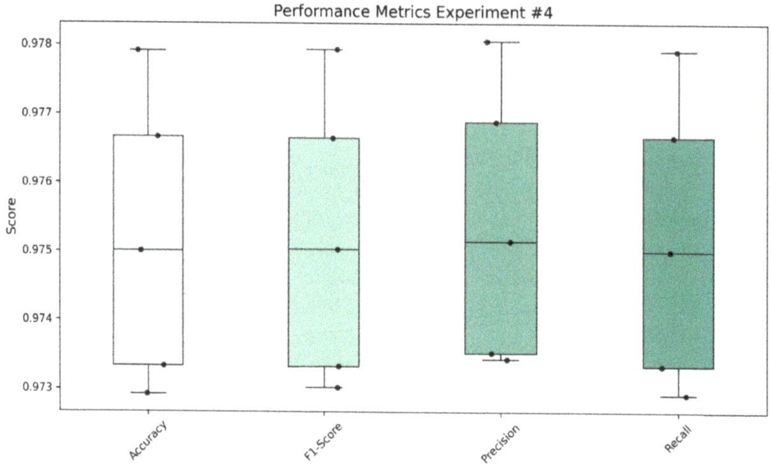

Figure 11. Box plot of different performance metrics from Experiment 4, showing the distribution of results across multiple runs. Performance reduction of less than 1%.

Figure 12 reveals a notable observation: despite reducing the dataset size by nearly half, the number of misclassifications remained nearly constant at around 70 errors, similar to Experiment 3. This indicates that the reduction in data volume had minimal impact on the model's performance or generalization ability. In comparison to traditional CNN-based models, which often experience significant performance drops with smaller datasets, ViTs exhibit superior generalization, highlighting their effectiveness in data-limited scenarios. Additionally, some of these errors could be attributed to incorrect labels, where the model's predictions were accurate, but the labels were flawed. Further investigation and retraining with a correctly labeled dataset could help validate these findings and enhance performance.

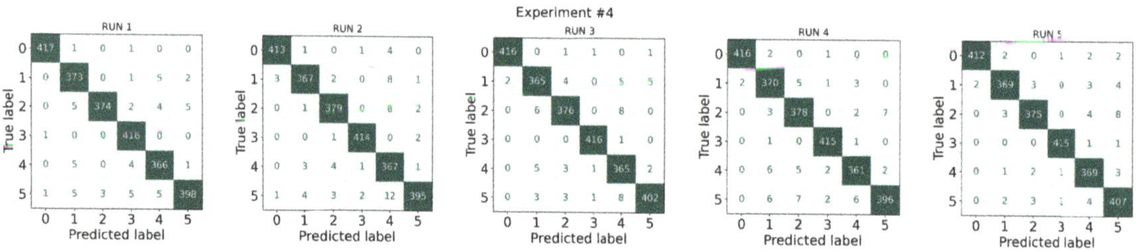

Figure 12. Confusion matrices for the 5 different runs of the dataset used in Experiment 4.

3.3.3. Experiment 5

In this final multiclass experiment, we reduced the dataset to just 2400 images, representing about 30% of the original data. As a result, as seen in Figure 13, we observed a performance drop of around 98% across most metrics in Experiment 3 (using the full dataset) to 95–96% in this experiment. This minimal performance decrease, despite a 70% reduction in data, demonstrates the model's efficiency and robustness with limited data. The model's ability to maintain high accuracy even with substantially less data suggests it can be effective in scenarios where data collection is costly or time-consuming, further supporting its viability for practical applications.

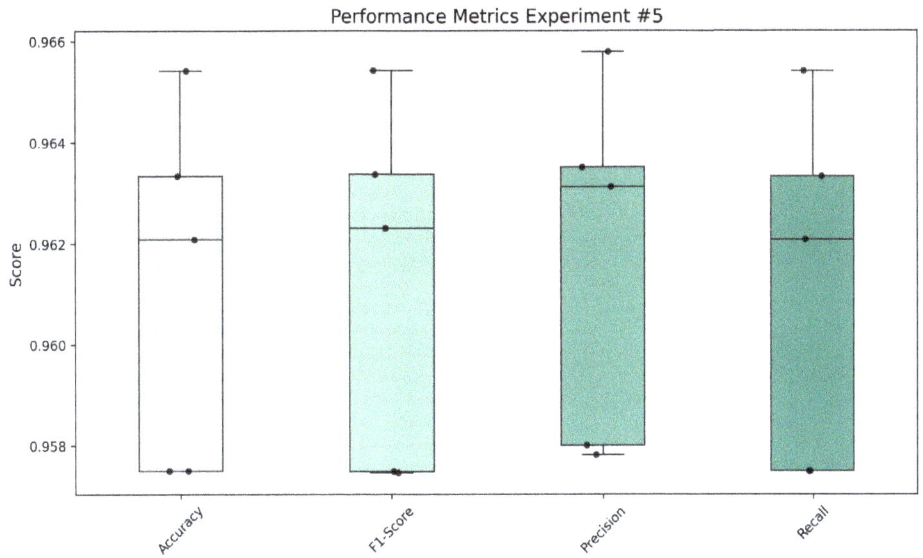

Figure 13. Box plot of different performance metrics from Experiment 5, showing the distribution of results across multiple runs. Performance metrics for multiclass classification with limited data.

In Figure 14, it can be observed that, even with the dataset reduced to about one-third of its original size, the model maintained solid performance. Although the number of misclassifications rose to around 90, this increase was minimal—only about 1.5 higher than when using the full dataset. This modest rise in errors, despite a significant reduction in training data, highlights the model's robustness in multiclass classification tasks, demonstrating its ability to generalize effectively even with limited data.

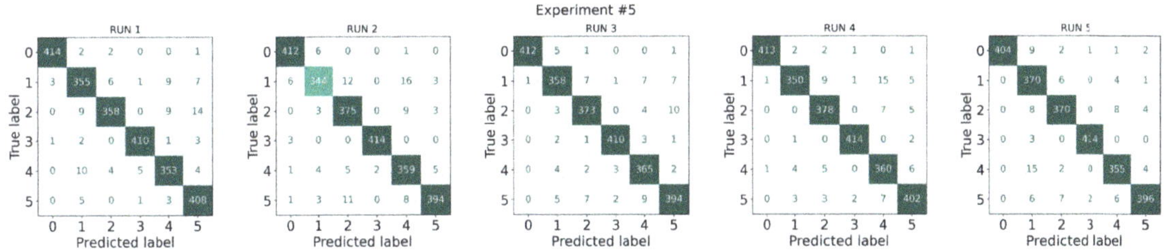

Figure 14. Confusion matrices for the 5 different runs of the dataset used in Experiment 5.

3.4. Inference Time

We evaluated the model's feasibility in an industrial setting by testing its prediction speed. In a simple experiment, we sequentially processed 2400 images from different classes (see Figure 15), simulating a camera sending data for evaluation. Using an NVIDIA Tesla T4, the model achieved a preprocessing speed of 25 images per second. While this performance is decent, it could likely be improved with a distilled model.

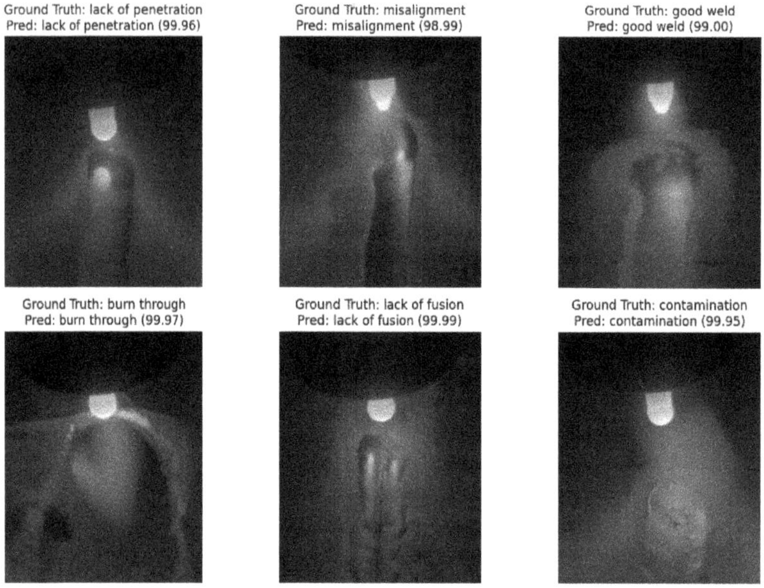

Figure 15. Each class with its associated prediction score.

3.5. Summary of All Experiments (Testing)

After evaluating each model's performance, we compiled the final results, summarized in Table 4. These results show that reducing the amount of training data can still yield strong performance when proper balancing techniques are applied. For instance, in Experiment 2, reducing the dataset size led to only a 1% decrease across all metrics. Similarly, in Experiments 3 to 5, where the multiclass dataset was reduced to just 2400 images, performance dropped by only 2%. This demonstrates the efficiency of ViTs and highlights their potential for real-world applications, particularly in scenarios where data collection and labeling are time-consuming or costly, such as in niche industrial use cases. In contrast, an additional experiment (not included in the summary) using the full 33,000 images without balancing the classes resulted in significantly lower metrics, ranging from only 65 to 70%. This underscores the critical importance of data balancing for model accuracy, especially in transformer-based models. These experiments highlight the ability of ViTs to generalize effectively with smaller, balanced datasets, paving the way for future research into their deployment in resource-constrained industrial environments. Furthermore, the findings emphasize the role of balanced datasets and preprocessing techniques, such as data augmentation and weighted loss functions, in enabling the model to learn effectively from all classes and maintain high performance even with reduced data. Given the good metrics, these models can be used for transfer learning in weld defect classification on other types of welding datasets [41].

Table 4. Final results for all experiments.

Experiment	Accuracy	Precision	Recall	F1-Score	Training Size
Exp. #1—binary	0.990	0.990	0.990	0.990	12,000
Exp. #2—binary	0.980	0.980	0.980	0.980	3000
Exp. #3—multiclass	0.980	0.980	0.980	0.980	8400
Exp. #4—multiclass	0.974	0.976	0.974	0.976	4800
Exp. #5—multiclass	0.962	0.962	0.962	0.962	2400

Table shows the final results for all experiments.

3.6. Comparison with SOTA

After evaluating the ViTs across various experiments, we compared the results with the state-of-the-art outcomes referenced in the introduction. Some teams addressing the same dataset developed 12 models named 1 to 12, categorized into two groups: the first six utilized CNNs, while the second group employed fully connected neural networks. In these experiments, both their models and ours were trained five times. However, a key difference lies in their approach, as described in [14], where each of the 12 models was trained five times using a different learning rate for each iteration, and the entire dataset of 33,000 images was utilized. In contrast, our experiments utilized a smaller subset of the dataset, with 12,000 images for Experiment 1 and 3000 images for Experiment 2. Despite the reduced data, the ViTs demonstrated superior performance, achieving accuracies approaching 99%. Table 5 below presents the best-performing fully connected and convolutional architectures from their experiments, emphasizing the remarkable effectiveness of the ViTs. Object detection models have been used in manufacturing for defect classification [25,26], but for weld defect classification using the AL5083 TIG dataset, only CNNs and dense neural networks have been applied. Thus, the performance of the proposed models is only compared to that of these models [14,34].

In Table 6, we present a summary of the multiclass experiments with a focus on accuracy. The table compares experiments conducted by Daniel Bac et al. [14] and a more recent study by Rongdi Wang et al. [34], which focused on smaller pretrained models. The experimental conditions for Daniel Bac's team remained consistent: five experiments with different learning rates using the full dataset of 33,000 images. In contrast, Rongdi Wang's team employed various models and implemented data augmentation techniques to balance the classes, resulting in a dataset exceeding 33,000 images. In comparison, our experiments prioritized reduced datasets. Notably, Experiment 5 utilized the smallest dataset of all, with just 2400 images, yet achieved an impressive accuracy of approximately 96%, even better than the WeldNet + FE method that uses over 33,000 images. Furthermore, we observed that using around 8000 images brought our accuracy close to 99%. These results demonstrate that our approach outperformed state-of-the-art (SOTA) models, and also highlighted the efficiency of our methodology in achieving high accuracy with limited data.

Table 5. Binary experiments comparison with the SOTA (average accuracy).

Model Name	Accuracy
BeiT-ViT (Experiment 1)	**0.990**
Best Convolutional Architecture (#5) [14]	0.899
Best Fully Conected Architecture (#9) [14]	0.727

The best results are highlighted in bold.

Table 6. Multiclass experiments comparison with the SOTA (average accuracy).

Model Name	Accuracy
BeiT-ViT (Experiment 3)	**0.980**
Best Convolutional Architecture (#1) [14]	0.699
Best Fully Conected Architecture (#12) [14]	0.468
WeldNet + FE [34]	0.891

The best results are highlighted in bold.

Experiment #5 stands out as the most challenging test, yet our model still delivered remarkable results, outperforming all existing models. It achieved an impressive accuracy of 96.20%, which is a substantial improvement over the best convolutional model [14] (69.9%) and fully connected architectures [14] (46.8%), with relative performance gains of 37.63% and 105.56%, respectively. Even compared to WeldNet + FE [34], the previous top-performing model with 89.1% accuracy, our model showed a 7.97% increase. These results underscore the robustness of our approach, proving its effectiveness even when faced with the hardest experimental conditions. Despite the dataset's limited size (2400 images), our model indicated superior defect classification capabilities, reinforcing its efficiency in learning from minimal data. Furthermore, similar patterns of improvement are evident across all other experiments, solidifying the reliability and generalizability of our methodology.

Relative improvement was calculated as follows:

$$\text{Relative Improvement} = \frac{\text{New Score} - \text{Baseline Score}}{\text{Baseline Score}} \times 100 \quad (5)$$

4. Conclusions

In this study, we implemented and evaluated ViT classifiers for industrial welding defect detection, demonstrating their ability to outperform state-of-the-art performance in binary and multiclass scenarios using the AL5083 TIG dataset. This research focused on the development of five experiments: two aimed at binary classification and the other three focused on multiclass classification. Each experiment was designed to robustly validate and train a model using cross-validation. All models achieved an accuracy exceeding 95%. Our best binary classification model reached an accuracy of 99%, surpassing the state-of-the-art by 9%. Similarly, our top-performing multiclass model achieved an accuracy of 98%, outperforming the state-of-the-art by 8%. These metrics not only exceed the state-of-the-art but also demonstrate a key advantage: our models were trained using significantly fewer images. For instance, in Experiment 2, we used only 3000 images to achieve 98% accuracy, and in Experiment 5, we used 2400 images to reach 96.2%. In contrast, state-of-the-art methods often rely on the full dataset or extensive data augmentation, while we employed downsampling techniques. The advantage of our work is that it achieves high accuracy with a small amount of data, allowing us to slightly reduce accuracy while significantly reducing the dataset through downsampling. We performed a comprehensive performance analysis of various ViT configurations, comparing their effectiveness against existing models in the defect detection domain, with a particular focus on the challenges posed by multiclass classification. Multiclass classification, which is inherently more complex than binary classification due to visual similarities between defect types and class imbalances, was effectively addressed in this study, achieving high accuracy even with reduced datasets. Additionally, we found that reducing the dataset size from 20,000 to 3000 images resulted in a small performance drop, demonstrating the potential for deploying ViTs in data-limited scenarios, such as early-stage defect detection workflows or industries with restricted data availability. The experimental results demonstrated strong performance, with ViTs achieving up to 8% higher accuracy compared to existing models, particularly excelling in multiclass defect classification. Although our study has achieved high accuracy, there are still certain limitations that need to be addressed. One such limitation is that the ViT architecture does not offer fast inference times.

Future research could establish a benchmark by evaluating different ViT models, such as DeiT (Data-efficient Image Transformer) or Swin Transformer (Shifted Window Transformer), while also focusing on improving model performance by identifying the minimum dataset size required for high accuracy, enhancing inference speed for real-time applications, and adapting the model to handle various types of welding defects and datasets. When it comes to exploring the deployment of the model in diverse manufacturing environments, we are considering the use of specialized hardware, such as embedded systems. Additionally, to gain deeper insight into the model's decision-making in defect detection, we intend to explore interpretability techniques like attention maps, Gradient-weighted Class Activation Mapping (Grad-CAM), and Shapley Additive Explanations (SHAP) to visualize the regions influencing the model's decisions. We will extend our training process to other welding defect classification problems and datasets, with the aim of generalizing the approach to broader industrial applications, such as pipeline inspection or automotive assembly. Another aspect we intend to explore is the incorporation of transfer learning strategies, such as fine-tuning pre-trained models, and exploring domain adaptation techniques. Additionally, we will investigate data augmentation methods, including the generation of synthetic data, to minimize misclassification between highly similar classes. Finally, we will optimize prediction inference time by prioritizing efficient model selection, such as faster ViT models, to ensure rapid inference and CPU-only feasibility. This is crucial for real-time quality control and decision-making in industrial workflows, especially high-throughput manufacturing environments. In industrial applications, rapid inference is essential for real-time decision-making and operational efficiency, making AI solutions more practical and impactful. By demonstrating the effectiveness of ViTs for welding defect detection, this work lays the foundation for faster, more reliable, and fully automated inspection workflows, significantly reducing operational costs while enhancing quality assurance in manufacturing.

Author Contributions: Conceptualization, A.C.O. and R.R.S.; Data curation, A.C.O. and R.R.S.; Formal analysis, A.C.O. and R.R.S.; Investigation, A.C.O. and R.R.S.; Methodology, A.C.O., R.R.S. and M.L.-M.; Project administration, M.L.-M.; Resources, A.C.O. and R.R.S.; Software, A.C.O. and R.R.S.; Supervision, D.E.H. and M.L.-M.; Validation, A.C.O. and R.R.S.; Visualization, A.C.O. and R.R.S.; Writing—original draft, A.C.O. and R.R.S.; Writing—review&editing, A.C.O., R.R.S., D.E.H. and M.L.-M. All authors have read and agreed to the published version of the manuscript.

Funding: This research received no external funding.

Data Availability Statement: The data presented in this study are available in the Kaggle repository at https://www.kaggle.com/datasets/danielbacioiu/tig-aluminium-5083/data (accessed on 22 January 2025), reference number [14].

Acknowledgments: This research was made possible by ITJUANA, S. DE R.L. DE C.V. through its ITJ Labs department. Furthermore, we extend our gratitude to the Tecnológico Nacional de México for the invitation to the Numerical and Evolutionary Optimization workshop. This work was partially supported by TecNM through the project "Modelos predictivos para clasificación de series de tiempo usando aprendizaje de máquina". We are truly grateful for the contributions and support of the ITJ Labs team, which has played a crucial role in successfully completing this endeavor.

Conflicts of Interest: The authors declare no conflicts of interest. This research was conducted as a Proof of Concept (PoC) within the R&D activities of ITJUANA, S. DE R.L. DE C.V. and is not currently associated with any commercial product or financial interest. While there is a possibility that this research could contribute to a commercial product in the future, no commercialization efforts or financial agreements related to this work exist at the time of publication. The findings presented in this study are independent and solely for academic and scientific dissemination.

References

1. Li, B.; Hou, B.; Yu, W.; Lu, X.; Yang, C. Applications of artificial intelligence in intelligent manufacturing: A review. *Front. Inf. Technol. Electron. Eng.* **2017**, *18*, 86–96. [CrossRef]
2. Lee, J.; Davari, H.; Singh, J.; Pandhare, V. Industrial Artificial Intelligence for industry 4.0-based manufacturing systems. *Manuf. Lett.* **2018**, *18*, 20–23. [CrossRef]
3. Nti, I.K.; Adekoya, A.F.; Weyori, B.A.; Nyarko-Boateng, O. Applications of artificial intelligence in engineering and manufacturing: A systematic review. *J. Intell. Manuf.* **2022**, *33*, 1581–1601. [CrossRef]
4. Arinez, J.F.; Chang, Q.; Gao, R.X.; Xu, C.; Zhang, J. Artificial Intelligence in Advanced Manufacturing: Current Status and Future Outlook. *J. Manuf. Sci. Eng. Trans. ASME* **2020**, *142*, 110804. [CrossRef]
5. United States Department of Transportation. *Federal Effort Allocation Overview*; United States Department of Transportation: Washington, DC, USA, 2024.
6. Wang, Y.J. A Robust FOPD Controller That Allows Faster Detection of Defects for Touch Panels. *Math. Comput. Appl.* **2024**, *29*, 29. [CrossRef]
7. Jäntschi, L.; Louzazni, M. Accelerating Convergence for the Parameters of PV Cell Models. *Math. Comput. Appl.* **2024**, *29*, 4. [CrossRef]
8. Chung, S.B.; Venter, M.P. Analysis and Design for a Wearable Single-Finger-Assistive Soft Robotic Device Allowing Flexion and Extension for Different Finger Sizes. *Math. Comput. Appl.* **2024**, *29*, 79. [CrossRef]
9. Chen, H.; Zendehdel, N.; Leu, M.C.; Yin, Z. Fine-grained activity classification in assembly based on multi-visual modalities. *J. Intell. Manuf.* **2024**, *35*, 2215–2233. [CrossRef]
10. Ramatlo, D.A.; Wilke, D.N.; Loveday, P.W. Digital Twin Hybrid Modeling for Enhancing Guided Wave Ultrasound Inspection Signals in Welded Rails. *Math. Comput. Appl.* **2023**, *28*, 58. [CrossRef]
11. Frixione, M.G.; Roffet, F.; Adami, M.A.; Bertellotti, M.; D'Amico, V.L.; Delrieux, C.; Pollicelli, D. Integrating Deep Learning into Genotoxicity Biomarker Detection for Avian Erythrocytes: A Case Study in a Hemispheric Seabird. *Math. Comput. Appl.* **2024**, *29*, 41. [CrossRef]
12. Zhang, Y.M.; Li, L.; Kovacevic, R. Neurofuzzy model based control of weld fusion zone geometry. In Proceedings of the 1997 American Control Conference (Cat. No.97CH36041), Albuquerquem, NM, USA, 6 June 1997; Volume 4, pp. 2483–2487. [CrossRef]
13. Zhang, W.; Liu, Y.; Wang, X.; Zhang, Y. Characterization of three-dimensional weld pool surface in GTAW. *Weld. J.* **2012**, *91*, 195s–203s.
14. Bacioiu, D.; Melton, G.; Papaelias, M.; Shaw, R. Automated defect classification of Aluminium 5083 TIG welding using HDR camera and neural networks. *J. Manuf. Process.* **2019**, *45*, 603–613. [CrossRef]
15. Vaswani, A.; Shazeer, N.; Parmar, N.; Uszkoreit, J.; Jones, L.; Gomez, A.N.; Łukasz Kaiser; Polosukhin, I. Attention is All you Need. In Proceedings of the 31st Conference on Neural Information Processing Systems (NIPS 2017), Long Beach, CA, USA, 4–9 December 2017; Volume 30.
16. Tunstall, L.; Werra, L.V.; Wolf, T. *Natural Language Processing with Transformers*, Revised ed.; O'Reilly Media: Sebastopol, CA, USA, 2022; Volume 19, pp. 11–15.
17. Shen, Y.; Song, K.; Tan, X.; Li, D.; Lu, W.; Zhuang, Y. HuggingGPT: Solving AI Tasks with ChatGPT and its Friends in Hugging Face. In Proceedings of the 37th Conference on Neural Information Processing Systems (NeurIPS 2023), New Orleans, LA, USA, 10–16 December 2023; Volume 36.
18. Dosovitskiy, A.; Beyer, L.; Kolesnikov, A.; Weissenborn, D.; Zhai, X.; Unterthiner, T.; Dehghani, M.; Minderer, M.; Heigold, G.; Gelly, S.; et al. An Image is Worth 16x16 Words: Transformers for Image Recognition at Scale. In Proceedings of the ICLR 2021—9th International Conference on Learning Representations, Virtual, 3–7 May 2020.
19. Redmon, J.; Divvala, S.; Girshick, R.; Farhadi, A. You only look once: Unified, real-time object detection. In Proceedings of the IEEE Conference on Computer Vision and Pattern Recognition, Las Vegas, NV, USA, 27–30 June 2016; pp. 779–788.
20. Carion, N.; Massa, F.; Synnaeve, G.; Usunier, N.; Kirillov, A.; Zagoruyko, S. End-to-End Object Detection with Transformers. In Proceedings of the 16th European Conference on Computer Vision—ECCV 2020, Glasgow, UK, 23–28 August 2020; Volume 12346, pp. 213–229. [CrossRef]

21. Zhao, Y.; Lv, W.; Xu, S.; Wei, J.; Wang, G.; Dang, Q.; Liu, Y.; Chen, J. DETRs Beat YOLOs on Real-time Object Detection. In Proceedings of the 2024 IEEE/CVF Conference on Computer Vision and Pattern Recognition (CVPR), Seattle, WA, USA, 16–22 June 2023.
22. Bacioiu, D.; Melton, G.; Papaelias, M.; Shaw, R. Automated defect classification of SS304 TIG welding process using visible spectrum camera and machine learning. *NDT E Int.* **2019**, *107*, 102139. [CrossRef]
23. Xia, C.; Pan, Z.; Fei, Z.; Zhang, S.; Li, H. Vision based defects detection for Keyhole TIG welding using deep learning with visual explanation. *J. Manuf. Process.* **2020**, *56*, 845–855. [CrossRef]
24. Hou, W.; Wei, Y.; Guo, J.; Jin, Y.; Zhu, C. Automatic Detection of Welding Defects using Deep Neural Network. *J. Phys. Conf. Ser.* **2018**, *933*, 012006. [CrossRef]
25. Zhang, K.; Shen, H. Solder joint defect detection in the connectors using improved faster-rcnn algorithm. *Appl. Sci.* **2021**, *11*, 576. [CrossRef]
26. Oh, S.J.; Jung, M.J.; Lim, C.; Shin, S.C. Automatic detection of welding defects using faster R-CNN. *Appl. Sci.* **2020**, *10*, 8629. [CrossRef]
27. Yang, Y.; Pan, L.; Ma, J.; Yang, R.; Zhu, Y.; Yang, Y.; Zhang, L. A high-performance deep learning algorithm for the automated optical inspection of laser welding. *Appl. Sci.* **2020**, *10*, 933. [CrossRef]
28. Guo, W.; Huang, L.; Liang, L. A weld seam dataset and automatic detection of welding defects using convolutional neural network. In Proceedings of the Advances in Intelligent Systems and Computing, Shanghai, China, 17–19 August 2018; Springer: Cham, Switzerland, 2020; Volume 905, pp. 434–443. [CrossRef]
29. Ghimire, R.; Selvam, R. Machine Learning-Based Weld Classification for Quality Monitoring. *Eng. Proc.* **2023**, *59*, 241. [CrossRef]
30. Varshney, D.; Kumar, K. Application and use of different aluminium alloys with respect to workability, strength and welding parameter optimization. *Ain Shams Eng. J.* **2021**, *12*, 1143–1152. [CrossRef]
31. Bendikiene, R.; Sertvytis, R.; Ciuplys, A. Comparative evaluation of AC and DC TIG-welded 5083 aluminium plates of different thickness. *Int. J. Adv. Manuf. Technol.* **2023**, *127*, 3789–3800. [CrossRef]
32. Ramana, M.V.; Krishna, M.; Kumar, B.V.R.; Kumar, E.P.; Dilip, Y.; Teja, V.N.S. Investigation on Joint Properties of AA5083 Aluminium Alloy Welded using A-TIG Process. *AIP Conf. Proc.* **2023**, *2754*, 030004. [CrossRef]
33. Kumar, J.D.; Thangaraj, K.; Kaliyaperumal, G.; Gogulan, C.; Kalidas, A.; Yokesvaran, K.; Liyakat, N.A.; Raj, L.S. Mechanical properties and fracture behaviour of varying filler rod composition in TIG welding of 5083 alloys. *Mater. Today Proc.* 2024, *in press*. [CrossRef]
34. Wang, R.; Wang, H.; He, Z.; Zhu, J.; Zuo, H. WeldNet: A lightweight deep learning model for welding defect recognition. *Weld. World* **2024**, *68*, 2963–2974. [CrossRef]
35. Wolf, T.; Debut, L.; Sanh, V.; Chaumond, J.; Delangue, C.; Moi, A.; Cistac, P.; Rault, T.; Louf, R.; Funtowicz, M.; et al. HuggingFace's Transformers: State-of-the-art Natural Language Processing. *arXiv* **2019**, arXiv:1910.03771.
36. Maurício, J.; Domingues, I.; Bernardino, J. Comparing Vision Transformers and Convolutional Neural Networks for Image Classification: A Literature Review. *Appl. Sci.* **2023**, *13*, 5521. [CrossRef]
37. Cuenat, S.; Couturier, R. Convolutional Neural Network (CNN) vs Vision Transformer (ViT) for Digital Holography. In Proceedings of the 2022 2nd International Conference on Computer, Control and Robotics, ICCCR 2022, Shanghai, China, 18–20 March 2022. [CrossRef]
38. Himel, G.M.S.; Islam, M.M.; Al-Aff, K.A.; Karim, S.I.; Sikder, M.K.U. Skin Cancer Segmentation and Classification Using Vision Transformer for Automatic Analysis in Dermatoscopy-Based Noninvasive Digital System. *Int. J. Biomed. Imaging* **2024**, *2024*, 3022192. [CrossRef]
39. Bao, H.; Dong, L.; Piao, S.; Wei, F. BEiT: BERT Pre-Training of Image Transformers. In Proceedings of the ICLR 2022—10th International Conference on Learning Representations, Virtual, 25–29 April 2022.
40. Géron, A. *Hands-On Machine Learning with Scikit-Learn, Keras, and TensorFlow: Concepts, Tools, and Techniques to Build Intelligent Systems*; O'Reilly Media: Sebastopol, CA, USA, 2019.
41. Kumaresan, S.; Aultrin, K.S.; Kumar, S.S.; Anand, M.D. Transfer Learning with CNN for Classification of Weld Defect. *IEEE Access* **2021**, *9*, 95097–95108. [CrossRef]

Mathematical and Computational Applications

MDPI

Article

An Experimental Study of Strategies to Control Diversity in Grouping Mutation Operators: An Improvement to the Adaptive Mutation Operator for the GGA-CGT for the Bin Packing Problem

Stephanie Amador-Larrea *, Marcela Quiroz-Castellanos * and Octavio Ramos-Figueroa

Artificial Intelligence Research Institute, Universidad Veracruzana, Xalapa 91000, Mexico; oivatco.rafo@gmail.com

* Correspondence: stephanieamadorlarrea@gmail.com (S.A.-L.); maquiroz@uv.mx (M.Q.-C.)

Abstract: Grouping Genetic Algorithms (GGAs) are among the most outstanding methods for solving NP-hard combinatorial optimization problems by efficiently grouping sets of items. Their performance depends on problem-specific heuristics and a balance between exploration and exploitation. The mutation operator plays a crucial role in exploring new solutions, but improper mutation control can lead to premature convergence. This work introduces adaptive mutation strategies for the GGA-CGT applied to the One-Dimensional Bin Packing Problem (1D-BPP). These strategies control the level of change that will be introduced to each solution dynamically by using feedback on population diversity, enabling better exploration. The proposed approach resulted in a 4.08% increase in optimal solutions (2227 across all classes) and a severe reduction in the average of individuals with equal fitness (from over 50% to less than 1%), enhancing diversity and avoiding local optima. The adaptive strategies were particularly effective in problem instances with larger item weights, where improvements were the most significant. Furthermore, statistical analysis confirmed the adaptive mutation approach's superior performance compared with the original one. These findings demonstrate the potential of adaptive mechanisms to improve genetic algorithms, offering a robust strategy for tackling complex optimization problems.

Keywords: bin packing problem; grouping genetic algorithm; mutation operator; adaptive control

1. Introduction

Combinatorial optimization seeks to discover the optimal configuration from a bounded set of potential solutions in complex problem spaces. Many of these cases belong to the family of grouping problems since they are related to efficiently partitioning a set of items into groups to optimize a specific criterion, usually to minimize or maximize a particular function. The specialized literature includes many NP-hard grouping problems, like the One-Dimensional Bin Packing Problem (1D-BPP), which is considered particularly challenging in such a way that no polynomial-time algorithms can solve all the instances optimally [1]. The 1D-BPP consists in packing a set of n items into m bins with a fixed capacity c, ensuring that the sum of the weights w_j of the items within each bin does not exceed this capacity [2]. The objective is to minimize the total number of bins used while satisfying the capacity constraints. Due to the fact that it is a combinatorial problem, obtaining the solution requires evaluating many possible combinations, and the number of possible solutions can be greater than $(n/2)^{n/2}$, making exact approaches impractical for large instances.

As a result, heuristic and metaheuristic algorithms have become fundamental tools for obtaining high-quality solutions within reasonable computational time. This has led to the extensive use of metaheuristic techniques, such as local search, swarm intelligence, and evolutionary algorithms, to address grouping problems [3,4]. While these algorithms do not guarantee finding the optimal solution, they can deliver high-quality results within an acceptable execution time, even for large and complex problems. Among the various metaheuristic approaches in the literature for addressing grouping problems, the Grouping Genetic Algorithm (GGA) stands out for its strong performance and adaptability to different problem characteristics and conditions. GGAs extend standard genetic algorithms by integrating a group-based representation scheme and variation operators that function at the group level, making them particularly effective for grouping problems [5]. As a result, the specialized literature includes several variants of the GGA, such as the Grouping Genetic Algorithm with Controlled Gene Transmission (GGA-CGT), designed to solve the 1D-BPP. Its particularity is that it uses genetic components, namely, population initialization strategy, crossover, mutation, and reproduction techniques, designed to operate in a controlled way, favoring the transmission of the best genes while balancing selective pressure and population diversity. In other words, these genetic components are derived from properties and characteristics fundamental to the 1D-BPP and the behavior of the solutions throughout the search process [6].

The success of GGAs lies in their ability to efficiently explore the solution space while addressing specific constraints by exploiting the search directions induced by their variation operators. Among them, crossover and mutation are the most commonly used. However, the literature highlights other operators, such as inversion, cloning, translocation, and injection. These operators play a key role in exploring the solution space, fostering diversity, and guiding convergence towards high-quality solutions. Their efficiency can be significantly improved through the incorporation of adaptive control strategies. Specifically, by integrating multiple mutation or crossover operators, the algorithm gains the ability to dynamically regulate the sequence of variation operations. This adaptability improves the balance between exploration and exploitation and reduces the risk of premature convergence. Consequently, the algorithm becomes more robust and effective in addressing diverse problem instances. The state of the art in GGAs includes methods that exploit the strengths of multiple operators or strategies in their variation operators. Some proposals employ various mutation operators, while others use probabilistic mechanisms to determine the appropriate strategy within the operator to use. This area remains an active field of research, presenting significant opportunities for improving the robustness and performance of GGAs and highlighting the advantages of exploiting multiple operators within variation mechanisms. Integrating various operators or strategies within these, whether for crossover, mutation, or both, improves the algorithm's ability to promote greater diversity in the search process, improves the balance between exploration and exploitation, and increases the chance of reaching optimal solutions across various problem instances. These adaptive strategies represent a promising direction for advancing the robustness and effectiveness of genetic algorithms in solving challenging optimization problems.

Despite its effectiveness and its accreditation as one of the best state-of-the-art algorithms for the 1D-BPP, experimental studies with new instances have shown that the GGA-CGT can converge prematurely to suboptimal solutions when working with some high-difficulty test cases [7]. The effectiveness of the GGA-CGT is caused mainly by the mutation operator, which is the main factor influencing this behavior and applies to more than 80% of the population in each generation. However, the lack of adaptability of this operator leads to a rigid mutation process that does not adjust to the current state of the population, which either disrupts promising solutions or does not introduce sufficient diversity. This rigidity limits the algorithm's ability to balance exploration and exploitation, making it susceptible to premature convergence. To address this limitation, this work proposes an online adaptive control mechanism for the mutation operator in the GGA-CGT. The objective is to dynamically select the most appropriate mutation strategy for each solution based on population diversity indicators, thus avoiding premature convergence and improving the algorithm's ability to successfully navigate complex search spaces. The fundamental scientific contribution of this work is introducing an adaptive control strategy for mutation in the GGA-CGT, which dynamically selects heuristic strategies based on real-time feedback regarding population diversity. Unlike traditional approaches that apply mutation statically or probabilistically, this method ensures a self-regulating balance between exploration and exploitation, improving adaptability and robustness in different problem instances. Experimental results demonstrate that this approach significantly improves the maintenance of population diversity, reduces the risk of premature convergence, and enhances performance on high-difficulty 1D-BPP test cases.

The structure of this work is organized as follows: Section 2 presents the literature review, providing a comprehensive overview of the related works. Section 3 introduces the GGA-CGT, detailing the components' procedure. Section 4 describes the proposed Adaptive_Mutation + RP operator, describing the methodologies and strategies employed in detail. Section 5 presents the experimentations of the proposed strategies, the results, and the analyses that were performed. Section 6 includes the experimental results of the GGA–CGT with the proposed strategy for controlling diversity in the Adaptive_Mutation + RP operator for the 1D-BPP and the comparison with the original GGA–CGT, demonstrating their effectiveness. Finally, Section 7 concludes the paper by summarizing the essential findings, highlighting the main contributions, and discussing potential directions for future research.

2. Related Work

The 1D-BPP has been extensively studied over recent decades due to its many applications in logistics, telecommunications, manufacturing, and transportation. Its complexity places it in the NP-hard class of problems, meaning that solving it optimally requires significant computational resources as the problem size increases [1]. Despite these challenges, the 1D-BPP remains a highly relevant problem, driving the development of heuristic and metaheuristic approaches that efficiently balance solution quality and computational effort. Among these, Genetic Algorithms (GAs) have been proven particularly effective [8]. The Grouping Genetic Algorithm (GGA), introduced by Falkenauer in 1996, is a significant breakthrough in treating the 1D-BPP and similar grouping problems [9]. The GGA introduced several important innovations in his proposal: (1) a representation scheme that encodes bins as genes rather than individual elements, (2) specialized variation operators for crossover and mutation that are suited to the representation of group structures, and (3) a fitness function adapted to evaluate the efficiency of bin utilization [10]. These features allowed GGA to explore the solution space more effectively, achieving solutions of higher quality and computational efficiency than traditional methods. Building on this foundation, Cruz-Reyes et al. [11] proposed the Hybrid GGA for Bin Packing (HGGA-BP), which introduced heuristics for generating the initial population and hybrid reinsertion strategies for items. This hybrid approach improved the performance of the GGA, in challenging instances of the 1D-BPP. Quiroz-Castellanos et al. [6] further advanced this field by developing the GGA with Controlled Gene Transmission (GGA-CGT), which integrated search heuristics to preserve high-quality groups at the same time as maintaining a critical balance between selective pressure and population diversity. This balance was essential to preventing premature convergence and ensuring robust solution space exploration. In 2021, González-San-Martín [12] introduced the GGA-CGT/D, which incorporated a problem reduction technique and a diversification mechanism. These enhancements significantly improved performance on the most challenging benchmark instances of the 1D-BPP. Similarly, in 2022, Amador Larrea [13] proposed a new crossover operator, FI-GLX-1, designed for the GGA-CGT to solve the 1D-BPP. This operator, developed through an experimental study, enhanced the algorithm's performance on challenging benchmarks. The study also highlighted the importance of adapting numerical and categorical parameters to the specific characteristics of the problem.

Maintaining diversity within the population is essential to avoiding premature convergence to suboptimal solutions, particularly in the early stages of the search process. A diverse population enables more uniform solution space exploration, increasing the likelihood of discovering high-quality solutions. Parameters such as mutation and crossover rates directly influence diversity, as they control the degree of variation introduced during the evolutionary process [8]. For example, disruptive mutations can explore previously unexplored regions of the solution space, improving diversity, whereas adaptive strategies balance the exploration and improvement of solutions [8]. Parameter management plays a crucial role in the effectiveness of GAs, with two primary approaches: parameter tuning (offline) and parameter control (online). Parameter tuning involves setting static values before execution [14], which, while being straightforward, often leads to suboptimal results for dynamic or complex landscapes [13]. In contrast, parameter control dynamically adjusts values during execution, allowing the algorithm to adapt to the problem's evolving characteristics [15]. Parameter control can be implemented through deterministic methods, which depend on predefined rules; adaptive methods, which use feedback from the search process; or self-adaptive methods, where parameters are part of the genetic representation [14]. Parameters can be categorized into numerical and symbolic types. Numerical parameters include crossover and mutation rates, while symbolic parameters enclose choices such as the type of operator used or the parent selection mechanism [8]. Despite the fact that numerical parameter control has been extensively studied, working symbolic parameters remain unexplored, presenting significant research opportunities.

Several studies illustrate adaptive control strategies for mutation operators. For instance, Bhatia and Basu [16] proposed an adaptive mutation probability regulation strategy for the 1D-BPP, which adjusts probabilities based on the bin space and the algorithm's progress. Singh and Gupta [17] applied a similar approach to the Multiple Knapsack Problem, dynamically calculating the probability of removing an item based on the number of knapsacks and items. Another study [18] introduced an evolutionary process adaptation strategy, starting with a low mutation probability that gradually increased as the algorithm progressed. Chaurasia and Singh [19] focused on parent solution transmission, determining the extent of inheritance based on fixed parameters and parent solution sizes. Jawahar and Subhaa [20] presented a dynamic operator adjustment strategy, calculating crossover and mutation percentages based on a generational gap factor. Inter-operator strategies have also shown promise. Rossi et al. [21] applied such a strategy in a GGA for Fixed Job Scheduling, dynamically adjusting two mutation operators to balance exploration and exploitation. Peddi and Singh [22] alternated between crossover and mutation operators in a GGA for data clustering, using feedback from the population's state to adaptively guide the search. Similarly, Balasch-Masoliver et al. [23] introduced multiple mutation types in a GGA for Multivariate Microaggregation, improving adaptability and diversity.

The literature also includes experimental approaches to designing GGAs. Ramos-Figueroa and Quiroz-Castellanos [24] proposed an experimental framework for high-performance GGAs, focusing on the individual analysis of algorithm components. In this line of research, the specialized literature includes works such as the study presented in [25], where the authors evaluate and design mutation operators in a GGA for the Parallel-Machine Scheduling Problem $R||C_{max}$, emphasizing problem-specific adaptation. Fernández-Solano [26] and Zavaleta-García [27] applied experimental methodologies to design mutation and crossover operators for image segmentation. Although most works focus on numerical parameter control, adapting these strategies to symbolic parameters is critical to future research. Feedback-driven approaches offer a promising path for dynamically adjusting symbolic parameters and optimizing evolutionary algorithms to meet the demands of increasingly complex problems.

Table 1 presents various applications of the GGA-CGT across different combinatorial optimization problems, including the One-Dimensional Bin Packing Problem (1D-BPP), the U-Shaped Assembly Line Balancing Problem (UALBP), the Parallel-Machine Scheduling Problem $R||C_{max}$, the Image Segmentation Problem (ISP), and the Variable Decomposition in Large-Scale Constrained Optimization Problem (VD-LSCOP). The table highlights key modifications made to the original GGA-CGT to adapt it to each problem, such as changes in crossover operators, mutation strategies, evaluation functions, repair mechanisms, and initial solution generation techniques. These modifications aim to improve solution quality, enhance the balance between exploration and exploitation, and optimize algorithmic efficiency for each specific problem.

Table 1. Applications and modifications of the GGA-CGT for different optimization problems.

Author	Year	Problem	Main Modification
Yorgancılar [28]	2020	1D-BPP, UALBP	Comparison of the techniques used in the GGA-CGT and GGA algorithms is performed, with the objective of measuring and analyzing performance when these techniques are modified one by one or combined.
González-San-Martín [12]	2021	1D-BPP	Problem reduction and a diversification technique are proposed.
Amador-Larrea [13]	2022	1D-BPP	The FI-GLX-1 crossover operator is proposed.
Ramos-Figueroa et al. [25]	2023	$R\|\|C_{max}$	Use of the adapted Gene-Level Crossover (AGLX) operator and the Download Mutation Operator, both specifically adapted to the problem being solved.
Zavaleta-García [27]	2023	ISP	The use of a new evaluation function focused on intracluster distance, random initial population generation, a repair method adapted to the problem, and the use of the FI-GLX-1 crossover operator is proposed.
Fernández-Solano [26]	2023	ISP	The use of a new evaluation function focused on intracluster distance, a random initial population generation, a repair method adapted to the problem, and the use of the Item Elimination operator is proposed.
Perez et al. [29]	2024	1D-BPP, $R\|\|C_{max}$	Part of the coevolutionary cooperation algorithm in one of its phases. Use of the BF-ñ technique for initial solution generation and the Grouping Mutation Operator.
Carmona-Arroyo [30]	2024	VD–LSCOP	Application of the replacement operator techniques and controlled selection from the GGA-CGT to the proposed GGA.

3. Grouping Genetic Algorithm with Controlled Gene Transmission

As was mentioned, one of the best state-of-the-art algorithms for solving the 1D-BPP is the GGA-CGT. In this algorithm, the variation operators and the strategies for selection and replacement are applied in a controlled way, promoting the transmission of the best genes. The GGA-CGT was first proposed by Quiroz-Castellanos et al. [6] and later improved by Amador-Larrea [13], who proposed a new crossover operator. This algorithm aims to maximize the filling of the bins, which also seeks to maximize the fitness values of the population. The fitness function for the GGA-CGT is presented in Equation (1).

$$F_{BPP} = \frac{\sum_{i=1}^{m}(S_i/c)^2}{m} \tag{1}$$

In this equation, m is the number of bins, S_i is the sum of the weights of the items in bin i, and c is the capacity of the bins. The GGA-CGT begins generating an initial population of solutions with the FF-$\tilde{n}$ heuristic. For each new solution in the population, first, the $\tilde{n}$ items with weights greater than $c/2$ are placed in separate bins; then, the remaining items are arranged by using the classical First Fit (FF) heuristic on a random permutation of this subset [6]. The variation operators include a crossover operator and a mutation operator. The crossover operator, Fullness_Items-Gene_Level_Crossover-1 [13], generates one child by sorting both parents in descending order according to the filling of each gene (bin), giving priority to the gene with fewer items when two genes have the same filling. The mutation operator used is Adaptive_Mutation+RP [6]. It includes an adaptive function to calculate the number of bins to be eliminated from each solution, eliminating the least full bins of the solution and reinserting the free items with the Rearrangement by Pairs heuristic. The GGA-CGT implements a controlled reproduction technique [6], where all the individuals have a chance to contribute to the next generation by forcing the survival of the best individuals through an elite group of solutions.

Figure 1 shows an example of the controlled selection and replacement for crossover and mutation, illustrating the main concepts. In Figure 1, in the first panel of the image, an instance of the 1D-BPP is presented, consisting of 16 objects with weights ranging from 1 to 10, where each bin has a capacity c of 10. On the left side, the population of solutions is displayed as individuals x_i, showing their genotype with group representation, the fullness of each bin, and the corresponding fitness values. On the right side, the population is shown after being arranged by using the sorting strategy, which arranges solutions from best to worst fitness. Additionally, groups are assigned: the elite group (x_5, and x_4), individuals with repeated fitness values (x_7), and those eligible to be selected as parents for the random group (R) ($x_6, x_1, x_2, x_3, x_8, x_7$) and the good group ($G$) ($x_5, x_4, x_6, x_1, x_2$).

Figure 1 (i) illustrates the crossover process. First, parents are selected for the random group, R (x_2, x_8, and x_7), and the good group, G (x_6, x_4, and x_1). Once the parent groups have been determined, the crossover begins by selecting one parent from each group. The parent solutions P_1 (x_6) and P_2 (x_2) are first sorted based on bin fullness, prioritizing bins with higher fullness. If two bins have the same fullness, preference is given to the bin containing fewer items. After sorting, genes are transmitted by comparing the parents gene by gene: the gene from the bin with the highest fullness is prioritized, and if two bins have the same fullness, the one with fewer items is transmitted first. If both bins are identical, P_1's gene is transmitted first, followed by P_2's. Once the child is finally formed, bins containing duplicate items are removed, and items that are not yet in the solution are freed. In this case, items 0 and 2 are freed. Finally, these items are reinserted into the solution, generating a new child solution with a fitness value of 0.763. This offspring and others produced during the crossover are then reintroduced into the population, replacing the parents from the random group.

Figure 1 (ii) illustrates the mutation operator. Once the children generated through crossover have been reinserted, the population is sorted from the best to the worst solution based on fitness, placing individuals with repeated fitness values at the end of the order. The elite group individuals (x_8 and x_5) and the individuals selected for mutation (x_2, x_6, x_1) are then identified. Mutation occurs in two steps: first, the elite group of individuals are cloned. Once this step is completed, the mutation process begins for the selected individuals. In this case, solution x_1 is mutated, where the last three genes are removed, freeing the items $(3, 15, 13, 7)$. Subsequently, they are reinserted into the solution by using the repair heuristic.

After mutation, solution x_1 improves its fitness from 0.748 to 0.945. This process is applied to each selected solution. Finally, the cloned solutions are reinserted into the population, replacing individuals with repeated fitness values and the worst-performing solutions. The details of the mutation operator are described in Section 3.1.

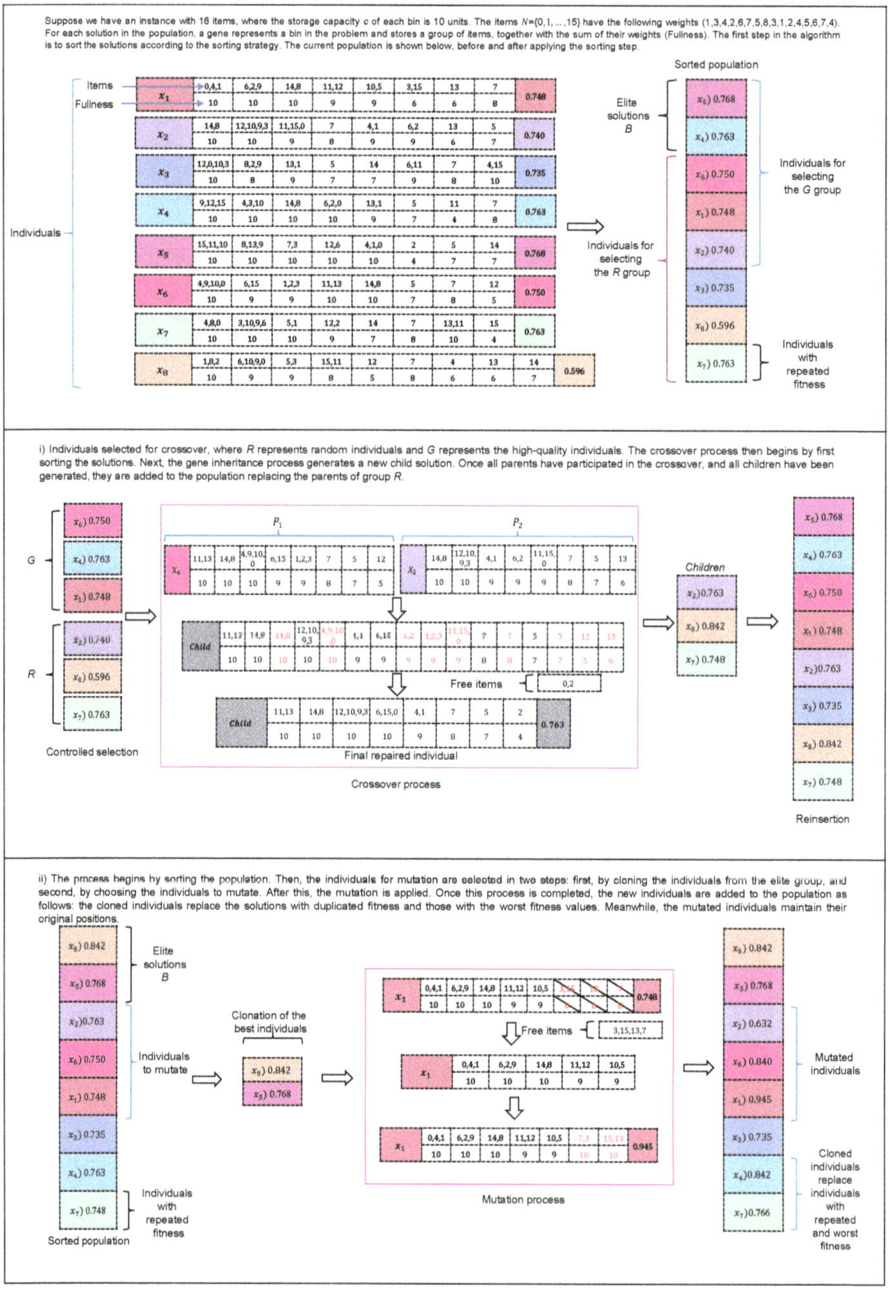

Figure 1. Operation of the GGA-CGT reproduction technique, illustrating the control applied within it for the selection and replacement of solutions for crossover and mutation.

As illustrated in Figure 1, the GGA-CGT applies a controlled reproduction technique in which selection, crossover, and mutation work together to maintain diversity while preserving high-quality solutions. A crucial component of this process is the mutation operator, which introduces selective modifications to improve exploration and avoid premature convergence. The mutation operator used in the GGA-CGT is Adaptive_Mutation + RP, designed to restructure solutions efficiently while maintaining a balance between exploration and exploitation. The following section details its mechanism, including the strategy for determining the number of bins to be removed and the reinsertion process using the Rearrangement by Pairs heuristic.

3.1. Adaptive Mutation Operator

The mutation operator aims to introduce minor, random changes into the solutions. The operator used in the GGA-CGT is Adaptive_Mutation + RP, which works at the gene level to promote the transmission of the best genes in the chromosome [31]. Before the mutation process, the genes in the solution are sorted in descending order according to their bin fill levels. If the solution being mutated is a cloned solution, a random permutation is applied first; then, the genes are sorted in descending order by bin fill. This operator then eliminates genes (bins) from the solution. The number of bins to eliminate in a solution with m bins is based on the relationship between the solution size and the number of incomplete bins, as defined by the following equation:

$$n_b = \lceil \imath \cdot \epsilon \cdot p_\epsilon \rceil, \tag{2}$$

where $\imath$ represents the number of incomplete bins, ϵ is the elimination proportion, and p_ϵ is the elimination probability:

$$\epsilon = \frac{(2 - (\imath/m))}{\imath^{(1/k)}} \tag{3}$$

$$p_\epsilon = 1 - U(0, \frac{1}{\imath^{1/k}}) \tag{4}$$

Equations (3) and (4) incorporate the parameter $k > 0$, responsible for defining the rate of change of ϵ and p_ϵ with respect to $(\imath/m)$; this parameter must be configured offline, and the larger the value of k, the larger the values of ϵ and p_ϵ. Furthermore, the elimination proportion is inversely proportional to the number of incomplete bins $\imath$ and the percentage of incomplete bins $(\imath/m)$. This implies that the smaller the solution, the higher the percentage of bins to be eliminated, and vice versa.

Rearrangement by Pairs

When the mutation is applied, some bins (genes) are removed from the solution, eliminating items that must be reinserted with the Rearrangement by Pairs heuristic to create a feasible solution. Rearrangement by Pairs (RP) is performed in two stages: (1) all bins are scanned for pairs of packed and free items to be exchanged, to improve the filling of bins, and (2) the free items are reinserted by using a First Fit (FF) heuristic. If there are no free items with a weight greater than half the bin capacity, a random permutation is applied to these items before reinsertion. Otherwise, the free items are sorted in descending order by weight before being added to the solution. This process helps to improve solutions while maintaining a balance between the exploration and exploitation of the search space.

Figure 2 presents an example of the operation of the Adaptive_Mutation + RP operator, illustrating step by step the procedure to correctly perform the mutation.

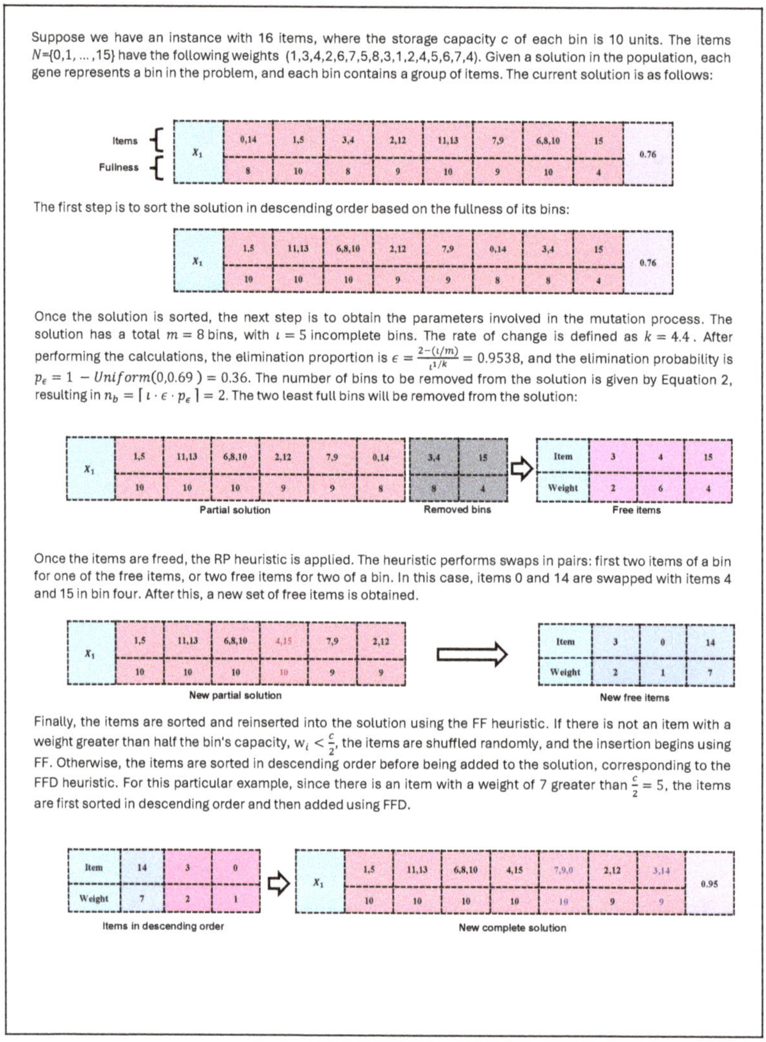

Figure 2. An example of the Adaptive_Mutation + RP operator proposed by Quiroz-Castellanos et al. [6].

3.2. Analysis of GGA-CGT Performance

As mentioned earlier, this research study focuses on an empirical experimental study motivated by observations that selective pressure in the GGA-CGT can lead to variation operators generating solutions with repeated fitness as the evolutionary process advances. This repetition negatively impacts diversity and results in convergence to suboptimal solutions. To address this issue and analyze the GGA-CGT's behavior, we performed the detailed execution of the algorithm to collect relevant behavioral data. It is important to emphasize again that we have worked with the version of the GGA-CGT that includes the crossover operator Fullness_Items-Gene_Level_Crossover-1 (FI–GLX–1) [13]. For the experimentation, the execution of the algorithm was run once for each instance, with the initial seed for random number generation set to 1. The parameter values were configured with the experimental approach used by Quiroz-Castellanos et al. [6]. The values for the parameters were as follows: population size $|P| = 100$, maximum number of generations

$max\_gen = 500$, number of individuals selected for the crossover $n_c = 84$, number of individuals selected for the mutation $n_m = 97$, rate of change for non-cloned solutions $k_{ns} = 4.4$, rate of change for cloned solutions $k_{cs} = 5.6$, and finally, maximal age for an individual to be cloned $life\_span = 10$.

The BPPv_{u}_c benchmark set proposed by Carmona-Arroyo et al. [7] was chosen for this research study due to its high difficulty level. This set consists of 2800 instances divided into four classes, each containing seven subclasses. Each subclass includes 100 test cases, with bin capacities c ranging from 10^2 to 10^8. The first class is BPP.25, where the number of items n ranges within [110, 154]; the item weights are uniformly distributed within the $(0, 0.25c]$ range, and the optimal solution requires 15 bins. The second class is BPP.5, where the number of items n ranges within [124, 167]; the weights are distributed within $(0, 0.5c]$, and the optimal solution uses 30 bins. The third class is BPP.75, where n ranges within [132, 165]; the item weights are distributed within $(0, 0.75c]$, and the optimal solution requires 45 bins. The final class is BPP1, where n ranges within [148, 188]; the weights are distributed within $(0, c]$, and the optimal solution requires 60 bins.

From the execution, the data collected for analysis included the number of instances in which the algorithm obtained an optimal solution and the average number of individuals with repeated fitness after mutation. The second measure was calculated by counting the number of individuals with identical fitness values for each generation after applying the mutation operator. Subsequently, this count was averaged across all generations to calculate a representative measure of the algorithm's performance regarding population diversity for each instance. Finally, we estimated the average for each class. This information provides insights into the algorithm's efficiency, population diversity, and capacity to converge to optimal solutions.

Table 2 presents the results achieved by the GGA-CGT with the FI-GLX-1 operator for solving the 1D-BPP. The optimal solutions obtained are displayed by problem class and bin capacity. The first column lists the problem class, the second indicates the bin storage capacity, the third shows the number of optimal solutions obtained for each class and bin capacity, and the final column presents the average of the maximum count of individuals with repeated fitness values within the population after applying the mutation operator.

The algorithm successfully obtains 2113 optimal solutions out of the 2800 available in the complete benchmark, equivalent to 75.4% of the entire benchmark. Additionally, regarding the number of individuals with repeated fitness values, an increase is observed for instances belonging to the BPP.75 and BPP1 classes with larger bin capacity values. In these cases, it has been observed that in some generations, more than fifty percent of the individuals in the population have repeated fitness values, which could indicate a lack of diversity in the population.

Furthermore, Figure 3 illustrates how the number of individuals with repeated fitness values changes after mutation, indicating whether it increases, decreases, or remains unchanged. To create this graph, the percentage of generations showing an increase, a decrease, or no change in the number of individuals with repeated fitness was calculated for each instance within each class and storage capacity. Then, for each class and capacity (comprising 100 instances each), the relative proportions of these percentages were averaged. This process provided the overall averages for each case: increase, decrease, and no change. The graph consists of four subplots, each corresponding to one of the following classes: BPP.25, BPP.5, BPP.75, and BPP1. Each subplot contains seven lines, one for each storage capacity. These lines represent the following percentages: the decrease in the number of individuals (shown in blue), the increase (shown in melon), and no change (shown in pink). The x-axis displays the percentage values, while the y-axis indicates the different storage capacities for each class. For those cases where the bar appears blank, the optimal solution was obtained in the initial population, so it did not enter the mutation process. Likewise, the cases where only a portion of the bar is visible indicate that the solution of some instances was found in the initial population (blank part of the bars), while the remaining ones went through the mutation process in at least one generation as part of the evolutionary search (visible segment of the bar). Similar to the results presented in Table 2, in Figure 3, it is graphically shown that in the instances belonging to the BPP.75 and BPP1 classes, the number of individuals with repeated fitness values fluctuates continuously after mutation, increasing and decreasing. However, it is also clear that for the instances in the BPP.25 and BPP.5 classes, a higher percentage of cases show a decrease in the number of individuals with repeated fitness values following the mutation process.

Table 2. Results obtained by the GGA-CGT for each subclass of the BPPv_u_c instances. The effectiveness of the algorithm is shown in terms of the number of optimal solutions found and the diversity of the population in terms of the number of individuals with repeated fitness after we applied the original Adaptive Mutation Operator proposed by Quiroz-Castellanos et al. [6].

Class	Bin Capacity	Optimal Solutions	Average Number of Individuals with Repeated Fitness After Mutation
BPP.25	100	100	1.26
	1000	100	0.11
	10,000	100	0.04
	100,000	100	0.49
	1,000,000	100	1.08
	10,000,000	30	2.11
	100,000,000	0	0.11
	Total	530	0.74
BPP.5	100	100	0.08
	1000	100	0.11
	10,000	99	0.30
	100,000	99	0.96
	1,000,000	2	1.64
	10,000,000	3	0.24
	100,000,000	32	0.28
	Total	435	0.52
BPP.75	100	100	0.15
	1000	100	1.58
	10,000	95	5.86
	100,000	20	65.93
	1,000,000	58	68.46
	10,000,000	78	70.67
	100,000,000	78	70.33
	Total	529	40.43
BPP1	100	100	0.42
	1000	100	1.99
	10,000	58	35.13
	100,000	74	55.75
	1,000,000	94	53.24
	10,000,000	97	55.58
	100,000,000	96	54.78
	Total	619	36.70
TOTAL		2113	19.60

As observed and mentioned, on average, the cases in which the number of individuals with repeated fitness values decreases the most are the instances of the classes in which fewer optima are obtained, that is, the subclasses of BPP.25 and BPP.5 with larger bin capacity values, which also have the characteristic of not containing items with weights larger than fifty percent of the bin capacity. In these classes particularly, it is observed that the impact of the mutation operator on diversity is positive, since in all but a minimal percentage, individuals with repeated fitness decrease in most cases. However, it is essential to note that in the BPP.75 and BPP1 classes there is a high proportion of individuals with repeated fitness values, with the increase being higher when the capacity of the bins is increased. This supports the conjecture that the algorithm may be converging to non-optimal areas within the search space. In addition, selective pressure and control may not allow more regions to be explored, decreasing diversity in the population, and resulting in suboptimal solutions.

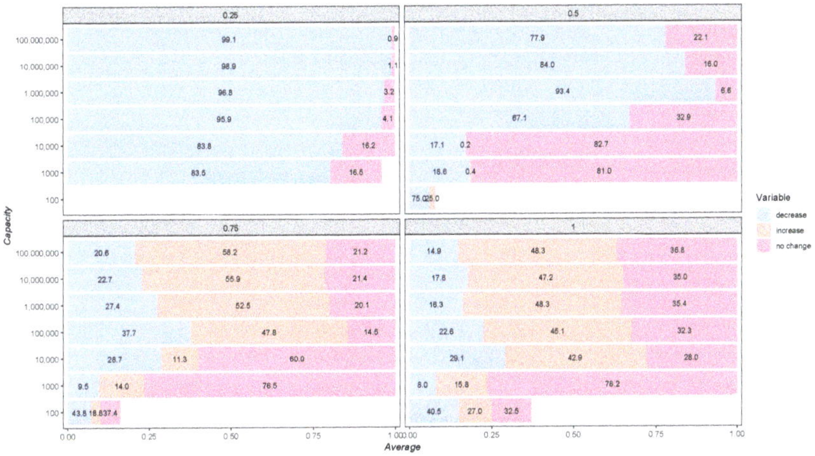

Figure 3. Proportion of generations in which the number of repeated fitness individuals increased, decreased, or remained unchanged during the execution of the GGA-CGT.

Based on the above, an experimental study is proposed to explore strategies that could be used as part of the mutation operator to control diversity within the algorithm and address the issue of multiple individuals with repeated fitness values. This would result in greater diversity within the population, allowing better exploration of the search space by covering a more significant portion of the search space and ultimately converging towards the global optima of the problem's solutions.

4. Experimental Study to Control the Diversity of the Adaptive Mutation Operator

Based on observations and analyses of the execution of the GGA-CGT and insights into selective pressure and control, this study proposes exploring potential areas for improvement within the algorithm. These enhancements focus specifically on strategies within the mutation variation operator. Two critical processes within the mutation operator are proposed for study. The first proposal focuses on how to consider the sort of the genes of solutions to be mutated, which impacts the selection of the bins (genes) to be eliminated from each solution. Doing so facilitates and promotes diversity through the adaptive management of selective pressure. The second proposal centers on the reinsertion of the free items within the Rearrangement heuristic.

4.1. Diversity Control with Adaptive Sorting Strategies of Mutation Operators

We analyzed two approaches for managing population diversity through the Adaptive_Mutation + RP of the GGA-CGT. The first involves the sorting of solution genes before mutation. Initially, as described, this was performed by sorting them in descending order based on gene filling. A random permutation was applied for cloned solutions first, followed by their sorting in descending order. This method aimed to allow solutions to retain the best genes (since mutation removes the worst, or least filled, genes); however, this approach may limit the search space by not enabling more significant disruption. The second process concerns the reinsertion of items in the second stage of the Rearrangement by Pairs heuristic, where the final free items are introduced to the solution by using the FF packing heuristic to ensure valid solutions. Exploring alternative strategies for sorting these free items before reinsertion could improve the mutation process, allowing for a more expansive range of potential solutions and facilitating escape from local optima. The potential strategies for controlling the sorting within the mutation operator are detailed in the following subsections. To evaluate the impact of these sorting strategies, we conducted experiments comparing different approaches to gene and item sorting. The results of these experiments are presented in Section 5.

4.2. Proposal 1: Gene Sorting

The proposals focus on determining how to sort the genes before removing the last genes in the sorted solution. Strategy 1 involves performing only a random permutation of the genes, regardless of the population's characteristics. Strategy 2, on the other hand, uses the value obtained from a threshold function and a random number r to decide whether to sort the genes by using a random permutation or the algorithm's predefined sorting method. The proposed function for the threshold is as follows:

$$threshold = \frac{1}{g} \sum_{i=1}^{g} \frac{n_i^*}{|P|}, \tag{5}$$

where $|P|$ is the population size, n_i^* is the number of individuals with unique fitness values, and g is the number of the current generation. The threshold function is updated at each generation by calculating the average proportion of individuals with unique fitness values relative to the total population, accumulated over all previous generations. When the number of individuals with repeated fitness values increases, the threshold value decreases more rapidly. Conversely, when it is lower, the threshold decreases more slowly. Additionally, this threshold rule ensures that when the number of individuals with unique fitness values is high (and the number of repeated values is low), the threshold remains stable or decreases very little. The opposite occurs when the number of unique fitness values is low.

For sorting strategy 2, the rule is as follows: Once the threshold value is calculated for a given generation, the next step is to generate a random value r with a uniform distribution, such that $r \sim \mathcal{U}(0,1)$. If r is greater than the threshold, then only a random permutation is performed on the genes of the solution. If the opposite case occurs, the genes are sorted by using the default operator within the algorithm, which consists of sorting the genes of the solution in descending order concerning the filling of the bins. The objective is that as the number of individuals with repeated fitness values increases, the threshold value decreases, giving a higher probability that the generated random value will be greater than the threshold. As a result, the genes of the parents will be randomly sorted, to introduce more diversity in the mutated solution. In Equation (6), the function representing the adaptive sorting strategy is presented:

$$\text{Sorting\_genes} = \begin{cases} \text{Random permutation} & \text{if } r > threshold, \\ \text{Sort the bins in descending order of their filling} & \text{if } r \leq threshold. \end{cases} \tag{6}$$

The proposals were implemented within the GGA-CGT, performing an online inter-operator parameter control.

4.3. Proposal 2: Item Sorting

The second proposal focuses on how free items are selected to be added to the solution by using the First Fit (FF) heuristic to generate a complete solution in the second stage of the Rearrangement by Pairs heuristic. As previously explained, the original strategy for sorting free items operates as follows: If at least one of the free items has a weight greater than half the bin's storage capacity, the items are sorted in descending order by weight. Otherwise, when there are no large items, they are arranged through a random permutation.

The strategies proposed for sorting the items are similar to those explored for gene sorting, as follows:

1. In the first approach, all the items are arranged by using a random permutation.

$$\text{Sorting\_items} = \text{Random permutation.} \tag{7}$$

2. The second method leverages the threshold defined in Equation (5). The rule is as follows: A random number is generated with a uniform distribution, such that $r \sim \mathcal{U}(0,1)$. If $r > threshold$, a random permutation is applied to the items; otherwise, they are sorted in descending order according to their weights.

$$\text{Sorting\_items} = \begin{cases} \text{Random permutation} & \text{if } r > \textit{threshold}, \\ \text{Sort items in descending order of their weights} & \text{if } r \leq \textit{threshold}. \end{cases} \tag{8}$$

Similar to the previous approach, this rule was implemented in the GGA-CGT, within the RP heuristic. After performing the swaps between items, the missing items in the solution are added by using the FF method. The critical difference lies in how these items are sorted before being selected, where the proposed strategy was introduced.

5. Experimental Results and Analysis

This section presents the proposed evaluation of the ordering strategies within the mutation operator. The analysis focuses on the number of optimal solutions, the diversity of the population, and the individuals with repeated fitness. The experimental conditions for this study were consistent with those detailed in Section 3.2: we use the same configuration for the parameters, and for each instance, the execution of the algorithm was run once with the initial seed for the random number generation set to 1. All the graphs that depict the percentages of individuals whose fitness values increase, decrease, or remain unchanged presented in this work were generated as follows: For each instance within each class and storage capacity, the percentage of generations showing an increase, a decrease, or no change in the number of individuals with repeated fitness was calculated. These percentages were then averaged across 100 instances per class and capacity to compute the relative proportions for each case: increase, decrease, and no change. Each graph consists of subplots corresponding to the four classes analyzed: BPP.25, BPP.5, BPP.75, and BPP1. Within each subplot, lines are presented for the seven storage capacities, illustrating the proportion of individuals whose fitness values decreased (shown in blue), the increase (shown in melon), and no change (shown in pink). The x-axis represents percentage values, while the y-axis displays the different storage capacities for each class. These results ensure a comprehensive representation of how mutation impacts the fitness of individuals across various scenarios.

5.1. Performance of Gene Sorting Strategies

The results of applying the gene sorting strategies are summarized in Table 3. This table shows the number of optimal solutions obtained by class and bin capacity with the implementation of the rule in strategies 1 and 2. We observe that a greater number of optimal solutions is achieved when utilizing the proposed adaptive strategy to control diversity by obtaining feedback on the population diversity during the evolutionary process (strategy 2). In the case of the strategy that always applies a random permutation to sort the genes, a total of 675 optimal solutions are obtained, representing only 24.1% of the total instances in the benchmark dataset. Compared with the original version, this means a 51.3% decrease in the number of optimal solutions, with reductions observed across all classes. Additionally, the number of individuals with repeated fitness values shows a significant decrease, particularly in the classes and capacities where fewer optimal solutions are achieved. The last is primarily due to the drastic reduction in selective pressure, which leads to such a high level of diversity among solutions that they fail to converge effectively.

For the second strategy, where the threshold rule is used, we observe that with the proposed rule, 78.6% of cases are solved optimally, compared with 75.4% in the original version; this represents an increase of 3.2% in optimally solved instances, equivalent to 88 additional cases. Furthermore, we measured the number of individuals with repeated fitness to assess the impact of exploring more regions within the search space. In the original version, as mentioned earlier, certain instances with specific storage capacities show that the BPP.75 and BPP1 classes exhibit more individuals with repeated fitness after the mutation process. This behavior persists when the algorithm is executed with the proposed threshold rule. However, despite this, a significant reduction in the average number of individuals with repeated fitness is achieved, especially in the BPP.75 class, where there is a decrease of over 30 individuals with repeated fitness in the population, and in the BPP1 class, with a reduction of up to 20 individuals. For the BPP.25 and BPP1 classes, there is an average increase of only one individual in the population, yet there is a noticeable rise in the number of optimal solutions found.

Additionally, Figures 4 and 5 illustrate the average of individuals with repeated fitness by class and capacity for the proposed gene-sorting strategies, namely, random sorting and sorting based on the threshold rule. These proportions are shown as the percentage of individuals whose fitness increased decreased or remained unchanged after the mutation process. Starting with the version that applies a random permutation before mutation, in Figure 4, it is observed across all four classes that for capacities greater than 10^3, the proportion of generations where the number of individuals with repeated fitness increases is less than one or even zero. There is a higher proportion for capacities of 10^2 and 10^3, even exceeding 50% in the BPP.5 class. This could be attributed to the randomness introduced, shifting from a fully controlled process favoring the best genes to allowing randomness to prioritize any gene. Furthermore, this is the class where the fewest optimal solutions were obtained.

For the version that employs the threshold rule to decide whether to sort by using a random permutation or in descending order, Figure 5 illustrates how proportion of generations in which individuals with repeated fitness changed, increased, decreased, or remained the same. In this proposal, the BPP.25 and BPP.5 classes exhibit very similar behavior, with a negligible or nearly negligible proportion of repeated fitness increases in instances with large storage capacities. However, for capacities of 10^2 and 10^3, there is a noticeable increase in individuals with repeated fitness. A distinct tendency occurs in the BPP.75 and BPP1 classes; most instances show either a significant increase or no change in the number of individuals with repeated fitness after mutation, and reductions are observed in only a small proportion of cases. This could suggest that while the proposal initially reduces the number of individuals with repeated fitness, the selective pressure eventually drives the search back into regions where premature convergence occurs, resulting in reduced diversity.

Table 3. Number of optimal solutions found and the diversity of the population in terms of number of individuals with repeated fitness after the execution of the GGA-CGT with the Adaptive Mutation Operator proposed by Quiroz-Castellanos et al. [6] with the random sorting of bins and with an adaptive strategy to control diversity. The table highlights in bold the case where the highest number of optimal solutions was achieved.

Class	Bin Capacity	GGA-CGT + Adaptive_Mutation + RP with Random Sorting of Genes		GGA-CGT + Adaptive_Mutation + RP with Adaptive Strategy to Sort Genes	
		Optimal Solutions	Average Number of Individuals with Repeated Fitness After Mutation	Optimal Solutions	Average Number of Individuals with Repeated Fitness After Mutation
BPP.25	100	100	1.26	100	1.26
	1000	100	0.11	100	0.07
	10,000	5	0.06	100	0.04
	100,000	0	0.00	100	0.38
	1,000,000	0	0.00	100	1.00
	10,000,000	0	0.00	37	2.55
	100,000,000	0	0.00	0	0.07
	total	205	0.20	**537**	0.77
BPP.5	100	100	1.20	100	1.26
	1000	44	0.79	100	0.19
	10,000	0	0.02	100	0.31
	100,000	0	0.00	99	0.99
	1,000,000	0	0.00	1	2.07
	10,000,000	0	0.00	6	0.14
	100,000,000	0	0.00	33	0.29
	total	144	0.29	**439**	0.75
BPP.75	100	100	8.62	100	3.22
	1000	2	1.06	100	1.88
	10,000	0	0.06	92	3.90
	100,000	0	0.00	26	28.95
	1,000,000	0	0.00	67	33.49
	10,000,000	0	0.00	94	34.33
	100,000,000	0	0.00	95	34.63
	total	102	1.39	**574**	20.06
BPP1	100	100	10.26	100	6.52
	1000	4	1.03	100	2.00
	10,000	9	0.56	65	18.67
	100,000	24	0.61	87	29.95
	1,000,000	31	0.66	100	31.90
	10,000,000	29	0.67	100	31.26
	100,000,000	27	0.61	99	32.55
	total	224	2.06	**651**	21.84
Total		675	0.99	**2201**	10.85

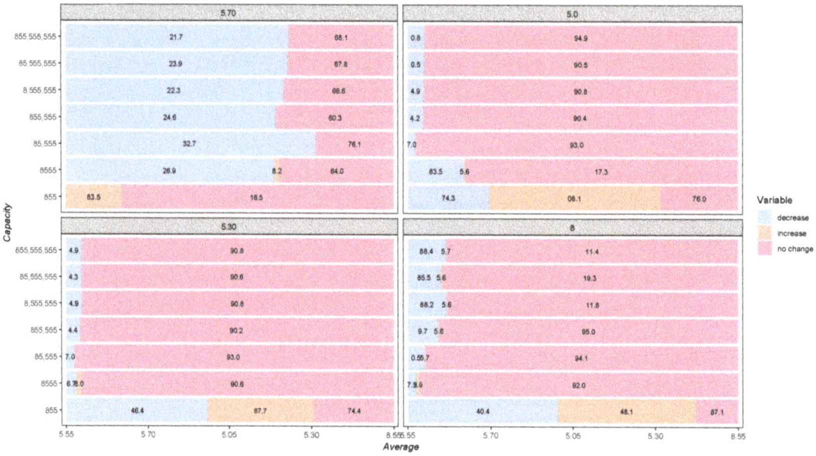

Figure 4. Proportion of generations in which the number of repeated fitness individuals increased, decreased, or remained unchanged during the execution of the GGA-CGT with random gene sorting.

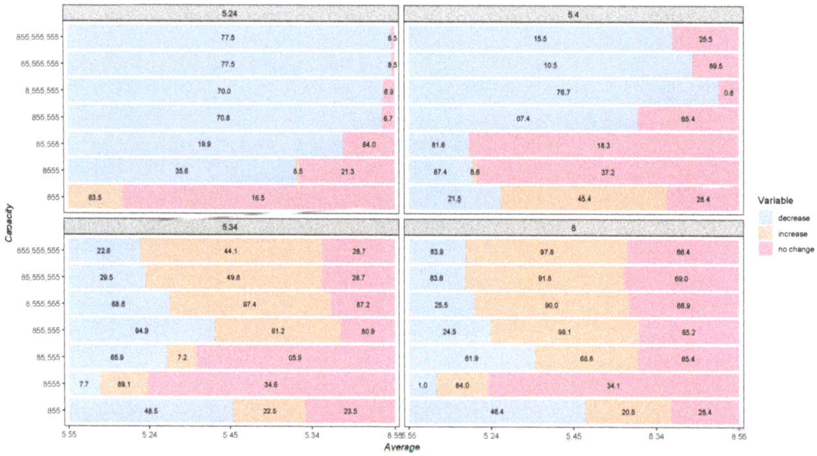

Figure 5. Proportion of generations in which the number of repeated fitness individuals increased, decreased, or remained unchanged during the execution of the GGA-CGT with the threshold rule.

5.2. Performance of Free Item Sorting Strategies

The results of the experimental approach for sorting free items are presented in Table 4. This table includes a column showing the number of optimal solutions obtained for each class and bin capacity, along with an additional column displaying the number of individuals with repeated fitness values after mutation. Notably, the random permutation method achieves the highest number of optimal solutions, totaling 2192, which corresponds to 78.3% of the overall results. In comparison, the strategy that sorts items based on the threshold condition produces 2171 optimal solutions, representing 77.5% of the total cases solved optimally. These findings underscore that consistently applying random permutation sorting yields the best outcomes, delivering 2.82% more optimal solutions than the original method, equivalent to an additional 79 optimal solutions.

Table 4. Number of optimal solutions found and the diversity of the population in terms of number of individuals with repeated fitness after the execution of the GGA-CGT with the Adaptive Mutation Operator proposed by Quiroz-Castellanos et al. [6] with the random sorting of items and with an adaptive strategy to sort the items in the second stage of the RP heuristic. The table highlights in bold the case where the highest number of optimal solutions was achieved.

Class	Bin Capacity	GGA-CGT + Adaptive_Mutation + RP with Random Sorting of Items		GGA-CGT + Adaptive_Mutation + RP with Adaptive Strategy to Sort Items	
		Optimal Solutions	Average number of Individuals with Repeated Fitness After Mutation	Optimal Solutions	Average Number of Individuals with Repeated Fitness After Mutation
BPP.25	100	100	0.00	100	0.00
	1000	100	0.03	100	0.02
	10,000	100	0.04	100	0.02
	100,000	100	0.48	100	0.39
	1,000,000	100	1.07	100	0.96
	10,000,000	30	2.11	21	1.94
	100,000,000	0	0.11	0	0.06
	total	**530**	4.15	521	0.48
BPP.5	100	100	0.12	100	0.12
	1000	100	0.11	100	0.09
	10,000	99	0.29	99	0.20
	100,000	99	0.96	97	1.00
	1,000,000	2	1.64	3	1.53
	10,000,000	3	0.24	5	0.30
	100,000,000	32	0.28	29	0.30
	total	**435**	5.05	433	0.51
BPP.75	100	100	0.92	100	0.92
	1000	100	0.21	100	0.22
	10,000	95	0.94	94	1.13
	100,000	27	3.95	22	5.65
	1,000,000	65	2.32	66	2.54
	10,000,000	96	0.56	96	0.54
	100,000,000	99	0.24	100	0.21
	total	**582**	9.33	578	1.60
BPP1	100	100	1.13	100	1.08
	1000	100	0.46	100	0.32
	10,000	65	4.13	59	4.47
	100,000	81	2.15	82	2.79
	1,000,000	99	0.37	98	0.42
	10,000,000	100	0.10	100	0.12
	100,000,000	100	0.11	100	0.12
	total	**645**	6.25	639	1.33
Total		**2192**	6.19	2171	0.98

Regarding the number of individuals with repeated fitness, we observe a significant decrease in these proposals, particularly in the BPP.75 and BPP1 classes. In some cases, the number of individuals with repeated fitness drops from over 55 in the original version to less than 1 (for the BPP1 class, instances with a capacity of 1,000,000) when using the threshold conditional proposals. This trend is also evident in the proposal that employs the random number conditional. As in the original version, a graph illustrates how the number of individuals with repeated fitness changes after mutation. The graph shows whether the number increases, decreases, or remains the same for the random gene sorting proposal and the threshold rule approach. The results are displayed in Figures 6 and 7.

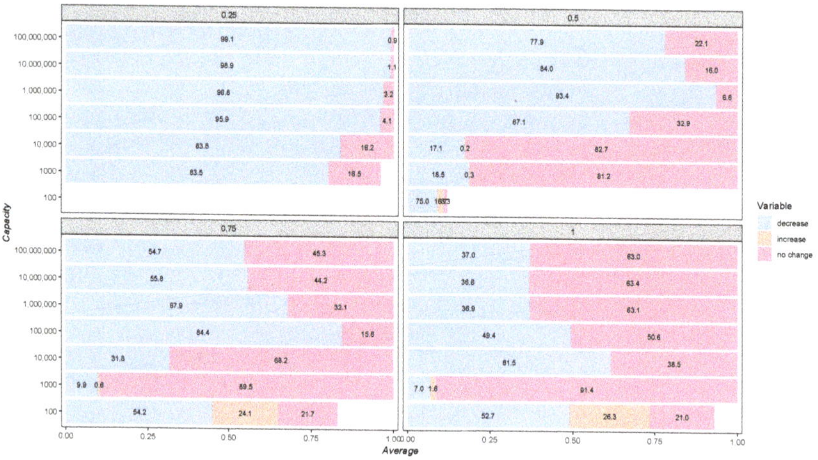

Figure 6. Proportion of generations in which the number of repeated fitness individuals increased, decreased, or remained unchanged during the execution of the GGA-CGT with random sorting for free item sorting.

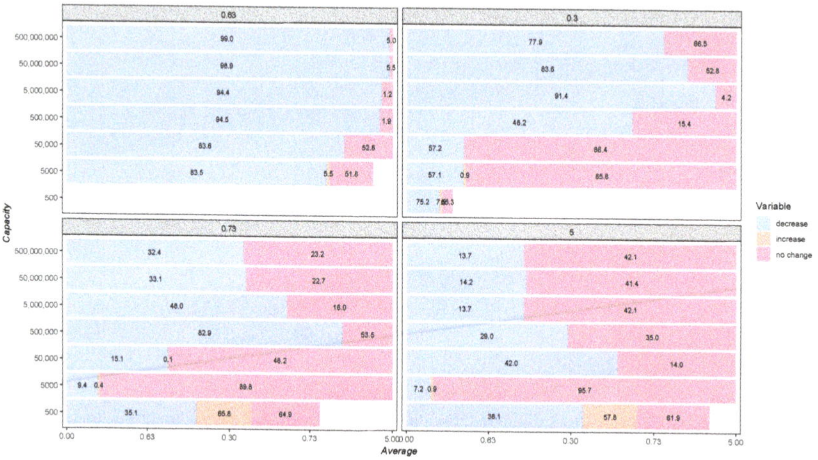

Figure 7. Proportion of generations in which the number of repeated fitness individuals increased, decreased, or remained unchanged during the execution of the GGA-CGT with the threshold rule for free item sorting.

For these sorting proposals, in the BPP.25 class, there is a consistent decrease or no change in the proportion of individuals with repeated fitness in all instances. In the BPP.5 class, it is observed that for instances with a capacity greater than 10^4, the number of repeated fitness individuals decreases or remains unchanged after mutation. For capacities of 10^3 and 10^4, the proportion of individuals showing an increase in repeated fitness is minimal, less than 1%. In the BPP.75 and BPP1 classes, for instances with a capacity of 10^2, there is an observed increase in the number of individuals with repeated fitness after mutation, though this accounts for only about 25% of the total. The proportion of instances showing a decrease or no change is higher, indicating better diversity retention overall. Specifically, in the version that yields the most optimal solutions, we observe the most notable decrease in individuals with repeated fitness after mutation, achieving the lowest recorded counts.

6. Comparison Between the Original GGA-CGT and the Adaptive Strategy for Mutation Control

In this comparison, we examine the differences between the original GGA-CGT and the version enhanced with the adaptive strategy for mutation control. To begin, we present the adaptive strategy for mutation control. The strategy, derived from this study, focuses on control for sorting mutation strategies using an adaptive approach. It employs a gene sorting method based on a threshold rule before the elimination of genes in the mutation operator. A random number r is drawn from a uniform distribution in each generation. If r exceeds the threshold, a random permutation of the genes in the solution is performed. Conversely, if r is less than or equal to the threshold, the genes are sorted according to the default operator within the algorithm. This sorting is combined with arranging free items through a random permutation in the second stage of the Rearrangement by Pairs heuristic.

The experimental conditions for this study were the same as detailed in Section 3.2: we used the same configuration for the parameters, and for each instance, the execution of the algorithm was run once with the initial seed for the random number generation set to 1.

The results obtained for the adaptive strategy are presented in Table 5. Regarding the number of optimal solutions found, it was 2227, 114 more than the original GGA-CGT, representing 79.54% of the total set, a 4.08% improvement over the original version. Additionally, concerning the number of individuals with repeated fitness, a significant decrease in the average number of individuals with repeated fitness after the mutation process is observed. This decreased from over 50% of individuals with repeated fitness to less than 1% in the classes with the highest incidence of repeated fitness, namely, the BPP.75 and BPP1 classes (see Figure 8). A slight reduction is also noted for classes with smaller items, that is, BPP.25 and BPP.5. However, since fewer individuals were in the original GGA-CGT, the change is less pronounced. Regarding the number of optimal solutions, the increase is particularly notable in the BPP.75 and BPP1 classes, which coincides with the most significant decrease in individuals with repeated fitness. This suggests that introducing more diversity allows the algorithm to explore other regions within the search space, enabling it to converge to optimal solutions. For the BPP.25 and BPP.5 classes, there is also an increase in the number of optimal solutions.

The proposed adaptive strategy for sorting free items in the 1D-BPP demonstrates promising results in enhancing solution diversity and optimality. By reducing the occurrence of repeated fitness after the mutation process, this approach helps strike a balance in selective pressure, allowing for greater diversity within the population and effectively mitigating the risk of premature convergence. The increase in optimal solutions, particularly in instances involving larger items, highlights the effectiveness of incorporating random and conditional sorting strategies. Overall, the adaptive sorting approach not only achieves a higher proportion of optimal solutions but also strengthens the robustness of the Genetic Algorithm by expanding its search capabilities and potential for generalization.

6.1. Limitations of the Adaptive Strategies for the Mutation Operator

The implementation of the control strategies showed an improvement in the algorithm's performance in terms of the number of optimal solutions obtained and the reduction in individuals with repeated fitness, suggesting an increase in diversity. However, the proposed approach has some limitations that should be considered.

First, this study focused exclusively on the 1D-BPP, using a single benchmark with specific features, such as the condition that the optimal solution requires all bins to be full. This could affect the generalization of the method to other variants of the problem or packing problems with more flexible constraints.

Second, diversity was estimated only by the number of individuals with repeated fitness. Although this indicator provides relevant information on the convergence of the population, other metrics could be considered in future studies to obtain a more complete picture of the algorithm's behavior.

Finally, the study was limited to modifying two processes within the mutation operator: (1) the ordering of genes in the solutions before mutation and (2) the reinsertion of the remaining items to complete the solution. Other variations in the mutation strategy were not explored, which could open new opportunities for improvement in future research.

Table 5. Number of optima and average of individuals with repeated fitness obtained by the GGA-CGT and GGA-CGT with sorted control. The table highlights in bold the case where the highest number of optimal solutions was achieved.

Class	Bin Capacity	GGA-CGT		GGA-CGT with Adaptive Strategies	
		Optimal Solutions	Average Number of Individuals with Repeated Fitness After Mutation	Optimal Solutions	Average Number of Individuals with Repeated Fitness After Mutation
BPP.25	100	100	1.26	100	1.26
	1000	100	0.11	100	0.07
	10,000	100	0.04	100	0.04
	100,000	100	0.49	100	0.38
	1,000,000	100	1.08	100	1.00
	10,000,000	30	2.11	37	2.55
	100,000,000	0	0.11	0	0.07
	total	530	0.60	**537**	0.62
BPP.5	100	100	0.08	100	0.88
	1000	100	0.11	100	0.13
	10,000	99	0.3	99	0.29
	100,000	99	0.96	98	1.06
	1,000,000	2	1.64	5	1.98
	10,000,000	3	0.24	8	0.19
	100,000,000	32	0.28	27	0.36
	total	435	0.52	**437**	0.84
BPP.75	100	100	0.15	100	1.58
	1000	100	1.58	100	0.24
	10,000	95	5.86	94	0.97
	100,000	20	65.93	33	3.53
	1,000,000	58	68.46	77	2.05
	10,000,000	78	70.67	98	0.43
	100,000,000	78	70.33	100	0.52
	total	529	40.28	**602**	1.33
BPP1	100	100	0.42	100	1.42
	1000	100	1.99	100	0.54
	10,000	58	35.13	65	3.22
	100,000	74	55.75	86	2.07
	1,000,000	94	53.24	100	0.46
	10,000,000	97	55.58	100	0.28
	100,000,000	96	54.78	100	0.20
	total	619	36.13	**651**	1.17
Total		2113	19.38	**2227**	0.99

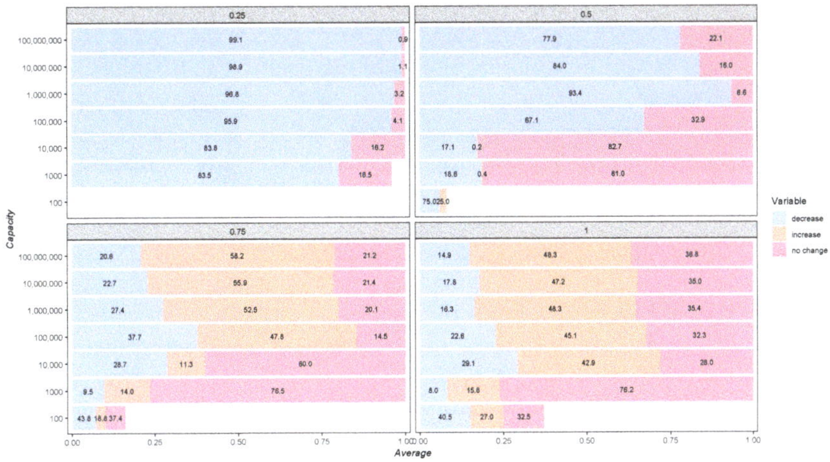

Figure 8. Proportion of generations in which the number of repeated fitness individuals increases, decreases, or remains unchanged after the mutation process during the execution of the GGA-CGT with adaptive strategies.

6.2. Statistical Test

Additionally, we conducted a statistical test to validate our results, comparing the original version of the GGA-CGT with the FI-GLX-1 operator and the GGA-CGT version with the proposed adaptive strategy for online mutation control. The statistical test applied is the Wilcoxon Rank Sum test, a non-parametric test. This test was performed on a sample of 31 runs with different seeds. For each run, we calculated the error as the relative difference, defined by the formula $(y - x)/x$, where x represents the number of bins in the optimal solution, and y represents the number of bins obtained by the algorithm. The average error for each instance was then calculated across the 31 runs. The statistical test was conducted by class and for the entire test set.

Table 6. Statistics for comparing the GGA-CGT and the GGA-CGT that includes sorted control in mutation. p-Values are presented by class and for the entire test suite, together with the average, standard deviation, and maximum value of optima obtained for each class.

Class	GGA-CGT			GGA-CGT with Adaptive Strategies			p-Value
	Average	Stdev	Max	Average	Stdev	Max	
BPP.25	525.87	4.56	536	537.39	4.43	548	9.77×10^{-1}
BPP.5	431.55	3.64	439	437.81	4.32	447	7.43×10^{-1}
BPP.75	516.97	7.16	530	599.58	4.25	608	9.54×10^{-12}
BPP1	618.23	3.81	629	652.77	2.89	659	4.32×10^{-5}
BPPv_{u_c}	2092.61	11.25	2115	2227.55	8.42	2246	7.58×10^{-7}

The results shown in Table 6 indicate that the adaptive mutation control in the GGA-CGT significantly improves performance compared with the original version of the GGA-CGT across various classes of the 1D-BPP. The adaptive strategy consistently achieves higher averages, standard deviations, and maximum values across 31 independent runs with different seeds.

For instance, in the BPP.75 class, the average number of optimal solutions increased from 516.97 in the original version to 599.58 with adaptive control, with a p-value of 9.54×10^{-12}. Similarly, in the BPP1 class, the average increased from 618.23 to 652.77, with a p-value of 4.32×10^{-5}. The overall improvement across all classes ($BPPv_{u}\_c$) is also evident, with an average increase from 2092.61 to 2227.55, supported by a highly significant p-value of 7.58×10^{-7}.

These statistically significant differences (p-values < 0.05) indicate that the adaptive mutation control mechanism enhances the diversity and convergence of the algorithm, achieving a higher number of optimal solutions in independent runs and effectively leveraging variations introduced by different seeds. This improvement highlights the potential of adaptive mutation control to enhance the quality and robustness of solutions in evolutionary algorithms applied to complex optimization problems.

7. Conclusions and Future Work

This work presented an experimental analysis of strategies integrated into the mutation operator, resulting in the development of online control strategies for mutation in the GGA-CGT. The findings demonstrated that adaptive mutation significantly improved the algorithm's performance. Additionally, the results suggested that the proposed approach promoted greater population diversity, mitigated premature convergence, and enabled the algorithm to explore a broader solution space.

Furthermore, the methods and analyses presented in this study can be applied to other combinatorial and real-world problems, which could lead to a deeper understanding of algorithmic performance and, potentially, to significant improvements in Genetic Algorithms for solving them. This study could be replicated and adapted to analyze other optimization problems and techniques, providing a broader understanding of the application of adaptive strategies in evolutionary algorithms.

The proposed strategy could be extended to real applications such as warehouse optimization, logistics planning, and industrial scheduling. However, since each problem has its characteristics, a specific study would be necessary to adapt the methodology and define an appropriate diversity metric. For example, the machines are predefined in the Parallel-Machine Scheduling Problem, and the process times are variable. In this case, the machines can be considered analogous to bins in the 1D-BPP, while the process times correspond to the bin's capacity. An adjustment of the adaptive approach and a redefinition of the diversity criteria would be required for its correct implementation.

Also, adapting the proposed mechanism to continuous optimization problems, such as economic dispatch and optimal energy flow, would demand modifications in the mutation operator to effectively manage the continuous variables. It would also be necessary to review the strategies within the mutation operator that could be controlled by the proposed method.

On the other hand, other combinatorial problems, such as the vehicle routing problem and flow-shop scheduling, could benefit from adaptive mutation to improve exploration and avoid premature convergence. Future research should explore integrating adaptive mechanisms into other genetic operators, enabling a dynamic balance between exploration and exploitation. In addition, extending the adaptive approach to control parameter settings between operators could further strengthen the robustness of the GGA-CGT, offering new perspectives for improving evolutionary algorithms in solving complex optimization problems.

Author Contributions: Conceptualization, S.A.-L., O.R.-F. and M.Q.-C.; methodology, S.A.-L. and M.Q.-C.; software, S.A.-L. and M.Q.-C.; validation, S.A.-L., O.R.-F., and M.Q.-C.; formal analysis, M.Q.-C.; investigation, S.A.-L. and M.Q.-C.; resources, S.A.-L.; writing—original draft preparation, S.A.-L. and O.R.-F.; writing—review and editing, S.A.-L., O.R.-F., and M.Q.-C.; visualization, S.A.-L., O.R.-F. and M.Q.-C.; supervision, O.R.-F. and M.Q.-C.; project administration, M.Q.-C. All authors have read and agreed to the published version of the manuscript.

Funding: This research received no external funding.

Data Availability Statement: The data supporting the reported results in this study are available upon request. Interested researchers may contact Stephanie Amador-Larrea at stephanieamadorlarrea@gmail.com and Marcela Quiroz-Castellanos at maquiroz@uv.mx to obtain access to the data.

Conflicts of Interest: The authors declare no conflicts of interest. The funders had no role in the design of the study; in the collection, analyses, or interpretation of data; in the writing of the manuscript; or in the decision to publish the results.

References

1. Garey, M.R.; Johnson, D.S. *Computers and Intractability: A Guide to the Theory of NP-Completeness*; W. H. Freeman & Co.: New York, NY, USA, 1979.
2. Martello, S. *Knapsack Problems: Algorithms and Computer Implementations*; J. Wiley & Sons: Chichester, NY, USA, 1990.
3. Ramos-Figueroa, O.; Quiroz-Castellanos, M.; Mezura-Montes, E.; Schütze, O. Metaheuristics to solve grouping problems: A review and a case study. *Swarm Evol. Comput.* **2020**, *53*, 100643. [CrossRef]
4. González-San-Martín, J.; Cruz-Reyes, L.; Gómez-Santillán, C.; Fraire, H.; Rangel-Valdez, N.; Dorronsoro, B.; Quiroz-Castellanos, M. Comparative Study of Heuristics for the One-Dimensional Bin Packing Problem. In *Hybrid Intelligent Systems Based on Extensions of Fuzzy Logic, Neural Networks and Metaheuristics*; Springer: Berlin/Heidelberg, Germany, 2023; pp. 293–305.
5. Ramos-Figueroa, O.; Quiroz-Castellanos, M.; Mezura-Montes, E.; Kharel, R. Variation Operators for Grouping Genetic Algorithms: A Review. *Swarm Evol. Comput.* **2021**, *60*, 100796. [CrossRef]
6. Quiroz-Castellanos, M.; Cruz-Reyes, L.; Torres-Jimenez, J.; Gómez, S.C.; Huacuja, H.J.F.; Alvim, A.C. A Grouping Genetic Algorithm with Controlled Gene Transmission for the Bin Packing Problem. *Comput. Oper. Res.* **2015**, *55*, 52–64. [CrossRef]
7. Carmona-Arroyo, G.; Vázquez-Aguirre, J.B.; Quiroz-Castellanos, M. *One-Dimensional Bin Packing Problem: An Experimental Study of Instances Difficulty and Algorithms Performance*; Springer: Cham, Germany, 2021; Volume 940.
8. Eiben, A.E.; Smith, J.E. *Introduction to Evolutionary Computing*, 2nd ed.; Springer Publishing Company, Incorporated: Berlin/Heidelberg, Germany, 2015.
9. Falkenauer, E. A hybrid grouping genetic algorithm for bin packing. *J. Heuristics* **1996**, *2*, 5–30. [CrossRef]
10. Falkenauer, E.; Delchambre, A. A genetic algorithm for bin packing and line balancing. In Proceedings of the 1992 IEEE International Conference on Robotics and Automation, Los Alamitos, CA, USA, 19–23 May 1992; pp. 1186–1192. [CrossRef]
11. Cruz-Reyes, L.; Quiroz-Castellanos, M.; Alvim, A.C.F.; Huacuja, H.J.F.; Gómez, C.; Torres-Jiménez, J. Heurísticas de agrupación híbridas eficientes para el problema de empacado de objetos en contenedores. *Comput. Sist.* **2012**, *16*, 349–360.
12. González-San-Martín, J.E. Un Estudio Formal de Heurísticas Para el Problema de Empacado de Objetos de una Dimensión. Master's Thesis, Instituto Tecnológico de Ciudad Madero, Ciudad Madero, Mexico, 2021.
13. Amador-Larrea, S. Un Estudio de Operadores Genéticos de Cruza para el Problema de Empacado de Objetos en Contenedores. Master's Thesis, Universidad Veracruzana, Veracruz, México, 2022.
14. Eiben, A.; Hinterding, R.; Michalewicz, Z. Parameter control in evolutionary algorithms. *IEEE Trans. Evol. Comput.* **1999**, *3*, 124–141. [CrossRef]
15. Karafotias, G.; Hoogendoorn, M.; Eiben, A.E. Parameter Control in Evolutionary Algorithms: Trends and Challenges. *IEEE Trans. Evol. Comput.* **2015**, *19*, 167–187. [CrossRef]
16. Bhatia, A.; Basu, S.K. Packing bins using multi-chromosomal genetic representation and better-fit heuristic. In *Proceedings of the International Conference on Neural Information Processing*; Springer: Berlin/Heidelberg, Germany, 2004; pp. 181–186.
17. Singh, A.; Gupta, A.K. Two heuristics for the one-dimensional bin-packing problem. *OR Spectr.* **2007**, *29*, 765–781. [CrossRef]
18. Agustín-Blas, L.E.; Salcedo-Sanz, S.; Ortiz-García, E.G.; Portilla-Figueras, A.; Pérez-Bellido, Á.M.; Jiménez-Fernández, S. Team formation based on group technology: A hybrid grouping genetic algorithm approach. *Comput. Oper. Res.* **2011**, *38*, 484–495. [CrossRef]
19. Chaurasia, S.N.; Singh, A. A hybrid evolutionary approach to the registration area planning problem. *Appl. Intell.* **2014**, *41*, 1127–1149. [CrossRef]
20. Jawahar, N.; Subhaa, R. An adjustable grouping genetic algorithm for the design of cellular manufacturing system integrating structural and operational parameters. *J. Manuf. Syst.* **2017**, *44*, 115–142. [CrossRef]
21. Rossi, A.; Singh, A.; Sevaux, M. A metaheuristic for the fixed job scheduling problem under spread time constraints. *Comput. Oper. Res.* **2010**, *37*, 1045–1054. [CrossRef]
22. Peddi, S.; Singh, A. Grouping Genetic Algorithm for Data Clustering. In *Proceedings of the Swarm, Evolutionary, and Memetic Computing*; Panigrahi, B.K., Suganthan, P.N., Das, S., Satapathy, S.C., Eds.; Springer: Berlin/Heidelberg, Germany, 2011; pp. 225–232.

23. Balasch-Masoliver, J.; Muntés-Mulero, V.; Nin, J. Using genetic algorithms for attribute grouping in multivariate microaggregation. *Intell. Data Anal.* **2014**, *18*, 819–836. [CrossRef]
24. Ramos-Figueroa, O.; Quiroz-Castellanos, M. An experimental approach to designing grouping genetic algorithms. *Swarm Evol. Comput.* **2024**, *86*, 101490. [CrossRef]
25. Ramos-Figueroa, O.; Quiroz-Castellanos, M.; Mezura-Montes, E.; Cruz-Ramírez, N. An Experimental Study of Grouping Mutation Operators for the Unrelated Parallel-Machine Scheduling Problem. *Math. Comput. Appl.* **2023**, *28*, 6. [CrossRef]
26. Fernández-Solano, O. Un Estudio Experimental de Operadores de Mutación para el Algoritmo Genético de Agrupación Para la Segmentación de Imágenes en RGB. Master's Thesis, Universidad Veracruzana, Veracruz, México, 2023.
27. Zavaleta-García, N.A. Estudio Experimental de Operadores Genéticos de Cruza Para el Problema de Segmentación de Imágenes en Color. Master's Thesis, Universidad Veracruzana, Veracruz, México, 2023.
28. Yorgancılar, G. Effects of Evolutionary Operators in Grouping Genetic Algorithms on Diversity and Result Quality. Master's Thesis, TED University, Ankara, Turkey, 2020.
29. Perez, A.; Michelt, J. Optimización Generalizada para Problemas de Agrupación: Un Enfoque Coevolutivo Auto-Adaptativo. 2024. Available online: https://rinacional.tecnm.mx/jspui/handle/TecNM/7945 (accessed on 12 December 2024).
30. Carmona-Arroyo, G. A Grouping Genetic Algorithm for Variable Decomposition in Large-Scale Constrained Optimization Problems. Master's Thesis, Universidad Veracruzana, Veracruz, Mexico, 2024.
31. Ramos-Figueroa, O.; Quiroz-Castellanos, M.; Carmona-Arroyo, G.; Vázquez, B.; Kharel, R. Parallel-Machine Scheduling Problem: An Experimental Study of Instances Difficulty and Algorithms Performance. In *Recent Advances of Hybrid Intelligent Systems Based on Soft Computing*; Melin, P., Castillo, O., Kacprzyk, J., Eds.; Springer International Publishing: Cham, Switzerland, 2021; pp. 13–49. [CrossRef]

Mathematical and Computational Applications

MDPI

Article

MASIP: A Methodology for Assets Selection in Investment Portfolios

José Purata-Aldaz [1], Juan Frausto-Solís [1,*], Guadalupe Castilla-Valdez [1], Javier González-Barbosa [1] and Juan Paulo Sánchez Hernández [2]

[1] Graduate Program Division, Tecnológico Nacional de México/Instituto Tecnológico de Ciudad Madero, Ciudad Madero 89440, Mexico; luis.pa@cdmadero.tecnm.mx (J.P.-A.); gpe_cas@yahoo.com.mx (G.C.-V.); jjgonzalezbarbosa@hotmail.com (J.G.-B.)

[2] Dirección de Informática, Electrónica y Telecomunicaciones, Universidad Politécnica del Estado de Morelos, Boulevard Cuauhnáhuac 566, Jiutepec 62574, Mexico; juan.paulosh@upemor.edu.mx

* Correspondence: juan.frausto@itcm.edu.mx

Abstract: This paper proposes a Methodology for Assets Selection in Investment Portfolios (MASIP) focused on creating investment portfolios using heuristic algorithms based on the Markowitz and Sharpe models. MASIP selects and allocates financial assets by applying heuristic methods to accomplish three assignments: (a) Select the stock candidates in an initial portfolio; (b) Forecast the asset values for the short and medium term; and (c) Optimize the investment portfolio by using the Sharpe metric. Once MASIP creates the initial portfolio and forecasts its assets, an optimization process is started in which a set with the best weights determines the participation of each asset. Moreover, a rebalancing process is carried out to enhance the portfolio value. We show that the improvement achieved by MASIP can reach 147% above the SP500 benchmark. We use a dataset of SP500 to compare MASIP with state-of-the-art methods, obtaining superior performance and an outstanding Sharpe Ratio and returns compared to traditional investment approaches. The heuristic algorithms proved effective in asset selection and allocation, and the forecasting process and rebalancing contributed to further improved results.

Keywords: investment portfolios; optimization; forecast; heuristics

1. Introduction

In financial asset management and investment decision-making, portfolio creation and optimization represent a fundamental paradigm and essential study area. Building investment portfolios addresses the complex task of allocating limited resources to a diverse range of financial assets, as well as forecasting them [1]. This process requires careful analysis of the risks and returns associated with each asset and understanding their interaction in the context of a diversified portfolio. Optimization involves the search for asset combinations that maximize expected returns or minimize risk, all subject to specific constraints and objectives. Similarly, there are diversification and asset selection methodologies, such as the Markowitz portfolio theory and its extensions, value at risk (VaR) models, and asset allocation strategies [2]. The Markowitz Mean-Variance model, also known as the Modern Portfolio Theory (MPT), is presented in Equations (1)–(4), which is a theoretical framework developed by Harry Markowitz [3]. This study is relevant because it allows portfolio optimization and considers risk-taking based on performance and diversification. This quantitative approach provides an objective framework for making investment decisions and lays the foundations for investment management.

$$Max\ E(X_i) = \sum_{i=1}^{N} X_i \mu_i \tag{1}$$

$$Min\ V(X_i,\ X_j) = \sum_{i=1}^{n} \sum_{j=1}^{n} X_i X_j \sigma_{ij} \tag{2}$$

$$\sum_{i=1}^{n} X_i = 1 \tag{3}$$

$$0 < X_i \leq 1 \tag{4}$$

where μ_i is the expected return of the ith asset, X_i is the relative value invested in the asset i, σ_{ij} represents the covariance between assets, and $E(X_i)$ is the expected return of the portfolio. Notice that the invested amount is normalized to one in the constraint defined by Equation (3), and no negative investments X_is are permitted according to Equation (4). The decision variables are the expected value $E(X_i)$ and the variance V, and the objectives are to achieve maximum profit and minimum risk.

In the model defined by Equations (1)–(4), the variance V of the portfolio is introduced as a risk measure, which is essential for constructing diversified portfolios. Covariance and correlation are also key elements for good diversification [3]. From the mean-variance model, two objectives must be met: firstly, to maximize the portfolio return and secondly, to minimize the risk. These two objectives become the main difficulty for solving the problem. The Sharpe Ratio (ς) is a popular and straightforward function [4]; ς is commonly used in modern portfolio methods [5–7], requiring only the expected value of the portfolio, its volatility, and a reference rate [8]. They are called Sharpe Ratio models and offer several advantages, including the facility to reduce the portfolio evaluation problem from a bi-objective scheme to a mono-objective optimization problem with Equations (5)–(7), thereby focusing solely on maximizing the ς function.

$$\max\varsigma(X_i) = \frac{\sum_{i=1}^{N} X_i \mu_i - R_B}{\sum_{i=1}^{n} \sum_{j=1}^{n} X_i X_j \sigma_{ij}} \tag{5}$$

$$\sum_{i=1}^{n} X_i = 1 \tag{6}$$

$$0 < X_i \leq 1 \tag{7}$$

where the $\varsigma(X_i)$ function indicates the expected differential return per unit of risk associated with different assets conforming to the portfolio, and R_B is the return on a benchmark or the risk-free rate. This portfolio optimization model has only one decision variable, defined by the Sharpe Ratio $\varsigma(X_i)$; this variable involves the relation between $E(X_i)$ and V. Thus, Equations (5)–(7) represent a mono-objective problem.

The composition of an investment portfolio entails more than merely selecting assets with the most significant potential to achieve stated objectives and satisfy constraints. Moreover, the optimal weights of the assets that satisfy the constraints and optimize the objective function must be determined. Also, the future value of the assets and their weighting in the portfolio for the periods during which the investment is planned must be known. These future values lead to a realistic portfolio being obtained for the investment plans. Therefore, it is necessary to forecast every asset as accurately as possible for the investment period. There are many successful forecasting methods, such as classical and modern machine learning, and combining forecast methods has proven valuable in enhancing accuracy and reducing the variance of forecasting errors [9]; they can generate more accurate forecasts than those from individual methods [10]. Optimization methods can obtain the best weights of the individual forecasting methods participating in an ensemble. One of these methods is Threshold Accepting (TA). This iterative algorithm controls the number of iterations by a temperature parameter and accepts some poor solutions when they fulfill an acceptance criterion defined in terms of a threshold parameter [11]. TA is an efficient tool in many applications [12].

The associated literature about how to conform an initial portfolio is scarce. However, most of the existing literature focuses on a single aspect of portfolio construction: (a) The identification of the candidate assets to be integrated in the portfolio [13,14]; (b) The optimization of the portfolio [15–17]; or (c) The application of a forecast strategy to predict the portfolio return [18–20]. This observation underscores a promising area of opportunity, which we seek to address in this study by implementing a novel approach. MASIP integrates three fundamental components: the preselection of assets, considering a minimum acceptable return; the forecasting of the selected assets based on an assembly method optimized by an enhanced threshold acceptance algorithm; and the rebalancing of assets to create the final portfolio. The above-mentioned points are performed by MASIP, an innovative methodology incorporating an asset pre-selection stage, an objective function based on the median-variance and Sharpe Ratio model, and asset diversification. This methodology also implements an acceptance scheme for a minimum return and a maximum permissible risk rate. The modern combining forecast method and the well-tested portfolio optimization heuristic algorithm accomplish MASIP.

The rest of this paper is organized as follows. Section 2 reviews the background of portfolios, including financial time series forecast modeling and portfolio integration. Section 3 introduces the details of MASIP methodology for portfolio integration models and describes the construction process of an ensemble forecasting model. Section 4 presents the experimentation and results from the studies using financial time series data and the application of MASIP. Detailed analysis and result interpretations are provided. Section 5 presents the conclusions.

2. Background

The construction of investment portfolios has been a central topic in the financial literature for decades, driven by the search for effective strategies that maximize return and minimize risk. From Markowitz's pioneering work on portfolio theory to the most recent advances in financial forecasting, research has generated a vast body of knowledge that underscores the importance of a meticulous approach to portfolio construction and management. This section considers the modern aspects related to portfolio construction: asset selection, forecasting stock methods, and portfolio optimization.

2.1. Asset Selection and Portfolio Construction

Cesarone et al. (2020) provided theoretical results for the risk-parity approach to general risk measures. They proposed an efficient and accurate Multi-Greedy heuristic to solve the portfolio selection problem [21]. Empirical results on real-world data show that diversified optimal portfolios are only slightly suboptimal in sample and generally exhibit better out-of-sample performance than purely diversified or optimized portfolios [22]. The authors analyzed NASDAQ100, FTSE100, and SP500 assets, finding good Sharpe values returns ratios for the SP500 index. The combined diversification–optimization approach is better than pure diversification or optimization.

A framework proposed by Puerto et al. (2023) adopts a bi-criteria mean-risk optimization2 approach to model the portfolio optimization problem, where the average portfolio rate of return is maximized while minimizing a given risk measure. These authors propose the use of mixed-integer linear programming (MILP) formulation to solve portfolio clustering and selection problems in a unified phase [23]. The authors focus on using Conditional Value at Risk (CVaR) as the objective function for portfolio optimization and evaluate the performance of the proposed integer linear programming formulation for the SP500 market. In another work performed by Kaczmarek and Perez, they propose the mean-variance model and hierarchical risk parity (HRP) for SP500, achieving a Sharpe Ratio ς in the range of 0.76 to 0.97 [24]. Unfortunately, they consider assets from an ancient dataset. In 2023, Zhang used datasets from the Chinese market based on the mean-variance model and Sharpe Ratio; therefore, a Monte Carlo process is proposed for asset selection to adjust the portfolio within the efficient frontier [25].

Ma et al. (2020) analyze the efficacy of a combination of machine learning and deep learning algorithms in predicting returns [26]. The analysis utilizes two machine learning models: Random Forest (RF) and Support Vector Regression (SVR), as well as three deep learning models: Long Short-Term Memory (LSTM), Deep Multilayer Perceptron (DMLP), and Convolutional Neural Network (CNN). These models were utilized for the purposes of stock preselection and portfolio optimization through the mean-variance (MV) and omega models for the Chinese market. The findings of this study indicated that RF + MV demonstrated superior performance.

For the European markets, LSTM neural networks were recently applied to predict asset returns and incorporated into the mean-variance optimization model to determine the optimal asset allocation. The average return prediction accuracy in this case was 95.8%, and the integrated portfolio outperformed the benchmark [27]. Table 1 presents a summary of related studies. The present study is focused on SP500; thus, we will compare our results with the references [21,23].

Table 1. Related studies.

Approach	Asset Preselection	Asset Forecasting	Portfolio Optimization
Cesarone [21]			X
Puerto [23]			X
Kaczmarek [24]			X
Zhang [25]			X
Ma [26]		X	X
Martinez [27]		X	X
MASIP	X	X	X

2.2. Asset Portfolio Prediction Methods

Nowadays, it is well known that combining several methods has advantages over individual forecasting models, and several ensemble methods are available [28]. For portfolio design and asset selection, an efficient way to analyze and predict risky assets is required, as well as a comparative analysis of the structure of a complete portfolio, including risky and risk-free assets, depending on the risk aversion coefficient [29,30]. Combinational forecasting has several approaches, such as the Simple Average, which is the sum of the forecasts divided by the number of forecasts. The Weighted Average model assigns weight to each forecast. The combined forecast is calculated as the sum of the forecasts multiplied by their respective weights, divided by the sum of the weights [31].

The Aggregation Model employs a machine learning algorithm to integrate forecasts derived from multiple models. It utilizes a range of algorithms, including neural networks, support vector machines, and decision trees, among others [32–34].

An ensemble model integrates various forecasting methodologies, employing a criterion to determine their contribution to the ensemble. These methodologies may encompass machine learning algorithms, autoregressive integrated moving average (ARIMA) model variants, exponential smoothing techniques, and other approaches. The mathematical formulation for this integration is dependent on the specific models being combined [30,35–38]. Metaheuristic methods, such as simulated annealing and genetic algorithms, can be applied to analyze the datasets while integrating the assets to determine their participation in a portfolio. Neural networks and support vector machines have also been used [39]. In other words, it is crucial to acknowledge that the models employed in forecast combination methodologies may differ depending on the specific approach utilized and the characteristics of the forecasting issue at hand. When there are a limited number of forecasting methods, assigning weights or balances to them is relatively straightforward. However, as the number of models increases, the task becomes significantly more complex. Thus, it is necessary to rely on optimization tools.

The integration of machine learning, deep learning, and probabilistic techniques into stock market analysis has the potential to enhance the precision of predictions and provide novel insights into portfolio management. However, it is imperative to recognize the limitations and challenges inherent in these approaches, such as data quality and model complexity, to ensure optimal effectiveness in practical applications [40,41].

2.3. Portfolio Optimization Methods

The potential for employing diverse forecasting techniques and integrating them into a portfolio introduces a novel challenge: identifying the optimal approach or discerning the most effective combination of forecasting methods that minimizes error and yields superior outcomes compared to individual forecasts. To address this, combinatorial and optimization strategies must be employed.

Optimization algorithms, such as TA, have become important in several areas [11]. The last reference highlights the superior performance of TA over other methods, such as Simulated Annealing, for combinatorial challenges. A new version of TA is used to improve portfolio asset weights by exploiting its convergence property [42], its applicability to econometrics applications [43], and in financial sector problems [44]. Gilli and Kellezi showed the effectiveness of TA in selecting a subset of assets that closely replicate the performance of the benchmark index, achieving high performance in investment portfolio design. Liu and Tadesse used the LSTM machine learning method to forecast assets from the SP500 market [45]; they applied the Sharpe Ratio and Equations (3)–(5) to integrate the investing portfolio. However, they considered only four assets of this market without a method for selecting the assets to incorporate into the portfolio. In addition, no risk-free reference rate was considered. When this value is considered, the yield and the Sharpe Ratio performance would be different. As shown in another section, this modification is performed in the model we propose in the present study.

These approaches permit investors to anticipate forthcoming market changes, optimize resource allocation, and mitigate potential risks. By employing classical, advanced, or combined models, it is possible to make more accurate predictions of market trends and asset behavior. This work, through the combination of various methods and their optimization, is intended to be a powerful tool for decision-making.

3. MASIP Methodology

This section presents the MASIP methodology for asset selection in investment portfolios and its related algorithms. A step-by-step approach addressing the issues relating to the following complex tasks is presented: asset selection, initial portfolio creation, asset forecasting, and final rebalancing or optimization. Furthermore, MASIP involves acquiring, analyzing, segmenting, and evaluating extensive time series data to achieve investment goals. Figure 1 presents the elements of MASIP and their interactions.

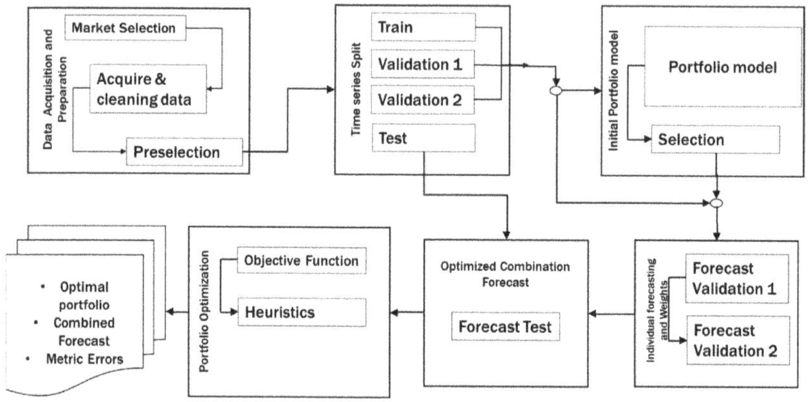

Figure 1. The MASIP methodology.

3.1. Data Acquisition and Preparation

Market selection is the first step to integrate the investment portfolio. At this point, the target market to be analyzed is selected; among the candidate markets to be evaluated, we could mention those listed in the New York Stock Exchange, the European region, the Asian region, or those belonging to developing countries such as BRICS or Latin America. In this case, the analyst must perform a previous analysis of the target market and establish the period to be analyzed. In acquisition and cleaning of data, the information on each of the candidate assets is obtained. We consider the daily closing prices to conform to the time series in a determined time. As we know, it is important to have quality, cleanliness, and order of information since, if erroneous or missing information is entered, we will have inaccurate conclusions that can lead to severe losses of resources. Likewise, the data must be complete; thus, the necessary techniques must be applied to enable the data series cleaning process. Some of the methods used may be normalization, other data transformation, inter- or extrapolation, application of simple averages, moving averages, or any other variation that allows us to have complete information without altering the quality of the data [46–48].

The selection of assets should be made before integrating the portfolio. At this point, the analyst knows which assets of the selected marketplace are already the candidates for integrating the optimal portfolio. Thus, only the corresponding time series would be used to design it, and the effort for this task can be reduced. For example, if we analyze SP500, one of the largest financial exchanges in the world, we will have to analyze more than 500 time series, which consumes a lot of time and resources. To streamline the evaluation process, we apply the discrimination of assets that meet a percentage of profit or return by the Minimum Acceptable Rate of Return or MARR [49], which could be used to select an investment that assures us a fixed return with minimal utility and no risk. Thanks to this scheme, we can significantly reduce the number of candidate assets that need to be evaluated [8].

3.2. Time Series Split

The segmentation of time series is an important practice since it allows for evaluating the performance of models using known data and comparisons with previously unseen data. When a model is trained with a time series, it is expected to be able to make accurate predictions. However, if the model is evaluated with the same training data, there is a risk of overfitting, and only training data are stored.

When the time series data are divided into training, validation, and test sets, it is possible to adjust the model parameters to minimize the error in the validation set and then evaluate the model performance in the test set, which contains data that are "new" to the model and not previously analyzed.

Segmentation also allows for the identification of problems with the data, such as outliers or anomalous values, which allows for treating and correcting the input data before training the models and obtaining a final model [35,50,51].

In this implementation, the training portion of each time series is initially employed to forecast the validation section, applying all available individual forecast methods. The outcomes of this process are used to determine the most effective methods for forecasting the validation section and to establish the distribution of weights for the combination of these optimal methods. Subsequently, the forecast is evaluated using the validation data, and the most effective methods are identified, along with the previously calculated weight distribution. The resultant forecasts are stored, and the optimization process with TA is initiated using Validation 1 to forecast Validation 2. The weights are adjusted until no further improvement is observed [30].

3.3. Initial Portfolio Model

Building a portfolio is not a simple task today since several factors can influence the selection, integration, and weighting of the assets of the investment portfolio. That is why it requires robust and reliable tools which allow us to analyze many candidate assets and the data that integrate them. As we present in Section 1, the MPT model (Equations (1)–(4)), proposed by Markowitz, is the pioneering work in portfolio management which considers two main objectives: maximizing the expected return and minimizing the risk [3,52]. The Sharpe model represented by Equations (5)–(7) is an MPT extension considering the two objective functions in only one. In the present work, we propose a like-Markowitz Model of MASIP defined by Equations (8)–(12).

$$max\ SR = \frac{E\left(R_{pf}\right) - \text{MARR}}{\sqrt{\sum_{i=1}^{n} \sigma_i{}^2 x_i{}^2 + 2\sum_{i=1}^{n} \sum_{j\neq i}^{n} \sigma_i \sigma_j x_i x_j \rho_{ij}}} \tag{8}$$

subject to

$$\sum_{i=1}^{n} x_i = 1 \tag{9}$$

$$\sum_{i=1}^{n} E\left(\text{R}_{\text{pf}}\right) > \text{MARR} \tag{10}$$

$$0 < x_i \leq 1 \tag{11}$$

$$\sum_{i=1}^{n} \sigma_i{}^2 x_i{}^2 + 2\sum_{i=1}^{n} \sum_{j\neq i}^{n} \sigma_i \sigma_j x_i x_j \rho_{ij} \leq \Phi_{max}{}^2 \tag{12}$$

where $E\left(R_{pf}\right)$ is the expected return, i and j are the assets, σ_{pf} is the portfolio risk, x_i, x_j are the contribution of i and j assets, σ_i, σ_j are the standard deviation of the assets, ρ_{ij} are the correlations between assets i and j, and $\Phi_{max}{}^2$ is the Maximum Risk Rate Allowable.

The MASIP model defined by Equations (8)–(12), as we mentioned above, is an MPT extension with the following differences:

- In the objective function, we used the parameter MARR to represent the Minimum Acceptable Rate of Return that the investor could accept; instead, the Markowitz model uses a risk-free rate;
- Constraint (10) uses the MARR parameter for weighting assets into the portfolio;
- Constraint (12) uses the ${\Phi_{max}}^2$ risk parameter to define the assets integrating the portfolio. This parameter binds the risk associated with the asset.

We use the parameter MARR, representing a common aspect used by decision-makers to accept an investment; however, the MPT and other models use a risk-free rate. Moreover, the average is sensitive to outliers and can be affected by extreme fluctuations. Conversely, the Expected Value (mathematical expectation) is calculated by multiplying each possible value by its probability of occurrence and adding up the results. In short, while the average is a basic measure of central tendency, the expected value more accurately represents the inherent uncertainty in the data [53].

In the state-of-the-art methods, a genetic algorithm is used for initial portfolio selection (GENPO), as well as algorithms based on threshold acceptance (TAIPO) and simulated annealing (SAIPO), which were tested and showed superior performance to other models, mainly focused on risk management (Yu, Gilli, and Masese) [16,17,54]. The last models are cited in the present study as YGM models. The individual YGM models versus MASIP are analyzed. The results of this analysis demonstrate the performance of the 12 combinations obtained between the YGM models and the hybridized algorithms. These results show that the YGM models using the maximum SR objective function applying the TAIPO algorithm obtain superior performance [8].

3.4. Individual Forecasting and Weights

Forecasting the performance of an investment portfolio is crucial for various reasons, such as identifying and quantifying the risks associated with the portfolio, financial planning, and evaluating strategy. Forecasting enables financial analysts to anticipate and adjust to changing market conditions, allowing them to make more timely and effective decisions.

The forecasting section of the Methodology is supported by FCTA (Forecasting Combined Method with Threshold Accepting) [30], which consists of assembling 14 forecasting methods and an enhanced threshold-accepting algorithm for weighting individual predictions. The forecast methods involved are Autoregressive integrated moving average ARIMA, exponential smoothing state space (ETS), Neural network forecasting method NNetar, Exponential smoothing state space-Box-Cox-ARMA (ETS), STLM with AR method, Naive and random walk forecasts with drift activated, Decomposing Time Series with Smoothing Methods (Theta), Naïve method, Autoregressive fractionally integrated moving average (ARFIMA), Bootstrapped method (Bagged), SplinesCubic, Holt, Feature-based forecast model averaging (FFORMA), and Jaganathan.

Notably, at least two of the models incorporated, FFORMA [36] and Jaganathan [55], obtained second and fourth place, respectively, in the competition, as both FFORMA and Jaganathan are ensemble and hybridization methods in certain cases [56]. Within this framework, it is essential to underscore the more than 24,500 financial series within the dataset, which suggests that the application of FCTA for the analysis of financial series can be considered reliable.

The FCTA employs a set of criteria to evaluate the efficacy of forecasting methods, including the uniform distribution, the normalized weights criterion, the Bayesian information criterion (BIC), and the Akaike information criterion (AIC). These criteria have been demonstrated to yield superior performance in the analysis and forecasting of the M4 competition series, which encompasses financial series.

The forecasting process is presented in Figure 2. In this stage, individual forecasts are made for each asset and its time series, using each forecasting technique considered in the algorithm, that is, the n assets selected in the initial portfolio are evaluated and forecasted by the M forecasting models. This is performed for each split of the time series, that is, each of the forecasting models is fed with the training section of each of the asset's data and compared with the validation section. The error is measured for all the best methods, and the ones with the lowest error evaluation are selected and compared against the test section.

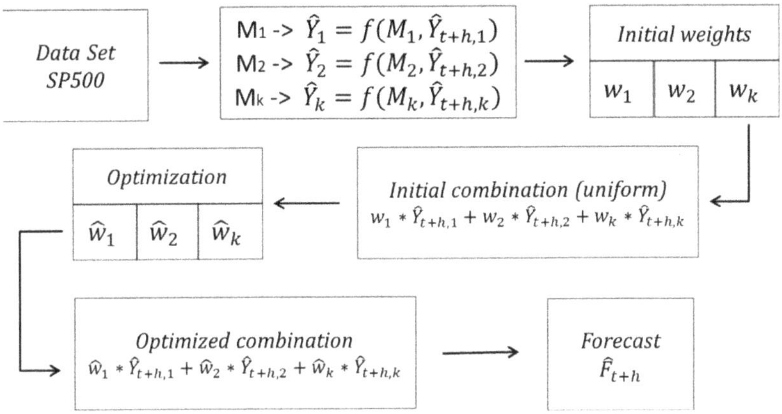

Figure 2. Combination of forecasting methods.

The initial combination of forecasting methods is established through the selection of the best-evaluated ones.

Through experiments involving enhanced weights, it was observed that a collection of methods yielded superior results compared to individual methods. They identified which features were crucial for selecting methods for the combination. Notice that the most effective combination is typically achieved when the best individual methods are employed.

Since the mid-20th century, ensemble methods for forecasting have been applied [57]; specifically, Bates [9] proposed using several forecasting methods in a weighted combination with weights w to be determined in terms of the error of each of the individual methods. This method is commonly known as the Bates method and was initially tested using Brown's exponential smoothing methods. The way to compare $n \geq 2$ methods in the Bates method is to determine the Mean Squared Error (MSE) for all methods for the series considered in an evaluation period; then, the methods are combined so that each participates in the forecast in terms of the accuracy corresponding to its MSE. A final forecast ensemble or combination can be expressed as Equation (13):

$$\hat{Y}_f = \sum_{i=1}^{n} w_i \hat{y}_i \tag{13}$$

It can be said that the relationship between the combined forecast $\hat{Y}_f$ and the individual forecast $\hat{y}_i$ are linear, where we have n individual forecast models, and for each model $\hat{y}_i$ we have a weight w_i.

Moreover, the objective function is measured using the symmetric mean absolute percentage error (sMAPE), a variation of the classic mean absolute percentage error (MAPE) metric [58]. However, other convenient error metrics could be used. The metric used in this paper is expressed in Equation (14).

$$sMAPE = \left[\frac{2}{h} * \sum_{t=n+1}^{n+h} \frac{|Y_i - \hat{Y}_i|}{|Y_i| + |\hat{Y}_i|}\right] * 100\ (\%) \tag{14}$$

where Y_i is the current observation of the time series, $\hat{Y}_i$ represents the predicted value, n is the amount of data in the time series, and h represents the length of the forecast horizon. This measure has been the subject of analysis since the late 1980s [59], but it gained popularity with the M3 competitions [60,61] as a metric to compare various forecasting methods. sMAPE is simple to calculate and understand, and unlike other errors, sMAPE considers both overestimates and underestimates equally.

Other error metrics considered for evaluation are MSE (15), Root Mean Squared Error (RMSE) (16), and Mean Absolute Percentage Error (MAPE) in Equation (17).

$$MSE = \frac{1}{n}\sum_{t=1}^{n} (Y_i - \hat{Y}_i)^2 \tag{15}$$

$$RMSE = \sqrt{\frac{1}{n}\sum_{t=1}^{n} (Y_i - \hat{Y}_i)^2} \tag{16}$$

$$MAPE = \frac{100}{n}\sum_{t=1}^{n} \left|\frac{(Y_i - \hat{Y}_i)}{Y_i}\right| \tag{17}$$

$$Rsquared = 1 - \frac{\sum (Y_i - \hat{Y}_i)^2}{\sum (Y_i - \overline{Y}_i)^2} \tag{18}$$

In the context of machine learning forecasting, the evaluation of model performance is subject to selecting appropriate metrics. This study includes MSE, MAPE, and RMSE to evaluate forecasting models. MSE is used as an objective function because of its ability to penalize large errors, although its sensitivity to outliers can be a disadvantage. MAPE can overestimate small errors and generate problems when actual values are close to zero. sMAPE, on the other hand, is used to present results which are easy to understand, as it provides a more symmetric error penalty compared to MAPE [62]. While MSE is useful for optimizing models, its interpretation is not intuitive for users. Therefore, sMAPE is used for internal model fitting, while MSE is used for algorithm comparison. This approach is designed to ensure a comprehensive and equitable evaluation of forecasting models, aligning with practices employed in time series forecast competitions [63].

The RMSE is useful in cases where large errors are unacceptable because it takes the square root of the errors. This makes the RMSE the same size as the original variable, which makes it easier to understand the average size of the error. However, it can also be too harsh on big errors, which may be a problem if they are rare or not important in the analysis. Outliers can also have a big effect on the RMSE, which can change how we judge the model when there are outliers [35]. The coefficient of the determination Rsquared (R^2) measures how well a forecasting model explains the variability of the actual data [35].

3.5. Optimized Combination Forecast

Once an initial combination of forecasting methods is obtained using the best-evaluated models, it is necessary to establish the optimal combination of the forecast models, i.e., to weigh them in such a way that this combination delivers the lowest possible error. For this, it is necessary to use again the heuristics that allow solutions, which, in this case, try to minimize the error of the combined forecast against real test data. As mentioned above, it is possible to use any of the heuristic methods.

Figure 3 shows the general process of the FCTA methodology. In the first step, the training data of the n time series are used in the m forecasting methods, $\hat{Y}_{t+h}(M1)$. These methods generate a weighted combination of methods for each time-series. This combination is then compared to validation segment 1. The process is repeated with the data of Validation 1. These data are used to forecast into the periods of Validation 2. The results are then compared to Validation 2. The test split is applied to evaluate the optimized combined forecasting. The error analysis in stages 1 and 2 selects the best individual forecasting methods for each of the selected assets in the initial portfolio, thereby establishing an initial mix of forecasting methods. This combination is uniform when equal weights are assigned to each forecasting method.

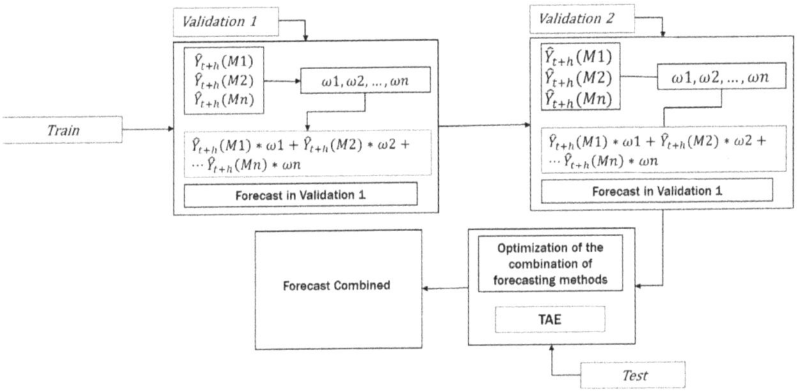

Figure 3. General outline of FCTA.

The initial combination of the best forecasting methods enters the optimization heuristic based on TA, which has been modified to obtain a superior performance to the classical TA algorithm. The algorithm Threshold Accepting Enhanced (TAE) is an algorithm including a modified cooling scheme, a reheating scheme, and thermal equilibrium. This algorithm was tested and showed a high performance in M4 time series [30].

The output of the TAE algorithm is the weighted combination of the best forecasting methods that obtain the minimum value of the sMAPE error metric. The weighted methods are then combined to generate forecasts, which are subsequently compared against the test section to calculate the error metric.

3.6. Portfolio Optimization

Following the conclusion of the forecasting process, it is imperative to perform a portfolio rebalancing or optimization. In this stage, the candidate assets' past and forecasted data are evaluated using the methodologies or algorithms outlined in Section 3.3. The anticipated outcome of this process is a portfolio that exhibits enhanced performance in comparison to its initial state. The logical sequence and steps of MASIP are shown in Algorithm 1.

Algorithm 1. Pseudo code for MASIP methodology

```
Market: Time series for portfolio forecasting (assets)
Result: Optimal Portfolio Forecast
  Initialization
   Market <- Get Market data
   Market <- Clear Market data
     For i in each asset in the Market
       If profit_asset >= MARR #Minimum Acceptable Rate Return
          Data <- asset
     Train, validation 1, validation 2, Test <- splitData()
     InitialPortfolio <- PortfolioSelection(Data[Train, validation 1, validation 2],
ObjectiveFunction)
     IndividualForecast <- IndividualMethods(Data[InitialPortfolio])
     InitialCombinationMethod <- IncludeProcess(IndividualForecast)
     ForecastCombination <- FCTA(InitialCombinationMethod, Test)
     OptimalPortfolio <- PortfolioSelection(Data[Train, validation 1, validation 2, Test],
ForecastCombination)
     Return (Optimal portfolio, Combined Forecast, Metric Errors)
```

4. Experimentation and Results

This section reviews the equipment and material resources used for developing this study. Since this is complex work, several computer parts equipment were required: Intel(R) Core(TM) i7-13620H, CPU @ 2.40 GHz, RAM: 16 GB, Intel(R) Xeon(R) Gold 6230 CPU @ 2.10 GHz, and 2.10 GHz, RAM: 12 GB (National Laboratory of Information Technology, Tamaulipas, Mexico).

The languages used for this experiment are Python 3.7 for initial and final portfolios and R 4.3.3 for forecasting.

A description of the time series utilized and an account of the experimentation conducted with this series through the portfolio integration and portfolio forecasting algorithms are also provided. A concise overview of both is included. The results and discussion of these experiments are subsequently presented.

4.1. Methodology Application

In this section the methodology application is described in detail and each step of our methodology is explained from the data acquisition and preparation to the final portfolio optimization.

4.1.1. Data Set Acquisition and Preparation

To evaluate the algorithms and the proposed solution methodology, a set of assets and their data series of the SP500 were used, consisting of 479 companies listed between January 2020 and the end of February 2025, giving a total of 262 observations (weekly prices). This period was chosen because there was significant variability in the index and the stocks due to the health emergency that affected the assets of the United States of America and the whole financial world.

Justifications for using the SP500 are the market representativeness and diversification (wide variety of companies of different sectors and sizes), access to historical data, liquidity which allows for easy and efficient trading, it being commonly used as a benchmark to compare the performance of an investment portfolio, and, finally, the U.S. equity market enjoys strong regulation and institutional stability.

A computer program was developed to obtain information on the SP500 market assets corresponding to the daily closing prices of the selected period from the Yahoo Finance portal. Once the initial set was formed, the amount of empty or non-numeric data (NAN) was checked, and those with missing data representing more than 50% of the total were discarded.

In the first phase of the proposed model, data validation and cleaning were performed, which consisted of verifying that all the assets had the complete and correct information for the period to be analyzed. The main tasks performed in this phase were detecting atypical data and data supplementation in case of missing data at the beginning, end, or within the series. The complementation process was performed by applying the extrapolation (missing data at the beginning or end of the series) and interpolation methods.

As a final step, the expected value of each asset was compared to the return of the SP500 index, as shown in Figure 4, where the assets with a return lower than this value were discarded, since the assets that do not exhibit a return higher than this limit would be of no interest to financial investors. As mentioned above, this helps to reduce computation time since the final set of data to be analyzed is reduced.

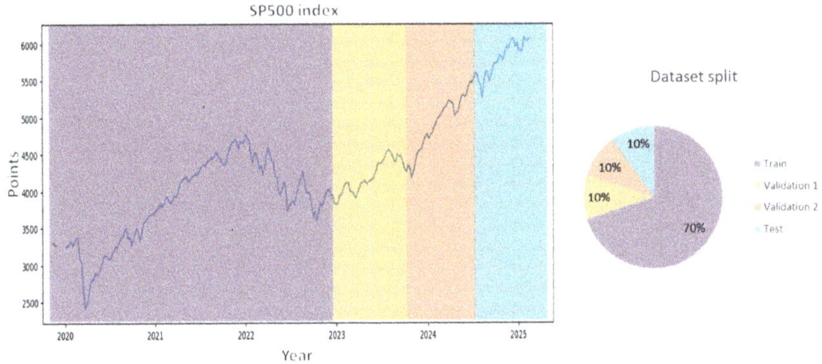

Figure 4. SP500 performance and split.

4.1.2. Segmentation of the Data Set

At this point, it was decided to segment the data series into four sections: the first segment is the training set, which was established as 70% of the data, i.e., 183 weekly observations, two validation segments, which consist of 10% of the observations each, i.e., 26 observations each, and finally a test segment equivalent to the remaining 10%. We decided to segment in this way to make the dataset compatible with the techniques applied in the FCTA methodology for data forecasting. Similarly, this same segmentation is applicable in constructing the initial portfolio where the training and validation sets are taken as input data. To summarize, the segmentation written above is in Figure 4.

4.1.3. Initial Portfolio Hyperparameters

For the construction of the initial portfolio, the TAIPO algorithm was used. This had already been tested and compared previously with other alternatives, demonstrating its advantages, such as obtaining high values in the Sharpe Ratio as well as a good return and a low correlation of assets.

The hyperparameters of the TAIPO metaheuristic were tuned by applying the analytical tuning method described in [8,64]. The values obtained for the parameters are: Initial Temperature: 273.8164, Final Temperature: 0.0000169, Cooling ratio: 0.96, while the equilibrium cycle length L, was dynamically calculated as a function of the number of assets and the temperature scheme.

Given that TAIPO is an approximate algorithm, 30 executions were carried out to obtain the average of the performance indicators, thereby enabling the selection of assets that maximize returns. The average values obtained were: Sharp Ratio (SR): 0.0221, Return: 0.0114, Risk: 0.2980, Correlation between assets: 0.2044, and the number of selected assets: 57, in weekly terms.

Compared to the performance of the SP500 during 2024, equivalent to 0.24 per year, the portfolio integrated in this stage had a performance of 147% above the index.

Figure 5 shows a sample of the graphs generated by the results of the algorithm. It can be observed that the Sharpe Ratio behaves as expected regarding the TAIPO algorithm; in the initial interactions, the system tolerates suboptimal solutions. However, as the iterations progress under the cooling scheme, the system becomes less tolerant of accepting bad solutions, ultimately identifying a higher SR value.

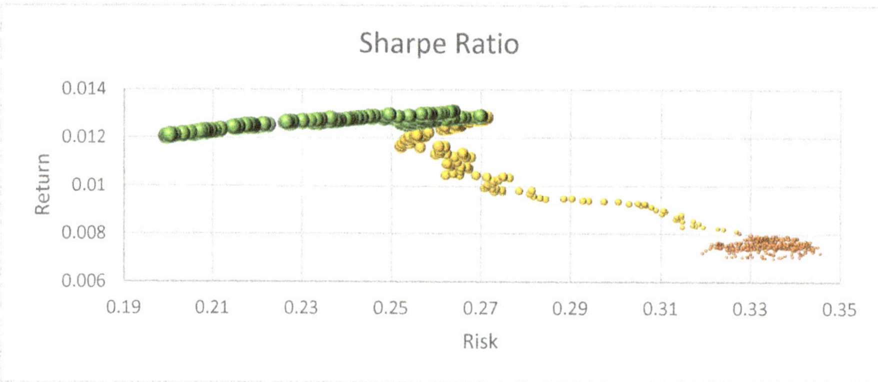

(**a**)

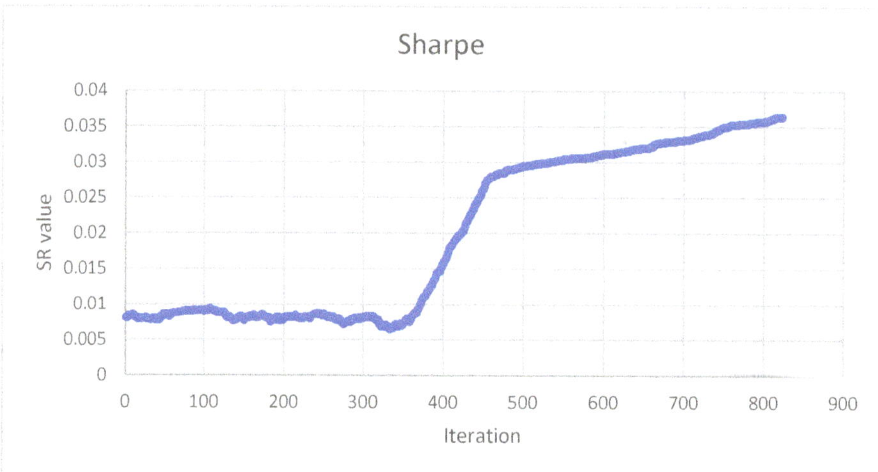

(**b**)

Figure 5. Algorithm performance in optimizing the Sharpe-based objective function value. (**a**) Efficient Frontier Risk vs. Return; (**b**) Sharpe Ratio values through algorithm execution.

Figure 5a presents the growth of the Sharpe Ratio value, which provides a representative example of the algorithm's performance. The size of the bubbles represents the SR, with smaller bubbles in the lower part corresponding to SR values between 0.0 and 0.01, indicating lower returns (below 0.008). In the intermediate section, the SR values range from 0.01 to 0.0299, with returns varying between 0.008 and 0.012. The larger bubbles at the top illustrate the efficient frontier, which is defined as a set of portfolios with high SR values above 0.03 and returns greater than 0.012.

The behavior of the SR value, which functions as the evaluation criterion for the TAIPO algorithm, is highlighted and shown in Figure 5b. Up to iteration 350, the values exhibit the anticipated behavior, accepting suboptimal solutions. However, after this initial phase, there is a decline in the acceptance of suboptimal solutions, leading to enhanced SR values. Notice that from iteration 500 to the conclusion of the process, the highest SR values correspond to the efficient frontier.

4.1.4. Forecasting: Weights and Optimization

For the initial portfolio forecasting, we applied the FCTA predicting tools [30]. Figure 6 shows the average result of the fourteen forecasting methods; we executed thirty executions according to the FCTA process. The sMAPE metric is presented for the initial combination using a uniform (sMAPE U) distribution for the weights. A 35.28% value was obtained for this initial solution, while the optimized combination (sMAPE O) yielded a sMAPE value of 23.60%. Notice that a large improvement of 49% of sMAPE was obtained. The forecast horizon for this case was set at 7 weeks ahead; nevertheless, MASIP works for any other short and medium forecast horizon.

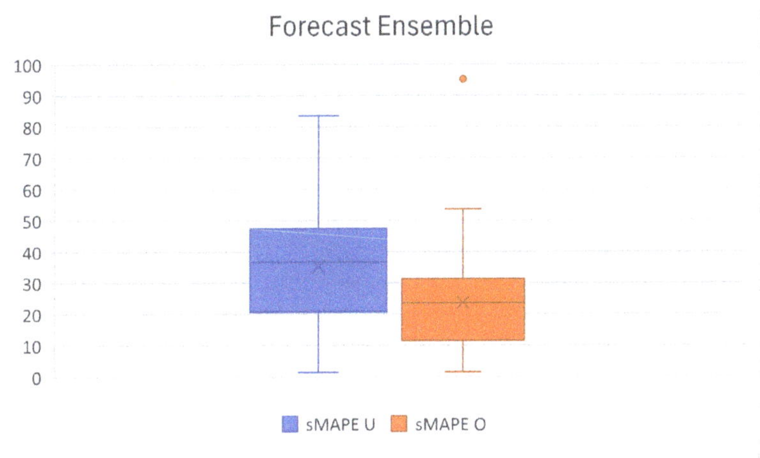

Figure 6. Forecast ensemble metric.

As illustrated in Table 2, a comparison is presented that integrates four performance error metrics.

Table 2. Forecasting performance of FCTA from 2020 to 2025.

	sMAPE	MSE	RMSE	MAPE	R^2
Mean	23.60	24.46	4.94	54.99	0.143
Std	16.5	16.22	9.03	83.20	0.178

4.1.5. Final Portfolio Optimization

In the final stage of the methodology, the combined forecasts of the assets were obtained. These were then refined using one of the integration algorithms and objective functions that have been previously mentioned. For this phase, the TAIPO algorithm was employed once more with the following metrics: Sharp Ratio: 0.0359, Return: 0.0126, Risk: 0.2170, Correlation between assets: −0.1061, and number of selected assets: 6 incorporating the complete time series as well as the forecasts calculated in the preceding stage. In terms of performance, the final portfolio presented an increase of 10.5% in returns and a reduction in correlation value compared to the initial portfolio. The graph of the evolution of the Sharpe-based objective function is shown in Figure 7.

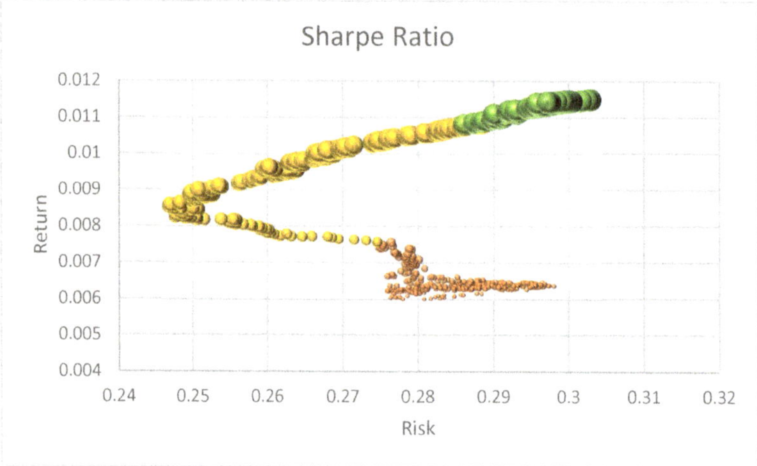

(a)

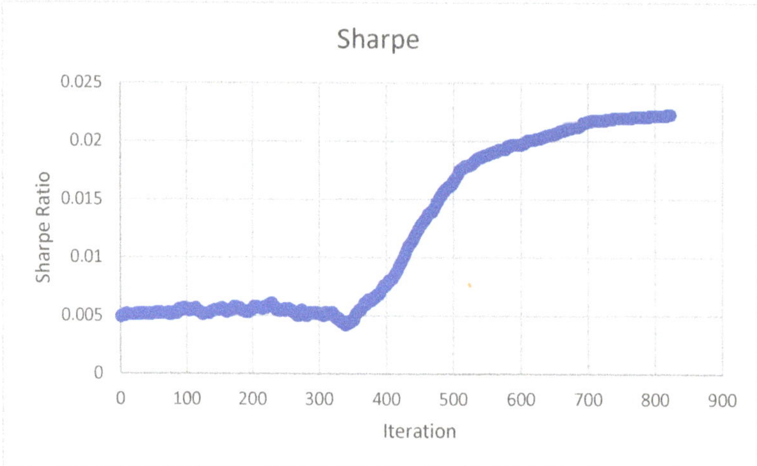

(b)

Figure 7. The Sharpe Ratio during final portfolio optimization: (**a**) Sample of Sharpe Ratio portfolios; (**b**) Sharpe Ratio through the TAIPO algorithm.

Figure 7a presents the growth of the Sharpe Ratio value, which provides a representative example of the algorithm's performance. The size of the bubbles represents the Sharpe Ratio (SR), with smaller bubbles in the lower part corresponding to SR values between 0.0 and 0.0099, indicating lower returns (below 0.00756). In the intermediate section, the SR values range from 0.01 to 0.02, with returns varying between 0.0075 and 0.0108. The larger bubbles represent the efficient frontier, which is defined as a set of portfolios with high SR values above 0.022 and returns greater than 0.011. The behavior of the SR value is presented in Figure 7b, which serves as the evaluation function for the TAIPO algorithm; up to iteration 250, the values behave as expected, accepting suboptimal solutions. However, in some periods ahead, the acceptance of suboptimal solutions decreases, resulting in better SR values. From iteration 310 to completion, the highest SR values form the efficient frontier. We can, therefore, conclude that the algorithm finds the best solutions and begins a stagnation process. This means that after 20 repetitions of the algorithm without finding an improvement in the solution, the algorithm stops [8].

By way of comparison and observing the results of the performance of the GENPO algorithm, an experiment was carried out for the optimization of the final portfolio, producing the following results: SR: 0.0308, Return: 0.0132, Risk: 0.2399, Correlation between assets: −0.0674, and number of selected assets: 10. In the results, we notice that, in practically all the items, there are similar values except for the assets where there are 10, as well as a slight increase in both return value and risk.

4.2. MASIP Comparison with State-of-the-Art Methods

To validate the proposed methodology, we compared MASIP with Puerto and Cesarone's model [21,23], and we used datasets from the SP500 market from 2005 to 2016. This period comprises the widest range of time common in both studies. For these datasets, we compared the average MASIP results with those reported by these authors in Table 3. In the last references, a portfolio for 52 weeks considering a zero risk-free rate was performed. The metrics used in the three alternatives were: the expected return, the risk value, the Sharpe Ratio, and the number of assets. These results are discussed in subsequent sections. We developed the programming codes necessary to emulate the former authors' works.

Table 3. MASIP results comparison with state-of-the-art methods for SP500.

Avg Results	Puerto [21]	Cesarone [23]	MASIP
Expected Return	0.002168	0.0026	0.0720
Risk	0.0231	0.0279	0.2988
Sharpe Ratio	0.094	0.0931	0.2218
Number of Assets	15	10	11

4.2.1. Data Acquisition, Preparation, and Split

We gathered datasets between 2005 and 2016 using Yahoo Finance, comprising, after data cleaning, a preparation dataset of 445 assets and 572 weekly observations. Observations were split into 70% for training, 10% each for Validation 1 and 2, and 10% for testing.

4.2.2. Initial Portfolio MARR Results

The TAIPO algorithm was used to build the initial portfolio, with the maxSharpe model. Table 4 shows a sensitivity analysis of the average results of the portfolios with different minimum acceptable rates of return (MARR). The value of the SR is directly related to the value of the MARR, since as the return value increases, the algorithm will look for a set of assets that delivers a higher return. We can also observe that as the minimum acceptable rate of return increases, the number of assets decreases; this is because, internally, the algorithm selects the assets with the highest return and that is close to the risk-free value. On the other hand, the results show a slight reduction in the correlation value.

Table 4. Initial portfolio applying MARR.

MARR %	Sharpe	Return	Risk	Correlation	Assets
15	0.1924	0.0383	0.1842	0.0305	45
20	0.2162	0.0572	0.2467	0.0142	17
25	0.2181	0.0636	0.2697	0.0091	10
30	0.2230	0.0723	0.2983	0.0135	11

The improvement in SR values achieved by the TAIPO algorithm and the target function can be seen in Figure 8.

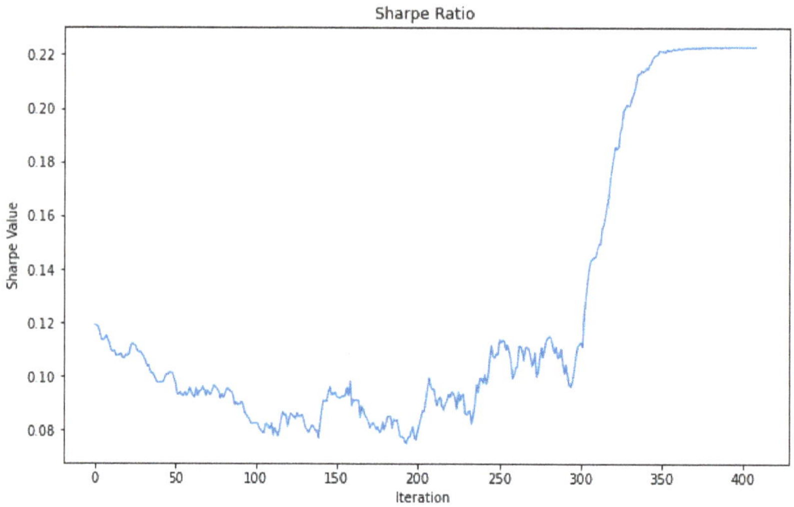

Figure 8. Sample of Sharpe Ratio in the initial portfolio.

4.2.3. Portfolio Forecasting

The initial portfolio was generated with the heuristic algorithm, and individual forecasts were obtained for each selected asset through the 14 forecasting methods within the FCTA methodology. A series of experiments were conducted, varying the forecast horizon within the range of 1 to 24 weeks.

Table 5 shows the differences in the error metric values of the forecast method combination models in a uniform distribution scheme, sMAPE U, and the contrast with the sMAPE O.

Table 5. Average sMAPE values at different forecasting horizons.

Horizon Weeks	sMAPE U	sMAPE O	Improvement
1	35.08	15.67	55%
4	10.64	7.74	27%
8	11.6	9.71	16%
24	12.59	11.68	7%

In conclusion, the application of the FCTA algorithm, once more, provides results that, at least in the case of the 1-week forecast horizon, demonstrate an improvement of up to 50%.

As is shown in Table 6, a comparison is presented that integrates four performance error metrics.

Table 6. Forecasting performance of FCTA from 2005 to 2016.

	sMAPE	MSE	RMSE	MAPE	R^2
Mean	13.15	27.47	4.62	74.93	0.241
Std	12.63	27.36	2.51	64.91	0.220

4.2.4. Final Optimized Portfolio

The subsequent section entails implementing rebalancing or optimization procedures. For this purpose, the TAIPO algorithm was used to obtain comparative results for each forecast horizon, as well as to contrast the advantages of each of the algorithms and the case study. It becomes evident that as the forecast horizon increases, there is a notable decline in the Sharpe Ratio and return values. Concurrently, the risk factor experiences a reduction, and the number of assets exhibits a decline. The performance shown in Figure 9 illustrates the evolution of the SR within the algorithm for the integration of the final portfolio. The final portfolios generated through MASIP methodology exhibit Sharpe values that are higher than those of the state-of-the-art methods in the short term, and this advantage is further enhanced once these portfolios have been forecasted. As we observe, they maintain or even increase the SR in the presented instances.

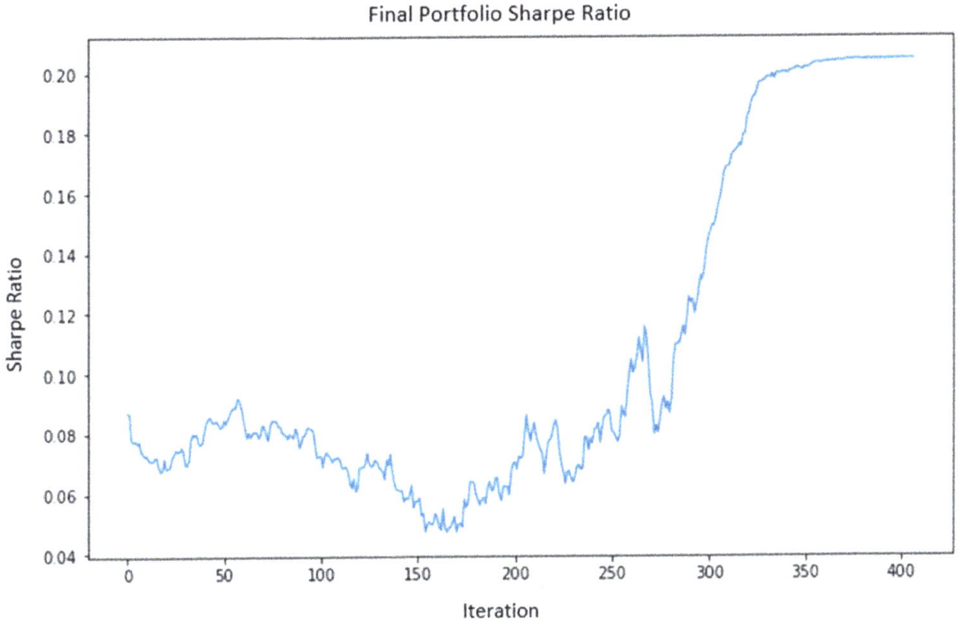

Figure 9. Sample of Sharpe Ratio in the final portfolio.

Table 7 shows the results obtained from these experiments, prioritizing the average of the highest Sharpe values found, identifying the algorithms that achieved superior performance in return terms. As shown in Table 7, the integrated portfolios with brief time horizons and the model MASIP yield outcomes surpass the contemporary standard. Conversely, in the medium-term horizons, between 8 and 24 weeks, Cesarone's model attains superior results in terms of SR. Conversely, when the focus is directed toward the variable yield, the MASIP algorithm demonstrates a markedly superior performance in comparison to the other two algorithms. This distinction can be attributed to the inherent characteristics of the MASIP algorithm and the mathematical models employed. The MASIP model is designed to optimize the SR by prioritizing the maximization of yield.

Conversely, the implementation of the MASIP methodology in the models under comparison yields analogous or superior outcomes. This is evident in the case of the Puerto approach, where predictions for the medium term (i.e., 8 to 24 weeks) indicate an SR considerably higher than those reported in the reference studied, reaching up to 415%. In contrast, Cesarone's model consistently shows superior performance, particularly in medium-term predictions, where significant improvements of up to 348% in terms of the SR are observed. As mentioned above, the MASIP methodology can be applied with various models, yielding favorable outcomes.

Table 7. Final comparison of forecasted values.

		Horizon Weeks					
		1	2	4	8	12	24
MASIP	Expected Return	0.072	0.0721	0.0721	0.06	0.0598	0.0609
	Risk	0.2991	0.2989	0.2986	0.2	0.1993	0.2091
	SR	**0.2215**	**0.2218**	**0.2222**	0.1752	0.1746	0.1717
	Num. Assets	11	11	11	7	7	7
Puerto [23]	Expected Return	0.0061	0.0061	0.0061	0.0339	0.0338	0.033
	Risk	0.0081	0.0081	0.008	0.0066	0.0065	0.0064
	SR	0.0829	0.0832	0.0839	**0.482**	**0.4845**	**0.477**
	Num. Assets	21	22	22	22	22	20
Cesarone [21]	Expected Return	0.0057	0.0056	0.0062	0.0068	0.0068	0.0067
	Risk	0.0282	0.0282	0.0321	0.0163	0.0163	0.0161
	SR	0.204	0.2018	0.195	**0.4167**	**0.4171**	**0.4148**
	Num. Assets	15	15	15	15	15	15

5. Conclusions

This paper proposes MASIP, an efficient methodology for investing portfolios. This methodology presents for the first time all the elements required for designing a portfolio: the selection of the best asset candidates for integrating the portfolio, the forecasting of the asset values in the investing horizon, including the best method selection for this task, the integration the best assets in this horizon and their participation, as well as the optimization model considering new constraints related to the maximum risk rate allowable and the minimum acceptable rates of return. MASIP was applied to SP500 and compared with state-of-the-art methods.

Regarding the initial phase of the portfolio construction process, it is imperative to identify the most suitable candidates. To this end, MASIP employs the TAIPO algorithm, a proposed mathematical model, and a constraint related to the lowest acceptable rate of return. This approach facilitates the identification of effective portfolios. In a subsequent stage, MASIP employs the FCTA to generate individual forecasts for each candidate asset and each of the prediction methods, selecting those methods with optimal performance and assembling them through the TAE algorithm to identify the optimal combination that minimizes the sMAPE and generates short- and medium-term forecast horizons. A rebalancing of the portfolio is subsequently implemented in the final stage, utilizing the TAIPO algorithm. This results in a portfolio comprising forecasted data and optimized weightings.

Experimental evaluations of the algorithms and data used yielded outstanding asset selection and evaluation results with SR, as well as expected return. The MASIP methodology was complemented by the forecast of the assets participating in the portfolio, thus collaborating in the final optimization, resulting in a final weighted, forecasted, and optimized portfolio.

A comparative analysis was conducted between MASIP and two alternative investment models, the Puerto and Cesarone models. The outcomes of this analysis were prioritized based on the meaning of the highest Sharpe values identified. The results of this study demonstrate that the MASIP algorithm exhibited robust performance in terms of returns. As the forecast horizon grows, the SR and return values drop significantly, concurrently with a drop in the risk factor and the asset number. The SR values from the created portfolios exceed those of the state-of-the-art methods, and forecasting enhances this advantage. The superiority of MASIP performance relative to the other two algorithms lies in its design, which prioritizes the maximization of short-term yield to optimize SR. Furthermore, the efficacy of the MASIP methodology in enhancing the performance of other methodologies is evident from the observed improvement in SR performance. This observation supports the potential for further experimentation with the same approach.

For future research, the application of other optimization methods is proposed, including other objective functions and metrics, both for asset preselection and portfolio rebalancing, the experimentation of novel methods or forecasting ensembles in financial time series, as well as the application of the MASIP methodology in other markets.

Author Contributions: Conceptualization, J.F.-S. and J.P.-A.; methodology and software, J.P.-A.; validation, G.C.-V., J.G.-B. and J.F.-S.; formal analysis, J.F.-S. and J.G.-B.; investigation, J.P.-A. and J.F.-S.; writing—original draft preparation, J.P.-A.; writing—review and editing, J.G.-B., J.F.-S., J.P.S.H. and G.C.-V.; supervision, J.F.-S. and J.G.-B.; project administration, J.F.-S. All authors have read and agreed to the published version of the manuscript.

Funding: This research received no external funding.

Data Availability Statement: https://github.com/DrJuanFraustoSolis/TAIPO.git (accessed on 30 January 2025).

Acknowledgments: The authors would like to acknowledge SECIHTI (Secretariat of Science, Humanities, Technology and Innovation), TecNM (National Technology of Mexico), Madero City Technological Institute, and the National Laboratory of Information Technologies (LaNTI) for access to the cluster.

Conflicts of Interest: The authors declare no conflicts of interest.

References

1. Guerard, J.B. *Introduction to Financial Forecasting in Investment Analysis*, 1st ed.; Springer: New York, NY, USA, 2013. [CrossRef]
2. Andersen, J.V. Investment Decision Making in Finance, Models of. In *Encyclopedia of Complexity and Systems Science*; Meyers, R.A., Ed.; Springer: New York, NY, USA, 2009; pp. 4971–4983. [CrossRef]
3. Markowitz, H. Portfolio Selection. *J. Financ.* **1952**, *7*, 77–91. [CrossRef]
4. Sharpe, W.F. Mutual Fund Performance. *J. Bus.* **1966**, *39*, 119–138. [CrossRef]
5. Scholz, H. Refinements to the Sharpe ratio: Comparing alternatives for bear markets. *J. Asset Manag.* **2007**, *7*, 347–357. [CrossRef]
6. Zakamouline, V.; Koekebakker, S. Portfolio performance evaluation with generalized Sharpe ratios: Beyond the mean and variance. *J. Bank. Financ.* **2009**, *33*, 1242–1254. [CrossRef]
7. Landete, M.; Monge, J.F.; Ruiz, J.L.; Segura, J.V. *Sharpe Portfolio Using a Cross-Efficiency Evaluation BT—Data Science and Productivity Analytics*; Charles, V., Aparicio, J., Zhu, J., Eds.; Springer International Publishing: Cham, Switzerland, 2020; pp. 415–439. [CrossRef]
8. Solis, J.F.; Aldaz, J.L.P.; Del Angel, M.G.; Barbosa, J.G.; Valdez, G.C. SAIPO-TAIPO and Genetic Algorithms for Investment Portfolios. *Axioms* **2022**, *11*, 42. [CrossRef]
9. Bates, J.M.; Granger, A.W.J. The Combination of Forecasts. *Oper. Res. Soc.* **1969**, *20*, 451–468. [CrossRef]
10. Winkler, R.L.; Makridakis, S. The Combination of Forecasts. *J. R. Stat. Soc. Ser. A* **1983**, *146*, 150–157.
11. Dueck, G.; Scheuer, T. Threshold accepting: A general purpose optimization algorithm appearing superior to simulated annealing. *J. Comput. Phys.* **1990**, *90*, 161–175. [CrossRef]
12. Winker, P.; Maringer, D. The Threshold Accepting Optimisation Algorithm in Economics and Statistics. In *Optimisation, Econometric and Financial Analysis*; Kontoghiorghes, E.J., Gatu, C., Eds.; Springer: Berlin/Heidelberg, Germany, 2007; pp. 107–125.
13. Choueifaty, Y.; Coignard, Y. Toward maximum diversification. *J. Portf. Manag.* **2008**, *35*, 40–51. [CrossRef]
14. Tella, R.; Rogel-Salazar, J. Portfolio Construction Based on Implied Correlation Information and VAR. *SSRN Electron. J.* **2013**, *12*, 125–144. [CrossRef]
15. Wang, S.; Xia, Y. *Portfolio Selection and Asset Pricing*, 1st ed.; Springer: Berlin/Heidelberg, Germany, 2002. [CrossRef]
16. Gilli, M.; Këlezi, E. A Heuristic Approach to Portfolio Optimization. Available online: https://citeseerx.ist.psu.edu/document?repid=rep1&type=pdf&doi=f6500f280e2c91a2ce11d0e46f90424aaac749f4 (accessed on 31 January 2025).
17. Masese, J.M.; Othieno, F.; Njenga, C. Portfolio Optimization under Threshold Accepting: Further Evidence from a Frontier Market. *J. Math. Financ.* **2017**, *7*, 941–957. [CrossRef]
18. Rangel-González, J.A.; Frausto-Solis, J.; González-Barbosa, J.J.; Pazos-Rangel, R.A.; Fraire-Huacuja, H.J. Comparative study of ARIMA methods for forecasting time series of the mexican stock exchange. *Stud. Comput. Intell.* **2018**, *749*, 475–485. [CrossRef]

19. Vijh, M.; Chandola, D.; Tikkiwal, V.A.; Kumar, A. Stock Closing Price Prediction using Machine Learning Techniques. *Procedia Comput. Sci.* **2020**, *167*, 599–606. [CrossRef]
20. Singh, P.; Jha, M. Portfolio Optimization Using Novel EW-MV Method in Conjunction with Asset Preselection. *Comput. Econ.* **2024**, *64*, 3683–3712. [CrossRef]
21. Cesarone, F.; Scozzari, A.; Tardella, F. An optimization–diversification approach to portfolio selection. *J. Glob. Optim.* **2020**, *76*, 245–265. [CrossRef]
22. Cesarone, F.; Mottura, C.; Ricci, J.M.; Tardella, F. On the Stability of Portfolio Selection Models. 2018, pp. 1–27. Available online: https://papers.ssrn.com/sol3/papers.cfm?abstract_id=3420081 (accessed on 31 January 2025).
23. Puerto, J.; Rodríguez-Madrena, M.; Scozzari, A. Clustering and portfolio selection problems: A unified framework. *Comput. Oper. Res.* **2020**, *117*, 104891. [CrossRef]
24. Kaczmarek, T.; Perez, K. Building portfolios based on machine learning predictions. *Econ. Res. Istraz.* **2021**, *35*, 1–19. [CrossRef]
25. Zhang, A. Portfolio Optimization of Stocks—Python-Based Stock Analysis. *Int. J. Educ. Humanit.* **2023**, *9*, 32–38. [CrossRef]
26. Ma, Y.; Han, R.; Wang, W. Portfolio optimization with return prediction using deep learning and machine learning. *Expert Syst. Appl.* **2021**, *165*, 113973. [CrossRef]
27. Martínez-Barbero, X.; Cervelló-Royo, R.; Ribal, J. Portfolio Optimization with Prediction-Based Return Using Long Short-Term Memory Neural Networks: Testing on Upward and Downward European Markets. *Comput. Econ.* **2024**, *65*, 1479–1504. [CrossRef]
28. Elliott, G.; Timmermann, A. Economic Forecasting. *J. Econ. Lit.* **2008**, *46*, 3–56. [CrossRef]
29. Centeno, V.; Georgiev, I.R.; Mihova, V.; Pavlov, V. Price forecasting and risk portfolio optimization. *AIP Conf. Proc.* **2019**, *2164*, 060006. [CrossRef]
30. Frausto-Solis, J.; Rodriguez-Moya, L.; González-Barbosa, J.; Castilla-Valdez, G.; Ponce-Flores, M. FCTA: A Forecasting Combined Methodology with a Threshold Accepting Approach. *Math. Probl. Eng.* **2022**, *2022*, 6206037. [CrossRef]
31. Zhang, Y.; Qu, H.; Wang, W.; Zhao, J. A Novel Fuzzy Time Series Forecasting Model Based on Multiple Linear Regression and Time Series Clustering. *Math. Probl. Eng.* **2020**, *2020*, 9546792. [CrossRef]
32. Rao, P.S.; Srinivas, K.; Mohan, A.K. *A Survey on Stock Market Prediction Using Machine Learning Techniques BT—ICDSMLA 2019*; Kumar, A., Paprzycki, M., Gunjan, V.K., Eds.; Springer: Singapore, 2020; pp. 923–931.
33. Becha, M.; Dridi, O.; Riabi, O.; Benmessaoud, Y. Use of Machine Learning Techniques in Financial Forecasting. In Proceedings of the 2020 International Multi-Conference on: "Organization of Knowledge and Advanced Technologies" (OCTA), Tunis, Tunisia, 6–8 February 2020; pp. 1–6. [CrossRef]
34. He, K.; Yang, Q.; Ji, L.; Pan, J.; Zou, Y. Financial Time Series Forecasting with the Deep Learning Ensemble Model. *Mathematics* **2023**, *11*, 1054. [CrossRef]
35. Hyndman, R.J.; Athanasopoulos, G. Forecasting: Principles and Practice. In *Forecasting: Principles and Practice*; Econometrics & Business Statistics: Monash, Australia, 2018; p. 504. Available online: https://otexts.com/fpp2/ (accessed on 31 January 2025).
36. Montero-Manso, P.; Athanasopoulos, G.; Hyndman, R.J.; Talagala, T.S. FFORMA: Feature-based forecast model averaging. *Int. J. Forecast.* **2020**, *36*, 86–92. [CrossRef]
37. Petropoulos, F.; Apiletti, D.; Assimakopoulos, V.; Babai, M.Z.; Barrow, D.K.; Ben Taieb, S.; Bergmeir, C.; Bessa, R.J.; Bijak, J.; Boylan, J.E.; et al. Forecasting: Theory and practice. *Int. J. Forecast.* **2022**, *38*, 705–871. [CrossRef]
38. Estrada-Patiño, E.; Castilla-Valdez, G.; Frausto-Solis, J.; González-Barbosa, J.; Sánchez-Hernández, J.P. A Novel Approach for Temperature Forecasting in Climate Change Using Ensemble Decomposition of Time Series. *Int. J. Comput. Intell. Syst.* **2024**, *17*, 253. [CrossRef]
39. Wasserbacher, H.; Spindler, M. Machine learning for financial forecasting, planning and analysis: Recent developments and pitfalls. *Digit. Financ.* **2022**, *4*, 63–88. [CrossRef]
40. Dharrao, D.S.; Bongale, A.M.; Deokate, S.T.; Doreswamy, D.; Bhat, S.K. Forecasting Stock Market Prices Using Machine Learning and Deep Learning Models: A Systematic Review, Performance Analysis and Discussion of Implications. *Int. J. Financial Stud.* **2023**, *11*, 94. [CrossRef]
41. Cheng, L.; Shadabfar, M.; Khoojine, A.S. A State-of-the-Art Review of Probabilistic Portfolio Management for Future Stock Markets. *Mathematics* **2023**, *11*, 1148. [CrossRef]
42. Althöfer, I.; Koschnick, K.-U. On the convergence of 'Threshold Accepting'. *Appl. Math. Optim.* **1991**, *24*, 183–195. [CrossRef]
43. Winker, P. *The Stochastics of Threshold Accepting: ANALYSIS of an Application to the Uniform Design Problem BT—Compstat 2006—Proceedings in Computational Statistics*; Rizzi, A., Vichi, M., Eds.; Physica: Heidelberg, HD, USA, 2006; pp. 495–503.
44. Gilli, M.; Kellezi, E. The Threshold Accepting Heuristic for Index Tracking. In *Financial Engineering, E-commerce and Supply Chain*; Springer: Boston, MA, USA, 2011; pp. 1–18. [CrossRef]
45. Ta, V.D.; Liu, C.M.; Tadesse, D.A. Portfolio optimization-based stock prediction using long-short term memory network in quantitative trading. *Appl. Sci.* **2020**, *10*, 437. [CrossRef]
46. Kelleher, J.D.; Tierney, B. *Data Science*; The MIT Press: Cambridge, MA, USA, 2018.

47. Provost, F.; Fawcett, T. *Data Science for Business: What You Need to Know About Data Mining and Data-Analytic Thinking*, 1st ed.; O'Reilly Media: Sebastopol, CA, USA, 2013.
48. Kuhn, M.; Johnson, K. *Feature Engineering and Selection: A Practical Approach for Predictive Models*, 1st ed.; CRC Chapman and Hall: Boca Raton, FL, USA, 2019.
49. SRoss, A.; Westerfield, R.W.; Jordan, B.D. *Fundamentals of Corporate Finance*, 13th ed.; McGraw Hill: New York, NY, USA, 2021.
50. Box, G.E.P.; Jenkins, G.M.; Reinsel, G.C. *Time Series Analysis: Forecasting and Control*; Prentice Hall: Englewood Cliffs, NJ, USA, 1994; Volume SFB 373, Chapter 5; pp. 837–900.
51. Goodfellow, I.; Bengio, Y.; Courville, A. *Deep Learning*; MIT Press: Cambridge, MA, USA, 2016; Available online: http://www.deeplearningbook.org (accessed on 31 January 2025).
52. Markowitz, H. Foundations of portfolio theory. In *Harry Markowitz: Selected Works*; World Scientific Publishing Company: Singapore, 2009; Volume 46, pp. 481–490. [CrossRef]
53. Markowitz, H.M. *Portfolio Selection*; Yale University Press: New Haven, CT, USA, 1959; Available online: http://www.jstor.org/stable/j.ctt1bh4c8h (accessed on 31 January 2025).
54. Yu, L.; Wang, S.; Lai, K.K. Multi-attribute portfolio selection with genetic optimization algorithms. *INFOR* **2009**, *47*, 23–30. [CrossRef]
55. Jaganathan, S.; Prakash, P.K.S. A combination-based forecasting method for the M4-competition. *Int. J. Forecast.* **2020**, *36*, 98–104. [CrossRef]
56. Cawood, P.; Van Zyl, T. Evaluating State-of-the-Art, Forecasting Ensembles and Meta-Learning Strategies for Model Fusion. *Forecasting* **2022**, *4*, 732–751. [CrossRef]
57. Barnard, G.A. New Methods of Quality Control New Methods of Quality Control. *J. R. Stat. Soc.* **1963**, *126*, 255–258.
58. Flores, B.E. A Pragmatic View of Accuracy Measurement in Forecasting. *Omega* **1986**, *14*, 93–98. [CrossRef]
59. Armstrong, J.S. Long-range Forecasting: From Crystal Ball to Computer. In *Wiley-Interscience Publication*; Wiley: Hoboken, NJ, USA, 1985; Available online: https://books.google.com.mx/books?id=t7V8AAAAIAAJ (accessed on 31 January 2025).
60. Makridakis, S. Accuracy measures: Theoretical and practical concerns. *Int. J. Forecast.* **1993**, *9*, 527–529. [CrossRef]
61. Makridakis, S.; Hibon, M. The M3-Competition: Results, conclusions and implications. *Int. J. Forecast.* **2000**, *16*, 451–476. [CrossRef]
62. Armstrong, J.S.; Franke, G. *Principles of Forecasting: A Handbook for Researchers and Practitioners*, 1st ed.; Springer: Boston, MA, USA, 2001.
63. Makridakis, S.; Spiliotis, E.; Assimakopoulos, V. The M4 Competition: 100,000 time series and 61 forecasting methods. *Int. J. Forecast.* **2020**, *36*, 54–74. [CrossRef]
64. Frausto-Solis, J.; Román, E.F.; Romero, D.; Soberon, X.; Liñán-García, E. *Analytically Tuned Simulated Annealing Applied to the Protein Folding Problem BT—Computational Science—ICCS 2007*; Shi, Y., van Albada, G.D., Dongarra, J., Sloot, P.M.A., Eds.; Springer: Berlin/Heidelberg, Germany, 2007; pp. 370–377.

Mathematical and Computational Applications

MDPI

Article

Internet-of-Things-Based CO_2 Monitoring and Forecasting System for Indoor Air Quality Management

Marya J. Marquez-Zepeda [1], Ildeberto Santos-Ruiz [1,*], Esvan-Jesús Pérez-Pérez [1,*], Adrián Navarro-Díaz [2] and Jorge-Alejandro Delgado-Aguiñaga [3]

1 TURIX-Dynamics Diagnosis and Control Group, I.T. Tuxtla Gutiérrez, Tecnológico Nacional de México, Carretera Panamericana S/N, Tuxtla Gutiérrez 29050, Mexico; m21271145@tuxtla.tecnm.mx
2 School of Engineering and Sciences, Tecnologico de Monterrey, Av. General Ramón Corona 2514, Zapopan 45138, Mexico; adrian.navarro@tec.mx
3 Centro de Investigación, Innovación y Desarrollo Tecnológico, CIIDETEC-UVM, Universidad del Valle de México, Tlaquepaque 45604, Mexico; jorge.delgado@uvmnet.edu
* Correspondence: ildeberto.dr@tuxtla.tecnm.mx (I.S.-R.); esvan.de.jesus.perez@upc.edu (E.-J.P.-P.)

Abstract: This study presents a low-cost and scalable CO_2 monitoring system that leverages NDIR sensors and a Long Short-Term Memory (LSTM) neural network to predict indoor CO_2 concentrations over both short- and long-term horizons. The proposed system aims to anticipate air quality deterioration in shared spaces, enabling proactive ventilation strategies. Various LSTM configurations were evaluated, optimizing the number of layers, neurons per layer, and input delays to enhance forecasting accuracy. The optimal model consisted of two LSTM layers with 128 neurons each and a time window of 10 previous observations. This model achieved an RMSE of approximately 57 ppm for an 8 h forecast in a classroom setting. Experimental results demonstrate the reliability of the proposed approach for CO_2 prediction and its potential impact on indoor air quality management.

Keywords: CO_2; air quality; remote monitoring; forecasting; artificial neural network

1. Introduction

In indoor environments, avoiding high concentrations of aerosols (microscopic particles exhaled when speaking or breathing) is critical, as these can degrade air quality and increase health risks. Poorly ventilated closed spaces exacerbate this issue, as the accumulation of aerosols and CO_2 rises with the number of occupants and the time spent in such environments [1]. This situation is common in classrooms during face-to-face sessions, where ventilation is often insufficient to maintain moderate CO_2 levels. Continuous CO_2 monitoring is essential to assess air quality, along with predicting or forecasting CO_2 concentration over time. This allows us to estimate how long it will take for a given space to reach CO_2 levels that could pose a significant risk, enabling proactive air quality management.

The clean air we breathe "outdoors", without pollution, contains approximately 400 parts per million (ppm) of CO_2. In the literature, minimum reference levels are reported between 412 ppm and 420 ppm, according to various sources [2]. Air with this concentration of CO_2 is considered to not have been breathed recently. CO_2 concentrations above the reference level indicate that the air has already been partially exhaled by someone, as shown in Table 1. For instance, when the CO_2 concentration reaches 1000 ppm, it is estimated that approximately 1.5% of the air has already been previously exhaled. Concentrations above 1000 ppm not only reflect reduced air quality but also pose a potential health risk, as elevated CO_2 levels can be toxic [3,4].

 https://doi.org/10.3390/mca30020036

Table 1. Relationship between CO_2 concentration and the fraction of breathed air [†].

CO_2 Concentration	Percentage of Breathed Air
400 ppm	0%
600 ppm	0.5%
700 ppm	0.7%
800 ppm	1.0%
1000 ppm	1.5%
2000 ppm	4.0%
3000 ppm	6.5%
4000 ppm	9.0%

[†] Based on IDAEA-CSIC-LIFTEC recommendations [5].

To assess air quality, the appropriate sensor must be selected. NDIR (non-dispersive infrared) sensors are suitable for measuring the concentration of CO_2 since the molecules of this gas are prone to absorbing infrared light. Evaluations have already been made of NDIR sensors as a low-cost option for CO_2 measurement. One of these was performed in a laboratory environment, demonstrating that, without any calibration or correction, NDIR sensors achieve RMS errors between 5 ppm and 21 ppm compared to a precision sensor [6]. CO_2 measurements can be managed and analyzed by various methods. Typically, as a complement to monitoring systems, diagnosis/prognosis applications are developed where the data are processed through specialized programs (e.g., MATLAB) or cloud computing services, such as ThingSpeak, Microsoft Azure, and Amazon Web Services, among others [7]. Cloud computing offers data storage and analysis services to forecast various physical variables using computational intelligence techniques like neural networks. In Robin et al. [8], convolutional neural networks were evaluated to monitor air quality; on the other hand, Altikat et al. [9] also used neural networks to predict the passage of CO_2 from the ground to the atmosphere. Recently, Kapoor et al. [10] designed a pilot monitoring system for CO_2 using neural networks and support vector machines. However, real-time CO_2 monitoring alone is insufficient for effective air quality management. Predictive modeling is necessary to estimate future CO_2 concentrations and optimize ventilation strategies.

Monitoring CO_2 levels is essential for ensuring indoor air quality (IAQ) and occupant well-being. Kwon et al. [11] classify CO_2 sensors into two main types: chemical sensors, which are energy-efficient and compact but suffer from short lifespan and low durability, and non-dispersive infrared (NDIR) sensors, which offer higher accuracy and are commonly used for air quality monitoring. The integration of Internet of Things (IoT) technologies has significantly improved CO_2 monitoring by enabling real-time data acquisition and remote accessibility. Marques and Pitarma [12] introduced iAQ WiFi, an IoT-based system that collects environmental data using low-cost sensors and transmits them via WiFi for real-time visualization and analysis. Marques et al. [13] expanded on this work with iAir CO_2, an advanced IoT solution designed for continuous CO_2 monitoring. Their study emphasizes the importance of real-time air quality tracking to anticipate and mitigate potential health risks.

Machine learning techniques have also been applied to CO_2 forecasting, allowing for more efficient and proactive air quality management. Kallio et al. [14] investigated multiple machine learning models, including ridge regression, decision trees, random forest, and multilayer perceptron, to predict indoor CO_2 concentration. Arsiwala et al. [15] developed a digital twin system integrating IoT, artificial intelligence, and Building Information Modeling (BIM) to automate CO_2 emissions tracking. Alsamrai et al. [16] provided a comprehensive review of IoT-based air quality monitoring systems, emphasizing the growing use of low-cost sensors and microcontrollers such as ESP8266 and ESP32. Their findings confirm that IoT applications offer a cost-effective and scalable alternative for pollution monitoring.

Building upon these advancements, this study integrates predictive modeling and real-time data collection to enhance CO_2 monitoring solutions. By leveraging machine learning and IoT technologies, our approach improves forecasting accuracy, supports proactive air quality management, and contributes to healthier indoor environments.

This study proposes an IoT-based CO_2 monitoring and forecasting system, integrating low-cost monitoring stations equipped with NDIR sensors and ESP32 microcontrollers to provide real-time CO_2 measurements. These devices are strategically deployed in classrooms, offices, and laboratories within Tecnológico Nacional de México campus Tuxtla Gutiérrez. The collected CO_2 data are processed using an LSTM autoregressive neural network, trained to predict future CO_2 concentrations up to eight hours in advance. Unlike traditional mathematical forecasting models, this approach allows the neural network to learn patterns directly from sensor data, enhancing adaptability to different environmental conditions. The results of this study suggest an alternative for scheduling classroom sessions to ensure safe air quality conditions. The main contributions of this study are summarized as follows:

- The LSTM network analyzes historical data to accurately forecast CO_2 levels up to four hours in advance, eliminating the need for explicit models.
- A network of affordable sensors and wireless transmitters enables cost-effective deployment and easy maintenance, making the system highly scalable.
- Predictive insights allow proactive ventilation control, improving air quality, health, and cognitive performance in indoor environments.
- Real-time monitoring and forecasting optimize space utilization and enhance safety in educational institutions, supporting data-driven decision making.

The remainder of this document is organized as follows: Section 2 presents the materials and methods used for the monitoring system and the configuration of the LSTM network for CO_2 concentration forecasting. Section 3 describes the results obtained in different configurations and the comparison with different methods reported in the literature. Finally, Section 4 presents the conclusions.

2. Materials and Methods

The CO_2 monitoring system to prevent COVID-19 infection involves using an NDIR sensor and an ESP32-Core2 microcontroller board to monitor CO_2 levels in indoor environments, as seen in Figure 1. The DNIR sensor is a kind of optical sensor that can detect the concentration of CO_2 in the air by measuring light absorption at a specific wavelength.

The ESP32-Core2 microcontroller board reads data from the DNIR sensor and sends them to the Thinhspeak cloud using WiFi connectivity. Thinhspeak is an IoT platform that provides data storage, analysis, and visualization tools. The collected CO_2 data are then analyzed in Matlab, a popular data analysis and modeling tool. Using these data, a CO_2 level prediction algorithm can be developed, which can estimate the CO_2 level in the near future based on current measurements.

The CO_2 level prediction algorithm based on LSTM results can be displayed to users on their mobile devices using an application. The application can show real-time monitoring graphs and alert users if the CO_2 level exceeds a certain threshold, indicating that the indoor environment may be poorly ventilated and potentially hazardous to human health.

This system has the potential to prevent the spread of airborne diseases, such as COVID-19, by providing a tool for monitoring indoor air quality and identifying poorly ventilated environments that could increase the risk of pathogen transmission. The following sections describe the method in detail, including device connections for collecting sensor data, the neural network used, the training process, and the tested configurations.

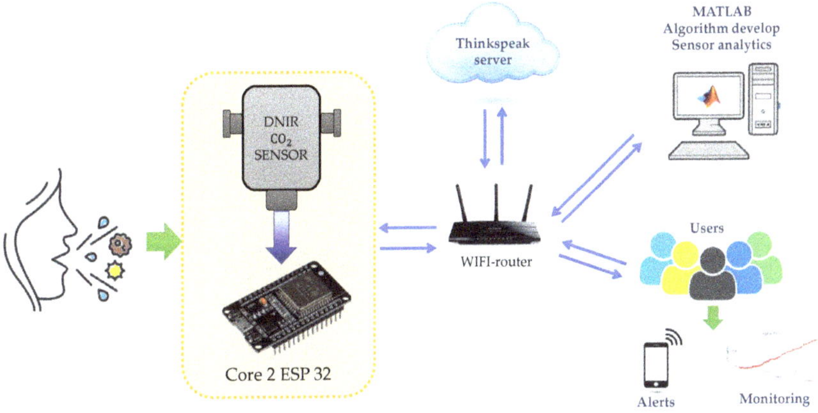

Figure 1. General schema of the proposed methodology.

2.1. Measurement of CO_2 Concentration

To measure the concentration of CO_2 in the air, an NDIR sensor is used, which is quite precise and easy to calibrate. This sensor consists of a tube, an optical filter, an emitter, and an infrared (IR) detector, as shown in Figure 2. The emitter produces IR light waves that travel through the air sample tube. The IR waves move toward the optical filter in front of the detector. The detector measures the amount of IR light that passes through the filter.

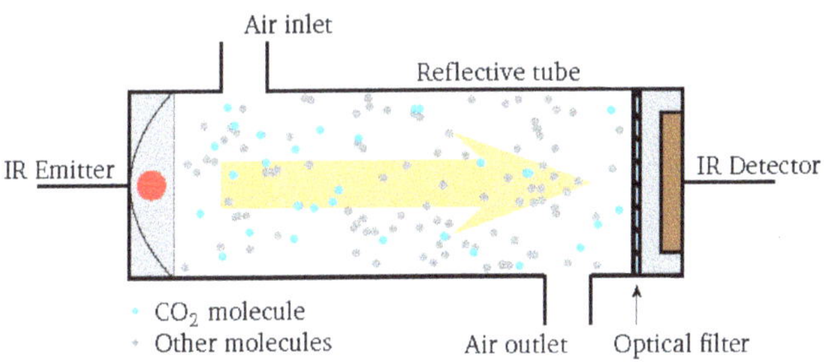

Figure 2. CO_2 sensor.

The radiation emitters band coincides with the CO_2's absorption band, located around 4.26 μm. The absorption spectrum is unique, so it is a signature or fingerprint to identify the CO_2 molecule.

As IR light travels through the tube, the CO_2 gas molecules absorb the characteristic 4.26 μm band while letting other wavelengths pass. At the detector end, the remaining light is incident on an optical filter that absorbs all wavelengths of light except the wavelength absorbed by the CO_2 molecules in the tube containing the air sample.

Finally, the detector receives the remaining amount of IR light not absorbed by the CO_2 molecules or the optical filter. To calculate the CO_2 concentration, the difference between the amount of IR light radiated by the emitter and the amount of IR light received by the detector is measured. Since this difference results from light absorption by the CO_2 molecules in the tube, it is directly proportional to the number of CO_2 molecules in the air sample.

In the monitoring station where the sensor is embedded, some aspects are taken into account so that the measurements are as accurate as possible; one of them is the warm-up time, which lasts approximately 60 s; during this time, the data are unreliable and are not recorded. In addition, the sensor must be calibrated using a process that references the lowest concentration recorded outdoors over some time. To verify the proper operation and accuracy of each NDIR sensor, its readings are compared to those of a precision CO_2 meter to validate the calibration.

The CO_2 concentrations recorded by the sensor vary depending on where it is placed within the monitored space; for a reliable reading, considering the influence of ventilation, the sensors are placed at least 120 cm from the ground, 60 cm from air flows (windows), and SI2m of the people inside the room, as suggested by previous studies [17].

2.2. Monitoring Stations

The monitoring stations capture the measurements from the sensors to transmit them in an IoT network where the measurements of all the monitored classrooms or offices converge. Each station is made up of the following elements:

- An ESP32 microcontroller module with WiFi, Bluetooth, and LoRa wireless connections; it also allows wired connections using I2C, UART, and SPI protocols.
- An NDIR sensor model MH-Z19D with UART-type serial interface; its detection range goes from 400 ppm to 10,000 ppm, with a maximum error of 50 ppm.
- Connection to an IoT network by WiFi or LoRa (long-range radio frequency), depending on the wireless connectivity available in each station.

The microcontroller captures the values the NDIR sensor detects and sends the data wirelessly for recording and processing in the cloud. The electrical diagram of the M5 tough device is shown in Figure 3, where the connections of the microcontroller with the sensor and the visual/sound indicators used as an alarm are specified when the concentration of CO_2 exceeds the safe values; and its specifications are presented in Table 2.

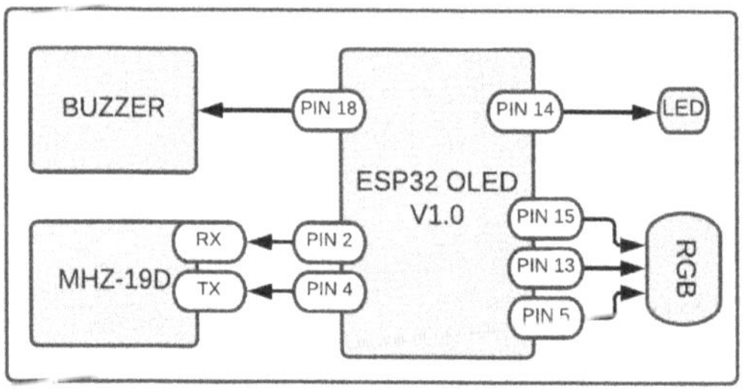

Figure 3. Connections at the monitoring station.

Table 2. Technical data for M5TOUGH.

Specifications	Parameters
ESP32-D0WDQ6-V3	240 MHz dual core, 600 DMIPS, 520 KB SRAM, WiFi
IPS LCD	Full-color display of 2.0″ 320 × 240 ILI9342C
Antenna	3D-WiFi
Speaker Configuration	NS4168 16-bit I2S amplifier + 1 W speaker
Voltage Input	USB (5 V at 500 mA) DC (24 V at 1 A)

The monitoring stations comprise an IoT network managed by ThingSpeak, a Cloud Computing service operated by MathWorks. The data are stored in the cloud, which can be updated and viewed using the ThingSpeak API, allowing them to be viewed on computers or mobile devices connected to the internet. The final prototype is presented in Figure 4, where the three concentration levels of CO_2 are shown, according to ppm and visual indicators (green, yellow, and red). The design includes a 2 cm hole which allows air to flow freely through it, thus facilitating readings from the CO_2 sensor. The marked levels are presented according to Table 3.

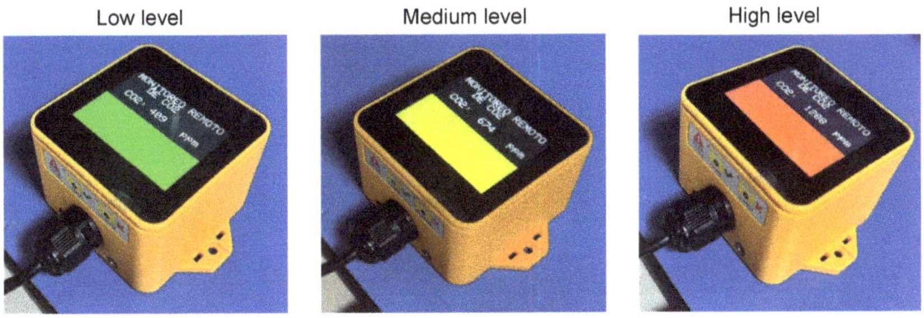

Figure 4. CO_2 concentration levels with indicator colors.

Table 3. Risk levels according to [18].

Risk Level	Color	Range (ppm)
Low	Green	400 to 700
Medium	Yellow	701 to 999
High	Red	1000 and above

The screenshot in Figure 5 shows the remote interface of one of the stations. The sensor sampling period is one second, although the cloud data are updated every 15 s due to limitations in bandwidth and available storage space; the communication latency is approximately one second, which is sufficient considering the frequency of data updating and the slow dynamics in the monitored area since there are no sudden changes in the concentration of CO_2.

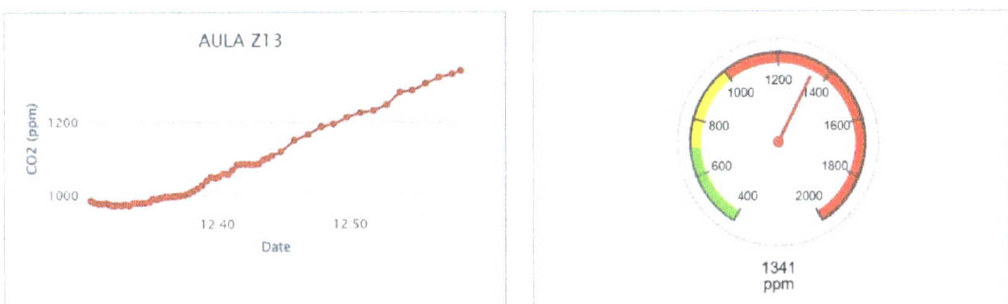

Figure 5. Monitoring interface in ThingSpeak.

2.3. *Prognosis of CO_2 Concentration with LSTM Network*

A type of neural network based on deep learning frequently used in time–series forecasting is autoregressive networks and those called Long Short-Term Memory (LSTM). From the CO_2 concentrations recorded by the stations, time–series are created for each monitored classroom, which are used to estimate future CO_2 concentrations based on the most recent measurements. LSTM networks work with time–series processing, using loops in their network diagram, and allowing them to remember/forget previous states and use this information to decide the next one. This LSTM comprises a status cell that transmits the data to be processed through the network. This gate allows us to decide what information is going to be discarded and another allows us to update the memory, as shown in Figure 6 and as expressed in Equations (1)–(6). Where x_t are the input data; f_t, i_t, and o_t, are the outputs of each gate, enabled by the activation function, σ or tanh. The subscripts f, i, and o are indicative of the gate that corresponds to them, *forget*, *input*, and *output*. In addition, there are short- and long-term memories, h_t and C_t.

$$f_t = \sigma(W_f[h_{t-1}, x_t] + b_f) \tag{1}$$
$$i_t = \sigma(W_i[h_{t-1}, x_t] + b_i) \tag{2}$$
$$o_t = \sigma(W_o[h_{t-1}, x_t] + b_o) \tag{3}$$
$$\widetilde{C}_t = \tanh(W_c[h_{t-1}, x_t] + b_c) \tag{4}$$
$$C_t = f_t C_{t-1} + i_t \widetilde{C}_t \tag{5}$$
$$h_t = o_t \tanh(C_t) \tag{6}$$

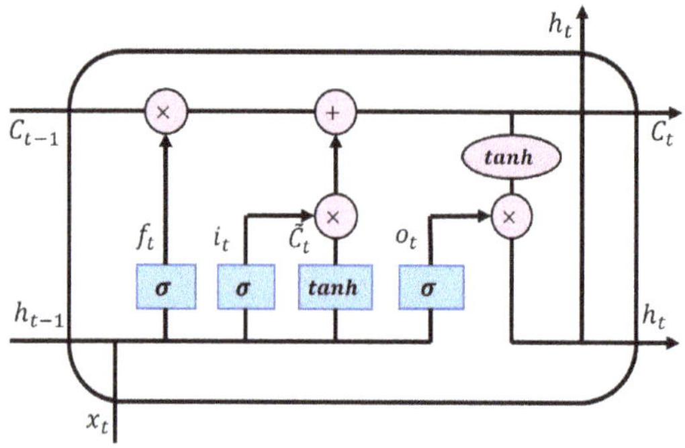

Figure 6. Structure of an LSTM network.

Conventional recursive networks are used to model short-term dependencies (i.e., close relationships in time–series), whereas LSTMs are useful for modeling long-term dependencies. The LSTM architecture is a block comprising three neural networks, better known as gates, which allow us to weigh the dataset to remember, discard, and update the information at the convenience of its application. This network will enable us to make a more extended prediction due to its long-term memory derived from the gates above. The LSTM block configurations are presented, which were proposed to analyze its performance with the dataset described above. The LSTM configurations were trained with 70% of the available data, and the other 30% were used to perform the forecast tests. The Adam learning algorithm was used with an initial learning rate of 0.005, and the iterations for the training were varied to know its impact on the performance of the network.

The number of hidden units and training times were varied in the neural architecture tuning. The first configuration selected took 200 epochs to train, having 128 hidden units. The second setup was 30 hidden units and trained in 1000 epochs. The third analysis case was of 208 hidden units and was trained in 1000 epochs.

3. Results and Discussion

The previously mentioned configurations of the LSTM block, varying the number of hidden units (128, 30, and 208 units), registered the best performances, obtaining accurate forecasts with a competitive RMSE, being lower concerning the results obtained with the NAR architecture. The first configuration with 128 hidden units took 200 epochs to train, obtaining an RMSE of 57.4396 ppm, an MAD of 27.67 ppm, and an MAPE of 0.026887%. Figure 7 shows the network output, Figure 8 contrasts the measured and forecast data, whereas Figure 9 presents the error between them.

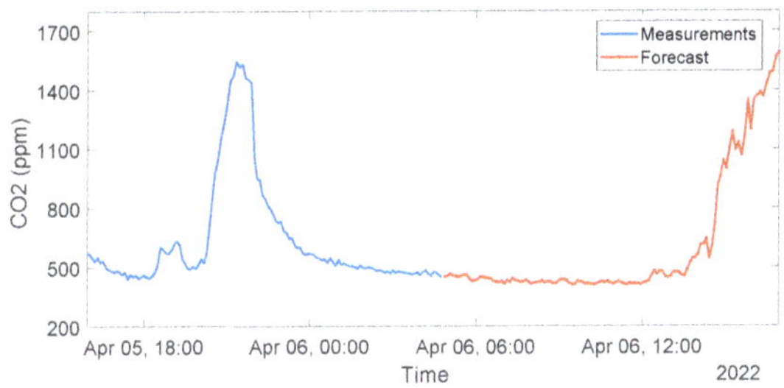

Figure 7. Neural network output, Case 1 LSTM.

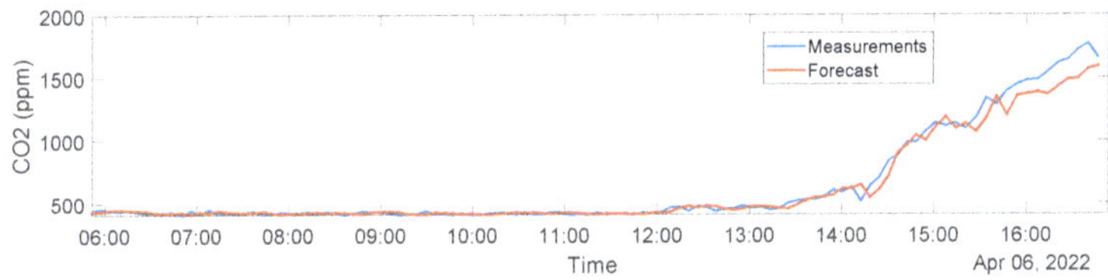

Figure 8. Measured data versus predicted data, Case 1 LSTM.

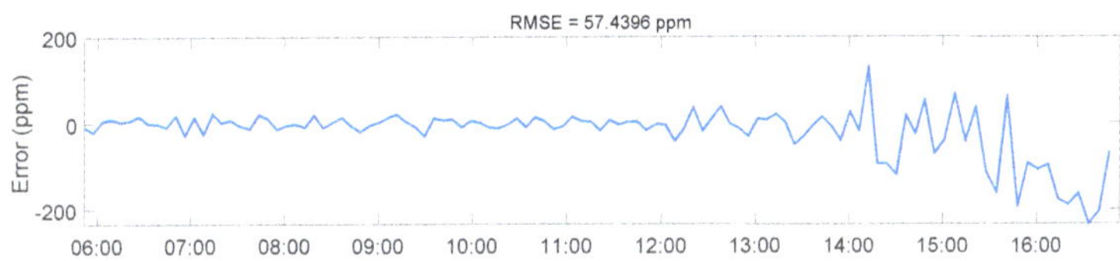

Figure 9. Forecast error, Case 1 LSTM.

The second configuration with 30 hidden units took 1000 epochs to train, obtaining an RMSE of 68.17 ppm, an MAD of 31.33 ppm, and an MAPE of 0.02871%. Figure 10 shows the network output, Figure 11 contrasts the measured and forecast data, and Figure 12 presents the error between them.

The third configuration, with 208 hidden units, took 1000 epochs to train, obtaining an RMSE of 69.86 ppm, an MAD of 29.1748 ppm, and an MAPE of 0.017992%. Figure 13 shows the network performance, Figure 14 contrasts the measured and forecast data, and Figure 15 presents the error between them.

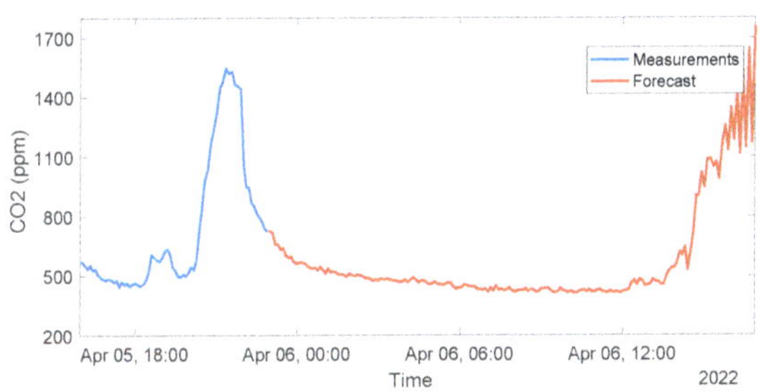

Figure 10. Neural network output, Case 2 LSTM.

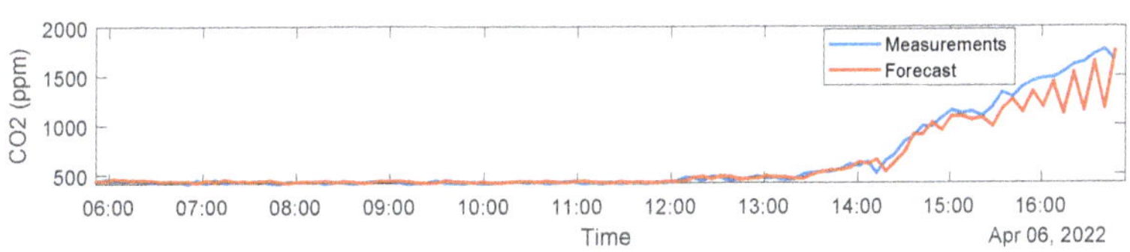

Figure 11. Measured data versus predicted data, Case 2 LSTM.

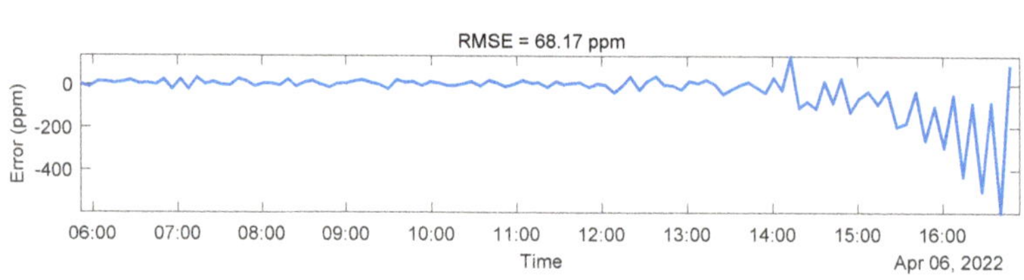

Figure 12. Forecast error, Case 2 LSTM.

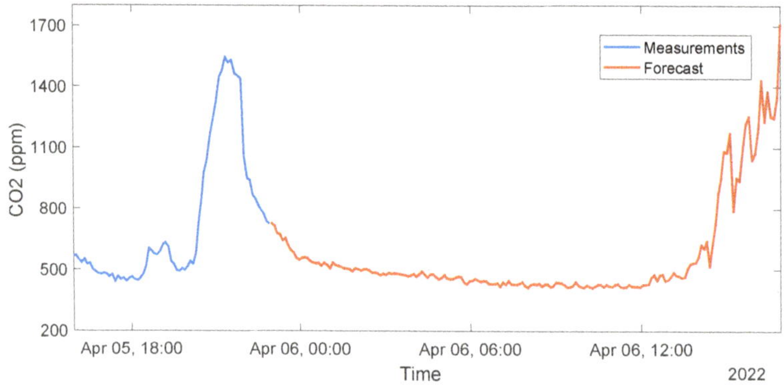

Figure 13. Neural network output, Case 3 LSTM.

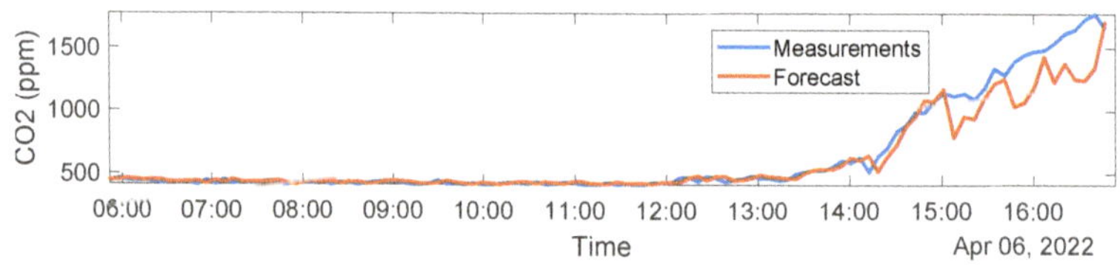

Figure 14. Measured data versus predicted data, Case 3 LSTM.

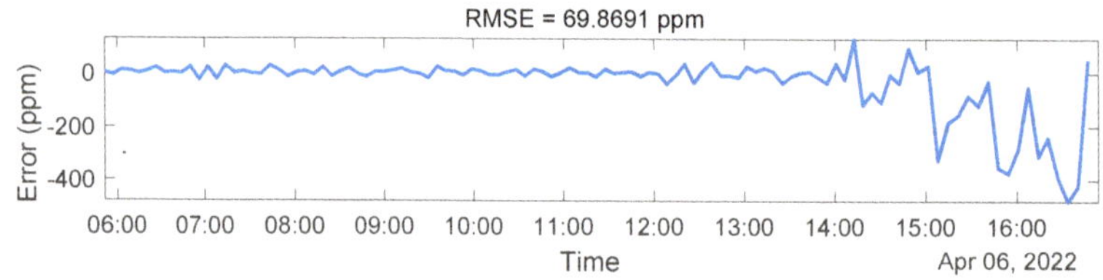

Figure 15. Forecast error, Case 3 LSTM.

Validation of the Proposed LSTM

Different architecture configurations for the validation of the LSTM were analyzed, which are presented below (Tables 4–7), highlighting the best performances. These tables are separated concerning selection percentages for training and test data to analyze their impact on configurations; in addition, they present the statistical indices to measure their performance. From this, it was concluded that the best results were obtained by selecting a data percentage of 70% for training and 30% for testing, maintaining the lowest RMSE on average.

Table 4. LSTM validation: 80% training, 20% testing.

Configuration	Epochs	RMSE (ppm)	MAD (ppm)	MAPE (%)
128	1000	63.86	34.04	0.040045
128	200	80.53	39.15	0.060933
100	800	105.82	46.28	0.045336
150	1000	84.03	41.40	0.043663
200	1000	67.25	34.33	0.029713
30	1000	76.77	38.76	0.05382
208	1000	81.2230	41.2516	0.051761
160	1000	64.7605	33.0171	0.041185
180	1000	104.5606	49.5327	0.082857
120	1000	80.5162	39.6079	0.038419

Table 5. LSTM validation: 70% training, 30% testing.

Configuration	Epochs	RMSE (ppm)	MAD (ppm)	MAPE (%)
128	1000	73.21	30.32	0.020867
128	200	57.4396	27.67	0.026887
100	800	70.88	29.35	0.019392
150	1000	76.48	32.65	0.027096
200	1000	75.91	34.88	0.032143
30	1000	68.17	31.33	0.028715
208	1000	69.8691	31.6915	0.028221
160	1000	86.5504	35.1914	0.033307
180	1000	67.4451	32.5854	0.0222
120	1000	68.1354	29.7362	0.028629

Table 6. LSTM validation: 60% training, 40% testing.

Configuration	Epochs	RMSE (ppm)	MAD (ppm)	MAPE (%)
128	1000	119.38	54.49	0.03586
128	200	92.37	42.60	0.030407
100	800	106.06	49.56	0.036635
150	1000	99.83	44.53	0.02597
200	1000	117.41	55.69	0.041978
30	1000	104.12	50.26	0.039152
208	1000	114.702	53.7591	0.038169
160	1000	112.7961	49.5180	0.038434
180	1000	105.6364	47.5359	0.033837
120	1000	123.9902	56.5427	0.044883

In relation to the results obtained, a comparison is made with previous research carried out by other authors, who have addressed the analysis of the concentration of CO_2 using various machine learning and deep learning approaches and techniques. These approaches and techniques are detailed in Table 8. A nonlinear autoregressive network (NAR) was tested, with a similar configuration presented for the LSMT of this work, the NAR obtained lower performance than the LSTM. On the other hand, the results were compared with the SVM, a linear regressive network (LR), and an artificial neural network (ANN) although its topology is not presented, reported in [19]. As can be seen, the proposed LSTM configuration obtained the lowest RMSE to predict CO_2. Table 9 expands this comparison by summarizing the key differences between the study by Liu et al. and the present research; whereas Liu et al. [19] achieved lower RMSE values (16.77 ppm), their model is limited to a 1 min prediction window. In contrast, the proposed approach extends the forecast to 8 h, making it more suitable for long-term air quality management. Additionally, the methodology provides a detailed description of the IoT implementation, specifying the use of NDIR sensors and ESP32 microcontrollers, whereas Liu et al. do not specify their hardware components; while Liu et al. validated their study in a residential environment, the present research was tested in various indoor spaces, such as classrooms, offices, and laboratories, demonstrating broader applicability.

Table 7. LSTM validation: second dataset.

Configuration	Epochs	RMSE (ppm)	MAD (ppm)	MAPE (%)
128	1000	114.3233	55.6462	0.037745
128	200	90.5694	40.2798	0.030747
100	800	104.3672	48.9134	0.039543
150	1000	99.83	44.53	0.02597
200	1000	111.1245	54.6689	0.04027
30	1000	101.3488	49.8211	0.03734
128–80	1000	158.3354	68.2418	0.05489
80–80	1000	129.8544	56.4882	0.04233
100–80	1000	118.695	52.658	0.03452
60–60	1000	120.4541	53.548	0.04065

Table 8. Summary of performances of different network architectures.

Architecture	RMSE (ppm)	MAPE (%)
NAR	66.56	0.022695
LSTM	57.4396	0.026887
SVM (*)	153.0833	1.9642
LR (*)	143.6322	1.9341
ANN (*)	111.5761	1.7404

(*) Results from [10].

Table 9. Comparison between Liu et al. [19] and the present study.

Aspect	Liu et al. [19]	Present Study
Main Model	LSTM	LSTM
Configurations	Single, Stacked, Bidirectional LSTM	Variation in layers, neurons, and input delays
Prediction Horizon	1 min	Up to 8 h
Best RMSE	16.77 ppm (Bidirectional LSTM)	57.44 ppm (128 neurons, 200 epochs)
Worst RMSE	21.96 ppm (Single-cell LSTM)	69.86 ppm (208 neurons, 1000 epochs)
Sensors Used	Not specified (generic IoT)	NDIR MH-Z19D
Microcontroller	Not specified	ESP32
IoT Platform	MQTT + Grafana	ThingSpeak
Test Environment	Residential	Classrooms, offices, and laboratories
Main Objective	Quick ventilation adjustment	Space optimization and mitigation strategies
Expected Impact	Immediate CO_2 prediction	Long-term air quality planning and management

4. Conclusions

The implementation of an IoT-based LSTM model for CO_2 monitoring has demonstrated high effectiveness in predicting CO_2 levels up to 8 h in advance. The ability of LSTM networks to capture long-term dependencies in time–series data allows for accurate and reliable forecasting, surpassing the performance of NAR neural networks. The proposed system provides a scalable and cost-effective solution for real-time CO_2 monitoring, offering valuable insights into air quality trends in shared indoor environments. These results highlight the potential of LSTM-based approaches to enhance air quality management by enabling proactive ventilation strategies and improving occupant well-being.

Future research could focus on optimizing the model's hyperparameters to further enhance predictive accuracy. Additionally, integrating other environmental factors such as temperature, humidity, and air quality indices could refine the system's performance. Developing a real-time alert mechanism for CO_2 threshold exceedance would further improve its practical applicability, allowing for immediate corrective actions. Advancements in this area will contribute to the development of more intelligent and efficient air quality monitoring systems, fostering healthier and safer indoor environments.

Author Contributions: Conceptualization, M.J.M.-Z. and I.S.-R.; Data curation, E.-J.P.-P., A.N.-D. and J.-A.D.-A.; Formal analysis, A.N.-D., J.-A.D.-A. and E.-J.P.-P.; Methodology, M.J.M.-Z. and I.S.-R.; Project administration, I.S.-R.; Software, M.J.M.-Z. and E.-J.P.-P.; Supervision, I.S.-R.; Validation, M.J.M.-Z. and I.S.-R.; Visualization, E.-J.P.-P., A.N.-D. and J.-A.D.-A.; Writing—original draft, M.J.M.-Z., I.S.-R. and E.-J.P.-P.; Writing—review and editing, J.-A.D.-A. and A.N.-D. All authors have read and agreed to the published version of the manuscript.

Funding: This research has been supported by the Consejo Nacional de Humanidades, Ciencias y Tecnologías (CONAHCyT) and by Tecnológico Nacional de México (TecNM).

Data Availability Statement: The raw data supporting the conclusions of this article will be made available by the authors on request.

Conflicts of Interest: The authors declare no conflicts of interest.

References

1. Moreno Grau, S.; Álvarez León, E.; García dos Santos Alves, S.; Diego Roza, C.; Ruiz de Adana, M.; Marín Rodríguez, I.; Rodríguez-Ba no, J.; Tomás Carmona, M.; Minguillón, M.C.; van der Haar, R. Evaluación del Riesgo de la Transmisión de SARS-CoV-2 Mediante Aerosoles. Medidas de Prevención y Recomendaciones. Documento Técnico. Ministerio de Sanidad. 2020. Available online: https://www.sanidad.gob.es/profesionales/saludPublica/ccayes/alertasActual/nCov/documentos/COVID19_Aerosoles.pdf (accessed on 30 December 2024).
2. Lahrz, T.; Bischof, W.; Sagunski, H.; Baudisch, C.; Fromme, H.; Grams, H.; Gabrio, T.; Heinzow, B.; Müller, L. Gesundheitliche Bewertung von Kohlendioxid in der Innenraumluft. *Bundesgesundheitsblatt Gesundheitsforschung Gesundheitsschutz* **2008**, *51*, 1358–1369.
3. Zemitis, J.; Bogdanovics, R.; Bogdanovica, S. The study of CO_2 concentration in a classroom during the COVID-19 safety measures. *E3S Web Conf.* **2021**, *246*, 01004.
4. Peng, Z.; Jimenez, J.L. Exhaled CO_2 as a COVID-19 infection risk proxy for different indoor environments and activities. *Environ. Sci. Technol. Lett.* **2021**, *8*, 392–397. [PubMed]
5. Minguillón, M.; Querol, X.; Riediker, M.; Felisi, J.; Garrido, T.; Alastuey, A.; Bekö, G.; Nehr, S.; Wiesen, P.; Carslaw, N. Guide for Ventilation Towards Healthy Classrooms. COST (European Cooperation in Science and Technology) Action CA17136 Report. 2020. Available online: https://scoeh.ch/wp-content/uploads/2021/01/Guide-for-ventilation_Indairpollnet.pdf (accessed on 30 December 2024).
6. Martin, C.R.; Zeng, N.; Karion, A.; Dickerson, R.R.; Ren, X.; Turpie, B.N.; Weber, K.J. Evaluation and environmental correction of ambient CO_2 measurements from a low-cost NDIR sensor. *Atmos. Meas. Tech.* **2017**, *10*, 2383–2395.
7. Tripathi, B.S.; Gupta, R.; Reddy, S. Cloud Architecture Based Learning Kit Platform for Education and Research—A Survey and Implementation. In Proceedings of the International Symposium on Ubiquitous Networking, Virtual, 19–21 May 2021; pp. 172–185.

8. Robin, Y.; Amann, J.; Baur, T.; Goodarzi, P.; Schultealbert, C.; Schneider, T.; Schütze, A. High-performance VOC quantification for IAQ monitoring using advanced sensor systems and deep learning. *Atmosphere* **2021**, *12*, 1487. [CrossRef]
9. Altikat, S.; Gulbe, A.; Kucukerdem, H.K.; Altikat, A. Applications of artificial neural networks and hybrid models for predicting CO_2 flux from soil to atmosphere. *Int. J. Environ. Sci. Technol.* **2020**, *17*, 4719–4732.
10. Kapoor, N.R.; Kumar, A.; Kumar, A.; Kumar, A.; Mohammed, M.A.; Kumar, K.; Kadry, S.; Lim, S. Machine Learning-Based CO_2 Prediction for Office Room: A Pilot Study. *Wirel. Commun. Mob. Comput.* **2022**, *2022*. [CrossRef]
11. Kwon, J.; Ahn, G.; Kim, G.; Kim, J.C.; Kim, H. A study on NDIR-based CO_2 sensor to apply remote air quality monitoring system. In Proceedings of the 2009 ICCAS-SICE, Fukuoka, Japan, 18–21 August 2009; pp. 1683–1687.
12. Marques, G.; Pitarma, R. Monitoring health factors in indoor living environments using internet of things. In *Recent Advances in Information Systems and Technologies*; Springer: Cham, Switzerland, 2017; Volume 570, pp. 785–794.
13. Marques, G.; Ferreira, C.R.; Pitarma, R. Indoor air quality assessment using a CO_2 monitoring system based on internet of things. *J. Med. Syst.* **2019**, *43*, 67. [CrossRef] [PubMed]
14. Kallio, J.; Tervonen, J.; Räsänen, P.; Mäkynen, R.; Koivusaari, J.; Peltola, J. Forecasting office indoor CO_2 concentration using machine learning with a one-year dataset. *Build. Environ.* **2021**, *187*, 107409.
15. Arsiwala, A.; Elghaish, F.; Zoher, M. Digital twin with machine learning for predictive monitoring of CO_2 equivalent from existing buildings. *Energy Build.* **2023**, *284*, 112851.
16. Alsamrai, O.; Redel-Macias, M.D.; Pinzi, S.; Dorado, M. A systematic review for indoor and outdoor air pollution monitoring systems based on Internet of Things. *Sustainability* **2024**, *16*, 4353. [CrossRef]
17. Nusseck, M.; Richter, B.; Holtmeier, L.; Skala, D.; Spahn, C. CO_2 measurements in instrumental and vocal closed room settings as a risk reducing measure for a Coronavirus infection. *medRxiv* **2020**. [CrossRef]
18. Mesa Silva, A.F. Evaluación de los Niveles de Riesgo Ocupacional Asociado a las Concentraciones de Gases Contaminantes en Atmosferas Confinadas en un Acueducto. Available online: https://ciencia.lasalle.edu.co/items/34b05257-d236-46f5-89ca-7c59d583c2d2 (accessed on 30 December 2024).
19. Liu, Z.; Ciais, P.; Deng, Z.; Lei, R.; Davis, S.J.; Feng, S.; Zheng, B.; Cui, D.; Dou, X.; Zhu, B.; et al. Near-real-time monitoring of global CO_2 emissions reveals the effects of the COVID-19 pandemic. *Nat. Commun.* **2020**, *11*, 5172. [CrossRef] [PubMed]

Mathematical and Computational Applications

MDPI

Article

An Evolutionary Strategy Based on the Generalized Mallows Model Applied to the Mixed No-Idle Permutation Flow Shop Scheduling Problem

Elvi M. Sánchez Márquez [1,*], Ricardo Pérez-Rodríguez [2], Manuel Ornelas-Rodriguez [1,*] and Héctor J. Puga-Soberanes [1]

1 National Technology of Mexico/Leon Institute of Technology, Leon 37290, Mexico; hectorjose.puga@leon.tecnm.mx
2 Secretaría de Ciencia, Humanidades, Tecnología e Innovación (SECIHTI), National Technology of Mexico/Aguascalientes Institute of Technology, Aguascalientes 20255, Mexico; dr.ricardo.perez.rodriguez@gmail.com
* Correspondence: malintzin041093@gmail.com (E.M.S.M.); manuel.ornelas@leon.tecnm.mx (M.O.-R.)

Abstract: The Mixed No-Idle Permutation Flow Shop Scheduling Problem (MNPFSSP) represents a specific case within regular flow scheduling problems. In this problem, some machines allow idle times between consecutive jobs or operations while other machines do not. Traditionally, the MNPFSSP has been addressed using the metaheuristics and exact methods. This work proposes an Evolutionary Strategy Based on the Generalized Mallows Model (ES-GMM) to solve the issue. Additionally, its advanced version, ES-GMMc, is developed, incorporating operating conditions to improve execution times without compromising solution quality. The proposed approaches are compared with algorithms previously used for the problem under study. Statistical tests of the experimental results show that the ES-GMMc achieved reductions in execution time, especially standing out in large instances, where the shortest computing times were obtained in 23 of 30 instances, without affecting the quality of the solutions.

Keywords: mixed no-idle permutation flow shop scheduling; evolutionary strategy; generalized Mallows model; estimation of distribution algorithm

1. Introduction

The Mixed No-Idle Permutation Flow Shop Scheduling Problem (MNPFSSP) is a mixed-integer scheduling problem in which both types of machines with idle time and no idle time are allowed, depicting a real-world scenario. This problem was first raised by Pan and Ruiz in 2014 and cataloged as NP-hard. Currently, the problem is solved by considering the only objective to be the maximum completion time or makespan, denoted as $Cmax\ (\pi)$, although new variants of the problem have emerged that include additional features and other target functions [1,2].

The MNPFSSP is challenging not only due to the combination of constraints imposed on machines, with and without idleness, but also because it guarantees the coexistence of both types of machines. It makes it more difficult and expensive to select feasible solutions to minimize the objective function.

In the state-of-the-art Iterated Greedy metaheuristics [3], exact methods, such as Benders Decomposition (BD), branch-and-cut (BC), and Automated Benders Decomposition (ABD), as well as metaheuristics [4], such as the Iterated Greedy algorithm (IGA) and Referenced Local Search (RLS), have been used to solve the aforementioned problem. In this study, the proposed algorithm is compared against the metaheuristics and some exact methods that decompose the MNPFSSP.

The decomposition methods are applied to problems that allow identifying structures in the constraints of linear programming problems. An example is the mixed-integer mathematical programming (MILP) problems, where their difficulty lies in the fact that they have some variables of an integer nature, causing the convexity property of the feasible region to be lost, making it difficult to solve the problem [5]. In these cases, the aim is to break down the original problem into subproblems that are more easily solvable than the original problem, taking advantage of their iterative solution. In particular, the Benders Decomposition Method (BD) identifies complicating variables as those that, when set as parameters, make the search process easier, breaking down the problem into two simpler ones, which are called the master problem and the subproblem [4].

In this scheme, the solution to the problem is derived by finding partial solutions to the master problem and subproblem using a feedback scheme. Convergence to the solution of the original problem is sought through the information on the dual problems associated with the subproblem, i.e., the master problem solves the complicating variables, and the subproblem is solved by setting such variables. Furthermore, the subproblem gives feedback to the master problem through the solution of the dual problem. The goal is to optimize a feasible region by increasing constraints (called cuts) to approximate the optimal solution [4,5]. The above description assumes that the subproblem is feasible and bound for any master problem, but this condition is not always satisfied; thus, the method is modified by constructing the feasibility cut. If the subproblem is feasible, its solution generates optimality cuts for the master problem. BD has the advantage of having a feasible solution: even if convergence is not achieved, it generates a quasi-optimal solution [6].

The branch-and-cut (BC) method is a combination of the cutting plane algorithm with the branch-and-bound method. A cut plane is a linear constraint that is added to the linear programming (LP) relaxation at any node in the search tree. Given an integer problem (IP), the BC method searches the feasible region by building a binary search tree, solving for the LP relaxation of the IP input, and then adding any number of cut planes [7].

Contrarily, the Estimation of Distribution Algorithms (EDAs) belong to the evolutionary computation field, specifically among the stochastic algorithms that simulate a natural evolution, known as evolutionary algorithms, which depend on parameters such as crossover and mutation operators, crossover and mutation probabilities, population size, number of generations, etc. Establishing appropriate values for these parameters has become an optimization problem, which has motivated the emergence of EDAs. They were first introduced in 1996 by Mühlenbein [8].

EDAs have been applied to solve combinatorial optimization problems [9–11], achieving competitive performance. Therefore, we justify their application for the MNPFSSP.

EDAs mainly replace crossover and mutation operators by estimating and sampling a probability distribution commonly learned from individuals (solution vectors) of the previous generation. An attempt is made to explicitly find correlations between the variables, i.e., the interactions between the variables (the genes of the individuals). Such interactions are expressed through the probability distribution associated with the selected individuals in each generation [12]. This is one of the most complicated features of the EDAs, i.e., estimating the probabilistic model that represents the interactions between the variables of the selected individuals. In addition, other factors such as individual representation and objective function are decisive factors in the process [13].

This study presents an evolutionary strategy implemented using the ES-GMM algorithm. The proposal is inspired by an EDA that uses the Generalized Mallows Distribution (GMM). The ES-GMM focuses on a practical proposal for estimating the parameters of the

GMM and the operating conditions. The ES-GMM performance is compared against the exact methods applied in [3] to solve the MNPFSSP.

The article is organized as follows: Section 2 presents the state of the art, including an overview of EDAs based on the Mallows Model. Section 3 provides the background, including a description of the MNPFSSP, the EDA based on the Mallows Model, and the GMM. Section 4 describes the proposed ES-GMM algorithm, including its advanced version ES-GMMc. Section 5 presents experimental development, while Section 6 presents the results and statistical analysis, comparing the performance of ES-GMM and ES-GMMc with BC, ABD, IGA, and RLS. Section 7 provides a discussion of the main findings, highlighting the contributions, practical implications, and comparative advantages of the proposed approach. Finally, Section 8 presents the conclusions and future research.

2. State of the Art

In recent years, EDAs have been proposed to solve problems similar to the MNPFSSP, such as the PFSSP [11] and NPFSSP [12]. These approaches adapt from discrete and continuous domains to the permutation's domain. Ceberio et al. [14,15] proposed the application of EDAs in permutation spaces and suggested probabilistic models such as those based on marginal, the Plackett–Luce and Mallows Models. Although the Mallows Model has shown competitive results, there has been a recognized need for a deeper study, particularly regarding parameter estimation [10,16].

Ceberio et al. [14] introduced a marginal k-order model that considers interactions among all problem variables through a matrix reflecting the joint probability of an index being in a specific position. This model, with a memory cost of $\mathcal{O}\left(\frac{n}{k}\right)^{2k}$, is practical only for small values of k (size of the marginal matrix) and n (problem size). The authors observed the necessity of modeling probability distributions over permutations and proposed to use the Mallows Model for this purpose [17], which is defined by two parameters: a central permutation π_0 (estimated by sampling and averaging permutations) and a spread parameter θ (defined via Maximum Likelihood Method (MLE) and calculated numerically), to balance the exploration and exploitation of the solution space. Specifically, with small values of θ, the probability assigned to π_0 was also small (exploration), but it increased rapidly once θ surpassed a certain threshold (exploitation).

Although the EDA working with the Mallows Model outperformed other algorithms for large instances of the Flow Shop Scheduling Problem (FSSP), despite the numerical challenge posed by the estimation of its parameters, a lack of control of the algorithm was observed, and the necessity for a better understanding of the spread parameter θ was indicated [15].

Ceberio et al. [3] detailed a generalization of the EDA working with the Mallows Model, facing similar instability issues, as previously observed. Empirical values and an upper bound for θ were defined to balance exploration and exploitation and avoid premature convergence [3]. To estimate the central permutation σ_0, the Maximum Likelihood Method was formulated, leading to σ_0 being given by the permutation that minimizes the sum of Kendall distances of the sample. This problem is known as the Kemeny Rank Aggregation Problem and is recognized as NP-hard. An alternative to estimating the central permutation was performing an exhaustive analysis to find an exact or approximate solution for the central permutation, and it was concluded that the Borda algorithm offers a balance between computation time and accuracy, providing an unbiased estimator of the Kemeny problem solution [18]. Some studies have focused on improving the quality of σ_0 by applying mutation methods [19], heuristics or metaheuristics [20], and Pareto front approaches [21].

Recently, several studies have explored the use of different distance metrics, such as Cayley and Ulam, for the GMM [3,13,18,22,23]. These advancements highlight the effectiveness of the Mallows Model in permutation optimization and suggest that implementation within an EDA framework should be promising for the MNPFSSP.

Despite the performance of previous implementations, the literature reveals a persistent challenge in systematically estimating GMM parameters. Studies continue to rely on empirical methods for parameter estimation, underscoring the need for improved precision in this process.

3. Background

3.1. The MNPFSSP

The MNPFSSP, denoted as $F|prmu, mixed\ no\text{-}idle|\ C_{max}$, belongs to the family of pure regular flow scheduling problems. It is a general case of the NPFSSP where some machines allow idle times while other machines do not: this problem is considered NP-hard. The objective is to find the optimal job sequence that minimizes the maximum completion time (makespan) [3].

The NPFSSP differs from the basic Permutation Flow Shop Scheduling Problem (PFSSP) because there is no idle time allowed between two consecutive jobs on the machines. This feature is established in the PFSSP by the condition that the completion time of a job on the machine is greater than or equal to the completion time of the previous job on the same machine plus its processing time. Meanwhile, in the case of NPFSSP, it is enforced as equality. Considering these two conditions, the mathematical model of the MNPFSSP can be modeled as a mixed-integer linear problem, where the objective function consists of the minimization of the makespan, i.e., the completion time at which the job that occupies the last position n of the sequence on the last machine m ends [3]. This is defined by the next equation.

$$min\ C_{max} = C_{m,n}. \tag{1}$$

It is subject to the following:

$$\sum_{k=1}^{n} x_{j,k} = 1, \qquad j = 1, \ldots, n; \tag{2}$$

$$\sum_{j=1}^{n} x_{j,k} = 1, \qquad k = 1, \ldots, n; \tag{3}$$

$$c_{1,k} \geq \sum_{j=1}^{n} x_{j,1} p_{1,j}, \qquad k = 1, \ldots, n; \tag{4}$$

$$c_{i,k} \geq c_{i-1,k} + \sum_{j=1}^{n} x_{j,k} p_{i,j}, \qquad k = 1, \ldots, n, \quad i = 2, \ldots, m; \tag{5}$$

$$\begin{cases} c_{i,k} = c_{i,k-1} + \sum\limits_{j=1}^{n} x_{j,k} p_{i,j}\,, & if\ i \in M' \\ c_{i,k} \geq c_{i,k-1} + \sum\limits_{j=1}^{n} x_{j,k} p_{i,j}, & in\ other\ case \end{cases} \qquad k = 2, \ldots, n, \ i = 1, \ldots, m; \tag{6}$$

$$c_{i,k} \geq 0, \qquad k = 1, \ldots, n, \ i = 1, \ldots, m; \tag{7}$$

$$x_{i,k} \in \{0,1\}, \qquad k = 1, \ldots, n, \ i = 1, \ldots, m; \tag{8}$$

where

$c_{i,k}$ is the completion time of work at position k on machine i.

$x_{j,k}$ indicates whether a job j occupies a position k in the sequence, taking the value 1 or 0.

$p_{i,j}$ is the processing time for job j on machine i.
M' is a no-idle machines subset of M.

The constraints (2) and (3) enforce that each job occupies a position in the permutation and that each position in any permutation is occupied by a job. Constraint (4) controls the completion time of the first job in the permutation. Constraint (5) imposes that the completion times of jobs on the second and subsequent machines are larger than the completion times of the previous tasks of the same job on previous machines plus its processing time, effectively avoiding the overlap of tasks from the same job position. Furthermore, the constraint (6) seeks to respect the existence or inexistence of inactive times [3,4].

Solution Representation

To generate a schedule for the MNPFSSP, a permutation-based representation is used in this study. In this representation, each element of the permutation, represented by an integer, corresponds to a specific job. The order of elements in the permutation defines the sequence in which the jobs will be processed in the machines from the production line. The solutions can be represented using Gantt charts. In Figure 1, we show the Gantt charts for the PFSSP, NPFSSP, and MNPFSSP, where m represents the machines and j the jobs in the sequence. We can notice the existence of no-idle and idle machines according to the characteristics of each problem.

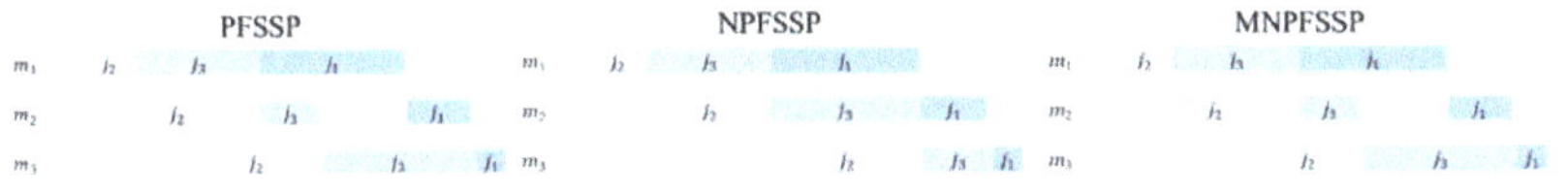

Figure 1. Gantt charts for the PFSSP, NPFSSP and MNPFSSP solutions.

3.2. EDAs Coupled with the GMM

The EDAs were first introduced in the field of evolution by Mühlenbein and Paaß [8], constituting a new tool in the field of evolutionary computation. They can be considered as a generalization of Genetic Algorithms (GAs), where the reproduction operators (crossover and mutation) are replaced by the estimation and sampling of the probability distribution of selected individuals [8].

The EDA constructs a probability distribution of the most promising solutions by leveraging the interrelationships between problem features. The distribution model is used to generate new individuals. The process is iterative, until a stopping criterion is met, and it returns the best-found solution.

Due to the need to solve permutation combinatorial problems, which are highly relevant in real-world situations in the industry, the EDA researchers have focused on developing new strategies to solve permutation-based problems. The Generalized Mallows Model on Estimation of Distribution Algorithm (GMMEDA) arises from a previous idea and has been applied to scheduling problems such as the PFSSP [3].

3.2.1. The GMM

The Mallows Model is a distance-based exponential probability model over permutation spaces that is based on the distance $d(\sigma_1, \sigma_2)$, which consists of counting the number of inversions necessary to order one permutation σ_1 compared to another σ_2. From this metric, the distribution is defined by Equation (9), with the parameter's central permutation σ_0 and the spread parameter θ.

$$P(\sigma) = \frac{exp(-\theta d(\sigma,\sigma_0))}{\psi(\theta)}, \quad (9)$$

where $\psi(\theta)$ is a normalization constant. When $\theta = 0$ assigns equal probability to every permutation, the larger the value of θ, the greater the probability assigned to σ_0. Thus, each permutation is assigned a probability that decays exponentially with respect to its distance from the central permutation [3].

The GM model was proposed as an extension of the Mallows Model [3]. This extension requires that the distance used can be decomposed into $n-1$ terms, such that the distance between permutations is expressed as $d(\sigma,\sigma_0) = \sum_{j=1}^{n-1} S_j(\sigma,\sigma_0)$, where S_j is the number of positions with smaller values to the right of j. When using the Kendall distance, the term S_j is denoted by V_j and is defined as $V_j(\sigma) = \sum_{i=j+1}^{n} I[\sigma(j) > \sigma(i)]$, and it ranges from 0 to $n-j$ for $1 \le j < n$. Then, Equation (9) is rewritten as follows:

$$P(\sigma) = \frac{exp\left(\sum_{j=1}^{n-1} -\theta_j V_j\left(\sigma,\sigma_0^{-1}\right)\right)}{\psi(\theta)}. \tag{10}$$

Considering V_j as independent variables, the probability distribution for the variables $V_j\left(\sigma,\sigma_0^{-1}\right)$ under the GM model is given by

$$P\left(V_j\left(\sigma,\sigma_0^{-1}\right) = r_j\right) = \frac{exp(-\theta_j r_j)}{\psi_j\left(\theta_j\right)}, r_j \in \{0,\ldots,n-j\}. \tag{11}$$

To introduce the GM model into an EDA concept, it is necessary to estimate its parameters σ_0 and θ. These are estimated using the principle of Maximum Likelihood Estimation (MLE). MLE starts from a sample of N permutations $\{\sigma_1,\ldots,\sigma_N\}$ where parameters σ_0 and θ maximize the likelihood function shown in Equation (12).

$$lnL(\sigma_1,\ldots,\sigma_N|\theta,\sigma_0) = lnP(\sigma_1,\ldots,\sigma_N|\theta,\sigma_0) = -N\sum_{j=1}^{N}\left[\theta_j\overline{V}_j - ln(\psi_j(\theta_j))\right], \tag{12}$$

where

$$\overline{V}_j = \frac{1}{N}\sum_{i=1}^{N} V_j\left(\sigma,\sigma_0^{-1}\right). \tag{13}$$

3.2.2. Position Parameter

Maximizing the log-likelihood with respect to σ_0 is equivalent to minimizing $\sum_{j=1}^{n-1} \theta_j \overline{V}_j\left(\sigma,\sigma_0^{-1}\right)$, and the central permutation is the minimum of the sum of the Kendall distances of a sample of permutations. Such estimation itself constitutes a problem classified as NP-hard, known as the Kemeny Rank Aggregation Problem [4]. In recent studies, the Borda algorithm has been used to compute the central permutation in the GMM, which has shown a good balance between time and quality of the solution [3].

3.2.3. Spread Parameter

The spread parameter for the GM model is estimated by using the likelihood function (12). For this purpose, the central permutation is taken, and the estimation is obtained by solving (14), which does not have a closed-form expression, but can be solved numerically using standard iterative algorithms for convex optimization. The Newton–Raphson method has been used for estimating the spread parameter in the literature [4].

$$\overline{V}_j = \frac{n-1}{exp(\theta_j) - 1} - \frac{n-j+1}{exp(\theta_j(n-j+1)) - 1}. \tag{14}$$

The pseudocode of the algorithm based on the EDA coupled with the GMM is presented in Algorithm 1, and this follows the procedure adopted in the state of the art. This approach is known in the literature as the GMEDA.

Algorithm 1: EDAs coupled with the GMM

Input : Population size N, maximum generations G, selection size S.
$g \leftarrow 0$ //generation counter.
Generate randomly or using heuristics, the initial population $Pop(0)$ of size N.
While Stop condition $\neq$ TRUE do :
  Compute the makespan C_{max} for each individual according [24].
  $S(g) \leftarrow$ Select a subset of the $N-1$ best individuals from $Pop(g)$.
  $\sigma_0 \leftarrow$ Estimate the central permutation with Borda algorithm.
  $\theta \leftarrow$ Estimate the spread parameter using Equation (14) and numerical methods.
  Implement the probabilistic GMM with the estimated parameters $P(\sigma_0, \theta)$.
  $X(g) \leftarrow$ Generate $N-1$ new candidate solutions through $P(\sigma_0, \theta)$ in Equation (10).
  $Pop(g) \leftarrow$ Update population by replacing with $X(g)$.
  $g \leftarrow g+1$.
End While
Output : Best solution in $Pop(g)$.

4. Evolutionary Strategy Based on the Generalized Mallows Model (ES-GMM)

The Differential Evolution Estimation Distribution Algorithm (DE-EDA) [4], the hybrid Estimation Distribution Ant Colony Search Algorithm (EDA-ACS) [5], and the Random Key-EDA [6] have been used for solving the Permutation Flow Shop Scheduling Problem (PFSP). In the case of PFSP, an EDA-GMM algorithm based on the ranking of distances has been applied [4]. This contribution was used as a reference to adapt and propose the version ES-GMM to solve the MNPFSSP.

In the proposed evolutionary strategy, each individual in the population is represented as a job permutation. This representation defines the sequence in which jobs are processed across machines. The population consists of many permutations, each evaluated based on its makespan, computed according to the methodology in [24].

To apply the GMM probabilistic model in our evolutionary strategy, the algorithm estimates the central and the spread parameters σ_0 and θ. σ_0 selects the individual with the smallest makespan from the selected population, and the spread parameter is proposed to be the inverse of the Kendall distances to the central parameter, according to Equation (15). These parameters are used to generate the offspring. This process is repeated until the predefined maximum generation (G) is reached. The structure of the algorithm is depicted in Figure 2.

4.1. Hyperparameters of Evolutionary Strategy

For the design of the proposed evolutionary strategy, the values of the hyperparameters were selected based on an exhaustive review of the literature and systematic experimentation. These hyperparameters are maximum number of generations, population size, selection percentage, and sampling percentage. The justification for each one is detailed below.

4.1.1. Maximum Number of Generations

The maximum number of generations was set at $G = 100$, in accordance with values reported in previous studies for similar problems [3]. According to the literature, this value ensures an adequate balance between computing time and the search space exploration capacity, allowing convergence towards high-quality solutions.

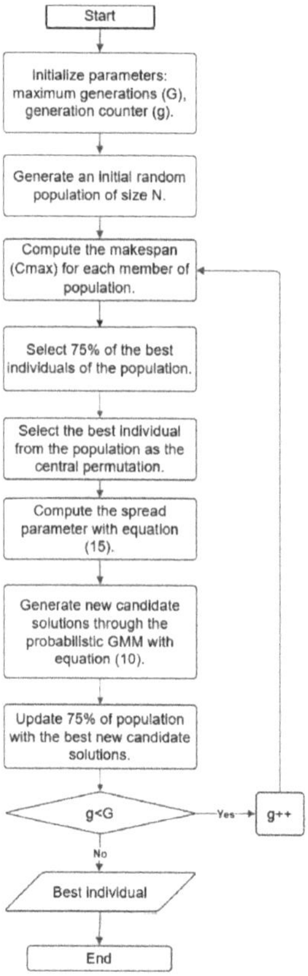

Figure 2. Flowchart of ES-GMM algorithm.

4.1.2. Population Size

The population size (N) was defined based on the number of jobs (n) in each instance of the MNPFSSP. During the experimentation, different configurations were evaluated $(N = n,\ N = 5n,\ N = 10n$, among others), and it was observed that the $N = 10n$ ratio offered better performance, maintaining population diversity throughout the optimization process, showing better adaptation to the complexity of the problem and producing higher-quality solutions.

4.1.3. Selection Rate

Three selection rate settings were tested, i.e., 50%, 75%, and 100%. Experimental results indicated that a rate of 75% was the most suitable for all instances, as it improved both solution quality and time efficiency. For small instances, this rate yielded significantly better solutions, while for large instances, it proved to be the most efficient setting in terms of runtime.

4.1.4. Sampling Rate

The sampling rate defines the proportion of individuals retained from the previous population. The sampling rates of 12%, 25%, 50%, and $\left(\frac{10n-1}{10n}\right) * 100\%$ were tested, using Relative Percentage Deviation (RPD) as the performance metric. Experimental analysis showed that a rate of 25% yielded the best results, with RPD values of -1.17% for small instances and -1.38% for large instances. This percentage not only outperformed the most common configurations in the literature but also ensured consistent and high-quality solutions.

4.2. Estimation of Proposed Spread Parameter

We propose estimating the spread parameter based on the inverse relationship between the Kendall distance V_j and the parameter value θ_j. This is proposed due to the difficulty of finding the spread value from Equation (14), as in the state of the art. Therefore, a practical proposal such as Equation (15) is proposed. When the average Kendall distance $\overline{V}_j$ is large, it indicates greater diversity among the permutations, promoting exploration. Conversely, when $\overline{V}_j$ is small, the sample of permutations converges toward the central permutation, favoring exploitation. Figure 3 illustrates the behavior of the $\overline{V}_j$ with respect to the values $\{1,2,3,4,5,6,7,8,9,10\}$ assigned for $\overline{\theta}_j$, for both Equation (14) and the proposed Equation (15). The proposed relation achieves higher values for the $\overline{V}_j$, with respect to $\overline{\theta}_j$, allowing for greater exploration and improving the diversity of solutions. Regarding Equation (15), it is not necessary to perform additional calculations.

$$\overline{V}_j = \frac{1}{\overline{\theta}_j}. \tag{15}$$

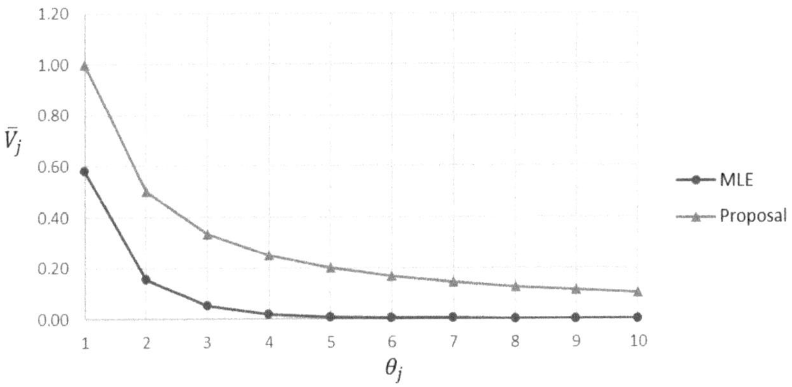

Figure 3. Behavior of the average Kendall distance as a function of the spread parameter for the MLE and the proposed approach.

4.3. Operating Conditions of ES-GMMc

To establish the minimum number of generations for the algorithm to generate satisfactory solutions, an experimental analysis was carried out on samples of small and large instances for the MNPFSSP. Figures 4 and 5 show the behavior of the algorithm's makespan over 100 generations in small and large instances, respectively. To propose a reference value for b, i.e., the smallest number of generations necessary so that the search process of the algorithm does not stop before approaching its convergence value, an experiment was carried out using a sample of the benchmark defined by Bektas, Hamzadayi, and Ruiz (2020) [4], which was built on the instance I_3_500_50_1, available at http://soa.iti.es/problem-instances, accessed on 1 January 2022. This benchmark classifies instances into two categories, i.e., small and large, according to the number of jobs. For the small instance category, the selected identifiers were $ID \in \{1,5,10,15,20,25\}$, while for the large instances, the selected identifiers were $ID \in \{1,5,10,15,20,25,30\}$. Tables 1 and 2 present the ID, the number of jobs (n), and the number of machines (m) for the small and large instances, respectively.

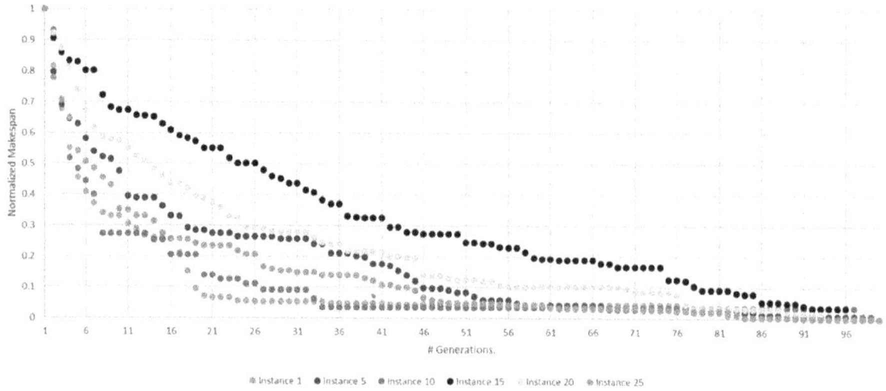

Figure 4. Evolution of normalized average makespan across generations for small instances.

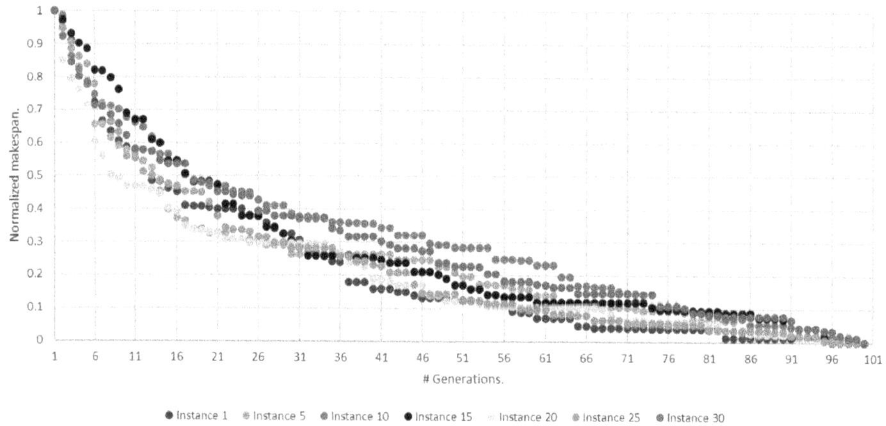

Figure 5. Evolution of normalized average makespan across generations for large instances.

Table 1. Characteristics of the small instances group [3].

ID Instance	n	m
1	10	3
5	15	6
10	25	3
15	30	9
20	40	6
25	50	3

Table 2. Characteristics of the large instances group [3].

ID Instance	n	m
1	50	5
5	100	10
10	200	5
15	250	15
20	350	10
25	450	5
30	500	15

The graphs in Figures 4 and 5 show the normalized average makespan of 30 experiments of 100 generations each for small and large instances, respectively.

From the observed behavior, the value l was obtained, which represents the minimum generation number in which the algorithm converges in each instance. This value represents the generation where no more changes are observed in the makespan per instance. Table 3 shows the l values, the average $\bar{l}$ values, and their standard deviation $\sigma_{\bar{l}}$ for small and large instances.

Table 3. Minimum number of generations l where the algorithm converges. The results were experimentally obtained for small and large instances.

Instances			
Small	l	**Large**	l
1	30	1	65
5	57	5	69
10	34	10	72
15	86	15	60
20	77	20	71
25	49	25	67
		30	79
$\bar{l}$	55.5	$\bar{l}$	69
$\sigma_{\bar{l}}$	22.6	$\sigma_{\bar{l}}$	5.9
b	33	b	63

Based on the previous information, b was defined as follows:

$$b = \left(\bar{l} - \sigma_{\bar{l}}\right). \tag{16}$$

This was established as the minimum number of generations required before applying an algorithm stop condition.

Stopping Condition

The stopping criterion considers two key aspects: (1) reaching the minimum number of generations b to avoid local minima and (2) confirming that the best makespan no longer improves after next generations.

Based on the calculated b value, a stopping condition was established to reduce the algorithm execution time. According to the flowchart in Figure 6, the stopping condition is defined as follows:

- The algorithm continues running for at least b generations, regardless of whether the makespan remains constant.
- After b generations, if the makespan of the best individual remains constant, the algorithm stops.

This stopping criterion ensures that the algorithm has enough time to explore solutions during the first b generations while allowing it to terminate early once it is clear that no further significant improvements are achievable. This approach accelerates the convergence process and reduces the execution time when the algorithm has reached a local or global optimum.

Figure 6 presents the flowchart of the ES-GMMc algorithm, highlighting the incorporation of a stopping criterion. The flowchart illustrates each stage of the algorithm, from the initial population generation to the selection of the best individuals and the updating of the parameters of the Mallows Distribution. In addition, the stopping conditions are detailed, where the algorithm verifies whether it has reached the threshold of generations b

or whether the makespan of the best individual has stabilized, in which case the execution is interrupted early.

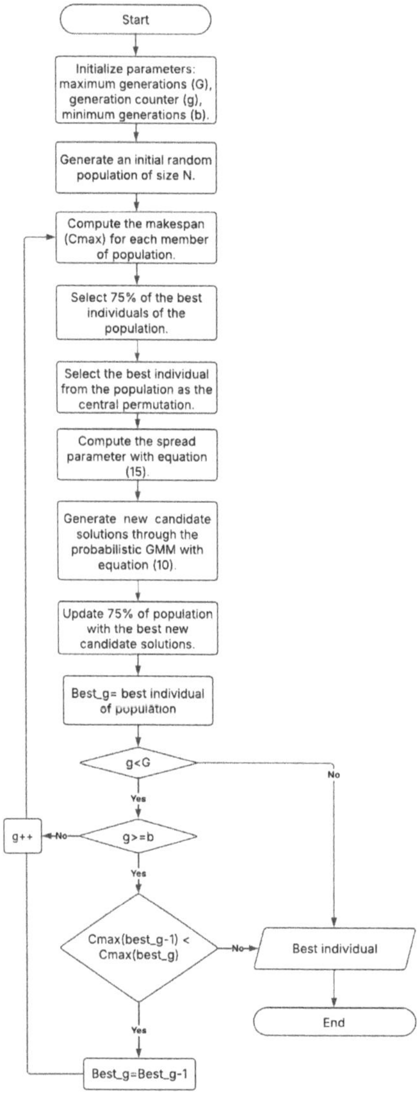

Figure 6. Flowchart of ES-GMMc algorithm.

5. Experimental Development

The proposed algorithm was executed in two versions with the same hyperparameters, without the stop condition (ES-GMM) and with the stop condition (ES-GMMc). To evaluate the results, they were compared to those of the exact methods (BC, BD, and ABD) reported in the study by Bektas, Hamzadayi, and Ruiz [4]. Since the results obtained by the BD method were not fully reported in the literature, they were re-evaluated experimentally. Additionally, the proposed algorithms were compared with the metaheuristics IGA and RLS.

Experiments were conducted on a desktop computer equipped with an Intel® Core™ i9-9900K CPU @ 3.60 GHz and 32 GB RAM. The exact methods were solved using the Cplex Studio IDE 12.9.0, while the metaheuristics were implemented in Java using the IDEA Community Edition 2018.

5.1. Instance Generation

The experimental instances were derived from a benchmark instance with 500 jobs and 50 machines, available at http://soa.iti.es. From this, the following subsets were generated: 27 small instances, with $n \in \{10, 15, 20, 25, 30, 35, 40, 45, 50\}$ and $m \in \{3, 6, 9\}$, and 30 large instances, with $n \in \{50, 100, 150, 200, 250, 300, 350, 400, 450, 500\}$ and $m \in \{5, 10, 15\}$.

5.2. Methodology

To compare makespan results with those reported in the state of the art, each algorithm was executed 30 experiments for each instance, and the minimum makespan obtained was recorded.

For the BD method, a limit time of 7200 s was imposed, consistent with the methodology in [1]. The reported time corresponds to the duration required to achieve the optimal solution or the best solution within the time limit. For the ES-GMM and ES-GMMc algorithms, the average execution time over the 30 experiments was reported.

This setup ensures a fair comparison of performance, focusing on both solution quality (makespan) and computational efficiency (execution time).

6. Results

This section presents the results obtained in terms of makespan and execution time for both small and large instances. The analysis focuses on comparing the performance of the different algorithms, evaluating their efficiency in terms of solution quality (makespan) and computational cost (execution time), incorporating a detailed statistic by evaluation and determining whether the observed differences are significant. The primary objective is to identify algorithms that achieve an appropriate balance between solution quality and computational cost, supported by statistical analysis.

6.1. Small Instances

6.1.1. Makespan Results

Figure 7 presents a comparative analysis of the makespan obtained by all the algorithms for small instances. Boxplots were used to visualize the distribution of results and evaluate the performance of each approach. It was observed that the ES-GMMc algorithm demonstrates competitive performance in terms of makespan, maintaining values close to the median of the most effective methods compared to the other approaches.

6.1.2. Statistical Analysis of Makespan

To formally assess whether there are statistically significant differences between algorithms regarding the makespan, normality and homogeneity tests were first applied with a significance level of $\alpha = 0.05$. The Shapiro–Wilk test confirmed that the makespan data follow a normal distribution ($p = 0.80$), and the Levene test confirmed the homogeneity of variances ($p = 0.75$). Therefore, an ANOVA test was applied with the null hypothesis that all algorithms exhibit equivalent performance.

The ANOVA test returned a p-value of 0.828, indicating that there are no statistically significant differences between the makespan performance in small instances. Although ES-GMMc shows competitive values in absolute terms, this difference is not statistically significant.

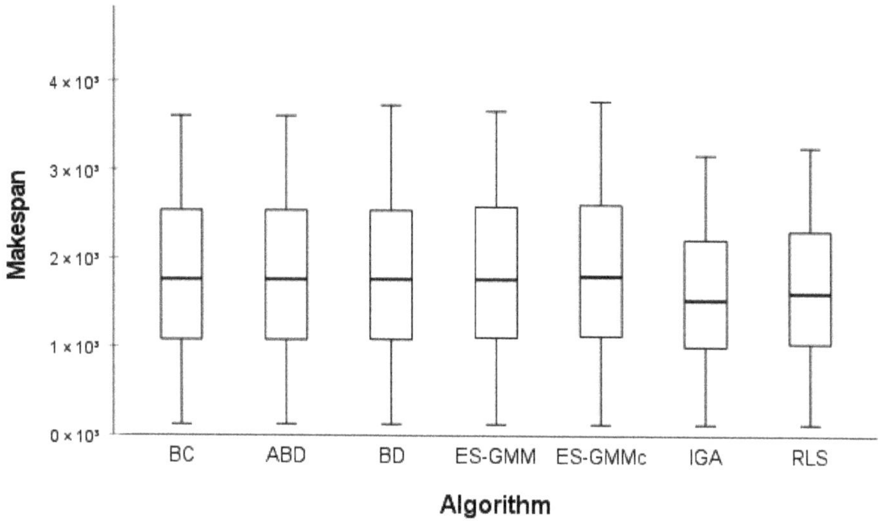

Figure 7. Makespan obtained by algorithms for small instances.

6.1.3. Execution Time Results

The execution times of the different algorithms for small instances are shown in Figure 8. It was observed that the BD algorithm exhibits a high variability, with significantly higher outlier values compared to the rest of the methods. Due to this variability, nine outlier extreme values were excluded from the BD algorithm to provide a clearer view of the performance of the remaining algorithms.

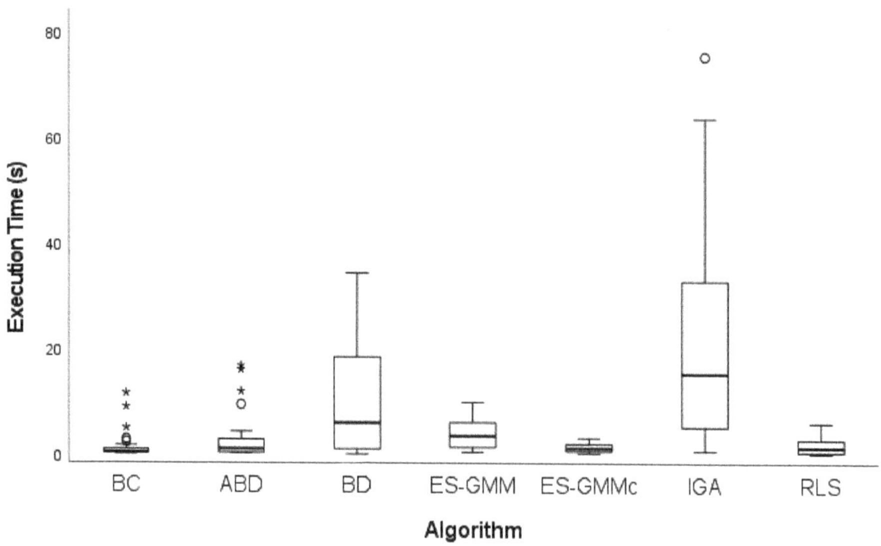

Figure 8. Execution times obtained by algorithms for small instances. ° represents moderate outliers, and ∗ represents extreme outliers.

In Figure 8, the IGA presents the highest dispersion and presence of outliers, indicating that its execution time is less predictable and, in some cases, considerably higher than that of the rest of the algorithms.

In contrast, the ES-GMMc algorithm maintains low and relatively stable execution times, without any outliers, reaffirming its computational efficiency. BC, ABD, and ES-GMM present similar execution times, although BC and ABD show a larger number of outliers, indicating that their performance may vary depending on the instance being solved. The RLS algorithm stands out for its low and consistent execution times, suggesting that it is an efficient option in terms of computational cost.

The combined analysis of makespan and execution time suggests that the ES-GMMc achieves a favorable balance between solution quality and computational efficiency, positioning it as a strong candidate for solving small instances. Although the IGA achieves competitive makespan values, its higher computational cost makes it less attractive for time-constrained environments. Finally, RLS emerges as an efficient option in terms of computational cost, although there is no significant difference in the quality of the solution, compared to the rest of the algorithms.

6.1.4. Statistical Analysis of Execution Times

Normality and homogeneity tests (Shapiro–Wilk and Levene) were applied with a significance level of $\alpha = 0.05$. Both tests indicated that execution times do not follow a normal distribution (see Table 4); thus, a non-parametric Friedman test was applied to evaluate differences between algorithms. The Friedman test calculated a p-value of 0.001, indicating significant differences in execution times.

Table 4. Statistical test results of the execution times for small instances.

Small Instances			
Algorithm	**Shapiro–Wilk**	**Levene**	**Friedman**
	***p*-Value**	***p*-Value**	**Average Ranges**
BC	0.001		1.52
ABD	0.001		3.20
BD	0.001		5.63
ES-GMM	0.048	0.001	5.00
ES-GMMc	0.057		3.07
IGA	0.003		6.30
RLS	0.003		3.26

The average ranks of Friedman (see Table 4) show that the BC method and the ES-GMMc achieve the shortest execution times, with the IGA being the least efficient. To further validate these findings, a Wilcoxon post hoc test was performed comparing the four best-ranked algorithms (see Table 5).

Table 5. Results of Wilcoxon post hoc test of the execution times for small instances.

Small Instances	
Comparison	**Wilcoxon**
	***p*-Value**
BC vs. ES-GMMc	0.013
BC vs. ABD	0.001
BC vs. RLS	0.002
ES-GMMc vs. ABD	0.737
ES-GMMc vs. RLS	0.049
ABD vs. RLS	0.885

The post hoc analysis (see Table 5) confirms that the BC method is better than the ES-GMMc, ABD, and RLS, which achieve statistically similar execution times.

6.2. Large Instances

6.2.1. Makespan Results

Figure 9 presents the makespan performance of the algorithms for large instances. In general, all algorithms present a similar range of values, with medians close to each other. However, slight differences were observed. In particular, the IGA presents a slightly lower median makespan, indicating that it marginally outperforms the other approaches in terms of solution quality.

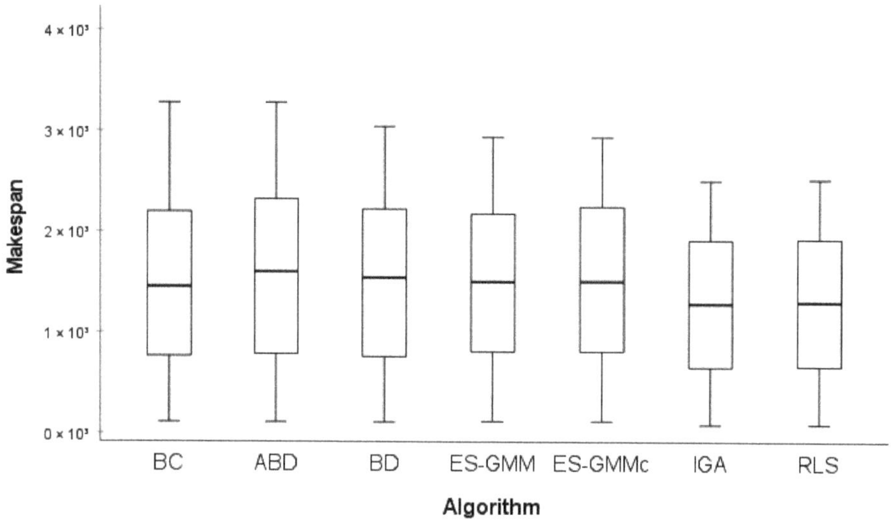

Figure 9. Makespan obtained by different algorithms for large instances.

Contrarily, algorithms such as BC, ABD, and BD show a wider range, indicating higher variability in their performance across different instances. The ES-GMMc algorithm maintains stable performance, with a competitive median and lower dispersion compared with BC and ABD. Similarly, RLS shows consistent behavior, with median and variability levels comparable to IGA, highlighting its ability to maintain stable solution quality across larger instances.

6.2.2. Statistical Analysis of Makespan

The Shapiro–Wilk and Levene tests were applied with a significance level of $\alpha = 0.05$. The Shapiro–Wilk (p-value: 0.209) and Levene tests (p-value: 0.507) confirmed the normality and homogeneity of variances, justifying the use of an ANOVA test. The ANOVA test returned a p-value of 0.762, indicating that there are no statistically significant differences between algorithms regarding makespan in large instances.

This result implies that, although the IGA achieves the best makespan values in absolute terms, these differences are not statistically significant and therefore cannot be used as conclusive evidence of its superiority in terms of solution quality.

6.2.3. Execution Time Results

The execution times for large instances are presented in Figure 10. As observed in small instances, the IGA presents an extremely high outlier, which implies that, in at least one instance, its execution time was significantly longer than the others. To provide a clearer view of the remaining algorithms, two extreme outliers were excluded from the IGA.

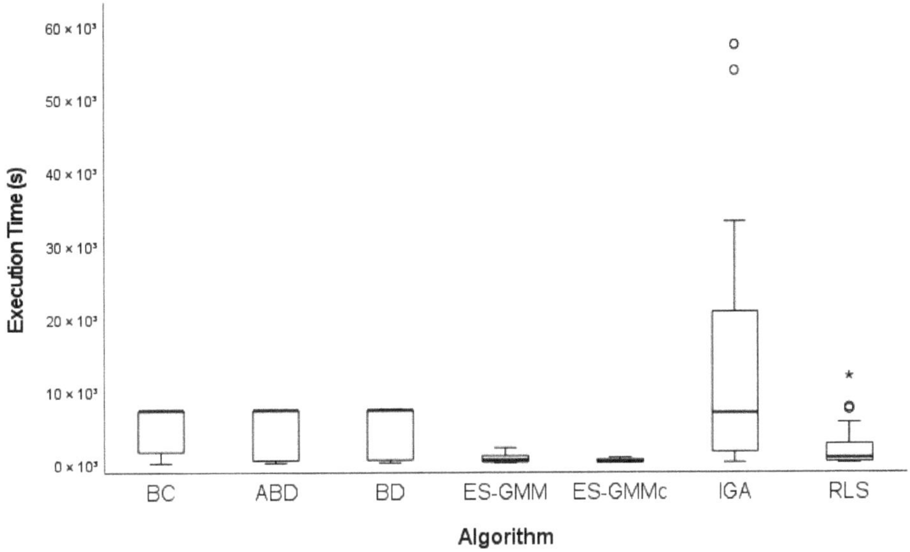

Figure 10. Execution times obtained by different algorithms for large instances. ° represents moderate outliers, and ∗ represents extreme outliers.

In this revised comparison, the ES-GMMc and ES-GMM stand out as the algorithms with the shortest execution times. Notably, the ES-GMMc reduced the execution time in 23 out of 30 instances. Both approaches not only achieved low execution times but also demonstrated very controlled variability, with compact boxes and short whiskers, indicating consistent performance for different instances.

In contrast, algorithms such as ABD, BC, and BD presented considerably longer execution times. Their boxes were more elongated, which suggested that these methods were more sensitive to the specific characteristics of each instance, causing highly variable resolution times.

The RLS algorithm showed relatively low execution times compared to the exact approaches (ABD, BC, and BD), but higher than those observed in the ES-GMM and ES-GMMc.

The combined analysis of makespan and execution times for large instances reinforces the conclusion that the ES-GMMc achieves a favorable balance between solution quality and computational efficiency, positioning it as a robust and competitive approach for solving larger problem instances. Although the IGA achieves competitive makespan values, its high computational cost limits its suitability for time-sensitive scenarios. Finally, RLS offers a balanced alternative, combining moderate computational cost with relatively stable solution quality, although it does not reach the same level of performance as the ES-GMMc.

6.2.4. Statistical Analysis of Execution Times

The Shapiro–Wilk and Levene tests confirmed that the execution times do not follow a normal distribution ($p = 0.001$), necessitating the use of the non-parametric Friedman test. The Friedman test calculated a p-value of 0.001, indicating significant differences between algorithms. Table 6 presents the average ranges showing that the ES-GMMc achieves the best (shortest) execution times, followed closely by the ES-GMM and RLS, while the IGA exhibits the longest execution times.

Table 6. Friedman test of execution times for large instances.

Large Instances	
Algorithm	Friedman
	Average Ranges
BC	4.62
ABD	4.17
BD	5.18
ES-GMM	2.70
ES-GMMc	1.43
IGA	6.17
RLS	3.73

A Wilcoxon post hoc test was applied to confirm the differences between the top four algorithms. The test confirmed that the ES-GMMc execution time is significantly lower than its competitors ($p < 0.001$), supporting the conclusion that the ES-GMMc is the most computationally efficient algorithm for large instances.

7. Discussion

This section discusses the performance of the proposals, mostly the key factors that contribute to the performance of the algorithm, with special emphasis on the stopping criterion, which allows an improvement in computational efficiency without compromising the quality of the solution. The strengths and weaknesses of the proposed method are identified and compared with those of the existing programming techniques. Finally, the practical implications of these findings for real-world production environments are examined, highlighting the potential advantages and applications of the proposed approach.

7.1. Stopping Criterion and Computational Efficiency

One of the key contributions of the ES-GMMc, the advanced version of the ES-GMM, is the incorporation of a stopping criterion. This criterion, experimentally defined and based on an evolution analysis of the normalized makespan through the generations, allows the algorithm to terminate once the convergence patterns are stable. Ending the search when there are no changes in the best individual's makespan avoids unnecessary evaluations in subsequent generations, substantially reducing the execution time without compromising the quality of the solution.

The statistical analysis conducted through the small and large instances consistently shows that the ES-GMMc achieves significantly lower execution times compared to the IGA, ABD, and BD, while maintaining competitive makespan values. Reducing the computational cost without affecting the quality of the solution is particularly relevant in real-world production scheduling environments, where minimizing both the processing time and the computational overhead is critical.

The ES-GMMc improves computational cost by assigning the best individual to the central permutation and the inverse of the mean distance with respect to the central permutation to the spread parameter, making parameter estimation easier and faster. Moreover, it refines the search for solutions, improving both exploration and exploitation, particularly in large instances, where the search space increases.

7.2. Comparison with Existing Approaches

The comparative analysis between algorithms highlighted that, although the IGA occasionally produced slightly better makespan values, these differences were not statistically significant in either small or large instances. This finding underscores that the ES-GMMc achieves equivalent or near-equivalent solution quality at a fraction of the computational cost, positioning it as a more efficient and scalable alternative for solving the MNPFSSP.

Furthermore, traditional approaches such as ABD and BC, while achieving reasonable makespan values, exhibited significantly higher variability in both execution time and performance. This variability limits their robustness, particularly in large instances with diverse job and machine configurations. In contrast, the ES-GMMc consistently maintained low dispersion in both makespan and execution time, showing a stable and reliable performance through the different problem sizes and complexities.

7.3. Implications for Real-World Programming

In production environments with limited computational resources and time constraints, the ES-GMMc offers an efficient solution. Its stopping criteria enables high-quality schedules to be delivered within reasonable computational budgets, making it particularly suitable for just-in-time systems, dynamic workshops, and frequent rescheduling scenarios.

8. Conclusions and Future Work

In this work, we developed an Evolutionary Strategy based on the Generalized Mallows Model (ES-GMM) and its advanced version, the ES-GMMc, which were designed to tackle the Mixed No-Idle Permutation Flow Shop Scheduling Problem (MNPFSSP). A key innovation of this approach is the definition of the central permutation as the individual with the best makespan within the population in each generation, effectively guiding the search toward the most promising solutions.

Additionally, the spread parameter is dynamically estimated based on the average Kendall distance of the solutions concerning the central permutation. This approach adjusts the balance between exploration and exploitation of the search space, promoting greater diversity in the early stages and efficient convergence toward best solutions in later stages. This dynamic adaptation prevents premature stagnation and significantly enhances algorithm efficiency.

The incorporation of a stopping criterion further strengthens the practical applicability of ES-GMMc, enabling the algorithm to detect convergence patterns and terminate the search when further improvement is unlikely. This feature significantly reduces computational costs, particularly in large instances, making the approach suitable for real-time and resource-constrained environments.

The experimental evaluation confirms that the ES-GMMc achieves an effective trade-off between solution quality and execution time, consistently outperforming the exact methods such as Benders Decomposition and branch-and-cut in larger instances. While the IGA achieved slightly better makespan values in some cases, the ES-GMMc demonstrated superior computational efficiency, making it a more practical alternative for large and complex scheduling problems. Particularly, we think that the ES-GMMc is suitable for industrial production scheduling, including flexible manufacturing systems, job shop scheduling with sequence-dependent setups, and just-in-time production environments where fast and high-quality scheduling decisions are essential.

In future research, a deeper study of the hyperparameter space should be carried out that can increase the efficiency of the algorithm performance. Furthermore, the ES-GMMc should be extended to multi-objective variants for the Flow Shop Scheduling

Problem, incorporating energy consumption and minimizing resource utilization as the additional criteria.

Author Contributions: Conceptualization, E.M.S.M.; methodology, R.P.-R. and M.O.-R.; formal analysis, H.J.P.-S.; investigation, E.M.S.M.; writing—original draft preparation, all authors; writing—review and editing, all authors; visualization, E.M.S.M. All authors have read and agreed to the published version of the manuscript.

Funding: This research received no external funding.

Data Availability Statement: The data presented in this study are available on request from the corresponding author, Elvi. M. Sánchez Márquez (malintzin041093@gmail.com).

Acknowledgments: The authors wish to thank the Secretariat of Science, Humanities, Technology and Innovation (SECIHTI) of Mexico, for the post-graduate study scholarship, 634738 (E. Sánchez), and the National Technology of Mexico/Leon Institute of Technology, for the support provided in this investigation.

Conflicts of Interest: The authors declare no conflicts of interest.

References

1. Li, Y.Z.; Pan, Q.K.; Li, J.Q.; Gao, L.; Tasgetiren, M.F. An Adaptative Iterated Greedy algorithm for distributed mixed no-idle permutation flowshop scheduling problems. *Swarm Evol. Comput.* **2021**, *63*, 100874.
2. Zhong, L.; Li, W.; Qian, B.; He, L. Improved discrete cuckoo-search algorithm for mixed no-idle permutation flow shop scheduling with consideration of energy consumption. *Collab. Intell. Manuf.* **2021**, *4*, 345–355.
3. Ceberio, J.; Irurozki, E.; Mendiburu, A.; Lozano, J.A. Extending Distance-based Ranking Models in Estimation of Distribution Algorithms. In Proceedings of the 2014 IEEE Congress on Evolutionary Computation (CEC), Beijing, China, 6–11 July 2014; pp. 6–11.
4. Bektas, T.; Dayi, A.H.; Ruiz, R. Benders decomposition for the mixed no idle permutation flowshop scheduling problem. *J. Sched.* **2020**, *23*, 513–523. [CrossRef]
5. Taskin, Z.C. *Benders Decomposition*; Wiley Encyclopedia of Operations Research and Management Science: Hoboken, NJ, USA, 2011.
6. Rojas, J.A. Descomposición Para Problemas de Programación Lineal Multi-Divisionales. Master's Thesis, CIMAT, Guanajuato, Mexico, 2012.
7. Cerisola, A.R.a.S. *Optimización Estocástica*; Universidad Pontificia, Technical Report: Madrid, Spain, 2016.
8. Balcan, M.F.; Prasad, S.; Vitercik, T.S.a.E. Structural Analysis of Branch-and-Cut and the learnability of Gomory Mixed Integer Cuts. de Conference on Neural Information Processing Systems. *Adv. Neural Inf. Process. Syst.* **2022**, *35*, 33890–33903.
9. Mühlenbein, H.; Bendisch, J.; Voigt, H.-M. From Recombination of Genes to the Estimation of Distributions II. Continuos Parameters. In *Parallel Problem Solving from Nature—PPSN IV*; Springer: Berlin/Heidelberg, Germany, 1996.
10. Ceberio, J.; Irurozki, E.; Mendiburu, A.; Lozano, J.A. A Distance-Based Ranking Model Estimation of Distribution Algorithm for the Flowshop Scheduling Problem. *IEEE Trans. Evol. Comput.* **2014**, *18*, 286–300. [CrossRef]
11. Ayodele, M.; McCall, J.; Regnier-Coudert, O.; Bowie, L. A Random Key based Estimation of Distribution Algorithm for the Permutation Flowshop Scheduling Problem. In Proceedings of the 2017 IEEE Congress on Evolutionary Computation (CEC), Donostia, Spain, 5–8 June 2017; pp. 2364–2371.
12. Sun, Z.; Gu, X. Hybrid algorithm based on an estimation of distribution algorithm and cuckoo search for the no idle permutation flow shop scheduling problem with total tardiness criterion minimizations. *Sustainability* **2017**, *9*, 953. [CrossRef]
13. Li, Z.C.; Guo, Q.; Tang, L. An effective DE-EDA for Permutation Flow-Shop Scheduling Problem. In Proceedings of the 2016 IEEE Congress on Evolutionary Computation (CEC), Vancouver, BC, Canada, 24–29 July 2016; pp. 2927–2934.
14. Ceberio, J.; Mendiburu, A.; Lozano, J.A. A preliminary study on EDAs for permutation problems based on marginal-based models. In Proceedings of the 13th Annual Conference on Genetic and Evolutionary Computation, Dublin, Ireland, 12–16 July 2011; pp. 609–616.
15. Ceberio, J.; Mendiburu, A.; Lozano, J.A. Introducing the Mallows Model on Estimation of Distribution Algorithms. In Proceedings of the ICONIP: International Conference on Neural Information Processing, Sanur, Bali, Indonesia, 8–12 December 2021; pp. 461–470.
16. Alnur, A.; Marina, M. Experiments with Kemeny ranking: What works when? *Math. Soc. Sci.* **2012**, *64*, 28–40.

17. Tsutsui, S.; Pelikan, M.; Goldberg, D.E. *Node Histogram vs. Edge Histogram: A Comparison of PMBGAs in Permutation Domains*; MEDAL Report No. 2006009; University of Missouri—St. Louis: St. Louis, MO, USA, 2006.
18. Irurozki, E. Sampling and Learning Distance-Based Probability Models for Permutation Spaces. Ph.D. Thesis, Universidad del País Vasco-Euskal Herriko Unibertsitatea, Leioa, Spain, 2014.
19. Pérez-Rodríguez, R.; Hernández-Aguirre, A. Un algoritmo de estimación de distribuciones copulado con la distribución generalizada de mallows para el problema de ruteo de autobuses escolares con selección de paradas. *Rev. Iberoam. Automática Informática Ind.* **2017**, *14*, 288–298.
20. Aledo, J.A.; Gámex, J.A.; Molina, D. Computing the consensus permutation in mallows distribution by using genetic algorithms. In Proceedings of the Recent Trends in Applied Artificial Intelligence: 26th International Conference on Industrial, Engineering and Other Applications of Applied Intelligent Systems, IEA/AIE 2013, Amsterdam, The Netherlands, 17–21 June 2013.
21. Pérez-Rodríguez, R.; Hernández-Aguirre, A. A hybrid estimation of distribution algorithm for flexible job-shop scheduling problems with process plan flexibility. *Appl. Intell.* **2018**, *48*, 3707–3734.
22. Irurozki, E.; Calvo, E.; Lozano, J.A.B. Permallows: An r package for mallows and generalized mallows models. *J. Stat. Softw.* **2016**, *71*, 1–30. [CrossRef]
23. Pérez-Rodríguez, R.; Hernández-Aguirre, A. A hybrid estimation of distribution algorithm for the vehicle routing problem with time windows. *Comput. Ind. Eng.* **2019**, *130*, 75–96. [CrossRef]
24. Quan-Ke, P.; Rubén, R. An effective iterated greedy algorithm for the mixed no-idle permutation flowshop scheduling problem. *Omega* **2014**, *44*, 41–50.

Mathematical and Computational Applications

MDPI

Article

Thau Observer for Insulin Estimation Considering the Effect of Beta-Cell Dynamics for a Diabetes Mellitus Model

Diana Gamboa †, Tonalli C. Galicia † and Paul J. Campos *

Posgrado en Ciencias de la Ingeniería, Tecnológico Nacional de México/I.T. Tijuana, Blvd. Alberto Limón Padilla s/n, Mesa de Otay, Tijuana 22454, BC, Mexico; diana.gamboa@tectijuana.edu.mx (D.G.); tonalli.galicia@tectijuana.edu.mx (T.C.G.)
* Correspondence: paul.campos@tectijuana.edu.mx
† These authors contributed equally to this work.

Abstract: In this work, a Thau observer is designed based on a nonlinear third-order mathematical model described by ODEs, which captures the dynamics among insulin levels, β-cells, and glucose concentration. The novelty of this research lies in its interdisciplinary approach to understanding a complex biological system. The observer's mathematical validation is established using the Localization of Compact Invariant Sets to determine the domain of attraction and global knowledge about the system's dynamic bounds. These bounds are used to compute the Lipschitz constant and the elements of the free gain matrix that satisfy the constraints for designing a Thau observer, such as the stability matrix and asymptotic stability. This analysis provides insights into how insulin levels evolve over time at various glucose concentrations, an essential step toward hardware implementation due to the system's chaotic behavior. It also establishes a mathematical background that could contribute to treatment planning in future Digital Twins studies. Numerical simulations demonstrate that the observer can accurately track the dynamic behavior of the Diabetes Mellitus model analyzed in this work through *in silico* methods.

Keywords: diabetes mellitus model; nonlinear analysis; Thau observer; insulin observation; compact invariant sets

MSC: 93C10; 93C15; 93D05

1. Introduction

Diabetes Mellitus (DM) is one of the most common metabolic disorders, and it is estimated that by the year 2045, the situation may be more alarming than previously envisaged [1]. Recently, it has been reported that diabetes prediction can be achieved by using supervised machine learning [2], leading to prediabetes surveillance [3], which, according to the authors, represents a growing global burden.

The main reason why diabetes is dangerous over a prolonged period is the lack of insulin produced by the pancreas in the presence of a high glucose concentration. This condition can be inherited or caused by any factor related to health conditions of the pancreas [4]. Insulin is a hormone secreted by β-cells located in response to glucose concentrations. The primary function of insulin is to distribute glucose to all the cells in the body via the bloodstream, where it is used as an energy source. Projections from the years 2010 to 2030 estimate that 154 million people with diabetes will be located in India and China alone [5], considering only population growth rates. Failing to take preventive measures for diabetes can lead to severe and irreversible health damage. This is why technological advances have introduced new techniques for administering insulin treatment, and one approach to studying DM is mathematical biology [6]. Developing and personalizing a mathematical model that accurately represents an entire or partial population is challenging when considering only factors such as insulin, glucose, β-cells, and insulin treatments, including rapid-acting, short-acting, intermediate-acting, and long-acting insulins. Nevertheless, these models further explain how each factor interacts with the immune system [7]. This paper focuses on the mathematical analysis of nonlinear dynamics described by ODEs [8], in which the design of a nonlinear observer is achieved by computing a Lipschitz constant based on the compact invariant set method for localization in open-loop scheme analysis. The purpose of our research is to continuously observe and monitor a patient's health state while minimizing additional costs and avoiding unnecessary invasiveness [9].

One approach to studying Type 1 Diabetes Mellitus (DM1) involves numerical methods and statistical techniques, which are commonly used to solve these models [10]. However, this approach is not entirely feasible for understanding biological processes due to their chaotic behavior [11]. In 2011, Little et al. [12] established that every diabetes model exhibits nonlinear properties and chaotic dynamics, even when designing a closed-loop glucose controller using a linear model derived from clinical data. Although the literature extensively covers the insulin–glucose relationship, previous research has overlooked the implementation of a nonlinear ODE-based biological model with chaotic behavior. In 2013 [13], the first comparison between the Van der Pol oscillator and the dynamics of DM1 was made, proving that the biological model is nonlinear and highly sensitive to initial conditions. This background allows us to discern a strategy for studying and analyzing DM1 by implementing nonlinear control techniques to address its chaotic behavior. Furthermore, a recent study [14] reported that the application of closed-loop blood glucose regulation in patients with Type 1 diabetes could be achieved through an artificial pancreas, thus improving the effectiveness of postprandial blood glucose control by extending the Bergman physiological model with a digestion kinetics equation. The ODE model of interest was introduced in [15] and has been mathematically studied using an adaptive controller [16] and a backstepping sliding mode controller [17]. The concept of Digital Twins [18] has a significant impact on innovation in the context of DM1. However, designing a nonlinear observer using the Lipschitz constant involves a mathematical estimation technique that is used to infer unmeasured states of a system based on a nonlinear model and limited observable data. Compared to Digital Twins, an observer is primarily used in real-time systems, such as artificial pancreas technologies, to enhance the precision of insulin delivery by estimating unmeasured physiological states. Digital Twins are broader, predictive, and simulation-focused, while nonlinear Thau observers are narrower, real-time estimation tools. Applying a Thau observer to a nonlinear model of DM1 provides a solid mathematical foundation to support treatment planning and serves as the groundwork for future Digital Twins research. As a result, our work makes a meaningful contribution to the field of biomathematics.

This paper presents a mathematical analysis of the complex interactions among insulin, glucose, and β-cells in the dynamic evolution of DM1. Using the LCIS, the minimum and maximum concentrations of each state variable in the system are determined, along with their evolution over time, demonstrating the boundedness of the system even in the presence of chaos. Additionally, conditions are established to support the design of the Thau observer. Using the maximum concentrations, the Lipschitz constant is computed to satisfy Thau's inequality, ensuring that the observer is asymptotically stable. Finally, a Thau observer is designed to accurately estimate insulin concentrations, even in the presence of unforeseen disturbances. This observer is particularly helpful for biological systems where measurements are often affected by noise, sensor limitations, and external perturbations. The proposed methodology ensures that the estimated insulin values remain stable and reliable under different physiological conditions. To validate the performance and robustness of the proposed observer, a series of in silico experiments are conducted, simulating various dynamic scenarios, including both normal and pathological conditions. These experiments demonstrate the observer's ability to track insulin levels with high precision and resilience against external perturbations. The results confirm that the Thau observer can be an effective tool for monitoring and analyzing insulin–glucose dynamics, contributing to the development of advanced control strategies for diabetes management.

The remainder of this paper is organized as follows. Section 2 presents the mathematical details of the model and how it describes the dynamical behavior of Diabetes Mellitus. Section 3 introduces the mathematical preliminaries of the Localization of the Compact Invariant Set method and its application to the model under study. Section 4 discusses Thau's observer preliminaries, its design, and the computation of the Lipschitz constant. Simulations and experimental results are presented in Section 5. Finally, Sections 6 and 7 present the discussion and conclusions of this work.

2. ODE Model

ODE models of diabetes typically aim to simulate the interactions among insulin secretion, glucose metabolism, and other physiological variables, such as insulin sensitivity and insulin resistance [19]. These models can range from simple representations of glucose–insulin dynamics to more complex systems incorporating multiple feedback loops, such as the effect of β-cells function, meal intake, and physical activity [20]. By incorporating both physiological principles and empirical data, ODE models offer valuable insights into the mechanisms of diabetes, help predict patient-specific outcomes, and support the development of therapeutic strategies.

The model under study describes the interaction of three population densities: insulin, glucose, and β-cells [15]. The significance of this model lies in the inclusion of β-cells as a regulatory agent, whereas the relationship between insulin and glucose exhibits characteristics similar to the predator–prey dynamics in the Lotka–Volterra model, where insulin acts as the predator and glucose as the prey. However, the presence of β-cells does not alter the chaotic behavior associated with the Lotka–Volterra model [21], even when β-cells are included. Therefore, the model is formulated using a set of nonlinear ordinary differential equations, as follows:

$$\begin{aligned}
\frac{dx_1}{dt} &= -a_1x_1 + a_2x_1x_2 + a_3x_2^2 + a_4x_2^3 + a_5x_3 + a_6x_3^2 + a_7x_3^3 + a_{20}, \\
\frac{dx_2}{dt} &= -a_8x_1x_2 - a_9x_1^2 - a_{10}x_1^3 + a_{11}x_2(1-x_2) - a_{12}x_3 - a_{13}x_3^2 - a_{14}x_3^3 + a_{21}, \\
\frac{dx_3}{dt} &= a_{15}x_2 + a_{16}x_2^2 + a_{17}x_2^3 - a_{18}x_3 - a_{19}x_2x_3,
\end{aligned} \tag{1}$$

where $x_1(t)$ is the population density of the predator (insulin), $x_2(t)$ is the population density of the prey (glucose), and $x_3(t)$ is the population density of the β-cells. The parameters a_{20} and a_{21} only have numerical values. If a_{20} and a_{21} are zero, the attractor remains. The numerical values of the system's parameters are presented below in Table 1.

Table 1. Dimensionless parameters proposed in [15].

Parameter	Description	Value
a_1	Indicates the normal decrease in concentration of insulin without glucose	2.04
a_2	Indicates the rate of propagation of insulin with existence of glucose	0.1
a_3, a_4	Indicate the increasing insulin rate once the concentration of glucose is raised	$1.09, -1.084$
a_5, a_6, a_7	Indicate the increasing insulin level rate independently excreted from different components by β-cells	$0.03, -0.06, 2.01$
a_8	Indicates the insulin effect on glucose	0.22
a_9, a_{10}	Indicate the rate of decrease in glucose in response to excretion of insulin	$-3.84, -1.2$
a_{11}	Indicates the normal increase of glucose without insulin	0.3
a_{12}, a_{13}, a_{14}	Indicate the rate of decreasing the concentration of glucose due to insulin excreted by β-cells	$1.37, -0.3, 0.22$
a_{15}, a_{16}, a_{17}	Represent the rate of increase in β-cells caused by the increase in glucose concentration	$0.3, -1.35, 0.5$
a_{18}, a_{19}	Indicate the rate of decreasing β-cells because of its existing level	$-0.42, -0.15$
a_{20}	Density rate of insulin constant	$-0.19 \leq a_{20} \leq 0.03$
a_{21}	Density rate of glucose constant	$-0.56 \leq a_{21} \leq 0.03$

3. Materials and Methods

3.1. Mathematical Preliminaries

This section addresses general theorems used to determine the bounds of a system. Starkov and Krishchenko outlined a general localization method for nonlinear systems in their works [22,23]. The proposed methodology focuses on identifying and analyzing compact invariant sets in nonlinear dynamical systems. These sets facilitate a localized examination of the system's behavior, reducing complexity while preserving essential dynamics. Recent research has shown that the compact invariant set localization method (LCIS) is valid and provides a deeper understanding of the nonlinear ODE model [24].

To understand the LCIS method, the methodology can be synthesized into the following key steps:

1. Define the nonlinear system. Represent the system with a set of ODEs in the following form:

$$x = f(x), \quad x \in \mathbb{R}^n, \tag{2}$$

where $f(x)$ is a C^∞-differental vector field.

2. Let $h(x)$ be a C^∞ differentiable function such that h is not the first integral of the system (2); therefore, the function $h(x)$ is used as a solution to the problem of localization of all compact invariant sets and is called the localizing function. Let $h\ |_B$ be the restriction of h to a set $B \subset \mathbb{R}^n$.
3. If the localization is set and all compact invariant sets are considered inside the domain $U \subset \mathbb{R}^n$, then the localization set $K(h) \cap U$ is valid, with $K(h)$ defined in Theorem 1. Let Q be a subset of $\mathbb{R}^n$. Then, the following theorem applies:

Theorem 1. *Each compact invariant set* Γ *of (2) is contained in the localization set* $K(h) = \{h_{\inf} \leq h(x) \leq h_{\sup}\}$ *[22].*

4. A refinement of the localization set $K(U)$ is achieved with the help of the iterative theorem, which is stated as follows:

Proposition 1. *If* $Q \cap S(h) = 0$*, then the system (2) has no compact invariant sets located in* Q *[22].*

5. The mathematical expression corresponds to the theorem defined in [25], known as the *iterative theorem*, which is described as follows:

Theorem 2. *Let* $h_m(x), m = 1, 2, \ldots$ *be a sequence of function of* $C^\infty(\mathbb{R}^n)$ *[23], then*

$$K_0 = K(h_0), \quad K_m = K_{m-1} \cap K_{m-1,m}, with \quad m > 0,$$

and

$$\begin{aligned} K_{m-1,m} &= \{x : h_{m,\inf} \leq h_m(x) \leq h_{m,\sup}\}, \\ h_{m,\sup} &= \sup_{S_{h_m} \cap K_{m-1}} h_m(x), \\ h_{m,\inf} &= \inf_{S_{h_m} \cap K_{m-1}} h_m(x), \end{aligned}$$

then, it contains any compact invariant set of the system (2) and $K_1 \supseteq K_2 \supseteq \cdots \supseteq K_m \supseteq \ldots$.

The advantage of this methodology lies in the fact that it can generate upper or lower bounds by combining several localizing functions. There is no limit to the number of localizing functions that can be combined to obtain a higher upper bound or a smaller lower bound. These criteria depend on the specific research; the key aspect of the method is to obtain one upper or lower bound for each state variable that is part of the nonlinear system.

6. This section reviews valuable results from these works. First, it is assumed that all state variables are positive and located in the positive orthant, $\mathbb{R}^3_+ = \{\, x_1 > 0; x_2 > 0; x_3 > 0; \}$, giving biological meaning. Also, consider $\mathbb{R}^3_{+,0}$ as the closed set $\mathbb{R}^3_+$:

$$\mathbb{R}^3_{+,0} = \{x_1 \geq 0; x_2 \geq 0; x_3 \geq 0\}.$$

The solution to the problem of localizing the compact invariant sets is limited to the positive invariant compact sets; therefore, the goal is to find the maximum densities of each variable to define the region.

3.2. Analyzing the Nonlinear ODE Model of DM1

Since a_{20} and a_{21} are free constant parameters that represent the instantaneous dose of insulin and glucose concentration, respectively, each depends on a specific moment of the day. If we assume the condition $a_{20} = a_{21} = 0$, this represents a steady state where no food is consumed, and, therefore, insulin inactivation is maintained. On the other hand, if $a_{20} \geq 0$, $a_{21} \geq 0$, this implies that insulin is activated due to glucose consumption at a given initial condition. In both cases, the system's set of equations in (1) has an attractor and retains its nonlinear properties. Therefore, nonlinear theory provides a broader understanding of the complex behavior of these types of systems. In this context, the localization method is applied. The system (1) is considered under the condition $a_{20} \geq 0$, $a_{21} \geq 0$, which ensures that the system operates within a positive domain. Comparing with Equation (2), the following localizing function is defined:

$$h_1 = qx_1 - x_2, \tag{3}$$

where q is a free positive parameter. Hence, considering β_1, the expression defined by $\beta_1 = (qa_2 - a_8)x_1x_2 + (qa_3 - a_{11})x_2^2 + qa_4x_2^3 + (qa_5 - a_{12})x_3 + (qa_6 - a_{13})x_3^2 + (qa_7 - a_{14})x_3^3 - a_{10}x_1^3$, and after applying the Lie derivative to the function (3), the set $S(h_1)$ is defined as the set containing the Lie derivative, i.e., $S(h_1) = \left\{ L_f h_1(x) = 0 \right\}$, which is represented by:

$$S(h_1) = \left\{ -x_2 = \frac{1}{a_{11}} \left(-\beta_1 - (qa_{20} + a_{21}) - a_9x_1^2 + qa_1x_1 \right) \right\}, \tag{4}$$

now, substituting Equation (4) into Equation (3) results in the set $h_1 \mid_{S(h_1)}$, which is defined as:

$$h_1 \mid_{S(h_1)} = qx_1 - \frac{\beta_1}{a_{11}} - \frac{(qa_{20} - a_{21})}{a_{11}} - \frac{a_9x_1^2}{a_{11}} + \frac{qa_1x_1}{a_{11}},$$

by grouping common terms with respect to x_1 and performing algebraic manipulation, $h_1 \mid_{S(h_1)}$ can be expressed as:

$$h_1 \mid_{S(h_1)} = -\frac{a_9}{a_{11}} \left(x_1 - \frac{qa_{11}}{2a_9} \left(\frac{a_1}{a_{11}} + 1 \right) \right)^2 + \frac{q^2a_{11}}{4a_9} \left(\frac{a_1}{a_{11}} + 1 \right)^2 - \frac{(qa_{20} - a_{21})}{a_{11}} - \frac{\beta_1}{a_{11}}.$$

Therefore, the compact invariant set K_1 is defined as follows:

$$K_1(h_1) = \left\{ h_1 \leq h_1 \mid_{S(h_1)} := \frac{q^2a_{11}}{4a_9} \left(\frac{a_1}{a_{11}} + 1 \right)^2 - \frac{(qa_{20} - a_{21})}{a_{11}} \right\}, \tag{5}$$

giving as a result the maximum concentration of insulin and the minimum concentration of glucose, which are defined as follows:

$$K_1(h_1) = \left\{ \begin{array}{l} x_1 \leq x_{1\max} := \frac{qa_{11}}{4a_9} \left(\frac{a_1}{a_{11}} + 1 \right)^2 - \frac{(qa_{20} - a_{21})}{qa_{11}} \\ x_2 \geq x_{2\inf} := \frac{q^2a_{11}}{4a_9} \left(\frac{a_1}{a_{11}} + 1 \right)^2 - \frac{(qa_{20} - a_{21})}{a_{11}} \end{array} \right\}. \tag{6}$$

Now, to establish an upper bound for the population density of glucose, consider a second localization function, h_2, defined as follows:

$$h_2 = x_2, \tag{7}$$

after applying the Lie derivative to h_2, setting $S(h_2) = \left\{L_f h_2(x) = 0\right\}$, and performing some algebraic rearrangement, $S(h_2)$ is given by:

$$S(h_2) = \left\{x_2 = 1 + \frac{a_{21}}{a_{11}x_2} - \frac{a_8}{a_{11}}x_1 - \frac{a_9 x_1^2}{a_{11}x_2} - \frac{a_{10}x_1^3}{a_{11}x_2} - \frac{a_{12}x_3}{a_{11}x_2} - \frac{a_{13}x_3^2}{a_{11}x_2} - \frac{a_{14}x_3^3}{a_{11}x_2}\right\},$$

this leads to the upper bound for the set $K_2(h_2)$, given by:

$$K_2(h_2) = \left\{h_2 \leq h_2 \mid_{S(h_2)} \cap K_1(h_1) := 1 + \frac{a_{21}}{a_{11}x_{2\,\mathrm{inf}}}\right\}, \tag{8}$$

this allows the maximum concentration of glucose to be defined by the set $K_* = \{K_2(h_2) \cap K_1(h_1)\}$, as follows:

$$K_* := \left\{x_2 \leq x_{2\,\mathrm{max}} := 1 + \frac{a_{21}}{\frac{q^2 a_{11}^2}{4a_9}\left(\frac{a_1}{a_{11}} + 1\right)^2 - (q a_{20} - a_{21})}\right\}. \tag{9}$$

Therefore, the lower and upper bounds for the glucose concentration x_2 are mathematically defined by the set:

$$\frac{q^2 a_{11}}{4a_9}\left(\frac{a_1}{a_{11}} + 1\right)^2 - \frac{(q a_{20} - a_{21})}{a_{11}} \leq x_2 \leq 1 + \frac{4a_9 a_{21}}{q^2 a_1^2 + 2q^2 a_1 a_{11} + q^2 a_{11}^2 - 4a_9 a_{20} q + 4a_9 a_{21}}. \tag{10}$$

Finally, the last localization function, in order to define an upper bound for the β cells, is:

$$h_3 = x_3, \tag{11}$$

after applying the previous steps to the proposed localization function (11), we obtain the set $K_3(h_3)$, given by:

$$K_3(h_3) = \left\{h_3 \leq h_3 \mid_{S(h_3)} \cap K_* := \frac{a_{15}}{a_{18}}x_{2\,\mathrm{max}} + \frac{a_{16}}{a_{18}}x_{2\,\mathrm{max}}^2 + \frac{a_{17}}{a_{18}}x_{2\,\mathrm{max}}^3\right\}, \tag{12}$$

This allows the maximum concentration of β-cells to be defined as follows:

$$K_3(h_3) := \left\{x_3 \leq x_{3\,\mathrm{max}} := \frac{u_{15}}{a_{18}}x_{2\,\mathrm{max}} + \frac{a_{16}}{a_{18}}x_{2\,\mathrm{max}}^2 + \frac{a_{17}}{a_{18}}x_{2\,\mathrm{max}}^3\right\}. \tag{13}$$

As a result of the previous analysis, the following theorem can be stated:

Theorem 3. *The model described by the set of equations in (1) contains all compact invariant sets within the polytope if the free parameter q satisfies the condition given in (10) and if $\mathbb{R}^3_{+,0}$ is a closed set, where $\mathbb{R}^3_+ : \mathbb{R}^3_{+,0} = \{x_1 \geq 0; x_2 \geq 0; x_3 \geq 0\}$. Under these conditions, the compact invariant set K is defined as:*

$$K = x_{1max} \cap x_{2max} \cap x_{3max}. \tag{14}$$

Defining a non-empty set in diabetes models is a compelling topic due to the mathematical complexity involved in representing nonlinear systems. However, on the basis of *in silico* experimentation, the results obtained are well-defined. Our findings contribute to demonstrating that the behavior of DM1 is inherently nonlinear, warranting further investigation from a mathematical perspective.

The work presented in [26] is particularly notable for introducing and evaluating the use of computational models in clinical trial design, an emerging methodology at the time. The authors propose a mathematical model that simulates the systemic inflammatory response in patients, allowing the prediction of therapeutic intervention outcomes without the need for traditional clinical trials. This approach has the potential to optimize clinical study designs, reduce costs, and improve patient safety by anticipating potential adverse effects.

In alignment with these advancements, we propose a threshold value for q such that $q > 46$.

4. Nonlinear Observer

The challenge of designing observers specifically tailored for nonlinear control systems was first introduced by Thau in his seminal work [27]. Over the past four decades, the control systems literature has evolved significantly, leading to the development of various methodologies for constructing observers that operate effectively in nonlinear frameworks [28]. These nonlinear observers have been applied across diverse fields, addressing key challenges, such as state estimation, parameter estimation, fault detection and isolation, disturbance estimation, and unknown input estimation. In 1998, a significant nonlinear observer was introduced to evaluate the validity of a biological model focused on phytoplanktonic growth [29]. However, subsequent research by Gabriele underscored the importance of parameter estimation in systems biology, highlighting its crucial role in generating meaningful predictions from computational models designed to represent biological systems accurately [30]. These insights have profound implications for computational biology, providing powerful tools to analyze and understand complex biological phenomena.

The use of state observers has also been explored in other biological processes, demonstrating their effectiveness in estimating variables that are difficult to measure directly. For instance, in [31], a robust observer was implemented to monitor ethanol fermentation in real time, enabling accurate inferences about key process variables. This approach is analogous to insulin and glucose estimation in this work, where the observer is used to infer non-measurable states in the glycemic regulation system.

Furthermore, mathematical modeling of glucose supply has been addressed in [32], where incomplete functions were used to describe system dynamics. This approach provides a relevant mathematical framework for formulating the model in this work, enhancing the accuracy of insulin and glucose estimation through the integration of observer techniques. These contributions underscore the growing role of control theory in computational biology, reinforcing the importance of state observers in developing reliable models for physiological processes.

In recent years, computational biology has turned to control theory to tackle the issue of parameter estimation, particularly through the use of state observers. Initially formulated for state estimation tasks, these algorithms aim to deduce the time evolution of unobserved components within a dynamical system. The existing literature on control theory in this context is extensive, with several classical techniques, such as Luenberger-like observers [33], being employed in biological or biochemical systems. A key advancement in nonlinear observer design for population biology models came with the application of Sundarapandian's theorem to the Lotka–Volterra model, which describes two-species competition with stable coexistence [34].

Despite the variety of available observer designs, Thau's observer has emerged as a particularly robust method. Its strength lies in its reliability in providing sufficient information about a system's trajectory over time, primarily due to its incorporation of a Lipschitz constant. This constant plays a crucial role in characterizing system behavior within chaotic regimes. In this context, it is possible to define a domain that encompasses the system's trajectories, often represented as an ellipsoid, thus facilitating the determination of the Lipschitz constant. This approach not only enhances the understanding of trajectory behavior but also strengthens the robustness of observer design in managing the complexities inherent in nonlinear dynamics.

Thau Observer

Based on the set of equations from system (1), the following conditions must be satisfied for the observer design. First, the pair (C, A) must be observable. Second, the nonlinear function $f(x)$ must be continuously differentiable and locally Lipschitz. If these two conditions are met, a nonlinear Thau observer can be constructed as:

$$\dot{z}(t) = Az + Bu(t) + F(z(t)) + kC(x(t) - z(t)). \tag{15}$$

Then, there exists a matrix $A_0 = A - kC$ that is symmetric and stable, with eigenvalues in the negative plane. In addition, there exists a positive matrix Q that is defined from a matrix P, which is determined by the Lyapunov equation as:

$$A_0^T P + PA_0 = -2Q, \tag{16}$$

hance, if k is chosen such that A_0 satisfies Equation (16), let γ be the Lipschitz constant expressed in the following inequality:

$$\gamma < \frac{\lambda_{\min} Q}{||P||},$$

where the Lipschitz constant must satisfy the following inequality:

$$||f(x_1) - f_2(x_2)|| \leq \gamma ||x_1 - x_2||,$$

for all x_1 and x_2, the error of the Thau observer will be globally asymptotically stable. By applying the theorem in [35], the observer can be obtained by satisfying the inequality:

$$||P|| \leq \frac{||A_0||^{n-1}}{|\det(A_0)|} \leq \frac{2}{\gamma}. \tag{17}$$

Since the matrix A_0 is defined by the linear values of the equations in (1), and considering the state variables of the glucose and β cells as potential variables needed to track the dynamical behavior of the observer over time, and to satisfy the symmetric restriction of A_0, the parameters of the matrix k are chosen as $k_1 = 0, k_3 = 0.3, k_4 = 0.03, k_5 = -1.37$; with $k_2 = 1$ and $k_6 = 100$ for a stable A_0. Hence, the extended form of A_0 is given by:

$$A_0 = \begin{bmatrix} -a_1 & 0 & a_5 \\ 0 & a_{11} & -a_{12} \\ 0 & a_{15} & -a_{18} \end{bmatrix} - \left(\begin{bmatrix} k_1 \\ k_2 \\ k_3 \end{bmatrix} \begin{bmatrix} 0 & 1 & 0 \end{bmatrix} + \begin{bmatrix} k_4 \\ k_5 \\ k_6 \end{bmatrix} \begin{bmatrix} 0 & 0 & 1 \end{bmatrix} \right), \tag{18}$$

Meanwhile, the nonlinear terms of system (1) are expressed as follows:

$$f'(x_1, x_2, x_3) = \begin{bmatrix} a_2 x_1 x_2 + a_3 x_2^2 + a_4 x_2^3 + a_5 x_3 + a_6 x_3^2 + a_7 x_3^3 \\ -a_8 x_1 x_2 - a_9 x_1^2 - a_{10} x_1^3 - a_{11} x_2^2 - a_{13} x_3^2 - a_{14} x_3^3 \\ a_{16} x_2^2 + a_{17} x_2^3 - a_{19} x_2 x_3 \end{bmatrix}. \tag{19}$$

Now, to satisfy inequality (17), the Frobenius norm is computed over the ellipsoidal domain where $|x_1| \leq 1.4$, $|x_2| \leq 3.0$ and $|x_3| \leq 2.0$. The numerical solution for the Frobenius norm gives:

$$\left\| f'(x_1, x_2, x_3) \right\|_\theta = 34.243,$$

meanwhile, solving the inequality (17) numerically yields:

$$1.4346 \times 10^{-2} = \frac{\|A_0\|^{n-1}}{|\det(A_0)|} \leq \frac{2}{\|f'(x, y, z)\|_\theta} = 5.8406 \times 10^{-2}.$$

Therefore, based on Equation (15), the Thau observer for the system (1) is defined as follows:

$$\begin{bmatrix} \dot{z}_1 \\ \dot{z}_2 \\ \dot{z}_3 \end{bmatrix} = \begin{bmatrix} -1.08z_2^3 + 1.09z_2^2 + 0.1z_1z_2 + 2.01z_3^3 - 0.06z_3^2 + 0.03x_3 - 2.04z_1 \\ 1.2z_1^3 + 3.84z_1^2 - 0.22z_1z_2 - 0.3z_2^2 - 0.7z_2 - 0.22z_3^3 + 0.3z_3^2 + x_2 - 1.37x_3 \\ 0.5z_2^3 - 1.35z_2^2 + 0.15z_3z_2 + 0.3x_2 + 100.0x_3 - 99.58z_3 \end{bmatrix}. \tag{20}$$

5. Numerical Simulation and Results

The Thau observer estimates the system's state variables based on the model's equations and the measured output. When simulated alongside the actual system, it allows us to validate that the observer accurately tracks the states and assesses the convergence of the estimated states to the actual states over time, as shown in Figure 1, considering the parameters in Table 1. Figure 1 also presents the phase plane comparison for each state variable.

Perturbations in the system, such as carbohydrate intake, have been extensively studied in the literature [36]. These perturbations are incorporated to simulate realistic scenarios in diabetic patients, where external factors significantly affect the dynamics of glucose and insulin. In this study, a nonlinear observer is used to estimate the system's states under these conditions by adjusting the population density by a finite factor at specific time intervals. The observer's estimation accuracy is demonstrated in Figure 2, which compares the system's actual states with those estimated by the Thau observer. The perturbations introduced in the initial stage are shown in Figure 3, which illustrates how the observer minimizes the estimation error and confirms its stability and robustness under various initial conditions or disturbances. Additionally, Figure 4 displays the error dynamics for each state variable, showing that all errors asymptotically converge to zero within a few seconds, even in the presence of perturbations.

The simulations were performed using the ode45 function in MATLAB® R2023b (The MathWorks, Inc., Natick, MA, USA), with a maximum step size of 0.001 [37]. This solver, which utilizes a fourth- and fifth-order Runge–Kutta algorithm, was employed to validate the derived conditions mathematically. In this research, direct comparisons with real-world measurements are not initially conducted, as the primary focus is to analyze the biological dynamic behavior of this type of model over time through mathematical modeling. Future work will compare these results with real-world conditions documented in the literature.

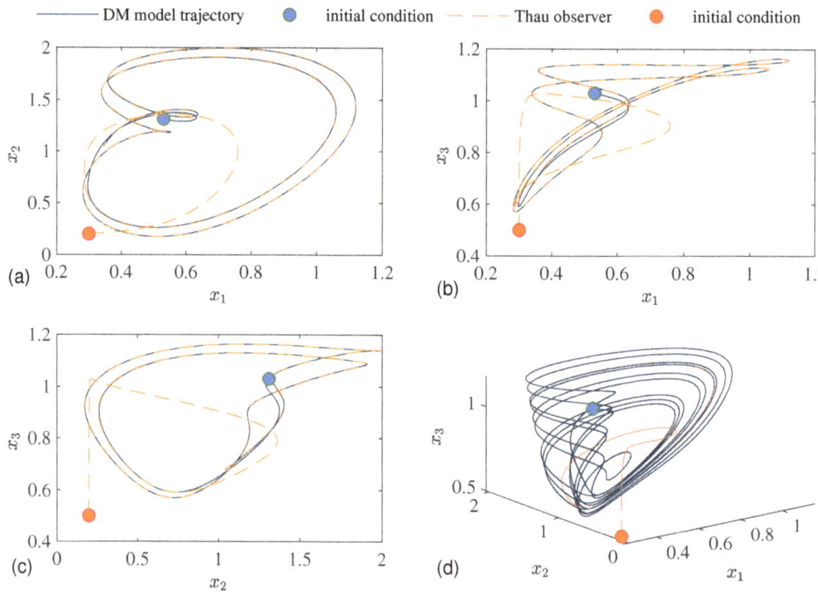

Figure 1. Comparison between the system (1) and Thau observer estimation in the absence of perturbations, with $X_0 = [0.53, 1.31, 1.03]$ and $Z_0 = [0.3, 0.2, 0.5]$. This comparison ensures the Thau observer is correctly implemented and performs as expected under given conditions. In (**a**), the phase planes $x_1(t)$ vs. $x_2(t)$ and $z_1(t)$ vs. $z_2(t)$ are depicted. In (**b**), the phase planes $x_1(t)$ vs. $x_3(t)$ and $z_1(t)$ vs. $z_3(t)$ are shown. In (**c**), the phase planes $x_2(t)$ vs. $x_3(t)$ and $z_2(t)$ vs. $z_3(t)$ are illustrated. Finally, in (**d**), the chaotic behavior is demonstrated.

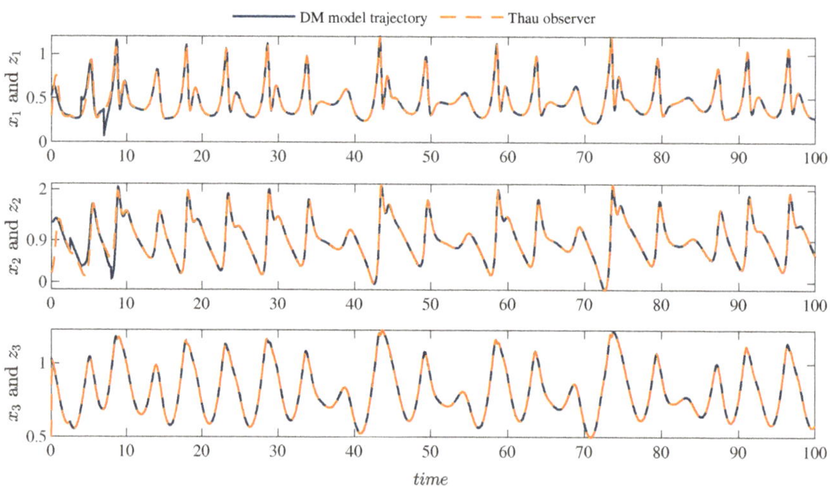

Figure 2. Comparison between the system (1) and the Thau observer. Solid lines indicate actual state trajectories, while dashed lines indicate estimated state trajectories from the observer. Perturbations are introduced prior to $t = 10$; nevertheless, the observer's trajectories consistently track the system's patterns.

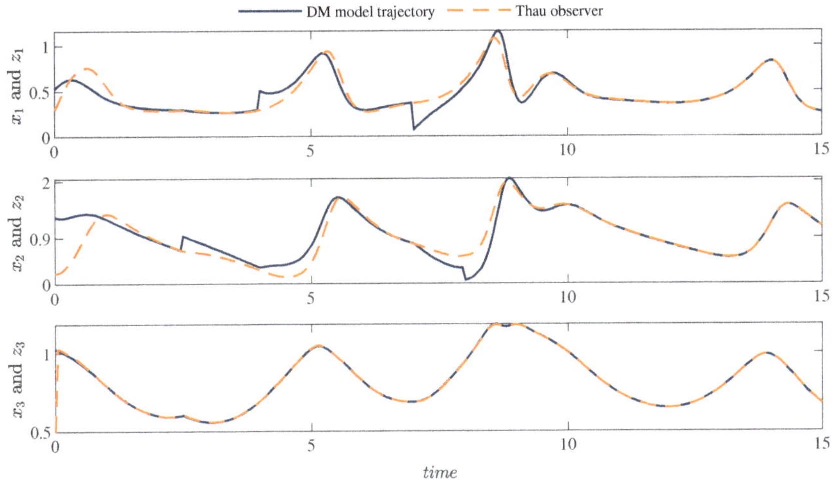

Figure 3. Comparison between the system (1) and the Thau observer. Perturbations are applied to the insulin population density, with values of $+0.2$ and -0.3 at $t = 4$ and $t = 7$, respectively. Similarly, perturbations of $+0.3$ and -0.25 are applied to the glucose population density at $t = 2.5$ and $t = 8$, respectively.

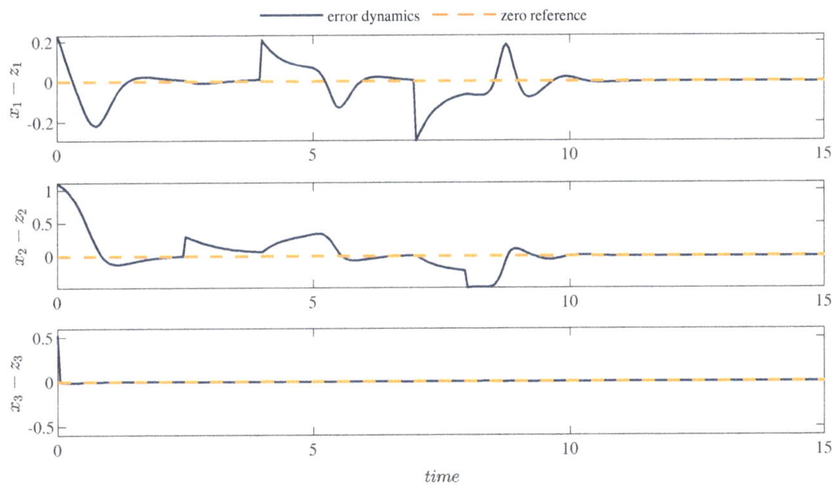

Figure 4. Error dynamics. Accurate estimation enhances control efficiency, robustness to disturbances, and overall system reliability, making error minimization essential for effective observer performance. As presented in this figure, at 10.5 seconds, the estimated error can be considered almost zero for $z_1(t)$, and $z_2(t)$ follows the dynamics of (1). Similarly, at 0.6 seconds, the state $z_3(t)$ also aligns with the system's dynamics.

6. Discussion

Designing a Thau observer for model (1) focuses on nonlinear DM1 systems, enabling the management of complex nonlinear behaviors and leading to more accurate trajectory tracking in nonlinear models [38], as shown in Figures 1 and 2. The dimensionless parameters in Table 1 exhibit chaotic behavior if and only if parameters a_{20} and a_{21} are less than or equal to 0.03. Other values of those parameters must be studied and analyzed for deeper biological implications. For example, in [39], computational and mathematical modeling approaches for drug development are discussed as advantageous due to their fast predictive capability and cost-effectiveness features. In future research, we will exploit the feasibility of implementing a comparative analysis considering our results with the Digital Twins predictions [40] to explore deeper biological implications. In addition, we will study the design of a nonlinear controller based on the designed observer, contributing to the approach of the implementation of new technology [41]. The LCIS method remains a key mathematical strategy for analyzing nonlinear ODE models with biological implications [24]. Therefore, the importance of nonlinear control theory applied in biomathematics can significantly enhance results. The Thau observer for the nonlinear DM model that considers β-cell dynamics allows accurate estimation of insulin levels, even under impaired β-cell functionality. It integrates the nonlinear interactions between glucose and insulin while reducing reliance on invasive measurements. The observer provides real-time insights into β-cell contributions, improving personalized diabetes management and treatment strategies. For instance, in Figure 3, the observer effectively manages uncertainties and disturbances in the system, ensuring reliable insulin level predictions. In Figure 4, the performance of the Thau observer is critical for ensuring accurate state estimation, system stability, and precise trajectory tracking. Significant errors can lead to unreliable state feedback, degraded control performance, and deviations from desired behavior.

7. Conclusions

This paper presents a Thau observer for a Diabetes Mellitus model that describes the dynamic interaction among insulin, glucose, and β-cells. The observer can estimate insulin levels by measuring the glucose and population of β-cells. However, designing a Thau observer is not easy; this type of observer requires certain conditions that must be satisfied, such as the Lipschitz constant γ and the stability matrix A_0. Therefore, there are established conditions for the free feedback matrix k such that all eigenvalues of the stability matrix defined by $A_0 = A - KC$ have a negative real part. Next, LCIS analysis is applied to determine the sets containing the dynamical behaviors of each state variable, defined by (14) considering the dimensionless parameters described in Table 1, to guarantee the stability condition established by inequality (17), needed to ensure the observer's asymptotic stability. Furthermore, the observer's performance and asymptotic stability are verified through MATLAB® simulations. Future work will focus on closed-loop control analysis and mathematical validation of the entire system, along with its implementation on a digital platform to develop a Digital Twin system. Both the conceptual Digital Twin and the closed-loop control system will enable the exploration of complex scenarios that would be dangerous or infeasible in real-life experiments due to validated treatment protocols. In addition, this approach will allow the study of perturbed cases influenced by factors such as caloric intake, exercise routines, and sleep patterns, facilitating the design of personalized treatment protocols. By integrating patient-specific data and real-time monitoring, the Digital Twin system has the potential to serve as an advanced simulation tool for predicting glucose–insulin dynamics under various physiological and pathological conditions. This will provide clinicians with a deeper understanding of the metabolic response of individuals, allowing for more precise treatment adjustments. Beyond individual treatment optimization, the implementation of a Digital Twin system could contribute to broader diabetes research by simulating large-scale population dynamics and testing novel therapeutic strategies before clinical trials. This would reduce the risks and costs associated with experimental treatments while improving the reliability of medical interventions. Ultimately, this research aims to improve the management of diabetes by offering a robust and patient-specific approach that minimizes the risks associated with hyperglycemia and hypoglycemia.

Author Contributions: Formal analysis and investigation, D.G.; software and validation, T.C.G.; writing—review and editing, P.J.C. All authors have read and agreed to the published version of the manuscript.

Funding: This research received funding from the Tecnológico Nacional de México/Instituto Tecnológico de Tijuana through the titled projects: ESTUDIO DE LA DIABETES MELLITUS TIPO 1 COMO UNA ENFERMEDAD MULTIFACTORIAL with ID 20136.24-P, and SISTEMA AUTÓNOMO DE REGULACIÓN DE INSULINA Y EXPERIMENTACIÓN IN-SILICO PARA TRATAMIENTOS DE LA DIABETES MELLITUS TIPO 1 with ID 23154.25-P.

Data Availability Statement: The detailed data supporting the findings of this article are available from the authors upon request. All original contributions presented in this study are included in the article. For further inquiries, please contact the corresponding authors.

Conflicts of Interest: The authors declare no conflicts of interest.

References

1. Saeedi, P.; Petersohn, I.; Salpea, P.; Malanda, B.; Karuranga, S.; Unwin, N.; Colagiuri, S.; Guariguata, L.; Motala, A.A.; Ogurtsova, K.; et al. Global and regional diabetes prevalence estimates for 2019 and projections for 2030 and 2045: Results from the International Diabetes Federation Diabetes Atlas, 9th edition. *Diabetes Res. Clin. Pract.* **2019**, *157*, 107843. [CrossRef] [PubMed]
2. Febrian, M.E.; Ferdinan, F.X.; Sendani, G.P.; Suryanigrum, K.M.; Yunanda, R. Diabetes prediction using supervised machine learning. *Procedia Comput. Sci.* **2023**, *216*, 21–30. [CrossRef]
3. Rooney, M.R.; Fang, M.; Ogurtsova, K.; Ozkan, B.; Echouffo-Tcheugui, J.B.; Boyko, E.J.; Magliano, D.J.; Selvin, E. Global prevalence of prediabetes. *Diabetes Care* **2023**, *46*, 1388–1394. [CrossRef] [PubMed]
4. Antar, S.A.; Ashour, N.A.; Sharaky, M.; Khattab, M.; Ashour, N.A.; Zaid, R.T.; Roh, E.J.; Elkamhawy, A.; Al-Karmalawy, A.A. Diabetes mellitus: Classification, mediators, and complications; A gate to identify potential targets for the development of new effective treatments. *Biomed. Pharmacother.* **2023**, *168*, 115734. [CrossRef]
5. Shaw, J.; Sicree, R.; Zimmet, P. Global estimates of the prevalence of diabetes for 2010 and 2030. *Diabetes Res. Clin. Pract.* **2010**, *87*, 4–14. [CrossRef]
6. Vallis, M.; Ryan, H.; Berard, L.; Cosson, E.; Kristensen, F.B.; Levrat-Guillen, F.; Naiditch, N.; Rabasa-Lhoret, R.; Polonsky, W. How continuous glucose monitoring can motivate self-management: Can motivation follow behaviour? *Can. J. Diabetes* **2023**, *47*, 435–444. [CrossRef] [PubMed]
7. Shi, M.; Zhou, J.; Cai, M. Multiple Physiological and Behavioural Parameters Identification for Dietary Monitoring Using Wearable Sensors: A Study Protocol. *medRxiv* **2024**. [CrossRef]
8. Lakshmanan, M. Nonlinear dynamics: Challenges and perspectives. *Pramana* **2005**, *64*, 617–632. [CrossRef]
9. Aliffi, G.E.; Nastasi, G.; Romano, V.; Pitocco, D.; Rizzi, A.; Moore, E.J.; De Gaetano, A. A system of ODEs for representing trends of CGM signals. *J. Math. Ind.* **2024**, *14*, 23. [CrossRef]
10. Savatorova, V. Exploring parameter sensitivity analysis in mathematical modeling with ordinary differential equations. *CODEE J.* **2023**, *16*, 4. [CrossRef]
11. Bortz, D.M.; Messenger, D.A.; Dukic, V. Direct estimation of parameters in ODE models using WENDy: Weak-form estimation of nonlinear dynamics. *Bull. Math. Biol.* **2023**, *85*, 110. [CrossRef]
12. Little, R.R.; Rohlfing, C.L.; Sacks, D.B. Status of hemoglobin A1c measurement and goals for improvement: from chaos to order for improving diabetes care. *Clin. Chem.* **2011**, *57*, 205–214. [CrossRef]
13. Wu, J.; Li, C.; Chen, W.; Lin, C.; Chen, T. Application of Van der Pol oscillator screening for peripheral arterial disease in patients with diabetes mellitus. *J. Biomed. Sci. Eng.* **2013**, *6*, 1143. [CrossRef]
14. Liu, S.; Song, R.; Lu, X. Research on Blood Glucose Nonlinear Controller Based on Backstepping Adaptive Control Algorithm. In Proceedings of the 2024 IEEE 13th Data Driven Control and Learning Systems Conference (DDCLS), Kaifeng, China, 17–19 May 2024; pp. 472–477. [CrossRef]
15. Shabestari, P.S.; Panahi, S.; Hatef, B.; Jafari, S.; Sprott, J.C. A new chaotic model for glucose-insulin regulatory system. *Chaos Solitons Fractals* **2018**, *112*, 44–51. [CrossRef]
16. Singh, P.P.; Singh, K.M.; Roy, B.K. Chaos control in biological system using recursive backstepping sliding mode control. *Eur. Phys. J. Spec. Top.* **2018**, *227*, 731–746. [CrossRef]
17. Saoussane, M.; Mohammed, T.; Mesaoud, C. Adaptive controller based an extended model of glucose-insulin-glucagon system for type 1 diabetes. *Int. J. Model. Simul.* **2023**, *43*, 282–293. [CrossRef]
18. Mosquera-Lopez, C.; Jacobs, P.G. Digital twins and artificial intelligence in metabolic disease research. *Trends Endocrinol. Metab.* **2024**, *35*, 549–557. [CrossRef]

19. Ackerman, E.; Rosevear, J.W.; McGuckin, W.F. A mathematical model of the glucose-tolerance test. *Phys. Med. Biol.* **1964**, *9*, 203. [CrossRef]
20. Dalla Man, C.; Breton, M.D.; Cobelli, C. Physical activity into the meal glucose—Insulin model of type 1 diabetes: In silico studies. *J. Diabetes Sci. Technol.* **2009**, *3*, 56–67. [CrossRef]
21. Vano, J.; Wildenberg, J.; Anderson, M.; Noel, J.; Sprott, J. Chaos in low-dimensional Lotka Volterra models of competition. *Nonlinearity* **2006**, *19*, 2391–2404. [CrossRef]
22. Krishchenko, A.P. Estimations of domains with cycles. *Comput. Math. Appl.* **1997**, *34*, 2–4. [CrossRef]
23. Krishchenko., A.P. Localization of invariant compact sets of dynamical systems. *Differ Equ* **2005**, *41*, 1669–1676. [CrossRef]
24. Starkov, K.E.; Krishchenko, A.P. On the Dynamics of Immune-Tumor Conjugates in a Four-Dimensional Tumor Model. *Mathematics* **2024**, *12*, 843. [CrossRef]
25. Krishchenko, A.P.; Starkov, K.E. Localization of compact invariant sets of the Lorenz system. *Phys. Lett. A* **2006**, *353*, 383–388. [CrossRef]
26. Clermont, G.; Bartels, J.; Kumar, R.; Constantine, G.; Vodovotz, Y.; Chow, C. In silico design of clinical trials: A method coming of age. *Crit. Care Med.* **2004**, *32*, 2061–2070. [CrossRef] [PubMed]
27. Thau, F.E. Observing the state of non-linear dynamic systems. *Int. J. Control* **1973**, *17*, 471–479. [CrossRef]
28. Besancon, G. *Nonlinear Observers and Applications*; Lecture notes in control and information sciences; Springer: Berlin/Heidelberg, Germany, 2007. [CrossRef]
29. Bernard, O.; Sallet, G.; Sciandra, A. Nonlinear Observers for a Class of Biological Systems: Application to Validation of a Phytoplanktonic Growth Model. *IEEE Trans. Autom. Control* **1998**, *43*, 1056 – 1065. [CrossRef]
30. Lillacci, G.; Khammash, M. Parameter Estimation and Model Selection in Computational Biology. *PLoS Comput. Biol.* **2010**, *6*, e1000696. [CrossRef]
31. Aguilar-López, R.; Alvarado-Santos, E.; Thalasso, F.; López-Pérez, P.A. Monitoring Ethanol Fermentation in Real Time by a Robust State Observer for Uncertainties. *Chem. Eng. Technol.* **2024**, *47*, 779–790. [CrossRef]
32. Bhatter, S.; Jangid, K.; Shyamsunder; Purohit, S.D. Determining glucose supply in blood using the incomplete I-function. *Partial. Differ. Equations Appl. Math.* **2024**, *10*, 100729. [CrossRef]
33. Hulhoven, X.; Wouwer, A.V.; Bogaerts, P. Hybrid extended Luenberger-asymptotic observer for bioprocess state estimation. *Chem. Eng. Sci.* **2006**, *61*, 7151–7160. [CrossRef]
34. Vaidyanathan, S. Nonlinear observer design for Lotka-Volterra systems. In Proceedings of the 2010 IEEE International Conference on Computational Intelligence and Computing Research, Coimbatore, India, 28–29 December 2010; pp. 1–5. [CrossRef]
35. Starkov, K.E.; Coria, L.N.; Aguilar, L.T. On synchronization of chaotic systems based on the Thau observer design. *Commun. Nonlinear Sci. Numer. Simulat.* **2012**, *17*, 17–25. [CrossRef]
36. Olay-Blanco, A.; Rodriguez-Linan, A.; Quiroz, G. Parameter and State Estimation of a Mathematical Model of Carbohydrate Intake. *IFAC-PapersOnLine* **2018**, *51*, 73–78. [CrossRef]
37. Xue, D.; Pan, F. Ordinary Differential Equation Solutions. In *MATLAB® and Simulink® in Action: Programming, Scientific Computing and Simulation*; Springer: Berlin/Heidelberg, Germany, 2024; pp. 283–321.
38. Gamboa, D.; Coria, L.N.; Cárdenas Valdez, J.R.; Ramírez Villalobos, R.; Valle Trujillo, P.A. Implementación en hardware de un observador no lineal para un modelo matemático de Diabetes Mellitus Tipo 1 (DM1). *Comput. Sist.* **2019**, *23*, 1475–1486. [CrossRef]
39. Hasan, M.R.; Alsaiari, A.A.; Fakhurji, B.Z.; Molla, M.H.R.; Asseri, A.H.; Sumon, M.A.A.; Park, M.N.; Ahammad, F.; Kim, B. Application of mathematical modeling and computational tools in the modern drug design and development process. *Molecules* **2022**, *27*, 4169. [CrossRef]
40. Cappon, G.; Facchinetti, A. Digital Twins in Type 1 Diabetes: A Systematic Review. *J. Diabetes Sci. Technol.* **2024**. [CrossRef]
41. Chan, P.Z.; Jin, E.; Jansson, M.; Chew, H.S.J. AI-Based Noninvasive Blood Glucose Monitoring: Scoping Review. *J. Med. Internet Res.* **2024**, *26*, e58892. [CrossRef]

Mathematical and Computational Applications

MDPI

Article

Bio-Inspired Multiobjective Optimization for Designing Content Distribution Networks

Gerardo Goñi [1], Sergio Nesmachnow [1,*], Diego Rossit [2], Pedro Moreno-Bernal [3,*] and Andrei Tchernykh [4]

1 Universidad de la República, Montevideo 11200, Uruguay; gerardo.goni@fing.edu.uy
2 Department of Engineering, Instituto de Matemática de Bahía Blanca (INMABB), Universidad Nacional del Sur-CONICET, Bahía Blanca B8000CPB, Argentina; diego.rossit@uns.edu.ar
3 Facultad de Contaduría, Administración e Informática, Universidad Autónoma del Estado de Morelos, Cuernavaca 62209, Morelos, Mexico
4 CICESE Research Center, Ensenada 22860, Baja California, Mexico; chernykh@cicese.mx
* Correspondence: sergion@fing.edu.uy (S.N.); pmoreno@uaem.mx (P.M.-B.)

Abstract: This article studies the effective design of content distribution networks over cloud computing platforms. This problem is relevant nowadays to provide fast and reliable access to content on the internet. A bio-inspired evolutionary multiobjective optimization approach is applied as a viable alternative to solve realistic problem instances where exact optimization methods are not applicable. Ad hoc representation and search operators are applied to optimize relevant metrics from the point of view of both system administrators and users. In the evaluation of problem instances built using real data, the evolutionary multiobjective optimization approach was able to compute more accurate solutions in terms of cost and quality of service when compared to the exact resolution method. The obtained results represent an improvement over greedy heuristics from 47.6% to 93.3% in terms of cost while maintaining competitive quality of service. In addition, the computed solutions had different tradeoffs between the problem objectives. This can provide different options for content distribution network design, allowing for a fast configuration that fulfills specific quality of service demands.

Keywords: content distribution networks; evolutionary algorithms; optimization; cloud computing

1. Introduction

A reliable internet connection has become an essential service for modern societies, enabling a growing number of daily activities. Internet providers face challenges not only in extending services to an increasing number of new users but also in meeting the demand for low-latency performance [1]. Content Distribution Networks (CDNs) were developed in response to the need for efficient internet service. A CDN is a system of distributed servers designed to deliver web content efficiently to users, who are served on a proximity basis [2].

CDN providers usually deploy servers as close to clients as possible in order to reduce latency and route requests so as to efficiently deliver content to end users [3]. Cloud-based CDNs rely on the Infrastructure-as-a-Service (IaaS) model, allowing CDN providers to lease infrastructure. The hardware of the servers is virtualized on the cloud, allowing the assigned resources to be adjusted according to computational needs. IaaS providers have data centers distributed around the world and provide storage, Virtual Machines (VMs), and network infrastructure to CDN providers. Cloud CDN providers maintain competitive Quality of Service (QoS), benefiting from the adaptability of cloud resources. However, implementing a cloud CDN presents the significant challenge of provisioning the necessary resources within the cloud to meet user demand while providing good QoS with an affordable cost.

The cloud broker is a business agent developed to act as an intermediary between users and cloud service providers while assisting users to find the best providers for their needs. Virtual brokers [4] represent a business model that handles the complex supply chain between content providers, IaaS providers, and users. A multi-tenancy approach is adopted to reduce costs by managing multiple content providers simultaneously, thereby leveraging discounted VM prices for large volumes. Resource-sharing strategies also enhance the QoS provided to users.

This article introduces an optimization model for finding strategies to design a cloud-based CDN by a virtual broker considering user demand, content providers, and IaaS providers. The model aims at simultaneously optimizing the cost for the virtual broker and the QoS provided to the users. Designing CDNs is an NP-hard problem, as it is a specific application of the combinatorial facility location problem [5]. Thus, to address this problem, a tailored Multiobjective Evolutionary Algorithm (MOEA) [6] is proposed using ad hoc solution representations and operators. The MOEA is compared to a set of three baseline heuristics and an exact solver based on a Mixed-Integer Linear Programming (MILP) model.

Experimental results on real-world instances demonstrate the effectiveness of the proposed bio-inspired multiobjective evolutionary approach. The MOEA was able to compute accurate and more diverse solutions than the exact solver for small problem instances, and the computed solutions significantly improved over greedy heuristics that optimize cost and QoS. The MOEA generated better Pareto fronts than greedy heuristics, which tended to concentrate on solutions with extreme objective values. MOEA computed significantly better results regarding the Relative Hypervolume (RHV) metric for multiobjective optimization, with an average improvement of 34.7%, as well as for the Empirical Attainment Surface (EAS) estimator. Improvements on the problem objectives ranged from 47.6% to 93.3% in terms of cost while maintaining very accurate QoS levels. The values provided by the EAS estimator also confirmed the superiority of the multiobjective evolutionary approach.

This problem was also studied in our previous work [7]. The current article presents the following new contributions: (i) an extended methodology applying specific search operators; (ii) a new set of realistic problem instances built using real information from internet and cloud providers; and (iii) a fully comprehensive experimental evaluation of the proposed MOEA approach, including a comparison with exact solutions and three baseline heuristics.

The rest of this article is organized as follows: Section 2 defines the cloud-based CDN problem, detailing its connection to cloud platforms and brokers, the mathematical formulation, and the QoS model; Section 3 reviews the key related work in the literature; Section 4 introduces the MOEA designed for the cloud-based CDN problem, covering solution representation, algorithm structure and stages, evolutionary operators, and implementation details; Section 5 describes the baseline optimization methods used for comparison, including the three heuristic approaches and the exact solver; Section 6 presents the experimental validation of the proposed MOEA and baseline methods via testing on realistic instances; finally, Section 7 highlights the main conclusions and outlines directions for future research.

2. The Cloud-Based CDN Design Problem

This section presents the cloud-based CDN design problem, including a conceptual description of the problem along with its main features and the mathematical formulation.

2.1. Cloud Platforms and Virtual Brokers

A CDN is a distributed system of servers hosted across multiple data centers on the internet. CDNs deliver online content, ensuring high availability and performance through efficient load balancing on servers and links. The geographic distribution of the servers on a CDN allows data to be served to users based on their geographic proximity to each server. However, it is prohibitively expensive for small content providers to compete on a large scale with conventional CDN service providers by implementing new data centers. Thus, the computing model based on cloud platforms [8] offers a cost-effective solution for small content providers based on a more efficient and dynamic allocation of resources.

Cloud platforms provide on-demand access to shared resources that are dynamically allocated and released [9], creating the perception of unlimited always-available resources for end users by automating allocation and scaling. If a cloud platform cannot meet resource demands, it can leverage additional resources from other platforms using the cloud bursting technique [10]. In a cloud-based CDN, it is possible to dynamically balance bandwidth, VMs, and storage resources based on content demand, resulting in reduced overall leasing costs while maintaining QoS for end users [11]. The cloud-based CDN must allow end users to access content shared by content providers and ensure that there is no interference between content providers. The previously-discussed IaaS business model in cloud platforms creates new opportunities for small content providers to use cost-effective and scalable CDNs without investing in the installation and maintenance of infrastructure; as such, there is a growing research trend around leveraging the advantages of developing cloud-based CDNs [12–15]. However, with the extension of cloud-based CDN, new managerial problems such as resource provisioning arise, representing a complex resource allocation problem [16].

To manage the diverse offerings of cloud service providers, a cloud broker acts as an intermediary between cloud service providers and cloud users [17]. The cloud broker assists users in identifying the providers that best meet their performance requirements, service level agreements, security, and costs. The cloud broker business model involves leasing VMs from various cloud service providers for extended periods in order to secure significant price discounts, allowing the cloud broker to offer VMs to cloud users on demand at a lower price than direct cloud service providers. The business model applied in this article considers a special type of cloud broker called a *virtual broker*, discussed in our previous publication [4]. A virtual broker focuses on minimizing the total cost of a cloud-based CDN built within IaaS provider data centers. It achieves this by enabling multiple content providers to share the cloud-based CDN while maintaining QoS for the end users who consume the content.

For two reasons, the virtual broker business model offers cost advantages for a cloud-based CDN used by multiple content providers when compared to a single-provider setup. The first is that IaaS service providers generally grant volume discounts when leasing resources on the same infrastructure. A cloud-based CDN with multiple content providers benefits from aggregating their collective resource demand, thereby qualifying for larger discounts. The second reason is that the virtual broker can maximize the utilization of reserved VM instances. These instances, which are more cost-effective when reserved for extended periods, can be shared among content providers based on fluctuating user demands, reducing reliance on more expensive on-demand VMs.

Figure 1 shows a diagram of how the virtual broker, content providers, IaaS service providers, and end users interact. Content providers contract the cloud-based CDN service offered by the virtual broker, which leases usage time and storage on VM instances from IaaS providers. End users then utilize the cloud-based CDN built by the virtual broker to consume the resources of the content providers. The virtual broker is responsible for managing and allocating resources to each content provider according to the demand of their end users. IaaS service providers grant discounts to the virtual broker for the volume of leased resources; in turn, the virtual broker is able to better utilize the usage time of the reserved VM instances by being able to use them with multiple content providers.

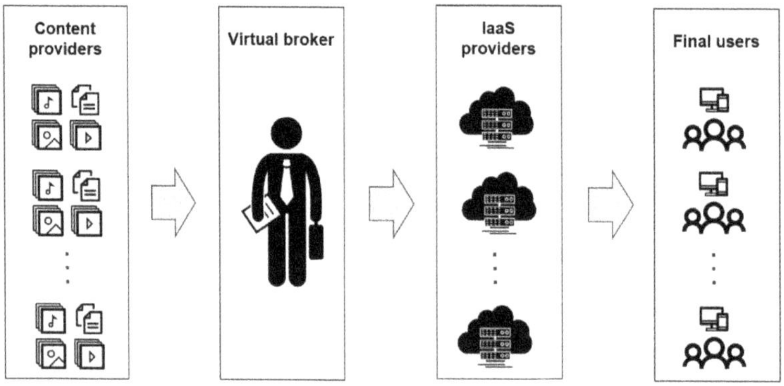

Figure 1. Interaction between content providers, virtual broker, IaaS providers, and final users.

2.2. Conceptual Model

Given a set of content, content providers, available VMs and VM instances, and a set of geographical regions demanding content, the design of a cloud-based CDN requires addressing a complex combinatorial optimization problem. The problem involves simultaneously solving two planning tasks over a given planning horizon:

- Assigning content shared by different providers to instances of VMs available in data centers at each time interval.
- Determining which data center should be used to fulfill the demand for content at each time interval from users located in different geographical regions.

Both planning tasks must be solved in order to minimize the overall cost (includes the cost of renting VMs, storing content, and transferring data from VMs to the internet) while maximizing the QoS provided to users. CDNs have several characteristics that make this a complex decision problem. First, the unit costs per data of content usually differ between data centers; thus, the assignment of content to data centers must be carefully planned. Additionally, data centers may have VMs that can be rented for the whole planning horizon as well as VMs that can be rented on-demand for a specific time interval. Moreover, the QoS provided to users is influenced by the selection of data centers to serve their requests.

2.3. Mathematical Formulation

The mathematical formulation of the cloud-based CDN design problem is developed using the sets, parameters, and variables specified in Table 1.

Three cost functions related to the costs of the CDN are defined:

1. $Cost^v(c) : crv_c \times rv_c + \overrightarrow{cdv_c} \times \overrightarrow{dv_c} \rightarrow \mathcal{R}_0^+$ returns the cost of reserving and renting on-demand VM instances in data center $c \in C$.

2. $Cost^s(c) : csd_c \times (\overrightarrow{bd} \times \vec{x_c}) \rightarrow \mathcal{R}_0^+$ returns the cost of storing content in data center $c \in C$ through a cost function $Cost^t(c) : \overrightarrow{ctd_c} \times (\overrightarrow{bd} \times \mathbf{Z_c}) \rightarrow \mathcal{R}_0^+$ that returns the cost of transferring data from the data centers to the internet.
3. $inv(QoS(c)) : \vec{q_c} \times \mathbf{Z_c} \rightarrow \mathcal{R}_0^+$ measures the inverse of the QoS provided to users.

Then, the objective functions of the cloud-CDN problem are defined as follows:

$$\min \quad \sum_{c \in C} (Cost^v(c) + Cost^s(c) + Cost^t(c)), \tag{1a}$$

$$\min \quad \sum_{c \in C} inv(QoS(c)). \tag{1b}$$

Equation (1a) proposes minimizing the overall cost of all the data centers in the CDN, while Equation (1b) proposes minimizing the inverse of the overall QoS provided to the users to guarantee maximization of the provided QoS (see the QoS model in the next subsection). The problem is bound by specific constraints. First, the demand for each content from each region and at each time interval must be fulfilled. A solution is only considered feasible if it satisfies the demand of every user. Second, a VM instance cannot host content from multiple providers at the same time. Third, the total content assigned to a VM must not exceed its processing capacity. Finally, a data center can only fulfill requests for content stored within it.

Table 1. Sets, parameters, and variables of the mathematical formulation.

	Sets
C	Set of data centers.
T	Set of time intervals within the planning horizon.
R	Set of geographical regions with content demand.
P	Set of content providers.
V	Set of VM providers.
$v \in V^c$	Set of instances of VMs that can be reserved at data center $c \in C$.
	Parameters
crv_c	Cost of reserving one VM instance in data center $c \in C$ for the planning horizon.
ctd_c	Cost of transferring a unit of data from data center $c \in C$ to the Internet.
$\overrightarrow{cdv_c} = \{cdv_t\}$	Cost to rent an on-demand VMs for each time interval $t \in T$ in data center $c \in C$.
csd_c	Data storage cost per byte stored in data center $c \in C$.
bd_k	Size in bytes of each content $k \in K$.
$\vec{q_c} = \{q_r\}$	Matrix indicating the QoS provided to users in region $r \in R$ by data center $c \in C$.
	Variables
rv_c	(integer) Number of instances of VMs that are reserved in data center $c \in C$ for the planning horizon.
$\overrightarrow{dv_c} = \{dv_t\}$	(integer) Number of on-demand VM instances that are used in time interval $t \in T$ in data center $c \in C$.
$\mathbf{Z_c} = \{z_{kr}\}$	(integer) Number of times content $k \in K$ is downloaded by a user in region $r \in R$ from data center $c \in C$.
$\vec{x_c} = \{x_k\}$	(binary) 1 if content $k \in K$ is stored in data center $c \in C$, and 0 otherwise.

2.4. Quality of Service Based on the Round-Trip Time

Proposing a metric to capture the service level provided to users from each region by every data center is a challenging task. This article uses a QoS metric based on the Round-Trip Time (RTT), that is, the time required for a user to send a request and receive a response from a server. The RTT is closely related to the effective network distance between the end users and cloud servers, as shown in the comprehensive analysis by Landa et al. [18]. A lower RTT corresponds to a shorter effective distance between the user and the content server.

To measure QoS based on RTT, we used data from RIPE Atlas [19]. RIPE Atlas is a platform that enables measurement of network parameters on the internet using a distributed worldwide network of sensors. The data center regions were chosen from locations where Amazon provides its Elastic Compute Cloud EC2 service, which allows users to rent VMs for executing applications while charging per second of active server usage [20]. The user regions were identified as the regions with the largest populations demanding Amazon EC2 services.

RTT measurements were collected every 2 h over two periods in 2016 and 2017. The data were downloaded from RIPE Atlas and processed using Python (version 3.10) libraries. The processing involved several stages. First, the Kolmogorov–Smirnov test [21] was applied to determine whether the RTT measurements followed a normal distribution. Because the results showed that normality could not be assumed, the median was selected as the measure of central tendency. Finally, the median RTT values for approximately 35,000 measurements between each user region and data center region were calculated in order to estimate the QoS provided by each data center region to the corresponding user region. The median RTT values provides a robust estimation of QoS, leveraging RTT as a precise and performance-focused metric. The smaller the median RTT values, the better the QoS provided to the users.

The collected measurements allowed us to design representative scenarios, especially considering that the proposed algorithmic models are robust against different values of the problem scenarios, as demonstrated in previous articles [7,22].

3. Literature Review

The design of CDNs on cloud platforms has been approached by considering a number of different optimization criteria for modeling the cloud resource provisioning problem. Related works have proposed optimization algorithms for efficient CDN design; however, few of these have followed multiobjective approaches. A review of related work is presented next.

Zheng and Zheng [23] proposed a Genetic Algorithm (GA) to optimize the delivery cost for static content delivery in cloud storage through a CDN. Costs were based on the network bandwidth and processing capability, allowing network peers to be classified within a local region based on the current peer load status. The evaluation was performed using the CloudSim simulator. Ten problem instances were built from the OR-Library based on set covering problem cases with weights, simulating content delivery for different groups of a static CDN topology. Experimental results showed improved CDN designs with reduced delivery costs.

Iturriaga et al. [24] studied multiobjective design and optimization of CDNs in the cloud while minimizing infrastructure costs and maximizing QoS for end users. An MOEA was applied to address the offline problem of provisioning infrastructure resources. The results indicated that the MOEA effectively optimized resource provisioning in cloud-based CDNs. The proposal was then extended by considering costs related to VMs, networking, and storage [22]. Three MOEAs were proposed for addressing the offline provisioning of cloud resources. Experimental results indicated that SMS-EMOA was the most accurate MOEA for all problem instances. In addition, the results demonstrated that the proposed approach reduced total costs by up to 10.6% while maintaining high QoS levels.

Stephanakis et al. [25] proposed a capacitated QoS network model to optimize the placement of surrogate media servers within the access portion of a CDN. The optimization objectives included minimizing costs and link delays based on capacity and utilization requirements as well as minimizing the network architecture. An MOEA was used to evaluate all possible connections between the first-level aggregation points in the access network and prioritize different types of traffic according to their class of service. The evaluation was performed on scenarios representing the traffic requirements of 20 access nodes aggregated on 20 digital subscriber line access multiplexers for live broadcast/IPTV and video-on-demand. The results indicated that the use of multiple cost-oriented strategies within the access network led to cost-effective implementations and improved quality of IPTV services.

Bektaş et al. [26] proposed a two-level Simulated Annealing (SA) to tackle the joint problem of object placement and request routing in a CDN under the constraints of server capacity and end-to-end object transfer delay. The objective function represented the total cost of transferring content when a client receives content through a proxy, reflecting the cost summed across all proxies, clients, and objects. Fifteen large-scale instances randomly generated using an internet topology generator were analyzed, where the number of proxy servers ranged from 20 to 60, the number of objects varied from 60 to 100, and the number of clients was set at 200. The results showed that SA consistently outperformed a greedy algorithm across all instances. Additionally, the proposed SA approach produced better solutions than the baseline algorithm for instances with more than 30 proxy servers.

Ellouze et al. [27] proposed a cross-layer framework to optimize traffic for a CDN media streaming service by using a random selection of a surrogate for each client and one-way transmission of delivery cost information from network operators to cloud service providers. The Cooperative Association for Internet Data Analysis (CAIDA) provided the topology and traffic model over 24 h, fluctuating between idle and busy times. The cross-layer optimization framework improved over the random selection, proportion map, and routing cost map techniques, yielding improved network utilization and higher perceived QoS.

Mangili et al. [28] proposed an optimization approach to assess the performance of Content-Centric Networking (CCN) compared to traditional CDNs. Their proposed approach included an optimization model designed to study the performance limits of a CCN by addressing the combined object placement and routing problem. The problem instances utilized topologies from the Abilene network (consisting of 11 routers and 14 links), the GEANT network (with 37 routers and 56 links), and a random-geometric graph (comprising 26 routers and 60 links) with 10 producers and 25 consumers. The results showed that the CCN provided significant performance improvements compared to a traditional IP-based network, even with small cache sizes; however, when the caching storage was large, the benefits of using advanced cache replacement policies were greatly diminished.

Coppens et al. [29] proposed a co-operative cost optimization algorithm for replica placement within a self-organizing hybrid CDN architecture using online content replication managed through control traffic. The proposed approach targeted the content distribution module within the CDN operation layer. The topology was based on a realistic European fiber-optic network with a throughput of 1 GB. The case study demanded an average of twelve movies per minute. Results showed that control traffic could be effectively regulated using a Control Ratio (CR). By applying an adaptive CR that measures the evolution of data traffic, the convergence time was reduced by up to 30%.

Cevallos et al. [30] proposed the multi-objective deployment of service function chains within virtualized CDNs for live streaming. Deep reinforcement learning was used on a real-world dataset from a video delivery operator. The dataset was limited to a five-day trace, with hosting nodes in Italy. The results demonstrated that the proposed method significantly improved QoS and QoE, as evidenced by a higher session acceptance ratio than the other tested algorithms while maintaining operational costs within acceptable limits.

Karaata et al. [31] proposed a multipath routing algorithm designed for CDN-P2P networks that enhances load balancing, scalability, and throughput while reducing buffer requirements and network bottlenecks. The proposed approach evaluates routing messages over all disjoint paths between two peers within a star P2P overlay network. The evaluation involved source peers ranging from 2000 to 5000 with a maximum network size of 5040 nodes, providing sufficient messages to keep the network channels active. The PeerSim simulator was employed to assess diverse network properties. The results indicated that an increase in network size slightly boosts the throughput achieved by the algorithm.

Bęben et al. [32] proposed a multi-criteria decision algorithm to address the problem of multisource content resolution within a CDN. First, the algorithm discovers multiple content delivery paths and gathers their respective transfer characteristics. Then, it triggers each content request to determine the best content server and delivery path combination. The evaluation used an internet-scale network model focusing on video content consumption. The topology included 36,000 domains with 103,000 inter-domain links. Results indicated that having better knowledge about links and servers enhances system efficiency. Specifically, integrating an appropriate routing algorithm with a suitable decision algorithm boosts the effectiveness of the system.

Neves et al. [33] addressed the replica placement and request distribution problem, taking into account server disk space, bandwidth, QoS requirements for requests, and variations in network conditions. Their approach combines exact methods with a heuristic hybrid strategy. Evaluation was conducted on 60 instances from the LABIC project using real data from Brazilian internet providers. The hybrid network heuristic achieved good solutions in significantly less computational time compared to the baseline greedy algorithm used for the offline version of the problem. The proposed method reduced the average gap by more than 1% and cut computational times by over 50%.

Khansoltani et al. [34] proposed an algorithm for selecting servers by evaluating both qualitative and quantitative features within a CDN. The authors used the Promethee decision-making algorithm to consider five key indicators: service price, server latency, CPU resource availability, memory, and server I/O capacity. The evaluation involved 50 servers with resource capacities determined by a normal distribution and 2000 applications submitted to the CDN. The proposed method improved over Promethee_RR (63.2%) and Promethee_MCT (68.7%) on average. The proposed method also demonstrated better performance in terms of service price, with averages of 59.8% and 58.4%.

Jabraili et al. [35] proposed a GA for allocating objects to servers based on their iteration frequency, load, and generated delay, with the goal of developing an effective policy for distributing outsourced content within a CDN infrastructure. The GA was evaluated over problem instances with a random transit-stub structure with 1464 nodes, 5000 objects, and 500,000 user requests. The proposed method significantly improved the mean response time and hit rate while maintaining a low load on the CDN servers.

Lai et al. [36] proposed the STARFRONT framework for optimizing cost and access latency in content distribution, considering dynamic network topology, workload distribution, and pricing policies from cloud operators. The minimum cost path was found by solving an integer linear programming problem. Then, a heuristic was employed for latency minimization by assigning cache servers to meet application-level requirements while minimizing bandwidth and storage costs. The problem instances considered 552,000 real-world flow records from a commercial cloud CDN operator and the Amazon AWS network topology. STARFRONT resulted in reduced round-trip times compared to a cloud-only strategy.

Jin et al. [37] introduced a randomized online edge-renting algorithm for content service providers with the aim of extending cloud-based CDNs into edge environments. The objective was to minimize the costs associated with renting edge services while accounting for bandwidth costs over the charging period without foresight into future demands. The evaluation included three real-world instances using actual charging prices for cloud and edge products where end users required 1 Mbps of bandwidth to access videos. The tested scenarios involved 600, 1200, and 1800 videos that generated visitor data over 150 days. The proposed algorithm was shown to represent an improvement over a standard solution.

Farahani et al. [38] proposed a hybrid P2P–CDN architecture for minimizing client latency and network costs during live video streaming. An unsupervised self-organizing map technique was designed to assist in decision-making for action selection. The approach was evaluated on a large-scale testbed with 350 players running an adaptive bitrate algorithm and five Apache HTTP servers (four CDN servers and an origin server) hosted in the CloudLab environment. Their approach significantly outperformed baseline schemes in regard to user QoE, latency, and network utilization. Later, the same authors presented a new framework for live video streaming that leverages the idle resources of peers to provide distributed video transcoding and super-resolution services [39]. This optimization model jointly optimizes end-to-end latency and network costs using a greedy algorithm while also considering resource limitations. The evaluation considered 350 HTTP Adaptive Streaming clients, and the results showed that their approach provided improvements over the traditional CDN and hybrid P2P–CDN baseline methods.

Yadav and Kar [40] proposed an optimization algorithm for minimizing the cost and delay of CDNs based on the placement of fog nodes within a specific region. Unsupervised Machine Learning (ML) was applied in two stages, namely, clustering and Voronoi. The algorithm was evaluated on datasets using real-time data from open public WiFi access points in Delhi and New York City, with link costs assigned to each link based on the CDN provider. The proposed approach reduced the overall cost of deployment and distribution.

Marri and Reddy [41] proposed an adaptive hybrid approach to enhance the performance of CDN–P2P by considering the serviceability of hosts and organizing peer groups in mesh and tree topologies. The approach was evaluated on a scenario with 1000 peers participating in a 60-min streaming session. The proposed hybrid approach improved over traditional CDN methods regarding upload capacity and startup delay.

The above analysis of related works allows us to identify a growing interest in optimization algorithms for problems involving efficient CDN design. Continuing in this line of work, the current article contributes to the field of multiobjective evolutionary approaches for cloud-based CDN design.

4. Multiobjective Evolutionary Approach for Cloud-Based CDN Design

This section describes the proposed MOEA to solve the CDN design problem.

4.1. Specification of the Proposed MOEA

To address the CDN design problem in cloud platforms, a specific MOEA based on the Nondominated Sorting Genetic Algorithm II (NSGA-II) was implemented. NSGA-II was introduced by Deb et al. [42] as an enhanced version of its predecessor, NSGA [43]. The new version integrates the following features, which enable improved evolutionary search compared to the original NSGA algorithm: (i) a non-dominated elitist sorting approach that uses an auxiliary subpopulation to reduce the complexity associated with dominance checking; (ii) a fitness assignment procedure based on nondominated rank ranges, incorporating crowding distance values to evaluate solution diversity; (iii) a new mechanism for preserving population diversity by using a crowding technique that does not require additional parameters, replacing the sharing technique used in the previous NSGA.

The implemented MOEA based on NSGA-II consists of three phases: population initialization, scheduling, and routing. Figure 2 presents the main activities comprising each phase of the proposed MOEA. The population initialization and scheduling phases determine the allocation of VMs to data centers in order to fulfill content requests, while the routing phase determines which VM serves each content request.

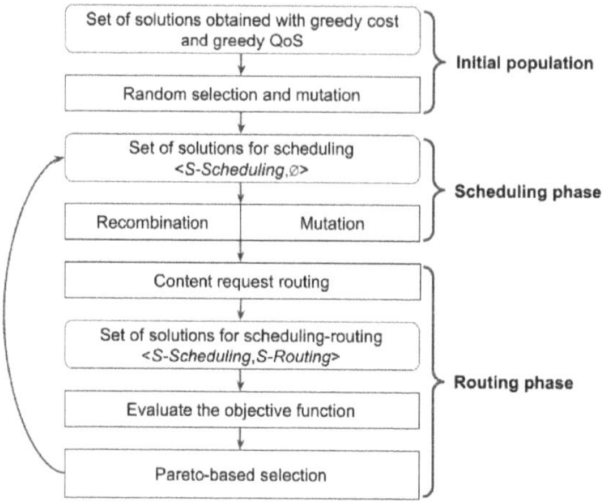

Figure 2. Stages of the implemented MOEA to solve the CDN design problem in cloud platforms.

The proposed MOEA represents solutions as <S-Scheduling, S-Routing>, where the S-Scheduling components is built in the first two phases and the S-Routing component is constructed in the routing phase (for a complete description of the applied representation, see Section 4.2).

4.1.1. First Stage: Population Initialization

The population initialization stage of the MOEA executes at the beginning of the algorithm. The initial population of the MOEA consists of solutions in the form <S-Scheduling,∅>. The S-Scheduling components of the initial solutions are randomly selected from a set of solutions G using a uniform distribution. The set G is constructed with solutions obtained through the greedy cost and greedy QoS heuristics described in Section 5.1. Random perturbations are performed by applying the mutation operator to each initial solution with a probability p_{MI} in order to introduce genetic diversity in the initial population.

4.1.2. Second Stage: Scheduling

In this phase, for each data center, the MOEA determines the number of each type of VM assigned in each planning hour considering the demand of content requests for each hour. The MOEA also determines which content resources are stored in each data center. Then, the MOEA constructs solutions in the form <S-Scheduling,∅>. Both crossover and mutation operators are applied to each solution with their respective probabilities.

4.1.3. Third Stage: Routing

In this stage, the MOEA determines which data center should serve each content resource request. For each solution, the S-Routing component is constructed using the content resource request routing algorithm described in Section 4.6. Finally, when each solution has both of the <S-Scheduling, S-Routing> components, the objective functions are evaluated. The evolutionary search continues until the termination criterion is met; after evaluating the objective functions, the MOEA selects parent solutions that generate offspring to be part of the new population, then returns to the beginning of the scheduling phase.

4.2. Representation of Solutions

In the proposed MOEA, solutions are represented by four matrices:

- Y: An integer matrix of dimension $H \times m$.
- X: A binary matrix of dimension $n \times m$.
- $\tilde{Y}$: An integer matrix of dimension $m \times 1$.
- Z: An integer matrix of dimension $s \times n \times m \times v \times (H \times 60)$.

where H is the number of hours to plan in the period $[0, T]$, m is the number of data centers, n is the number of content resources, s is the number of geographical regions from which users make requests, and v is the number of VMs allocated throughout the planning period.

The S-Scheduling component is represented by matrices Y, $\tilde{Y}$, and X. Matrix $\tilde{Y}$ indicates how many instances of reserved VMs are assigned to each data center throughout the planning period. Matrix X represents the allocation of content provider resources for each data center. Matrix Y indicates the number of VMs needed to meet the demand for content resource requests for each data center and in each planning hour $h \in [0, H-1]$. The number of VMs requested on-demand is expressed for each hour $h \in [0, H-1]$ and for each data center c_e as $\max 0, y_e^h - \tilde{y}_e$. The S-Routing component is represented by the matrix Z. The value $Z[l, i, e, v, t] = b$ indicates that users from region rl download content k_i a total of b times from VM v at data center c_e at time t, where time is expressed in minutes.

Equation (2) shows the encoding of the S-Scheduling component of a solution for an instance with three data centers C_0, C_1, C_2 and six resources $K_0, \ldots, K_5$ for an hour h within the planning period $[0, H-1]$. Table 2 displays the assignment of content resources to data centers and the number of VMs used at each data center in the given example.

$$
\tilde{Y} = \begin{pmatrix} 3 \\ 3 \\ 5 \end{pmatrix} \begin{matrix} C_0 \\ C_1 \\ C_2 \end{matrix} \qquad
X = \begin{matrix} & \begin{matrix} C_0 & C_1 & C_2 \end{matrix} & \\ & \begin{pmatrix} 1 & 0 & 1 \\ 0 & 1 & 0 \\ 0 & 1 & 1 \\ 1 & 0 & 0 \\ 0 & 1 & 0 \\ 0 & 0 & 1 \end{pmatrix} & \begin{matrix} K_0 \\ K_1 \\ K_2 \\ K_3 \\ K_4 \\ K_5 \end{matrix} \end{matrix}
$$
$$
Y = \begin{matrix} \begin{matrix} 0 & \cdots & h & \cdots & H-1 \end{matrix} & \\ \begin{pmatrix} \cdots & \cdots & 5 & \cdots & \cdots \\ \cdots & \cdots & 3 & \cdots & \cdots \\ \cdots & \cdots & 4 & \cdots & \cdots \end{pmatrix} & \begin{matrix} C_0 \\ C_1 \\ C_2 \end{matrix} \end{matrix} \tag{2}
$$

Table 2. VMs and resources allocated to each data center in the example representation of the S-Scheduling component of the solution.

Data Centers	$\tilde{Y}$	Y^h	$\max\{0, Y^h - \tilde{Y}\}$	Assigned Resources
C_0	3	5	2	K_0, K_3
C_1	3	3	0	K_1, K_2, K_4
C_2	5	4	0	K_0, K_5, K_2

4.3. Recombination Operator

The recombination operator is based on uniform crossover, with specific adaptations included to take the problem constraints into account. The recombination operator receives two feasible solutions, performs the crossover operation on the S-Scheduling component of each solution, and returns two feasible solutions.

For each data center, the recombination operator exchanges the content resource allocation, the total VM allocation, and the proportion of VMs that are assigned on demand. Each attribute is randomly selected with uniform distribution (probability 0.5) to be exchanged between the data centers of the solutions.

Algorithm 1 shows the pseudocode of the recombination operator. Given two solutions to recombine, the first loop (lines 1–6) exchanges the allocation of content resources to data centers. For each resource $k_i \in K$ selected with uniform distribution (probability 0.5) and for each data center $c_e \in C$, the loop exchanges the allocation of resource k_i to data center c_e between the two solutions participating in the crossover (line 4). Because the described exchange is performed for all data centers $c_e \in C$, no infeasible solutions are generated due to lack of allocation of any resource. In the second loop (lines 7–12), each scheduling hour $h \in [0, H]$ is selected with uniform distribution (probability 0.5) and the assignment of VMs for all data centers in C is swapped between the two solutions (line 10). Because the described swapping is performed for all the data centers $c_e \in C$, no infeasible solutions are generated due to not being able to cover the demand for VMs in a scheduling hour $h \in [0, H]$. The last loop of the algorithm (lines 13–16) selects each data center $c_e \in C$ with uniform distribution (probability 0.5) and exchanges the reserved VM assigned in c_e between the solutions involved in the crossing (line 15).

For all data centers $c_e \in C$, the following condition must hold: the total number of VMs assigned must always be greater than or equal to the number of VMs reserved. Therefore, the exchanges made in the last two loops of the algorithm can generate unfeasible solutions. Thus, line 17 executes the algorithm for correcting solutions, as described in Section 4.5.

Algorithm 1 Recombination operator

Input: solutions s_1 and s_2 to recombine
1: **for each** $k_i \in K$ **do**
2:     with probability 0.5
3:     **for each** $c_e \in C$ **do**
4:         exchange x_{ie} between solutions
5:     **end for**
6: **end for**
7: **for each** $h \in [0, H]$ **do**
8:     with probability 0.5
9:     **for each** $c_e \in C$ **do**
10:         exchange y_e^h between solutions s_1 and s_2
11:     **end for**
12: **end for**
13: **for each** $c_e \in C$ **do**
14:     with probability 0.5
15:     exchange $\tilde{y}_e$ between solutions s_1 and s_2
16: **end for**
17: execute algorithm for correcting solutions

4.4. Mutation Operator

The mutation operator receives a feasible solution, performs modifications on the S-Scheduling component, and returns a feasible solution. The mutation operator changes the allocation of content resources to data centers, taking into account that no content resources are left unallocated. Regarding the allocation of VMs to data centers, the mutation operator produces changes in the solution that vary the proportion of reserved and on-demand VMs that the data center uses to cover the demand for content resources.

Algorithm 2 describes the mutation operator. The first loop of the algorithm (lines 1–10) changes the allocation of resources to data centers. For each resource $k_i \in K$, data centers $c_e \in C$ are selected with uniform distribution (probability p_M). If the resource is not assigned to the selected data center, then the resource is assigned to the data center (line 5). If it is already assigned, the resource is unassigned to the data center after checking that the resource is assigned to at least one other data center in C (line 7). The check that the resource is assigned to another data center is performed to avoid generating an infeasible solution due to lack of resource allocation. The second loop of the algorithm (lines 11–23) modifies the allocation of VMs to data centers in the solution. For each scheduling hour $h \in [0, H]$, the loop selects data centers $c_e \in C$ with uniform distribution (probability p_M). If data center c_e has y_e^h VMs assigned to it, the loop selects a uniformly distributed $MVcant$ value in the set $[1, y_e^h]$ (line 15), then removes $MVcant$ number of VMs assigned to that data center at hour h (line 17) and assigns them at hour h to another data center C_f selected with uniform probability (line 18). If data center c_e has no VMs assigned, then the loop assigns a VM to it (line 20). Because the described mutation does not modify the total number of VMs assigned to the data centers in each planning hour h, no infeasible solutions will be generated due to not being able to cover the demand for VMs in hour h.

The third loop (lines 24–31) modifies the number of VMs reserved in each data center $c_e \in C$ selected with uniform distribution (probability p_M). If the selected data center has $\tilde{y}_e$ reserved VMs, a value with uniform distribution $cantMV \in [0, \tilde{y}_e - 1]$ is selected and assigned as the new value of reserved VMs of data center c_e (line 27). If the selected data center does not have a reserved VM, then one is assigned to it (line 28). Modifying the number of VMs assigned and reserved in a data center $c_e \in C$ can generate unfeasible solutions. Therefore, line 32 executes the algorithm for correcting solutions, as described in Section 4.5.

Algorithm 2 Mutation operator

Input: solution s to mutate
1: **for each** $k_i \in K$ **do**
2:   **for each** $c_e \in C$ **do**
3:     with probability p_M
4:     **if** $x_{ie} \leftarrow 0$ **then**
5:       $x_{ie} \leftarrow 1$
6:     **else if** $x_{ie} \leftarrow 1$ and $\exists\, c_f \in C / x_{if} \leftarrow 1$ **then**
7:       $x_{ie} \leftarrow 0$
8:     **end if**
9:   **end for**
10: **end for**
11: **for each** $h \in [0, H]$ **do**
12:   **for each** $c_e \in C$ **do**
13:     with probability p_M
14:     **if** $y_e^h > 0$ **then**
15:       select $cantMV \in [1, y_e^h]$
16:       select a data center $c_f \neq c_e \in C$
17:       $y_e^h \leftarrow y_e^h - cantMV$
18:       $y_f^h \leftarrow y_f^h + cantMV$
19:     **else**
20:       $y_e^h \leftarrow 1$
21:     **end if**
22:   **end for**
23: **end for**
24: **for each** $c_e \in C$ **do**
25:   with probability p_M
26:   **if** $\tilde{y}_e > 0$ **then**
27:     $\tilde{y}_e \leftarrow cantMV$, with $cantMV \in [0, \tilde{y}_e - 1]$
28:   **else**
29:     $\tilde{y}_e \leftarrow 1$
30:   **end if**
31: **end for**
32: execute Algorithm 3 for correcting solutions

4.5. Correction of Infeasible Solutions

According to the problem constraints, for all data centers $c_e \in C$, the total number of allocated VMs must always be greater than or equal to the number of reserved VMs. Therefore, applying the recombination or mutation operations may result in infeasible solutions. The correction algorithm operates on infeasible solutions, modifying the number of reserved VMs in the data centers to ensure that the solutions satisfy the feasibility condition.

Algorithm 3 presents the solution correction method. The loop modifies the number of reserved VMs in each data center to make the solution feasible. For each data center $c_e \in C$, the value of mc is computed as the minimum number of VMs used in the data center over the entire planning period (line 2). If mc is lower than the number of reserved VMs in the data center ($\tilde{y}_e$), then mc reserved VMs are assigned to the data center (line 4).

Algorithm 3 Correction of infeasible solutions

Input: infeasible solution s_U to mutate
1: **for each** $c_e \in C$ **do**
2: $\quad mc \leftarrow$ min. number of VMs used in c_e
3: $\quad$ **if** $\tilde{y}_e > mc$ **then**
4: $\quad\quad \tilde{y}_e \leftarrow mc$
5: $\quad$ **end if**
6: **end for**

4.6. Routing Algorithms

The Content Resource Request Routing (CRP) algorithm defines the routing phase of the MOEA. Considering the VM and content resource allocation specified in S-Planning, the CRP algorithm assigns each content resource request to a data center while respecting the problem constraints.

To assign resource requests to data centers, the CRP algorithm selects between two resource request allocation strategies with probability 0.5 and uniform distribution, choosing either a strategy that prioritizes cost (BestCostRoute) or a strategy that prioritizes QoS (BestQoSRoute). After selecting an allocation strategy, the CRP applies it in assigning all content resource requests.

The BestCostRoute and BestQoSRoute strategies apply more than one greedy method for allocating content resources. If a method fails to allocate all content resources, then the allocations made thus far are discarded and the next method with more relaxed restrictions is applied. In each strategy, the methods are ordered from the most restrictive to the least restrictive. In this way, the methods are ordered in decreasing order (measured by cost or QoS) of the solution they build.

4.6.1. BestCostRouting: Cost-Prioritized Routing Strategy

BestCostRouting has two methods for assigning content resources, namely, m1Cost and m2Cost. Both methods select the cheapest VM that meets the imposed conditions. m1Cost is the most restrictive strategy; as a necessary condition for assigning a content request, it considers that the data center must have the content resource to meet the request. In order to optimize the use of VMs, it prioritizes VMs that are already being used by the same content provider that has the resource associated with the request. If a VM used by the provider is not found, then a free VM is selected. m2Cost relaxes the above-mentioned restriction. Because m2Cost does not follow the allocation of resources to data centers in S-Planning, this component must be corrected.

Algorithm 4 presents the pseudocode of the BestCostRouting. At each time t, the function getVMBestCost $(t, p, add\_constraint)$ is applied to search for the VM with the best cost that meets the three constraints of the problem along with the additional constraint *add_constraint*.

The first loop of the algorithm (lines 1–17) implements the m1Cost assignment method. For each content request at each time t, m1Cost iteratively searches for the best cost VM using the function getVMBestCost (line 6). Here, *add_constraint* ensures that the VM returned by the function is being used by the content provider associated with the request that the VM will serve. Additionally, the constraint ensures that the data center has the resource associated with the request. If the function does not find a suitable VM, then the search is repeated (line 9) while relaxing *add_constraint*. If a suitable VM is found in either of the two searches, then the request is assigned to that VM. If no VM is found in the searches, then m1Cost ends and the m2Cost method is executed.

Algorithm 4 Best cost routing algorithm

1: **for each** $t \in [0, T]$ **do** ▷ m1Cost method
2:     **for each** $p \in Request(t)$ **do**
3:         k_p: resource of p
4:         pro_k: provider of k_p
5:         *add_constraint*← datacenter has resources k_p and pro_k using mv
6:         mv←getVMBestCost($t, p, add\_constraint$)
7:         **if** no mv found **then**
8:             *add_constraint*← datacenter has resource k_p
9:             mv←getVMBestCost($t, p, actual\_constrain$)
10:             **if** mv found **then**
11:                 assign request p to mv
12:             **else**
13:                 terminate m1Cost
14:             **end if**
15:         **end if**
16:     **end for**
17: **end for**
18: **if** $\exists p \in Request(t)$ not assigned **then** ▷ m2Cost method
19:     **for each** $t \in [0, T]$ **do**
20:         **for each** $p \in Request(t)$ **do**
21:             k_p: resource of p
22:             pro_k: provider of k_p
23:             *add_constraint*← pro_k using mv
24:             mv←getVMBestCost($t, p, add\_constraint$)
25:             **if** no mv found **then**
26:                 *add_constraint*←∅
27:                 mv←getVMBestCost($t, p, add\_constraint$)
28:             **end if**
29:             assign request p to mv
30:             assign k_p to data center of mv
31:         **end for**
32:     **end for**
33: **end if**

The second loop of the algorithm (lines 19–32) implements m2Cost. When applying m2Cost, none of the content request assignments made in m1Cost are considered; the assignment process starts anew from the first request. m2Cost iterates over the content resource requests for each time t. The initial search calls the getVMBestCost($t, p, add\_constraint$) function (line 24), with *add_constraint* requiring that the VM returned by the function is being used by the content provider associated with the request. In this case, the function may return a VM located in a data center that does not have the assigned content resource. If the function does not find a suitable VM meeting the imposed condition, then m2Cost searches for a VM without an associated *add_constraint*; the VM only needs to satisfy the problem constraints.

Due to the feasibility of the *S-Scheduling* solution component, the m2Cost method always finds a VM for each content request. Therefore, the request is assigned to the found VM in the last line of the algorithm, and the content resource associated with the request is allocated to the data center of the VM if it was not already assigned.

4.6.2. BestQoSRouting: QoS-Prioritized Routing Strategy

BestQoSRouting applies four methods for assigning content resource requests:

- m1QoS: Assigns a content request to a VM considering the constraint that the VM must be in a data center with the assigned content resource related to the request.
- m2QoS: Extends m1QoS by adding the constraint to prioritize VMs already in use by the content provider associated with the request.
- m3QoS: Removes the constraint from m1QoS
- m4QoS: Removes the constraint from m2QoS.

Methods m3QoS and m4QoS do not adhere to the resource allocation to the data centers as represented in the S-Scheduling component of the solution. Therefore, the BestQoS-Routing strategy must correct this component in the event of a content resource request being assigned to a data center that does not have the associated resource.

Algorithm 5 presents the pseudocode for the BestQoSRouting strategy. The getVMBestQoS($t, p, add\_constraint$) function is applied to search for the VM with the best QoS at each time t, satisfying the constraints of the problem along with the additional constraint *add_constraint*, which varies depending on which method is being used.

The first loop (lines 1–9) implements the m1QoS method. For each content request at each time t, it iteratively searches for the VM with the best QoS using the function getVMBestQoS (line 4) while requiring that the VM belongs to a data center with the resource associated with the request. If a VM is found, then the request is associated with that VM. If no VM is found, the m1QoS method ends and the m2QoS method is executed next.

The second loop (lines 12–23) implements the m2QoS method. This method does not consider any of the content request assignments made in m1QoS; the assignment process starts anew from the first request. Similar to m1QoS, m2QoS iterates over content resource requests for each time t and searches for the VM with the best QoS using the function getVMBestQoS (line 14), requiring that the VM is being used by the content provider associated with the request that will be served by that VM. Additionally, the constraint requires that the data center has the resource associated with the request. If no VM meeting the condition is found, then the search is repeated (line 16) while removing from *add_constraint* the requirement that the VM must be in use by the content provider associated with the request. If a VM is found in either of these two searches, the request is associated with it; if no VM is found in the searches, then the m2QoS method ends and the m3QoS method is executed next.

The m3QoS method has the same structure as m1QoS, with two differences: (i) during invocation of the getVMBestQoS (line 29) function, the requirement that the data center must have the resource associated with the request is not considered, and (ii) it modifies the S-Scheduling component of the solution when assigning the content resource request to a data center that does not have the associated resource (line 31). Similarly, the m4QoS method has the same structure as m2QoS, with two exceptions: (i) The getVMBestQoS function (lines 41 and 44) removes the requirement that the data center must have the resource associated with the request, and (ii) it modifies the S-Scheduling component of the solution when assigning the content resource request to a data center that does not have the associated resource (line 47).

Algorithm 5 Best QoS routing algorithm

```
1: for each t ∈ [0, T] and p ∈ Request(t) do                          ▷ m1QoS method
2:     k_p: resource of p
3:     add_constraint← datacenter has resource k_p
4:     if mv←getVMBestQoS(t, p, add_constraint) then
5:         assign request p to mv
6:     else
7:         terminate m1QoS
8:     end if
9: end for
10: if ∃p ∈ Request(t) not assigned then                               ▷ m2QoS method
11:     for each t ∈ [0, T] and p ∈ Request(t) do
12:         k_p: resource of p and pro_k: provider of k_p
13:         add_constraint← pro_k using mv and data center has resource k_p
14:         if not mv←getVMBestQoS(t, p, add_constraint then
15:             add_constraint ← data center has resource k_p
16:             mv ← getVMBestQoS(t, p, add_constraint)
17:         end if
18:         if mv found then
19:             assign request p to mv
20:         else
21:             terminate m2QoS
22:         end if
23:     end for
24: end if
25: if ∃p ∈ Request(t) not assigned then
26:     for each t ∈ [0, T] and p ∈ Request(t) do                      ▷ m3QoS method
27:         k_p: resource of p
28:         add_constraint← ∅
29:         if mv←getVMBestQoS(t, p, add_constraint) then
30:             assign request p to mv
31:             assign request k_p to data center of mv
32:         else
33:             terminate m3QoS
34:         end if
35:     end for
36: end if
37: if ∃p ∈ Request(t) not assigned then
38:     for each t ∈ [0, T] and p ∈ Request(t) do                      ▷ m4QoS method
39:         k_p: resource of p and pro_k: provider of k_p
40:         add_constraint← pro_k using mv
41:         mv←getVMBestQoS(t, p, add_constraint)
42:         if no mv found then
43:             add_constraint← ∅
44:             mv←getVMBestQoS(t, p, add_constraint)
45:         end if
46:         assign request p to mv
47:         assign request k_p to data center of mv
48:     end for
49: end if
```

4.6.3. Implementation Details

The proposed MOEA was implemented using the JMetal framework [44]. JMetal aims at the development, experimentation, and study of metaheuristics to solve multiobjective optimization problems. JMetal is developed in Java using the object-oriented paradigm. JMetal provides a set of classic and modern optimizers, a collection of benchmark problems, and a set of quality indicators for use in evaluating the quality and computational efficiency of algorithms. With JMetal, it is possible to configure and execute complete experimental studies that generate statistical information about the obtained results. The framework also takes advantage of multi-core processing to accelerate the execution time of experiments.

5. Baseline Optimization Methods for Comparison

In this section, we present the baseline optimization algorithms for comparison with the MOEA, which include a set of three heuristics methods and an exact resolution approach.

5.1. Heuristics

Three heuristics are used for the cloud-based CDN design problem: greedy cost, greedy QoS, and round robin. To clarify their implementations, the general procedure common to all heuristics is presented first, followed by a discussion of their specific features.

5.1.1. General Procedure

The general procedure of the heuristics is outlined in Algorithm 6. The heuristics iterate over content resource requests for each time instant. For every content request, assignment to a data center is made based on a predefined criterion (line 3). Then, the heuristics check for an available VM in the selected data center with sufficient capacity to assign the request. If no VM with sufficient capacity is found, a new VM is added to the selected data center. The type of VM added depends on whether the time instant corresponds to a demand peak; either an on-demand VM is rented for a demand peak (lines 8 and 9), or a VM is reserved for the entire planning horizon (lines 10 and 11).

A time instant is classified as a demand peak if the required number of VMs at that instant meets or exceeds a threshold number of VMs, denoted as VM_{peak}. By adjusting the value of VM_{peak}, different subsets of time instants are identified as demand peaks, resulting in varied heuristic solutions. The value of VM_{peak} ranges from zero to the maximum number of VMs needed to meet demand at any time within the planning horizon $[0, T]$. To determine the number of required VMs at each time instant, content requests are analyzed in accordance with the constraints outlined in Section 2.3, i.e., that no two providers can share the same VM simultaneously and that a single VM cannot process more than a predefined number of concurrent content requests. Finally, dominated or repeated solutions are removed to construct the Pareto front for each heuristic.

For a given set of input parameters, each heuristic produces a unique solution. To facilitate comparison between the proposed baseline heuristics and the MOEA, multiple solutions are generated by executing each heuristic with different input parameters. Specifically, the variation is applied to the set of time instants considered as demand peaks.

Algorithm 6 General procedure for the baseline heuristics

```
for each t ∈ [0, T] do
    for each p ∈ Request(t) do
        Assign p to data center ce according to a predefined criterion
        Increase QoS, Cost^s, Cost^t.
        if there is a VM in ce_minCost with available resource for p then
            Assign p to VM
        else
            if t is marked as demand peak then
                Rent a new on-demand VM' in ce_minCost for the hour that includes time t
            else
                Reserve a VM' in ce_minCost for the whole planning horizon [0, T].
            end if
            Assign p to VM'
            Increase Cost^v.
        end if
    end for
end for
```

5.1.2. Specific Features of Heuristics

The specific features of each baseline heuristic are as follows.

Greedy Cost. The predefined criterion of the greedy cost is to assign each content request to the data center that offers the lowest cost while satisfying a minimum QoS threshold. Input parameters include the subset of time instants identified as demand peaks and the minimum QoS threshold. Greedy cost aims at obtaining cost-efficient solutions that fulfill a required minimum QoS threshold. To generate a range of solutions, in addition to the subsets of demand peak time instants, the QoS thresholds for each subset of demand peak time instants are also varied.

Greedy QoS. The predefined criterion of the greedy QoS is to assign the content request to the data center that provides the best QoS for each content request. The input parameter is the subset of time instants identified as demand peaks. The greedy QoS heuristic aims to obtain solutions that prioritize high QoS for users.

Round Robin. The predefined criterion of the round robin heuristic is to iteratively assign content requests to data centers in a sequential manner. Data centers are numbered from 0 to n, where n is the total number of data centers. Each content request is assigned to the next data center in the sequence. When a content request is assigned to data center n, the next content request is assigned to data center 0. The goal of the round robin heuristic is to distribute content requests evenly across data centers, avoiding imbalances where some data centers handle too many requests while others handle too few.

5.2. Exact Resolution Approach

AMPL [45] was used as the modeling language for the CDN design problem. CPLEX version 12.6.3.0 was used as the exact solver. The mathematical formulation results in an MILP model driven by the inclusion of discrete variables rv_c, $\overrightarrow{dv_c}$, $\overrightarrow{x_c}$, and $\mathbf{Z}_c$.

The ε-constraint method was applied to solve the resulting bi-objective problem. The ε-constraint is effective for generating a set of Pareto-optimal solutions in complex problems [46]. Equation (1a) remains the primary objective function, whereas Equation (1b) is transformed into a constraint, serving as a restricted objective (Equation (3)).

$$\begin{aligned} \min \quad & \sum_{c \in C} (Cost^v(c) + Cost^s(c) + Cost^t(c)) \\ \text{subject to:} \quad & \\ & \sum_{c \in C} inv(QoS(c)) \leq \varepsilon_{QoS} \\ & \text{other constraints of the problem} \end{aligned} \tag{3}$$

By adjusting the parameter ε_{QoS}, the model yields different compromise solutions, allowing us to explore the tradeoff between cost and user QoS. Initially, εQoS is set to the total number of content requests in the instance multiplied by the worst QoS value between a user region and a data center. Then, εQoS is progressively reduced by one unit until it reaches a minimum value of one.

6. Experimental Validation

This section describes the experimental evaluation of the proposed bio-inspired multi-objective optimization approach for cloud-based CDN design.

6.1. Methodology for the Evaluation

Our evaluation of the proposed MOEA involved three stages: parameter tuning, empirical evaluation, and comparison with exact and heuristic baseline methods.

Several important solutions for multiobjective problems were considered: the best solution in terms of cost, the best solution in terms of QoS, and the compromise solution that most equally weights the two objectives (i.e., the solution that is closest to zero according to the Euclidean norm).

Regarding metrics, we computed two relevant indicators to determine the quality of solution found by each studied method: RHV and EAS. RHV is a relevant metric in multiobjective optimization that assesses the two main purposes of a multiobjective optimization algorithm, namely, the proximity to the Pareto front and the diversity of the obtained solutions [47]. The RHV metric is the ratio between the volume determined by the Pareto front calculated for the resolution method and the volume determined by the reference Pareto front. The ideal value of the ratio is 1 [47]. The RHV metric was computed regarding an approximation of the optimal Pareto front built by gathering the nondominated solutions found in all executions of the studied methods, which is a common approach when solving multiobjective optimization problems with an unknown optimal Pareto front [48]. In turn, the attainment surface represents the set of ideal tradeoffs between conflicting objectives. Points in the attainment surface indicate the best achievable performance for each objective [49]. EAS is a statistical estimator of the attainment surface; for each vector in the objective space, it determines the probability of its being dominated by the computed Pareto front in a given execution of the studied algorithm. Unlike RHV, EAS does not depend on the (unknown) real Pareto front. The analysis is based on the 50%-EAS [49], which considers the estimated attainment surface obtained by at least 50% of multiple executions, i.e., it is analogous to the median in single objective optimization. The ideal value of the 50%-EAS metric is 1.

The methodology of Knowles and Corne [50] was adopted to compare the EAS of each studied method. First, a statistical pairwise comparison was performed by computing the 50%-EAS with a statistical confidence level of 95%. Then, two key metrics were evaluated: (i) the proportion of the area where there is statistical confidence that the compared algorithm is undefeated by any other algorithm (U), and (ii) the proportion of the area where there is statistical confidence that the compared algorithm outperforms all other algorithms in terms of 50%-EAS values (B). Unlike the comparison relying on the

RHV metric, this does not rely on the optimal Pareto front constructed for each specific problem instance.

In our empirical evaluation, we solved ten realistic instances to compute approximations to the Pareto front considering different tradeoff solutions. Due to the stochastic nature of MOEAs, 50 independent executions using different seeds for the pseudorandom number generator were performed for each problem instance. The distributions of the results were examined using the Kolmogorov–Smirnov statistical test to determine whether the samples followed a normal distribution. The Kruskal–Wallis statistical test was applied to determine the statistical significance of the differences between results.

To avoid biased results, parameter tuning experiments were performed on a set of smaller problem instances different from the validation instances. A set of relevant parameters that impact the search capabilities of the proposed MOEA were studied: recombination probability (p_R), mutation probability (p_M), and population size (#P).

Finally, a comparison with baseline optimization methods is reported. The comparison considered an exact approach using CPLEX, a traditional business-as-usual round robin method, and greedy heuristics focused on optimizing cost (greedy cost) and QoS (greedy QoS). To guarantee the statistical significance of the MOEA results over the baseline heuristics, the median (*med*) RHV was used as estimator and the Interquartile Range (IQR) of the RHV values was used to evaluate the dispersion. Improvements were considered statistically significant if $|median(RHV(MOEA)) - RHV(greedy)| > IQR(RHV(MOEA))$.

6.2. Evaluation Instances

A set of realistic instances of the cloud-based CDN design problem were generated using the GlobeTraff tool [51]. GlobeTraff is a traffic generator that allows realistic mixes of internet traffic to be created. GlobeTraff is able to generate various types of application traffic based on models presented in the literature, allowing for detailed parameterization of both the generated models and the composition of the resulting traffic mix. In each generated instance, the traffic generated by GlobeTraff was divided by geographic regions. For a more realistic traffic division, we used information provided in the report made by the Cisco company [52], in which a division of traffic by types and geographic areas is proposed.

Different scenarios were generated by varying the size of the traffic simulations created with GlobeTraff based on the execution time and the number of independent executions of the evaluated MOEA. The instance sizes for the experimental evaluation were adjusted using reference data consisting of the number of video requests over time obtained from statistics of the OpenFING project, an online educational video service at Engineering Faculty of the Universidad de la República, Uruguay (https://open.fing.edu.uy/).

Smaller instances were used for the MOEA parameters setting experiments. For the MOEA validation, the size of the instances was adjusted into three groups: small, medium, and large. For experiments solving the problem using AMPL with CPLEX, the size of the instances was adjusted based on the execution time and the memory available in the infrastructure used to run AMPL. For all the generated instances, the number of user regions was kept fixed at 30 and the number of data centers at 10, as explained in Section 6.2. The other characteristics are reported in Table 3.

Table 3. Details of problem instances for each stage in the experimental evaluation.

Stage	Instance	#Videos	#Min	#Providers	#Requests
Exact resolution using CPLEX	AMPL1	500	240	6	48
	AMPL2	300	300	6	303
	AMPL3	400	300	5	249
	AMPL4	400	420	5	547
MOEA parameters setting	A1	500	300	6	134
	C1	1000	240	7	364
	D1	400	300	5	249
	D2	4000	480	7	600
MOEA validation	S1	11,500	240	6	8219
	S2	11,500	240	6	6744
	M1	18,600	240	6	13,000
	M2	18,600	240	6	12,304
	L1	30,000	240	6	27,700
	L2	30,000	240	6	22,827

6.3. Development and Execution Platform

The MOEA and baseline heuristics were developed in the Java programming language using the JMetal framework [44]. The exact resolution algorithm was implemented in AMPL, and the problem was solved using CPLEX version 12.6.3.0.

The experimental evaluation was performed using the National Supercomputing Center (Cluster-UY), Uruguay [53]. The computing resources included a DELL Power Edge R720 server, two Intel Xeon E52650 2.00 GHz processors with eight cores each, 64 GB of RAM, 600 GB of disk storage, and the Linux CentOS operating system.

6.4. Parameters Setting Experiments

Parameter setting experiments were performed over instances A1, C1, D1, and D2 using up to 4000 videos, a planning horizon of 480 min, seven providers, and 600 video requests. For each instance, 50 independent executions were performed, with a stopping criterion of 1000 generations for each combination of three relevant parameters of the studied MOEA: p_R, p_M, and #P. The candidate values were $p_R \in \{0.5, 0.7, 0.9\}$, $p_M \in \{0.001, 0.01, 0.1\}$, and $\#P \in \{50, 100, 200\}$. The RHV metric was considered in the analysis.

The Kolmogorov–Smirnov statistical test was applied to analyze the result distributions. Based on the obtained p-values, not all instances followed a normal distribution for the RHV metric. Therefore, the median was chosen as an estimator of the RHV results.

For the four instances we used, a classification of each parameter combination was made according to its median RHV and 50%-EAS value. When the nonparametric Kruskal–Wallis test did not allow us to determine statistical significance with a confidence level of 95%, which would ensure a significant difference between the distribution of results of two parametric configurations, the execution time was used to determine the classification. For each parameter combination, the number of times each position was occupied in the different classifications was determined. A total number of 27 different configurations were studied. Table 4 reports the configurations on the first four positions of the ranking.

The results in Table 4 show that configurations with a population size equal to 200 individuals occupied the first four positions of the ranking more often. In turn, the best configurations had the values $p_C = 0.5$ and $p_M = 0.001$. The Kruskal–Wallis test for different population sizes yielded p-values less than 1×10^{-2}; therefore, the differences are assured with a statistical confidence level greater than 95%. For configurations with different values of p_C and p_M where the p-values could not guarantee statistical significance, the median of the MOEA execution time was used as a tiebreaker. The best parameter configuration we found was $\#P = 200$, $p_C = 0.5$, and $p_M = 0.001$.

Table 4. Number of times each parameter configuration occupied the top four positions in the ranking of best results for RHV, 50%-EAS, and execution time.

Configuration			RHV Position				50%-EAS Position				Time Position			
p_C	p_M	#P	1st	2nd	3rd	4th	1st	2nd	3rd	4th	1st	2nd	3rd	4th
0.5	0.010	200	2	1	1	0	1	1	1	0	0	0	1	0
0.5	0.001	200	2	1	1	0	2	1	1	0	0	0	1	1
0.7	0.001	200	0	1	0	2	0	1	0	2	0	1	1	0
0.7	0.010	200	0	1	2	0	0	1	1	0	0	0	0	1

6.5. *Validation Results and Comparison with Baseline Optimization Methods*

This subsection reports the results of the experimental evaluation. A comparison is presented with solutions obtained by exactly solving small instances of the cloud-based CDN design problem and with solutions computed using the greedy cost, greedy QoS, and round robin techniques.

6.5.1. Comparison of Results for Instances That Are Solvable Using an Exact Method

Table 5 reports the median results of the RHV and 50%-EAS metrics computed for each problem instance. For the proposed MOEA, the Interquartile Range (IQR) is included as a measure to assess the robustness and spread of the results. The greedy QoS and round robin techniques are not included in the RHV and 50%-EAS comparison, as they compute a single solution.

Table 5. RHV (±IQR for the proposed MOEA) and 50%-EAS results.

	RHV			50%-EAS					
	CPLEX	MOEA	Greedy Cost	CPLEX		MOEA		Greedy Cost	
				U	B	U	B	U	B
AMPL1	1.00	0.98 ± 0.01	0.74	100%	100%	12%	3%	0%	0%
AMPL2	1.00	0.94 ± 0.01	0.69	100%	100%	7%	2%	0%	0%
AMPL3	1.00	0.94 ± 0.01	0.68	100%	100%	10%	4%	0%	0%
AMPL4	0.82	0.95 ± 0.01	0.75	20%	16%	82%	73%	0%	0%

The results in Table 5 demonstrate that the proposed MOEA computed high-quality solutions for small size problem instances. CPLEX solved the problem to optimality and computed the optimal RHV value on instances AMPL1 and AMPL3 (the ideal value of 1) and 0.99 in instance AMPL2. In those instances, the MOEA computed solutions with a maximum difference of only 0.06. Different results were computed for problem instance AMPL4, where the MOEA computed a RHV value 0.13 better than CPLEX. The greedy cost heuristic computed the worst RHV results for all instances. The MOEA improved over greedy cost by up to 0.26 in problem instance AMPL4. The RHV values computed by the proposed MOEA demonstrate strong convergence towards the reference Pareto front and show good diversity within the calculated solutions. The statistical significance of the difference between the median RHV of the proposed MOEA and the RHV values of the other methods is confirmed for all problem instances. A similar trend is observed for 50%-EAS values, showing that the proposed MOEA improved significantly over the greedy cost method and computed better values of the metric in problem instance AMPL4, with a large percentage of both undefeated and outperforming solutions.

Table 6 reports the compromise solution and best solutions regarding each problem objective computed by each studied method for the four problem instances that were solved exactly. The best values regarding cost and QoS are marked in bold.

Table 6. Comparison of results, showing the compromise solution, best cost solution, and best QoS solution using the studied methods for exactly solvable instances.

		Compromise Solution		Best Cost Solution		Best QoS Solution	
Instance	Method	Cost	inv(QoS)	Cost	inv(QoS)	Cost	inv(QoS)
AMPL1	CPLEX	0.41	3876.24	**0.13**	7380.25	0.98	**3036.11**
	MOEA	0.43	3922.67	0.15	7864.66	1.02	**3036.11**
	greedy QoS	1.07	3036.11	1.07	3036.11	1.07	**3036.11**
	greedy cost	0.51	5812.83	0.19	7380.25	0.67	5394.00
	Round Robin	1.64	10,642.48	1.64	10,642.48	1.64	10,642.48
AMPL2	CPLEX	0.59	24,504.13	**0.22**	45,745.74	1.38	18,048.38
	MOEA	0.61	29,993.89	**0.22**	45,745.74	2.00	17,789.23
	greedy QoS	2.35	17,232.70	2.35	17,232.70	2.35	**17,232.70**
	greedy cost	1.11	35,634.35	0.44	45,745.74	1.37	34,226.50
	Round Robin	2.62	67,629.22	2.62	67,629.22	2.62	67,629.22
AMPL3	CPLEX	0.59	20,047.25	**0.22**	37,782.11	1.50	13,649.30
	MOEA	0.77	20,372.47	**0.22**	37,782.11	1.83	13,970.86
	greedy QoS	2.20	13,474.58	2.20	13,474.58	2.20	**13,474.58**
	greedy cost	0.95	29,172.80	0.44	37,782.11	1.08	28,609.66
	Round Robin	2.64	56,029.40	2.64	56,029.40	2.64	56,029.40
AMPL4	CPLEX	0.52	59,557.83	**0.34**	83,139.64	0.52	59,557.83
	MOEA	1.04	51,339.93	**0.34**	83,139.64	2.77	32,958.34
	greedy QoS	2.98	31,802.58	2.98	31,802.58	2.98	**31,802.58**
	greedy cost	1.39	63,223.03	0.45	83,139.64	1.71	60,829.69
	Round Robin	3.54	121,857.86	3.54	121,857.86	3.54	121,857.86

Regarding the best cost solutions, Table 6 shows that the MOEA matched the value of CPLEX in both cost and QoS for instances AMPL2, AMPL3, and AMPL4. Only in instance AMPL1 did the MOEA reported a slightly worse value for the value of cost compared to CPLEX. Regarding the best QoS solutions, greedy QoS always obtained the best QoS value, as it is strongly biased towards that problem objective. The MOEA computed the same QoS as greedy QoS on AMPL1 instance, but the cost value was 4.8% better. In instances AMPL2 and AMPL4, the MOEA computed the second-best QoS value, but with significantly better cost values (up to 14.9% better in instance AMPL2). Overall, the MOEA computed accurate results with better tradeoff values between the problem objectives. In the particular case of AMPL4, CPLEX obtained a solution that was significantly worse than the MOEA solution in terms of QoS. This occurred because CPLEX failed to find solutions when the ε_{QoS} value was set to low levels. In fact, CPLEX was unable to find a feasible solution even after an entire day of computation.

The proposed MOEA simultaneously improved in terms of cost and QoS over greedy cost for all instances. The average cost improvement was 36.4%, with values between 21.1% (for AMPL1) and 50.0% (for AMPL2 and AMPL3). QoS improvements were even more significant, with an average of 47.2%, minimum of 43.7% (for AMPL1) and maximum of 51.2% (for AMPL3). Even larger improvements were obtained over the round robin technique, between three and four times in terms of QoS and reducing the cost up to 63.6%, with an average improvement of 45.7$. These results imply a significant improvement in both objective functions over a business-as-usual (BaU) strategy that does not apply an intelligent resource planning.

Table 7 reports the execution times of the studied methods (in seconds). The proposed MOEA computes solutions efficiently, with execution times of around two minutes for all instances. While the greedy methods are faster, their results are not accurate. The execution times of the exact method were over two orders of magnitude higher than the execution time of the MOEA for instances AMPL1 to AMPL3, and over three orders of magnitude higher for instance AMPL4. These results demonstrate the efficiency of the proposed MOEA for computing accurate solutions with reduced execution times.

Table 7. Execution times of the studied methods (in seconds).

Instance	CPLEX	MOEA				Greedy Cost/Greedy QoS
		Avg.	σ	Min.	Max.	
AMPL1	18,414.79	119.71	10.61	108.23	130.96	<1.0
AMPL2	30,332.62	158.39	13.33	144.09	171.96	<1.0
AMPL3	66,421.85	156.66	15.01	140.80	173.62	<1.0
AMPL4	120,650.66	227.98	17.20	210.18	246.81	<1.0

Figure 3 presents the Pareto fronts generated by the studied methods for instances AMPL1 to AMPL4. The Pareto fronts were generated by gathering the nondominated solutions computed by each proposed optimization method in all the performed executions.

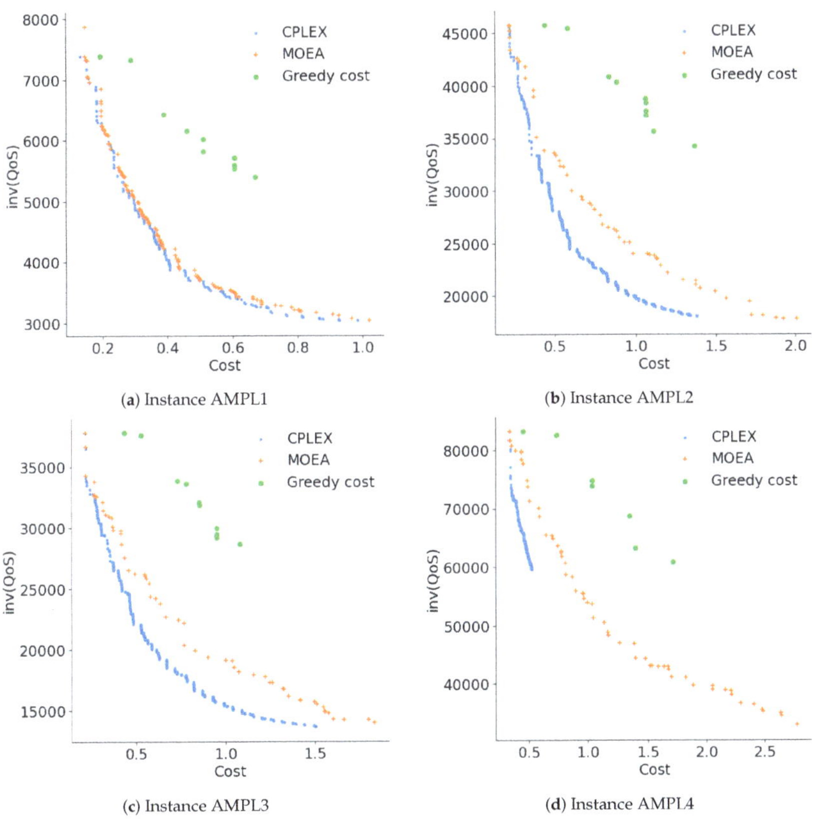

(**a**) Instance AMPL1 (**b**) Instance AMPL2 (**c**) Instance AMPL3 (**d**) Instance AMPL4

Figure 3. Pareto fronts computed using the studied methods.

Figure 3 shows that the proposed MOEA generated more diverse Pareto fronts compared to greedy cost, which tended to concentrate on solutions with extreme cost values. The overall Pareto front for each instance consists of solutions from CPLEX and MOEA. Except for instance AMPL4, a significant portion of the reference Pareto front comprises solutions from CPLEX, followed by the MOEA, and finally the greedy methods. The MOEA managed to cover a wider range of solutions in the QoS component, particularly on instance

AMPL4, where it formed the majority of the reference Pareto front. Figure 3d shows that CPLEX was unable to obtain feasible solutions with good QoS values for instance AMPL4.

Figure 4 presents the 25%, 50%, and 75% attainment surfaces of representative executions of the studied methods for instances AMPL1 to AMPL4. In general, the RHV and 50%-EAS results show that CPLEX struggled to cover a wide range of solutions in the QoS component, with greedy cost also facing this issue to a lesser extent. On the other hand, the MOEA stands out for its diverse solutions, being closer or directly forming the reference Pareto front when compared to the greedy method. In general, it can be concluded that the MOEA yields good results in the best solutions according to QoS.

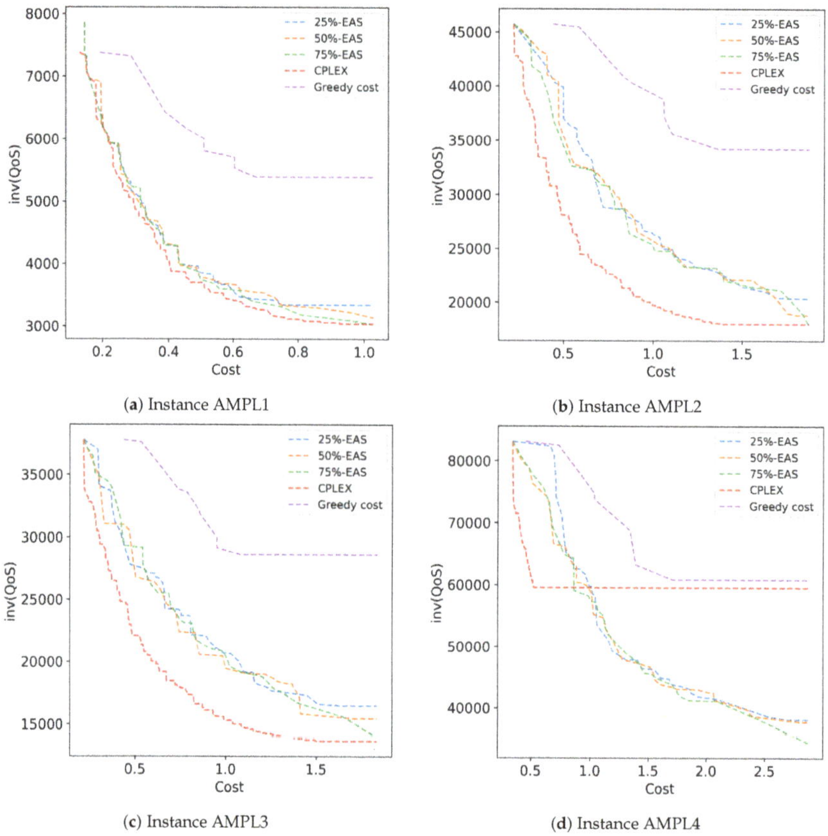

(**a**) Instance AMPL1 (**b**) Instance AMPL2 (**c**) Instance AMPL3 (**d**) Instance AMPL4

Figure 4. Attainment surfaces computed by the studied methods for instances AMPL1 to AMPL4.

6.5.2. Comparison of Results for Instances That Are Not Solvable Using an Exact Method

Table 8 reports the RHV and 50%-EAS results, cost/inv(QoS) values ($\times 10^3$) of the compromise solution, and best solutions regarding each problem objective computed by each studied method.

The best values regarding cost and QoS are marked in bold. RHV results are also reported for the MOEA and greedy cost. Greedy QoS and round robin are not included in the RHV comparison, as they compute a single solution.

Table 8. Comparison of results, showing RHV, 50%-EAS, compromise solution, best cost solution, and best QoS solution using the studied methods for instances that are not exactly solvable.

I	Method	RHV	50%-EAS	Compromise Solution		Best Cost Solution		Best QoS Solution	
				Cost	inv (QoS)	Cost	inv (QoS)	Cost	inv (QoS)
S1	MOEA	0.97 ± 0.02	85–79%	2.50	585.984	**0.93**	1232.226	4.20	466.690
	Greedy cost	0.65	4–1%	5.29	953.486	3.70	1232.226	5.77	906.182
	Greedy QoS	-	-	8.24	459.401	8.24	459.401	8.24	**459.401**
	Round Robin	-	-	11.58	1852.399	11.58	1852.399	11.58	1852.399
S2	MOEA	0.98 ± 0.01	88–82%	2.17	454.303	**0.77**	984.215	3.86	350.421
	Greedy cost	0.67	3–0%	2.92	772.895	1.47	984.215	3.26	745.020
	Greedy QoS	-	-	5.46	345.948	5.46	345.948	5.46	**345.948**
	Round Robin	-	-	7.88	1458.439	7.88	1458.439	7.88	1458.439
M1	MOEA	0.97 ± 0.01	82–75%	3.42	924.831	**1.32**	1911.590	5.31	739.137
	Greedy cost	0.64	4–2%	6.98	1485.437	4.86	1911.590	7.34	1427.574
	Greedy QoS	-	-	10.58	724.694	10.58	724.694	10.58	**724.694**
	Round Robin	-	-	15.62	2891.293	15.62	2891.293	15.62	2891.293
M2	MOEA	0.99 ± 0.01	90–86%	2.96	984.155	**1.26**	1820.476	5.47	682.330
	Greedy cost	0.67	1–0%	4.73	1418.478	3.00	1820.476	5.23	1364.557
	Greedy QoS	-	-	8.04	681.630	8.04	681.630	8.04	**681.630**
	Round Robin	-	-	10.90	2734.637	10.90	2734.637	10.90	2734.637
L1	MOEA	0.98 ± 0.01	91–87%	4.11	1857.605	**1.79**	4026.547	6.34	1510.223
	Greedy cost	0.58	0–0%	20.52	3150.290	18.62	4026.547	20.96	3041.886
	Greedy QoS	-	-	26.89	1491.477	26.89	1491.477	26.89	**1491.477**
	Round Robin	-	-	29.75	6135.782	29.75	6135.782	29.75	6135.782
L2	MOEA	0.99 ± 0.01	95–92%	3.45	1617.467	**1.57**	3396.253	5.67	1288.672
	Greedy cost	0.59	0-0%	11.77	2634.308	9.79	3396.253	12.34	2554.905
	Greedy QoS	-	-	16.33	1270.953	16.33	1270.953	16.33	**1270.953**
	Round Robin	-	-	20.13	5097.385	20.13	5097.385	20.13	5097.385

Regarding RHV values, the MOEA computed significantly better results than greedy cost; the average improvement was 34.7% and the best improvement in the large problem instances was 40.0%. Following the trend detected in the comparison between the MOEA and the CPLEX exact solver, the multiobjective evolutionary search showed proper convergence to the reference Pareto front; at the same time, the computed solutions showed good diversity. The RHV results were highly robust in all experiments, fulfilling the statistical significance criterion described in Section 6.1. Overall, the MOEA computed significantly better solutions than the greedy cost method.

The 50%-EAS results confirm that MOEA computed better solutions that the other methods, with significant improvements in both the proportion of the area where there is statistical confidence that MOEA solutions are undefeated by any other algorithm (U) and the proportion of the area where there is statistical confidence that the compared algorithm outperforms all other algorithms in terms of 50%-EAS values (B).

Regarding the best cost solutions, MOEA obtained the best cost for all instances. The MOEA was able to properly sample the region of the Pareto front associated with low cost values while maintaining accurate QoS values. The improvements provided by the MOEA were between 47.6% and 93.9%. Regarding the second-best greedy cost method, the improvements were between 47.6% and 90.4%. In absolute values, these improvements allow the cost to be reduced by between half (for small instance S2) and one sixth (for large instance L2) over the greedy cost heuristic. The best improvements were computed for the larger instances, demonstrating the scalability of the multiobjective evolutionary approach.

Regarding the best QoS solutions, the greedy QoS heuristic computed solutions with the same value of inv(QoS) and different cost values. Therefore, solutions with the same value of inv(QoS) turned out to be dominated and were removed from the solution set.

The improvements provided by the MOEA were between 48.5% and 75.9%. Regarding the greedy QoS method, the improvements in cost were between 84.3% and 93.3%, and MOEA was slightly lower in QoS (between 0.1% and 1.99%). The best QoS values were computed for the larger instances, again demonstrating the scalability of the proposed MOEA.

For the compromise solutions, the MOEA significantly improved in both cost and QoS over the greedy cost and round robin techniques for all instances. When compared with greedy QoS, the MOEA only improved regarding the cost objective.

Figure 5 shows the Pareto fronts generated for instances that are not solvable by exact methods. The graphics show that the round robin technique generated a solution that is very far from the reference Pareto front. The reference Pareto front is composed by solutions computed by the MOEA plus a single nondominated solution computed by greedy QoS. MOEA was the technique with the greatest diversity of solutions. Regarding diversity, the second-best method was greedy cost; however, the graphics show that the computed solutions were not close to the reference Pareto front. In general terms, Figure 5 indicates that the MOEA generated more diverse and better quality solutions than the rest of the techniques studied in the comparative analysis.

The diversity of the solutions computed by the MOEA captures different tradeoffs between the problem objectives. The diverse solutions provide different options for CDN design, allowing for fast deployment and configuration to fulfill specific QoS demands.

Overall, the comparative analysis allows us to conclude that the MOEA is able to compute accurate solutions that represent an improvement over baseline methods. The improvements on the cost objective stand out, where the MOEA is able to compute cost-effective solutions to the cloud-based CDN design problem. Improvements in QoS are also significant when considering that almost the same QoS of the greedy QoS method is obtained with a significantly lower cost. The most significant factor when comparing with other methods to solve the problem is that the MOEA shows a much greater diversity of solutions, with different tradeoffs between the problem objectives. These solutions provide different suggestions to decision-makers, allowing them to properly balance cost investments and inv(QoS) level according to specific needs.

Figure 6 presents the 25%, 50%, and 75% attainment surfaces of representative executions of the studied methods for the six instances solved in the comparative analysis.

The MOEA demonstrates its ability to compute competitive results compared to well-known heuristics and exact solvers, and is integrable into online systems used to process real-time data. This study represents an essential first step before integrating the MOEA into an online framework. In addition to this, the contributions of this article also include managerial insights. By leveraging recurring demand patterns, the offline solutions obtained from the MOEA can provide decision-makers with valuable information on key problem parameters. For instance, the MOEA can help to identify the extreme values of system costs, the level of QoS provided to users, and the approximate range of both objectives in tradeoff solutions.

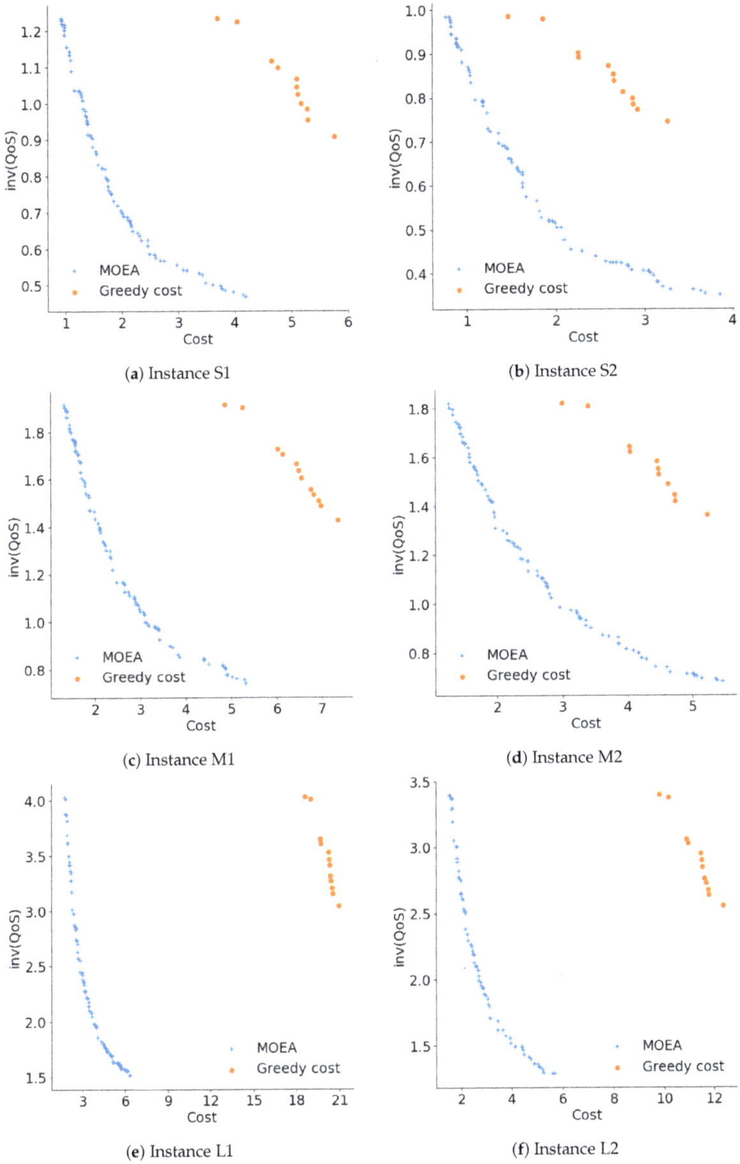

(**a**) Instance S1 (**b**) Instance S2

(**c**) Instance M1 (**d**) Instance M2

(**e**) Instance L1 (**f**) Instance L2

Figure 5. Pareto fronts computed by the studied methods; inv(QoS) values are expressed in millions.

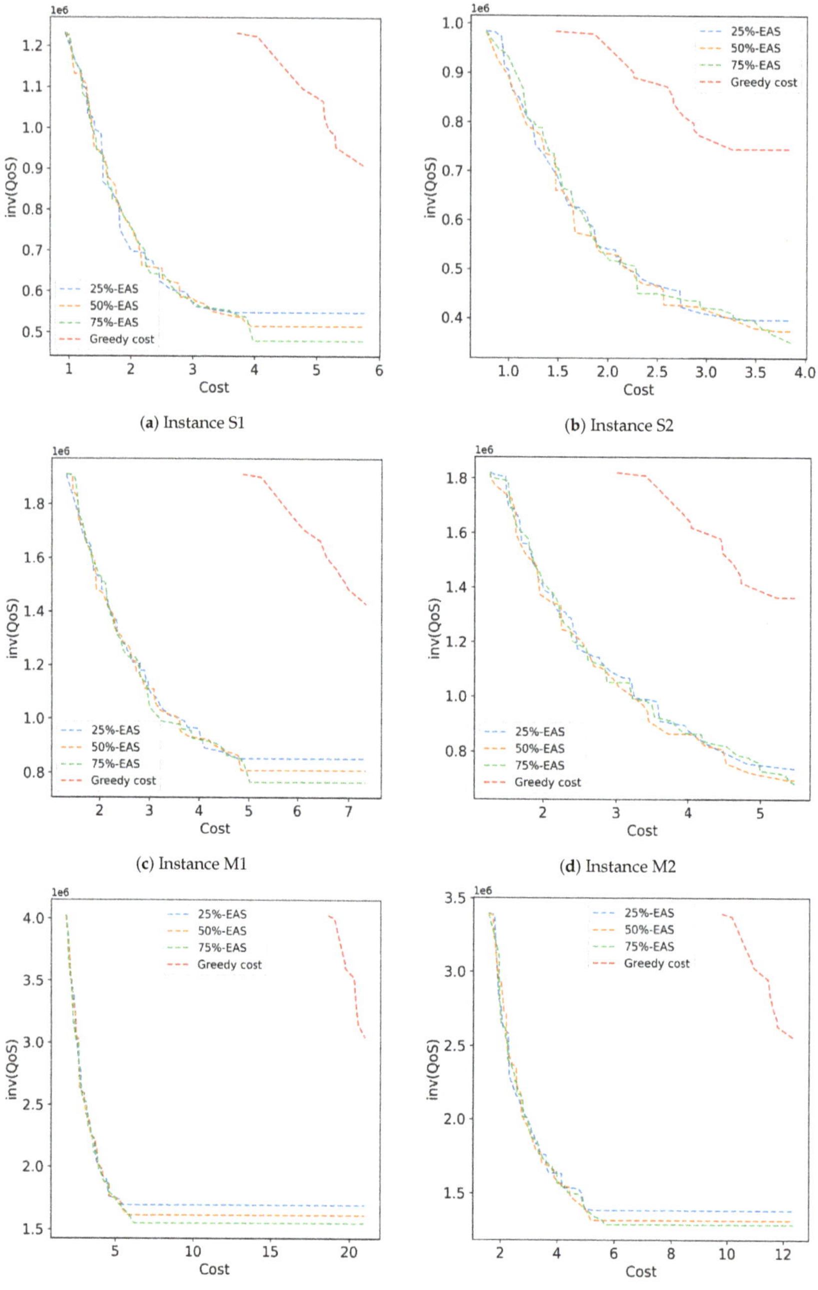

(**a**) Instance S1 (**b**) Instance S2 (**c**) Instance M1 (**d**) Instance M2 (**e**) Instance L1 (**f**) Instance L2

Figure 6. Attainment surfaces computed by the studied methods for problem instances S1 to L2.

6.6. Comparison with Recent Related Works

The proposed MOEA obtains significant improvements in CDN design costs and QoS as evaluated by the RTT for accessing content. The improvements are comparable and in certain cases even better than those reported for similar problems in recent related works, especially considering that few articles have followed multiobjective approaches.

The cost improvements of the proposed MOEA (47.6% to 93.3%) are significantly better than the randomized online edge-renting algorithm presented by Jin et al. [37] (10.5% to 17.6% over BaU), taking advantage of the intelligent evolutionary exploration of the search space. QoS improvements (47.2%) were lower than those reported for the STARFRONT framework by Lai et al. [36], mainly because STARFRONT is heavily focused on reducing round-trip times (90% improvement over BaU) and does not optimize for cost.

The improvements of MOEA regarding QoS are not at good as the improvements reported for the unsupervised self-organizing map technique presented by Farahani et al. [38], which improved by over 59% in Quality of Experience (QoE). However, the self-organizing map approach does not optimize for design costs, and higher costs will eventually be required to guarantee the reported QoE improvements. When taking cost optimization into account, the MOEA provides better improvements compared to the greedy algorithm for joint cost/QoS optimization presented by the same authors [39] in terms of both QoS/Quality of Experience (QoE) (22% over the traditional CDN and hybrid P2P–CDN baseline methods) and cost (34% over baseline).

The cost improvement provided by the proposed MOEA is also significantly better than the improvement reported for the two-step machine learning optimization method by Yadav and Kar [40] (30% of improvement in overall cost of deployment and distribution), where QoS was not optimized but was included as constraint. Finally, the QoS improvement of the proposed MOEA over BaU is higher than that provided by the adaptive hybrid approach of Marri and Reddy [41] (19% to 25% improvement regarding upload capacity and startup delay), where no cost optimization was proposed.

The above comparison with recent related works allows us to conclude that the proposed MOEA is effective for the simultaneous optimization of cost and QoS in cloud-based CDN design. Table 9 summarizes the main aspects and results of the comparison.

Table 9. Comparison with recent related works.

Author(s)	Algorithm	QoS Improvement	Cost Improvement
Lai et al. [36]	two-step heuristic	>90% in QoS	cost not optimized
Jin et al. [37]	randomized online	QoS not optimized	11.9% in edge service renting cost
Farahani et al. [38]	heuristic and machine learning	>59% in QoE of users	cost not optimized
Farahani et al. [39]	greedy	>22% in QoE of users	>34% in cost
Yadav and Kar [40]	two-step machine learning	QoS not optimized	>30% in overall cost
Marri and Reddy [41]	machine learning	>25% in upload capacity of peers, >19% in delay	cost not optimized
this article	MOEA	47.2% in Qos over BaU	47.6% to 93.3% in cost

7. Conclusions

A reliable internet connection is essential for modern societies, driving an increasing number of everyday activities, particularly through mobile and wireless devices. Internet providers face the dual challenge of expanding services to accommodate more users while ensuring low-latency performance. To meet these demands, cloud-based CDNs have emerged as a key solution for constructing cost-efficient networks. By deploying servers in strategic locations and intelligently allocating content, CDNs enhance the speed and reliability of web content delivery while maintaining network cost efficiency. However, CDN design and content allocation must be carefully planned in order to minimize operational costs while ensuring high service quality for users.

In this context, the present article introduces an optimization model aimed at designing cloud-based CDNs that account for the demands of end users, content providers, and IaaS providers. The model simultaneously optimizes the operational costs of the virtual broker that manages the system and the QoS delivered to users. A tailored MOEA incorporating ad hoc solution representations and evolutionary operators is proposed to solve the cloud-based CDN problem. The performance of the proposed MOEA was benchmarked against three baseline heuristics and an exact solver based on an MILP formulation. Experimental results on real-world instances highlight the effectiveness of the proposed MOEA, which consistently produced accurate and diverse solutions. The proposed MOEA outperformed the exact solver for small problem instances and yielded significantly better results than baseline heuristic algorithms focused solely on cost or QoS optimization. In addition, the MOEA generated more diverse Pareto fronts compared to the heuristics, which tended to concentrate on solutions with extreme objective values. In terms of the RHV metric, the MOEA achieved an average improvement of 34.7% over the baseline heuristics. Significant improvement were also obtained regarding the 50%-EAS metric. Cost reductions ranged between 47.6% and 93.3%, while QoS levels remained highly accurate, improving by 47.2% over BaU.

Future work will focus on expanding the problem instances to create a more diverse and complex test environment. The online variant of the problem should also be studied further in order to enhance the applicability of the proposed approach. Incorporating historical traffic data and refining the QoS function by including factors such as network bandwidth in addition to RTT could enhance the precision of the model. This would provide a deeper understanding of the effectiveness of the model and lead to more realistic solutions. Additionally, efforts will be directed towards improving the efficiency of the proposed algorithm to enable real-time online execution, complementing the current offline approach.

This research was developed under the replicable research paradigm. Source code and data are available in the GitHub repository https://github.com/gerardogoni20170819/CDN_CLOUD (accessed on 29 December 2024). Details about the research are also reported on the website https://www.fing.edu.uy/inco/grupos/cecal/hpc/DRCC/ (accessed on 29 December 2024).

Author Contributions: Conceptualization, S.N.; methodology, S.N. and G.G.; software, G.G.; validation, G.G., S.N., D.R., P.M.-B. and A.T.; formal analysis, S.N. and A.T.; investigation, G.G., S.N., D.R. and P.M.-B.; resources, G.G. and S.N.; data curation, G.G.; writing—original draft preparation, G.G., S.N., D.R., P.M.-B. and A.T.; writing—review and editing, G.G., S.N., D.R., P.M.-B. and A.T.; visualization, G.G., S.N., D.R. and P.M.-B.; supervision, S.N.; project administration, G.G. and S.N. All authors have read and agreed to the published version of the manuscript.

Funding: This research received no external funding.

Data Availability Statement: Data and results are available upon request.

Conflicts of Interest: The authors declare no conflicts of interest.

References

1. Zolfaghari, B.; Srivastava, G.; Roy, S.; Nemati, H.; Afghah, F.; Koshiba, T.; Razi, A.; Bibak, K.; Mitra, P.; Rai, B. Content delivery networks: State of the art, trends, and future roadmap. *ACM Comput. Surv.* **2020**, *53*, 34. [CrossRef]
2. Vakali, A.; Pallis, G. Content delivery networks: Status and trends. *IEEE Internet Comput.* **2003**, *7*, 68–74. [CrossRef]
3. Salahuddin, M.; Sahoo, J.; Glitho, R.; Elbiaze, H.; Ajib, W. A survey on content placement algorithms for cloud-based content delivery networks. *IEEE Access* **2017**, *6*, 91–114. [CrossRef]
4. Nesmachnow, S.; Iturriaga, S.; Dorronsoro, B. Efficient Heuristics for Profit Optimization of Virtual Cloud Brokers. *IEEE Comput. Intell. Mag.* **2015**, *10*, 33–43. [CrossRef]
5. Drezner, Z.; Hamacher, H.W. (Eds.) *Facility Location: Applications and Theory*; Springer: Berlin/Heidelberg, Germany, 2004.
6. Nesmachnow, S. An overview of metaheuristics: Accurate and efficient methods for optimisation. *Int. J. Metaheuristics* **2014**, *3*, 320–347. [CrossRef]

7. Goñi, G.; Nesmachnow, S.; Chernykh, A. Design of Content Distribution Networks for smart cities. In *Smart Cities*; Springer Nature: Cham, Switzerland, 2025; pp. 267–281.
8. Buyya, R.; Broberg, J.; Goscinski, A. *Cloud Computing: Principles and Paradigms*; Wiley Series on Parallel and Distributed Computing; Wiley: Hoboken, NJ, USA, 2010. [CrossRef]
9. Papagianni, C.; Leivadeas, A.; Papavassiliou, S. A cloud-oriented content delivery network paradigm: Modeling and assessment. *IEEE Trans. Dependable Secur. Comput.* **2013**, *10*, 287–300. [CrossRef]
10. Mattess, M.; Vecchiola, C.; Garg, S.K.; Buyya, R. Cloud Bursting: Managing Peak Loads by Leasing Public Cloud Services. In *Cloud Computing*; CRC Press: Boca Raton, FL, USA, 2017; pp. 343–367. [CrossRef]
11. Wang, M.; Jayaraman, P.; Ranjan, R.; Mitra, K.; Zhang, M.; Li, E.; Khan, S.; Pathan, M.; Georgeakopoulos, D. An overview of cloud based content delivery networks: Research dimensions and state-of-the-art. In *Transactions on Large-Scale Data-and Knowledge-Centered Systems XX*; Springer: Berlin/Heidelberg, Germany, 2015; pp. 131–158. [CrossRef]
12. Xiao, W.; Bao, W.; Zhu, X.; Wang, C.; Chen, L.; Yang, L.T. Dynamic request redirection and resource provisioning for cloud-based video services under heterogeneous environment. *IEEE Trans. Parallel Distrib. Syst.* **2016**, *27*, 1954–1967. [CrossRef]
13. Gao, G.; Zhang, W.; Wen, Y.; Wang, Z.; Zhu, W. Towards Cost-Efficient Video Transcoding in Media Cloud: Insights Learned from User Viewing Patterns. *IEEE Trans. Multimed.* **2015**, *17*, 1286–1296. [CrossRef]
14. Hu, M.; Luo, J.; Wang, Y.; Veeravalli, B. Practical resource provisioning and caching with dynamic resilience for cloud-based content distribution networks. *IEEE Trans. Parallel Distrib. Syst.* **2014**, *25*, 2169–2179. [CrossRef]
15. Jokhio, F.; Ashraf, A.; Lafond, S.; Lilius, J. A Computation and Storage Trade-off Strategy for Cost-Efficient Video Transcoding in the Cloud. In Proceedings of the 2013 39th Euromicro Conference on Software Engineering and Advanced Applications, Santander, Spain, 4–6 September 2013; pp. 365–372. [CrossRef]
16. Zhang, J.; Huang, H.; Wang, X. Resource provision algorithms in cloud computing: A survey. *J. Netw. Comput. Appl.* **2016**, *64*, 23–42. [CrossRef]
17. Baruwal Chhetri, M.; Chichin, S.; Bao Vo, Q.; Kowalczyk, R. Smart Cloud Broker: Finding your home in the clouds. In Proceedings of the 2013 28th IEEE/ACM International Conference on Automated Software Engineering (ASE), Silicon Valley, CA, USA, 11–15 November 2013; pp. 698–701. [CrossRef]
18. Landa, R.; Araujo, J.; Clegg, R.; Mykoniati, E.; Griffin, D.; Rio, M. The large-scale geography of Internet round trip times. In Proceedings of the 2013 IFIP Networking Conference, Brooklyn, NY, USA, 22–24 May 2013; pp. 1–9.
19. Candela, M.; Kisteleki, R.; Aben, E.; Homburg, P.; Beest, J.; Amin, C.; Strikos, A.; Queen, D.; Antony, A. RIPE Atlas: A Global Internet Measurement Network. *Internet Protoc. J.* **2015**, *18*, 3.
20. Agmon, O.; Ben, M.; Schuster, A.; Tsafrir, D. Deconstructing Amazon EC2 spot instance pricing. *ACM Trans. Econ. Comput.* **2013**, *1*, 16. [CrossRef]
21. Massey, F.J. The Kolmogorov-Smirnov test for goodness of fit. *J. Am. Stat. Assoc.* **1951**, *46*, 68–78. [CrossRef]
22. Iturriaga, S.; Nesmachnow, S.; Goñi, G.; Dorronsoro, B.; Tchernykh, A. Evolutionary Algorithms for Optimizing Cost and QoS on Cloud-based Content Distribution Networks. *Program. Comput. Softw.* **2019**, *45*, 544–556. [CrossRef]
23. Zheng, Z.; Zheng, Z. Towards an improved heuristic genetic algorithm for static content delivery in cloud storage. *Comput. Electr. Eng.* **2018**, *69*, 422–434. [CrossRef]
24. Iturriaga, S.; Goñi, G.; Nesmachnow, S.; Dorronsoro, B.; Tchernykh, A. Cost and QoS Optimization of Cloud-Based Content Distribution Networks Using Evolutionary Algorithms. In Proceedings of the High Performance Computing, Bucaramanga, Colombia, 26–28 September 2018; Meneses, E., Castro, H., Barrios Hernández, C.J., Ramos-Pollan, R., Eds.; Springer: Cham, Awitzerland, 2019; pp. 293–306. [CrossRef]
25. Stephanakis, I.; Logothetis, D. Evolutionary Algorithm Optimization of Edge Delivery Sites in Next Generation Multi-service Content Distribution Networks. In Proceedings of the Engineering Applications of Neural Networks, Corfu, Greece, 15–18 September 2011; Iliadis, L., Jayne, C., Eds.; Springer: Berlin/Heidelberg, Germany, 2011; pp. 192–202. [CrossRef]
26. Bektaş, T.; Cordeau, J.F.; Erkut, E.; Laporte, G. A two-level simulated annealing algorithm for efficient dissemination of electronic content. *J. Oper. Res. Soc.* **2008**, *59*, 1557–1567. [CrossRef]
27. Ellouze, S.; Mathieu, B.; Lemlouma, T. A bidirectional network collaboration interface for CDNs and Clouds services traffic optimization. In Proceedings of the 2013 IEEE International Conference on Communications (ICC), Budapest, Hungary, 9–13 June 2013; pp. 3592–3596. [CrossRef]
28. Mangili, M.; Martignon, F.; Capone, A. A comparative study of Content-Centric and Content-Distribution Networks: Performance and bounds. In Proceedings of the 2013 IEEE Global Communications Conference (GLOBECOM), Atlanta, GA, USA, 9–13 December 2013; pp. 1403–1409. [CrossRef]
29. Coppens, J.; Wauters, T.; de Turck, F.; Dhoedt, B.; Demeester, P. Design and Performance of a Self-Organizing Adaptive Content Distribution Network. In Proceedings of the 2006 IEEE/IFIP Network Operations and Management Symposium NOMS 2006, Vancouver, BC, Canada, 3–7 April 2006; pp. 534–545. [CrossRef]

30. Cevallos Moreno, J.F.; Sattler, R.; Caulier Cisterna, R.P.; Ricciardi Celsi, L.; Sánchez Rodríguez, A.; Mecella, M. Online Service Function Chain Deployment for Live-Streaming in Virtualized Content Delivery Networks: A Deep Reinforcement Learning Approach. *Future Internet* **2021**, *13*, 278. [CrossRef]
31. Karaata, M.; Al-Mutairi, A.; Alsubaihi, S. Multipath Routing Over Star Overlays for Quality of Service Enhancement in Hybrid Content Distribution Peer-to-Peer Networks. *IEEE Access* **2022**, *10*, 7042–7058. [CrossRef]
32. Bęben, A.; Batalla, J.M.; Chai, W.K.; Śliwiński, J. Multi-criteria decision algorithms for efficient content delivery in content networks. *Ann. Telecommun.—Ann. Télécommun.* **2013**, *68*, 153–165. [CrossRef]
33. Neves, T.; Ochi, L.; Albuquerque, C. A new hybrid heuristic for replica placement and request distribution in content distribution networks. *Optim. Lett.* **2015**, *9*, 677–692. [CrossRef]
34. Khansoltani, A.; Jamali, S.; Fotohi, R. A Request Redirection Algorithm in Content Delivery Network: Using PROMETHEE Approach. *Wirel. Pers. Commun.* **2022**, *126*, 1145–1175. [CrossRef]
35. Jabraili, H.; Yousefi, S.; Boukani, B.; Rad, M.B. Replication based on objects iteration frequency and load using a genetic algorithm under a content distribution network. In Proceedings of the 2013 21st Iranian Conference on Electrical Engineering (ICEE), Mashhad, Iran, 14–16 May 2013; pp. 1–6. [CrossRef]
36. Lai, Z.; Li, H.; Zhang, Q.; Wu, Q.; Wu, J. Cooperatively Constructing Cost-Effective Content Distribution Networks upon Emerging Low Earth Orbit Satellites and Clouds. In Proceedings of the 2021 IEEE 29th International Conference on Network Protocols (ICNP), Dallas, TX, USA, 1–5 November 2021; pp. 1–12. [CrossRef]
37. Jin, Z.; Pan, L.; Liu, S. Randomized online edge service renting: Extending cloud-based CDN to edge environments. *Knowl.-Based Syst.* **2022**, *257*, 109957. [CrossRef]
38. Farahani, R.; Bentaleb, A.; Çetinkaya, E.; Timmerer, C.; Zimmermann, R.; Hellwagner, H. Hybrid P2P-CDN Architecture for Live Video Streaming: An Online Learning Approach. In Proceedings of the GLOBECOM 2022—2022 IEEE Global Communications Conference, Rio de Janeiro, Brazil, 4–8 December 2022; pp. 1911–1917. [CrossRef]
39. Farahani, R.; Çetinkaya, E.; Timmerer, C.; Shojafar, M.; Ghanbari, M.; Hellwagner, H. ALIVE: A Latency- and Cost-Aware Hybrid P2P-CDN Framework for Live Video Streaming. *IEEE Trans. Netw. Serv. Manag.* **2024**, *21*, 1561–1580. [CrossRef]
40. Yadav, P.; Kar, S. Efficient Content Distribution in Fog-Based CDN: A Joint Optimization Algorithm for Fog-Node Placement and Content Delivery. *IEEE Internet Things J.* **2024**, *11*, 16578–16590. [CrossRef]
41. Marri, S.; Reddy, P.C. Deep Temporal LSTM Regression Network (DTLR-Net) Model for Optimizing Quality of Video Streaming Quality in CDN-P2P Model. *IEEE Access* **2025**, *13*, 42521–42529. [CrossRef]
42. Deb, K.; Agrawal, S.; Pratap, A.; Meyarivan, T. A Fast Elitist Non-dominated Sorting Genetic Algorithm for Multi-objective Optimization: NSGA-II. In Proceedings of the Parallel Problem Solving from Nature PPSN VI, Paris, France, 18–20 September 2000; Schoenauer, M., Deb, K., Rudolph, G., Yao, X., Lutton, E., Merelo, J.J., Schwefel, H.P., Eds.; Springer: Berlin/Heidelberg, Germany, 2000; pp. 849–858. [CrossRef]
43. Srinivas, N.; Deb, K. Muiltiobjective Optimization Using Nondominated Sorting in Genetic Algorithms. *Evol. Comput.* **1994**, *2*, 221–248. [CrossRef]
44. Durillo, J.; Nebro, A. JMetal: A Java framework for multi-objective optimization. *Adv. Eng. Softw.* **2011**, *42*, 760–771. [CrossRef]
45. Fourer, R.; Gay, D.M.; Kernighan, B.W. A Modeling Language for Mathematical Programming. *Manag. Sci.* **1990**, *36*, 519–554. [CrossRef]
46. Rossit, D.G.; Toutouh, J.; Nesmachnow, S. Exact and heuristic approaches for multi-objective garbage accumulation points location in real scenarios. *Waste Manag.* **2020**, *105*, 467–481. [CrossRef]
47. Deb, K.; Pratap, A.; Agarwal, S.; Meyarivan, T. A fast and elitist multiobjective genetic algorithm: NSGA-II. *IEEE Trans. Evol. Comput.* **2002**, *6*, 182–197. [CrossRef]
48. Coello Coello, C.A.; Lamont, G.B.; Van Veldhuizen, D.A. *Evolutionary Algorithms for Solving Multi-Objective Problems*; Springer: New York, NY, USA, 2007. [CrossRef]
49. Knowles, J. A summary-attainment-surface plotting method for visualizing the performance of stochastic multiobjective optimizers. In Proceedings of the 5th International Conference on Intelligent Systems Design and Applications, Warsaw, Poland, 8–10 September 2005; IEEE: Piscataway, NJ, USA, 2005. [CrossRef]
50. Knowles, J.; Corne, D. Approximating the Nondominated Front Using the Pareto Archived Evolution Strategy. *Evol. Comput.* **2000**, *8*, 149–172. [CrossRef] [PubMed]
51. Katsaros, K.V.; Xylomenos, G.; Polyzos, G.C. GlobeTraff: A Traffic Workload Generator for the Performance Evaluation of Future Internet Architectures. In Proceedings of the 2012 5th International Conference on New Technologies, Mobility and Security (NTMS), Istanbul, Turkey, 7–10 May 2012; pp. 1–5. [CrossRef]

52. Cisco. *Cisco Visual Networking Index: Global Mobile Data Traffic Forecast Update, 2015–2020*; Technical Report C11-738429-00; Cisco: San Jose, CA, USA, 2016.
53. Nesmachnow, S.; Iturriaga, S. Cluster-UY: Collaborative Scientific High Performance Computing in Uruguay. In *Supercomputing*; Torres, M., Klapp, J., Eds.; Communications in Computer and Information Science; Springer: Cham, Switzerland, 2019; pp. 188–202. [CrossRef]

Mathematical and Computational Applications

MDPI

Article

TAE Predict: An Ensemble Methodology for Multivariate Time Series Forecasting of Climate Variables in the Context of Climate Change

Juan Frausto Solís [1,*], Erick Estrada-Patiño [1], Mirna Ponce Flores [1], Juan Paulo Sánchez-Hernández [2], Guadalupe Castilla-Valdez [1] and Javier González-Barbosa [1,*]

[1] Graduate Program Division, Tecnológico Nacional de México/Instituto Tecnológico de Ciudad Madero, Ciudad Madero 89440, Mexico; estrada1792@gmail.com (E.E.-P.); mirna.pf@cdmadero.tecnm.mx (M.P.F.); guadalupe.cv@cdmadero.tecnm.mx (G.C.-V.)

[2] Dirección de Informático, Electrónica y Telecomunicaciones, Universidad Politécnica del Estado de Morelos, Boulevard Cuauhnáhuac 566, Jiutepec 62574, Mexico; juan.paulosh@upemor.edu.mx

[*] Correspondence: juan.frausto@gmail.com (J.F.S.); jjgonzalezbarbosa@hotmail.com (J.G.-B.)

Abstract: Climate change presents significant challenges due to the increasing frequency and intensity of extreme weather events. Mexico, with its diverse climate and geographic position, is particularly vulnerable, underscoring the need for robust strategies to predict atmospheric variables. This work presents TAE Predict (Time series Analysis and Ensemble-based Prediction with relevant feature selection) based on relevant feature selection and ensemble models of machine learning. Dimensionality in multivariate time series is reduced through Principal Component Analysis, ensuring interpretability and efficiency. Additionally, data remediation techniques improve data set quality. The ensemble combines Long Short-Term Memory neural networks, Random Forest regression, and Support Vector Machines, optimizing their contributions using heuristic algorithms such as Particle Swarm Optimization. Experimental results from meteorological time series in key Mexican cities demonstrate that the proposed strategy outperforms individual models in accuracy and robustness. This methodology provides a replicable framework for climate variable forecasting, delivering analytical tools that support decision-making in critical sectors, such as agriculture and water resource management. The findings highlight the potential of integrating modern techniques to address complex, high-dimensional problems. By combining advanced prediction models and feature selection strategies, this study advances the reliability of climate forecasts and contributes to the development of effective adaptation and mitigation measures in response to climate change challenges.

Keywords: climate change; multivariate time series; deep learning; principal component analysis; ensemble methods; particle swarm optimization

1. Introduction

Climate change is defined as a significant and prolonged alteration in atmospheric, geographic, and natural patterns [1]. Throughout history, climatic phenomena of various kinds have occurred. However, current evidence points to a strong anthropogenic influence on contemporary phenomena [2]. According to the Climate Risk Index 2023, extreme weather events such as hurricanes, severe droughts, and torrential rains, have increased in frequency and intensity, putting the economic, social, and environmental stability of affected regions at risk [3]. The Conference of the Parties (COP21) highlighted that temperature is a key variable in the understanding and modeling of climate change phenomena, as it highly correlates with other meteorological variables such as humidity, wind, and atmospheric pressure [4,5].

Due to its geographic location, climatic diversity, and border between two oceans, Mexico is particularly exposed to these extreme weather events [6]. This vulnerability demands the development of robust strategies to understand and accurately predict the behavior of atmospheric variables, thus facilitating the planning of mitigation and adaptation measures [7]. Accurate temperature prediction and other related variables are essential to prevent human and material losses and make informed decisions in agriculture, water management, and environmental protection [8].

In this context, current prediction models face multiple challenges, including the high dimensionality of multivariate time series and the difficulty of capturing the complex interactions between climate variables [7,9]. Moreover, no single model has proven universally superior regarding generalization and accuracy [10–12].

This paper presents TAE Predict (Time series Analysis and Ensemble-based Prediction with relevant feature selection), an innovative strategy based on feature selection and ensembles of machine learning models designed to address the inherent challenges in predicting relevant climate variables of climate change. This methodology focuses on integrating multiple forecasting models, optimizing their individual contributions through a weighting scheme based on combinatorial optimization techniques, such as Particle Swarm Evolutionary Algorithms (PSO). This approach allows us to identify and combine the strengths of each model, mitigating their weaknesses and obtaining more accurate and robust predictions.

One of the main features of this strategy is the ability to select and prioritize the most relevant variables through Principal Component Analysis (PCA), thus reducing the dimensionality of the problem without losing critical information. This approach not only improves computational performance but also increases the interpretability of the results, facilitating the understanding of the influence of each variable in the forecast. In addition, the methodology employs data remediation techniques, such as quadratic interpolation and singular value decomposition, to ensure the quality of the data set and minimize the impact of noise and outliers.

The results obtained from this strategy are evaluated experimentally, using multivariate time series from weather stations in key cities in Mexico. These data reflect diverse and complex climatic conditions and highlight the vulnerability of the country to extreme events resulting from climate change. By combining the predictive capacity of the ensembles with advanced feature selection and data remediation techniques, this work establishes a solid and replicable methodological framework for predicting climate variables.

With this contribution, we seek not only to advance the accuracy and reliability of forecasts, but also to provide a powerful analytical tool that allows decision-makers, researchers, and environmental managers to implement more informed and effective mitigation and adaptation strategies in the face of the challenges imposed by climate change. This methodology demonstrates that by integrating innovative approaches and leveraging modern machine learning techniques, it is possible to address complex, high-dimensional problems with high impact and applicability results.

This work is organized as follows: section two presents some works related to the problems presented and approaches that have addressed the forecasting of these variables. Section three describes the proposed methodology, detailing the models used, the remediation and feature selection strategies, and the ensemble scheme designed. Subsequently, section four analyzes the results obtained in the experimental process, comparing the performance of the proposed strategy against individual models and evaluating its generalization capacity in different contexts. Finally, section five presents the conclusions.

2. Related Works

Temperature forecasting is essential for understanding and mitigating the impacts of climate change. Traditionally, statistical models such as ARIMA (Autoregressive Integrated Moving Average) and its seasonal extension, SARIMA (Seasonal ARIMA), have been employed for this purpose [13–16]. These models effectively capture linear and seasonal patterns in the time series. However, they have limitations in dealing with nonlinear and complex relationships inherent in climate data, which can affect the accuracy of predictions in dynamic and variable scenarios [17].

Machine learning techniques have been incorporated into climate time series prediction to overcome these limitations. Models such as Long-Term Memory Neural Networks (LSTMs), Support Vector Regression (SVRs), and Random Forests (RFRs) have shown superior performance in capturing nonlinear and complex patterns [18–23]. For example, LSTMs can model long-term dependencies in sequential data, making them suitable for forecasting climate variables with high variability.

Likewise, Random Forests have been successfully applied in time series forecasting, showcasing their ability to handle large data sets and capture complex interactions between variables [24].

Ensemble methods, which combine multiple models to improve the accuracy and robustness of predictions, have gained relevance in this context [25]. Combining models such as LSTM, SVR, and RFR in an ensemble approach has been shown to improve prediction accuracy compared to individual models [11]. However, these methods also face challenges, such as computational complexity and the need for careful selection and weighting of component models to avoid overfitting and ensure generalization [12,26].

Despite the aforementioned advances, a gap persists in the integration of machine learning techniques and ensemble methods for temperature prediction in specific regions, such as Mexico, which present high susceptibility to extreme weather events. This work seeks to address this gap by developing an ensemble strategy that combines machine learning models and heuristic techniques to improve accuracy and robustness in the prediction of multivariate time series in the context of climate change.

3. Methodology

Multivariate time series forecasting represents a significant challenge due to multiple factors affecting its performance. This process requires rigorous preprocessing to ensure complete, consistent, high-quality data, eliminate outliers, and, in many cases, to solve the missing data problem. In addition, it is essential to implement feature selection strategies that reduce the problem's dimensionality, preserving as much explained variance as possible. Subsequently, forecasting models should be trained with modern and robust machine learning techniques, selecting them for their outstanding performance on similar problems, in part of the problem; in our case, we consider it more effective to select them by their performance in the validation section of each time series. In other words, using a heuristic ensemble ensures a broad exploration of the solution space, achieving high-performance predictive models.

In this context, this paper presents an ensemble methodology for multivariate time series forecasting with feature selection, whose modular architecture is illustrated in Figure 1. Each component of the proposed model is replaceable by different techniques, which endows the methodology with flexibility and adaptability.

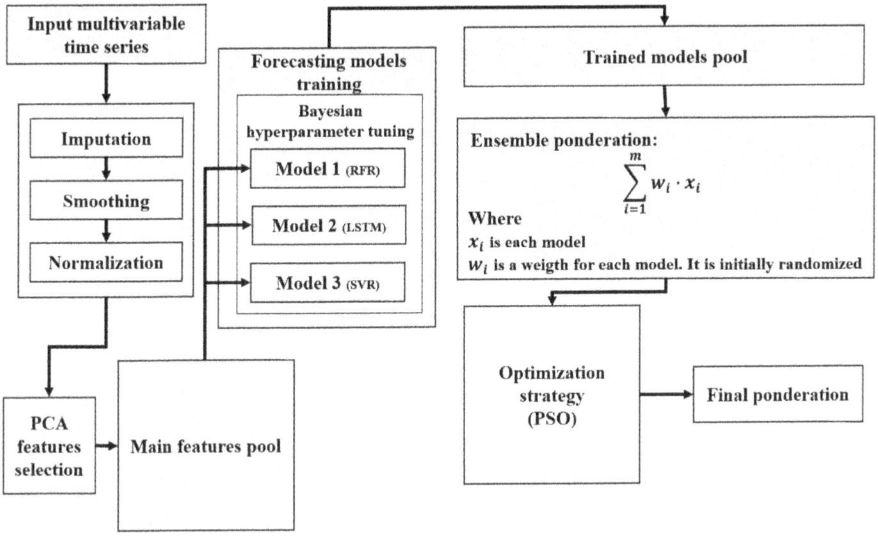

Figure 1. Forecasting methodology proposed.

3.1. Time Series Preprocessing

The data used in this work come from physical weather stations, which may face conditions that generate noise or missing data in the time series. This makes a remediation process prior to model adjustment indispensable, since cured data favor a better forecast performance and reduce training times. As a first step, an imputation based on quadratic interpolation was applied. This choice responds to the data analysis, where absences are usually isolated or span less than two consecutive periods. This method uses neighboring values to estimate missingness more accurately.

To mitigate the impact of noise and outliers, two smoothing strategies were implemented. First, a moving average with a window of size two was used, which allowed smoothing time series without significantly altering their essential patterns. Subsequently, Singular Value Decomposition (SVD) was applied [12]. This technique consisted of hankelization of the series with a seasonal periodicity of 365 periods and retaining 95% of the principal components after decomposition. The hankelized series reduced noise by eliminating low-relevance components. It is important to mention that SVD is applied to each variable individually, so that information is not mixed across variables. Therefore, n strategies are applied to the model, where n is the number of variables in the system.

These strategies not only ensure data integrity for model fitting, but are also critical for capturing seasonal patterns and underlying trends in the time series. By removing noisy elements, models can focus on learning relevant information, resulting in more consistent and robust performance, especially in critical applications such as weather forecasting.

Finally, the data were normalized to the interval [1,2]. This choice was motivated by the need to avoid potential problems with forecasting methods and evaluation metrics, ensuring consistency of subsequent analysis.

3.2. Selection of Relevant Characteristics

In the field of multivariate time series, the problem of high dimensionality acquires crucial relevance due to the issues generated in adjustment and forecasting [27–29]. Each variable or indicator incorporated in the time series adds an additional dimension to the data set. Although it could be assumed that a greater number of variables implies an improvement in the quality of the forecast due to the increase in the amount of information available, it is essential to know that not all the information added is useful for the model. In fact, many times, these variables can be redundant, irrelevant, or introduce noise, hindering the learning process.

Increasing dimensionality also significantly increases the time and computational resources required by the models [30]. This not only lengthens the fitting process, but can negatively affect model quality due to the complexity of exploring a solution space that grows exponentially with the number of dimensions.

In this context, one of the key approaches to addressing this problem is feature selection. Proper selection involves identifying those variables that have a significant impact on the target variable, eliminating those that are highly correlated with each other or that contribute little information to the model. This process not only improves the quality of the fit, but also reduces processing time and allows a simple analysis of the relationships between the selected variables and the target.

For this work, PCA is implemented as the principal feature selection strategy, a mathematical technique that transforms the original data set into a new system defined by principal components [28]. These components are calculated from the eigenvalues and eigenvectors of the covariance matrix of the data set, where each eigenvector defines a direction in the space of the variables, and its corresponding eigenvalue represents the variance explained in that direction.

The central idea of PCA is to rearrange the dimensions of the system so that the first components capture the largest proportion of the variability present in the data. In this work, a variance explained threshold of 92% is used, which means that only those principal components whose cumulative variance exceeds this percentage are selected. This approach allows for a significant reduction in dimensionality without losing relevant information for forecasting.

Although PCA generates new variables (components) that are linear combinations of the original variables, the components are used in terms of the initial variables. In this paper, the PCA process is not completed, and it is stopped when it reaches the phase of weighting the variables with the highest explained variance, returning the necessary variables until the threshold is reached.

This facilitates a subsequent analysis of how each variable contributes to the forecast, which is essential for evaluating its impact on the target variable.

The implemented strategy not only optimizes computational efficiency, but also provides a more manageable and relevant feature set, allowing forecasting models to be fitted more accurately and quickly.

3.3. Forecasting Models

The strategies of tuning hyperparameters are performed in order to fit the data. Each strategy does this differently and performs differently in both quality and fitting time. Although there is no universally superior strategy, the strategies selected for this work stand out for their high accuracy, robustness, and reliability. These strategies have also been successfully applied to problems relating to the forecast of atmospheric variables and to issues related to climate change.

3.3.1. Long Short-Term Memory

Neural networks are a powerful approach for generalizing information and extracting complex patterns from a data set [31,32]. These networks stand out for responding accurately, even to entirely unknown data. However, their performance in time series forecasting has shown limitations because they tend to process data individually and do not explicitly consider the temporal dependencies inherent in this type of information.

To address the last limitation, LSTM offers an innovative solution [33]. These networks incorporate a structure based on specialized cells to maintain relevant information over time and discard information that is useless for future learning. This ability "to remember" and "to forget" in a controlled manner is crucial for modeling temporal dependencies in time series data, facilitating complex learning of dynamic patterns.

In this work, an LSTM neural network architecture specifically designed to address the challenges of forecasting climate change related variables has been implemented [20]. This architecture consists of LSTM cells organized in stacked layers, which allows the efficient capture of complex multivariate patterns and improves the representation of temporal dependencies in the data.

Once the LSTM cells process the information, a fully connected network (or standard dense layer) is responsible for generalizing the learned representation and providing the final model output. This combination of structures ensures an effective integration between the capture of temporal dependencies and the generalization of the learned patterns.

Table 1 details the hyperparameters used in the LSTM network configuration employed in this work. These hyperparameters include the number of LSTM layers, the number of units in each layer, the learning rate, and the regularization of values applied, among others.

Table 1. LSTM configuration.

Hyperparameters *	Value
LSTM Cells	64
LSTM Layers	5
Dropout per layer	0.5
MLP layers	5
Loss function	Mean Squared Error
Early stopping patience	80
Epoch	2000

* Configuration of the LSTM model, including key hyperparameters such as the number of layers, dropout rate, and training epochs used in this study.

3.3.2. Random Forest Regression

This regression method is called RFR (that stands up Random Forest Regression). RFR belongs to the Decision Trees popular models in machine learning. RFR is widely applied in forecasting because it divides complex data into simpler subgroups [34–36]. RFR attractive features include a small number of tunable parameters, automatic calculation of generalization errors and handling of missing data, different types of data, and general resistance to overfitting [37]. RFR combines multiple decision trees to form a robust model using the bagging approach, where each tree is trained with random subsamples of the original data. In our case, the data consists of a time series, which allows us to identify both specific patterns and broader trends, as well as seasonal patterns.

Each tree generates an independent prediction, and, in the end, these predictions are combined by an operation (usually the arithmetic mean) to obtain the final result. This approach reduces variance and improves the model's ability to capture complex nonlinear relationships. Figure 2 shows this flow, while Table 2 details the hyperparameters used, such as the number of trees, maximum depth, and splitting criteria.

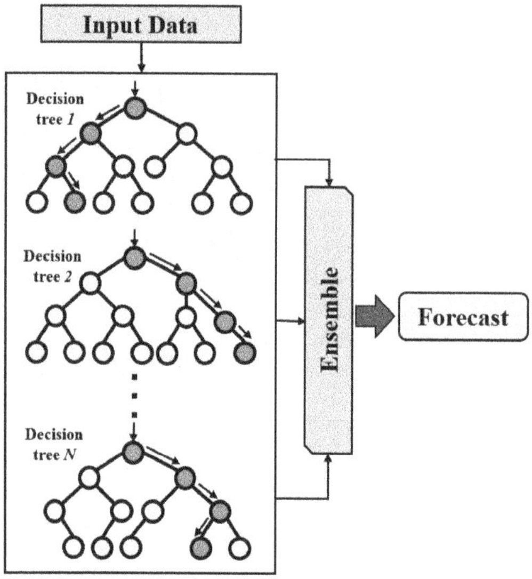

Figure 2. Random Forest Regression flow chart.

Table 2. Random Forest configuration.

Hyperparameter *	Value
Estimators	500
Max depth	10
Min samples split	10
Min samples leaf	15

* Hyperparameters used for the RFR model, detailing the number of estimators, maximum depth, and splitting criteria.

In this work, we apply the methodology to the analysis of climate data, a domain where nonlinear relationships and complex temporal dynamics are common. Random Forest proves to be effective in identifying patterns in key variables such as temperature, precipitation, and greenhouse gas concentration, providing accurate and reliable predictions.

3.3.3. Support Vector Regression

Support vector regression machines are a widely used strategy for function fitting in forecasting problems [38]. This approach is based on finding a function that minimizes the prediction error while maintaining a balance between model complexity and data generalization. The main objective is to provide an efficient forecast consistent with the information used during training.

Regarding time series, SVR captures complex patterns in the data using kernel functions, which allow modeling nonlinear relationships between variables. This method is especially useful in highly nonlinear problems, and represents typical behavior of atmospheric data [39]. Figure 3 shows a graphical representation of the function fitting process using SVR. The circles in the input and output of the boundary region represent events of a general process. Moreover, the width of this boundary delimited by broken lines represents the confidence interval, delimited by the parameter ϵ or controlled margin, allowing to focus on predicting relevant patterns. The hyperparameters listed in Table 3 follow the standard formulation commonly adopted in the literature for SVR models [38,39]. They are the penalty parameter C, the epsilon parameter ε, and the kernel coefficient γ. The C parameter controls the trade-off between the model complexity and the degree to which deviations greater than ε are penalized. The epsilon ε defines a margin of tolerance where no penalty is given to errors, effectively shaping the SVR loss function. The configuration of the hyperparameters used, including γ, ϵ, and C, is detailed in Table 3. These hyperparameters were obtained by Bayesian experimentation [40,41].

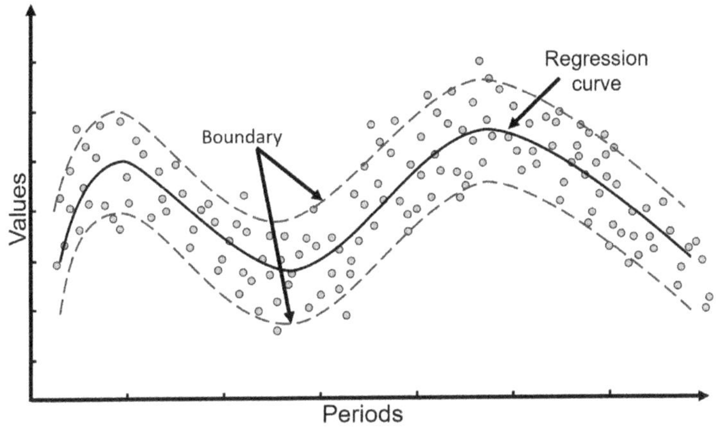

Figure 3. Graphical representation of the function fit and confidence intervals applied to the data.

Table 3. SVR configuration.

Hyperparameter	Value
Kernel	RBF
C	1.0
Epsilon	0.1
Gamma	Scale

3.4. Hyperparameter Tuning

The configuration of hyperparameters is critical to the performance of forecasting models, as it defines their ability to generalize and avoid problems such as overfitting or underfitting [40,41]. An overfitted model presents low errors in the training set, but poor performance on unknown data, while an underfitted model lacks the flexibility to capture relevant patterns.

In this work, Bayesian fitting was used to determine the optimal hyperparameters, ensuring a balance between accuracy and generalization. Bayesian fitting prioritizes well-performing configurations using an iterative probabilistic model, reducing the search space and increasing efficiency. Unlike traditional methods such as GridSearch, this approach dynamically adapts to previous results, maximizing the probability of finding optimal configurations with fewer evaluations [41].

3.5. Ensemble Strategy

Despite the tunability of forecasting models, there is no guarantee that they generalize information uniformly, since each strategy fits specific aspects of the training data. This implies that there is no universally superior model for all forecasting problems, due to the diversity and complexity of the data used for tuning [12].

Ensemble strategies address this problem by combining predictions from multiple models, taking advantage of their individual strengths and mitigating their fitting errors. In this approach, each model contributes a given weight to the final result. However, an inappropriate allocation of these weights can generate results inferior to the best individual model, making it essential to optimize this weighting.

In this work, PSO is used to determine the weights of the models in the ensemble. This heuristic technique, inspired by the behavior of natural systems, seeks efficient solutions to highly complex problems. The particle swarm begins by randomly initializing a population of candidate weight vectors, where each particle represents a possible combination of model weights. These weights are subject to the constraint that their sum equals one. During each iteration, particles evaluate their performance on the objective function, typically based on validation error, and adjust their positions in the search space based on their own best historical performance and the best performance observed in their neighborhood. This process allows the algorithm to iteratively refine the weight configuration toward an optimal or near-optimal solution [42,43].

Although PSO does not guarantee finding the globally optimal solution, its exploratory capability ensures obtaining combinations of models that at least equal the performance of the best individual model.

3.6. Error Metrics

Fitting forecast models and ensemble weighting require an accurate assessment of their performance. For this purpose, error metrics that quantify the variation between model predictions and the actual values of the time series are used. These metrics allow the models to be adjusted to minimize the prediction error, thus improving their forecasting capability. In the literature, several error metrics are found, each one focused on specific aspects of model performance [44]. However, there is no universally superior metric, as the choice depends on the type of problem, the characteristics of the data, and the purpose of the analysis. In this paper, the Mean Squared Error (MSE) is used as the main metric for model fitting. This metric is defined as $MSE = \frac{1}{N} \cdot \sum_{i=1}^{N} \left(y_i - \hat{y}_i \right)^2$ where y_i are the real values, $\hat{y}_i$ are the values predicted by the model, and N is the total number of observations. The MSE penalizes large errors more severely, which favors a more accurate fit by reducing significant deviations during training. However, the MSE generates results in quadratic units, which makes it difficult for users to interpret directly.

To facilitate the understanding of the results, the Mean Absolute Percentage Error ($MAPE$) is used as a reporting metric. This metric is defined as $MAPE = \frac{100}{N} \sum_{i=1}^{N} \left| \frac{y_i - \hat{y}_i}{y_i} \right|$. The $MAPE$, expressed in percentage terms, provides a more intuitive and straightforward interpretation, since it indicates the average percentage relative error between predicted and actual values. This feature makes it especially useful for comparing results between different models and time series, providing a more accessible view for end users.

4. Experimental Results

This section describes the data set used during the experimental process, as well as the results obtained after the implementation of the proposed method. In addition, a comprehensive comparison is made between the performance of the ensemble method and the individual models, highlighting the advantages and limitations of each approach. All models used in this study were executed a minimum of 30 times to ensure the stability and reliability of the results. The values reported correspond to the average of these runs, allowing for a more accurate representation of the performance of each model. In addition, the runs were performed without time constraints, allowing each model to reach its best possible fit under the experimental conditions. Nevertheless, the average execution time for each strategy is also reported to provide a comparative perspective on computational efficiency.

4.1. Data Description

The information used in the experimental process of this work is based on four multivariate time series corresponding to key cities in Mexico: Monterrey, Guadalajara, Tijuana, and Tampico. The selection of these cities responds to their high susceptibility to the effects of climate change, such as extreme weather events, and their relevance as densely populated metropolises with diverse climatic environments. According to Germanwatch's Climate Risk Index 2023, Mexico ranks 31st globally, with a high vulnerability score due to extreme weather events such as hurricanes, catastrophic storms, and prolonged droughts [3]. This situation is aggravated by its geography, with two coasts exposed to the Atlantic and Pacific Oceans, its vast territorial extension, and its complex diversity of climatic ecosystems. Furthermore, according to data from the National Institute of Statistics and Geography (INEGI), over the last 30 years, Mexico has registered an average increase of 0.85 °C in annual temperature and a notable increase in the frequency of extreme events, such as torrential rains and severe droughts [45]. In this study, the time series were formed from weather data collected daily from airport weather stations located within the selected cities. These stations, being operated under international standards, offer reliable and consistent measurements. The data set spans the period from January 2012 to December 2022, ensuring a comprehensive temporal coverage for the analysis. The variables analyzed include maximum, minimum, and average temperature; maximum average and minimum dew point; maximum and average relative humidity; maximum average and minimum wind speed; and finally, maximum, average, and minimum atmospheric pressure. The choice of these variables allows capturing a comprehensive representation of the daily weather conditions in each region. These variables are not only relevant for forecasting, but also have a direct impact on climate risk assessment. The information used in the experimental process was divided sequentially into four main blocks. The first block corresponds to the training set, which represents 55% of the time series and is used to adjust the models. The subsequent blocks are as follows: validation set 1 (15% of the data), used for hyperparameter tuning; validation set 2 (15% of the data), used to evaluate model performance under test-like conditions; and finally, the test set (15% of the data), which is used to assess the final performance of the models on data not seen during any stage of training or validation.

4.2. Experimentation

In the first instance, the proposed models were evaluated individually using the temperature target in its three representations: maximum, average, and minimum. The results obtained, presented in Table 4, have the evaluations performed on validation set 2, which was reserved exclusively for reporting results. The forecast horizon used in this work is 15 days for each prediction.

The values indicate that the SVR and Random Forest models performed better on average in terms of percentage error than the LSTM model. However, it is important to clarify that the SVR model is deterministic under the configuration used in this study. That is, given a fixed set of hyperparameters and training data, it produces identical results across executions, resulting in a standard deviation of zero. In contrast, the Random Forest and LSTM models include stochastic components in their training processes, which leads to variability in their outputs across runs. Among these stochastic models, LSTM exhibited the lowest standard deviation, reflecting greater stability and consistency in its predictions relative to Random Forest. However, Table 4 shows two clear outliers in the standard deviation values: Guadalajara (Max, RFR: $\sigma = 2.00$) and Tampico (Max, LSTM: $\sigma = 3.03$). These cases suggest a higher variability in the model's predictions when trained multiple times. This behavior can be explained by the randomness present in the training processes and the complexity of the data in those particular cities. Not all models respond the same way to a given data set, and a higher deviation does not necessarily mean poor performance. It may reflect that the model is more sensitive to certain features or to noise in the data. Although the average error values are relatively stable, the presence of these outliers indicates that some models may be overfitting or underfitting in specific scenarios.

Table 4. Forecasting models comparative.

Cities *	Temperature	SVR	RFR	LSTM	σ SVR	σ RFR	σ LSTM
Guadalajara	Max	2.78%	1.61%	3.27%	0	2	0.04
	Avg	2.94%	1.64%	3.73%	0	0.11	0.04
	Min	2.79%	1.58%	2.87%	0	0.01	0.11
Monterrey	Max	3.60%	2.86%	4.65%	0	0.95	0.65
	Avg	4.18%	5.71%	5.66%	0	0.74	0.07
	Min	3.08%	2.66%	6.09%	0	0.46	0.37
Tampico	Max	3.16%	3.04%	3.97%	0	0.38	3.03
	Avg	2.76%	2.89%	4.18%	0	0.05	0.30
	Min	3.16%	3.31%	3.97%	0	0.08	0.09
Tijuana	Max	5.02%	3.97%	4.07%	0	0.25	0.12
	Avg	3.56%	2.99%	4.41%	0	0.12	0.01
	Min	4.87%	3.31%	4.22%	0	0.76	0.64

* Comparison of the average error and standard deviation for the SVR, RFR, and LSTM models, evaluated on Validation Set 2 for maximum, average, and minimum temperature across cities.

Subsequently, the models were ensembled to generate a common forecast using a particle swarm algorithm, fitted with training and validation sets 1. Table 5 presents the results of this ensemble evaluated on validation set 2, completely invisible during the fitting of both the individual models and the ensemble. The results show that the ensemble method not only improves the average performance but, in all cases, achieves results at least as good as the best individual model, with consistent improvements. This is due to the exploratory nature of the swarm algorithm, which evaluates solutions in the search space by collaboratively combining the strengths of the individual models. Once the PSO algorithm determined the optimal weight configuration based on training and validation set 1, this fixed combination was subsequently applied to the test set without further adjustments. This ensures that the evaluation on the test data remained unbiased and reflects the true generalization capacity of the ensemble model. Figure 4 graphically illustrates an example of the forecasting behavior, evidencing an excellent fit to the original curve. Table 6 compares the results of the best model versus those of the ensemble. It is relevant to note that, in any case, the ensemble method has a forecasting performance at least as good as the best single method.

In order to quantify the improvement in forecast accuracy using the ensemble strategy, for each data set, we present in Table 6 the percentage improvement obtained from the single best method versus the ensemble. The results show that, in most cases, the ensemble outperformed the best base model, achieving improvements of up to 27.27% in MAPE, as observed in the average temperature series for Monterrey. On average, the ensemble reduced the MAPE by 9.13% compared to the best individual model, demonstrating greater generalization capacity and robustness when dealing with multivariate climate data variability.

To evaluate the stability of the ensemble model, data from the test set were invisible during the fitting processes. Table 7 reports the average results in this ensemble, which represents the tail of the time series and may include new patterns or abrupt changes. Despite these challenges, the ensemble demonstrated a remarkable generalization capability, as graphically observed in Figure 5.

Table 5. Ensemble results in validation 2 set.

Cities *	Temperature	Ensemble
Guadalajara	Max	1.61%
	Avg	1.60%
	Min	1.58%
Monterrey	Max	2.59%
	Avg	3.04%
	Min	2.52%
Tampico	Max	2.62%
	Avg	2.39%
	Min	2.53%
Tijuana	Max	3.28%
	Avg	2.87%
	Min	3.10%

* Performance of the ensemble model on Validation Set 2, showing average percentage error for maximum, average, and minimum temperature predictions across cities.

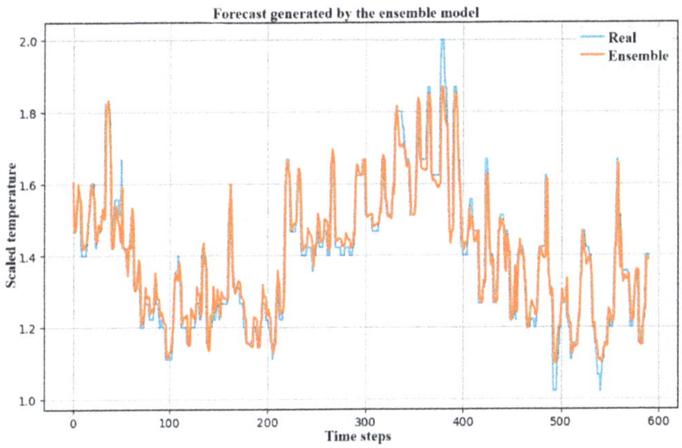

Figure 4. Forecast results for the validation set 2.

Table 6. Comparative results of the ensemble with the best of the individual methods.

Cities *	Temperature	Best Model	Best Model MAPE	Ensemble MAPE	Improvement
Guadalajara	Max	RFR	1.61%	1.61%	0%
	Avg	RFR	1.64%	1.60%	2.43%
	Min	RFR	1.58%	1.58%	0%
Monterrey	Max	RFR	2.86%	2.59%	9.44%
	Avg	SVR	4.18%	3.04%	27.27%
	Min	RFR	2.66%	2.52%	5.26%
Tampico	Max	RFR	3.04%	2.62%	13.81%
	Avg	SVR	2.76%	2.39%	13.26%
	Min	SVR	3.16%	2.53%	19.93%
Tijuana	Max	RFR	3.97%	3.28%	17.38%
	Avg	RFR	2.99%	2.87%	4.01%
	Min	RFR	3.31%	3.10%	6.34%

* Comparison showing the percentage improvement between the results of the ensemble method versus the best of the individual forecasting methods obtained in the data set of each cited city.

Table 7. Results in test set.

Cities *	Temperature	Ensemble
Guadalajara	Max	2.04%
	Avg	3.92%
	Min	1.93%
Monterrey	Max	5.57%
	Avg	4.32%
	Min	3.37%
Tampico	Max	2.44%
	Avg	4.55%
	Min	4.12%
Tijuana	Max	4.24%
	Avg	3.47%
	Min	5.31%

* Evaluation of the ensemble model's performance on the test set, including average percentage error for maximum, average, and minimum temperature predictions across cities.

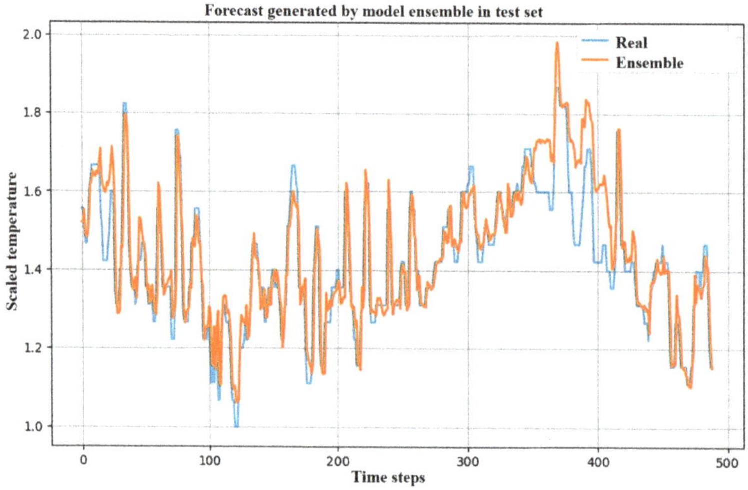

Figure 5. Forecast generated by the ensemble model for the test set.

Regarding the computational efficiency, although the primary objective of this study was to assess the predictive performance and generalization capabilities of the ensemble model, we also report the average training times of the individual components. The SVR model required approximately 3 s per series, while RFR averaged 6.2 min, and LSTM models executed with GPU acceleration took about 8.4 min. The ensemble optimization using PSO required an average of 1.2 min per configuration. All experiments were conducted on a workstation equipped with a Ryzen 7 5700X processor, 32 GB of RAM, and an NVIDIA GeForce RTX 4060 OC, leveraging multithreading for traditional models and GPU computing for neural networks.

The proposed method stands out for its ability to integrate multiple models, improving overall performance, robustness, and generalization in multivariate time series. This approach also highlights the relevance of the selected variables by prioritizing those that contribute significantly to the final model. However, some limitations were identified, such as the dependence on the quality and completeness of the data, as well as the need to evaluate its scalability in larger data sets. In future work, this methodology can be extended to other regions and variables, as well as to explore improvements in the ensemble algorithm.

5. Conclusions

This paper presents an innovative methodology based on an ensemble approach for multivariate time series forecasting applied to climate change. The methodology combines multiple machine learning techniques with a heuristic particle swarm algorithm to select relevant features and optimize the forecasting process. The results obtained demonstrate that the ensemble approach consistently outperforms the individual models. In all cases evaluated, the ensemble strategy produced results equal to or better than the best individual model and, on average, achieved significant improvements. This situation reflects the algorithm's ability to leverage the strengths of each model and explore optimal solutions in the search space. A highlight of the approach is its ability to generalize information. The methodology was evaluated on several test sets with distinct data patterns in each data set, with common features with observations related to climate change. Despite these difficulties, the ensemble model maintained robust performance, adapting to the inherent variability of the data and demonstrating its utility in complex stages. The inclusion of a feature selection process is also fundamental to the effectiveness of the model. This process not only reduces the dimensionality of the problem, but also retains most of the variance explained, improving the interpretability and efficiency of the model. This selection allows the model to work exclusively with the most significant variables contributing to optimal prediction performance.

As for the individual models, although SVR and Random Forest show a better average performance, the LSTM model stands out for its stability, evidenced by a lower standard deviation. This analysis confirms that the integration of complementary models within the ensemble is an effective strategy to improve both the accuracy and consistency of results. We proposed to refine the ensemble approach by incorporating Machine Learning and heuristic optimization techniques, as well as a new general architecture. In addition, the application of this methodology to other atmospheric variables and regions will be explored, deepening the understanding of climate change and its effects. With these improvements, the proposed approach has the potential to consolidate as a robust and efficient tool for global climate forecasting.

Author Contributions: Conceptualization, J.F.S., E.E.-P., M.P.F. and J.P.S.-H.; methodology, E.E.-P. and J.G.-B.; software, E.E.-P., J.P.S.-H. and M.P.F.; validation, J.F.S., E.E.-P., G.C.-V. and J.G.-B.; formal analysis, G.C.-V.; investigation, J.F.S. and E.E.-P.; resources, J.G.-B.; data curation, E.E.-P. and J.G.-B.; writing—original draft preparation, E.E.-P. and J.F.S.; writing—review and editing, J.F.S., E.E.-P., J.G.-B. and J.P.S.-H.; visualization, E.E.-P., J.G.-B., G.C.-V. and M.P.F.; supervision, J.F.S.; project administration, J.F.S. and J.G.-B. All authors have read and agreed to the published version of the manuscript.

Funding: This research received no external funding.

Data Availability Statement: https://github.com/DrJuanFraustoSolis/TAE-Predict.git (accessed on 22 January 2025).

Acknowledgments: In this section, the authors would like to acknowledge SECIHTI (Secretaria de Ciencia, Humanidades, Tecnologías e Innovación), TecNM/Instituto Tecnológico de Ciudad Madero, and the National Laboratory of Information Technologies (LaNTI).

Conflicts of Interest: The authors declare no conflicts of interest.

References

1. Abbass, K.; Qasim, M.Z.; Song, H.; Murshed, M.; Mahmood, H.; Younis, I. A Review of the Global Climate Change Impacts, Adaptation, and Sustainable Mitigation Measures. *Environ. Sci. Pollut. Res.* **2022**, *29*, 42539–42559. [CrossRef]
2. Al-Ghussain, L. Global Warming: Review on Driving Forces and Mitigation. *Environ. Prog. Sustain. Energy* **2019**, *38*, 13–21. [CrossRef]

3. Jan, B.; Thea, U.; Leonardo, N.; Christoph, B. The Climate Change Performance Index 2023: Results. 2022. Available online: https://www.germanwatch.org/en/87632 (accessed on 23 January 2025).
4. CMNUCC. COP 21. Available online: https://unfccc-int.translate.goog/event/cop-21?_x_tr_sl=en&_x_tr_tl=es&_x_tr_hl=es&_x_tr_pto=tc (accessed on 23 January 2025).
5. Lal, R. Beyond COP 21: Potential and Challenges of the "4 per Thousand" Initiative. *J. Soil Water Conserv.* **2016**, *71*, 20A–25A. [CrossRef]
6. Rodríguez-Aguilar, O.; López-Collado, J.; Soto-Estrada, A.; Vargas-Mendoza, M.d.l.C.; García-Avila, C.d.J. Future Spatial Distribution of *Diaphorina citri* in Mexico under Climate Change Models. *Ecol. Complex.* **2023**, *53*, 101041. [CrossRef]
7. Fildes, R.; Kourentzes, N. Validation and Forecasting Accuracy in Models of Climate Change. *Int. J. Forecast.* **2011**, *27*, 968–995. [CrossRef]
8. Hargreaves, J.C.; Annan, J.D. On the Importance of Paleoclimate Modelling for Improving Predictions of Future Climate Change. *Clim. Past* **2009**, *5*, 803–814. [CrossRef]
9. Yerlikaya, B.A.; Ömezli, S.; Aydoğan, N. Climate Change Forecasting and Modeling for the Year of 2050. In *Environment, Climate, Plant and Vegetation Growth*; Fahad, S., Hasanuzzaman, M., Alam, M., Ullah, H., Saeed, M., Ali Khan, I., Adnan, M., Eds.; Springer International Publishing: Cham, Switzerland, 2020; pp. 109–122, ISBN 978-3-030-49731-6.
10. Wolpert, D.H.; Macready, W.G. No Free Lunch Theorems for Optimization. *IEEE Trans. Evol. Comput.* **1997**, *1*, 67–82. [CrossRef]
11. Frausto-Solis, J.; Rodriguez-Moya, L.; González-Barbosa, J.; Castilla-Valdez, G.; Ponce-Flores, M. FCTA: A Forecasting Combined Methodology with a Threshold Accepting Approach. *Math. Probl. Eng.* **2022**, *2022*, e6206037. [CrossRef]
12. Estrada-Patiño, E.; Castilla-Valdez, G.; Frausto-Solis, J.; González-Barbosa, J.; Sánchez-Hernández, J.P. A Novel Approach for Temperature Forecasting in Climate Change Using Ensemble Decomposition of Time Series. *Int. J. Comput. Intell. Syst.* **2024**, *17*, 253. [CrossRef]
13. Dimri, T.; Ahmad, S.; Sharif, M. Time Series Analysis of Climate Variables Using Seasonal ARIMA Approach. *J. Earth Syst. Sci.* **2020**, *129*, 149. [CrossRef]
14. Zia, S. Climate Change Forecasting Using Machine Learning SARIMA Model. *iRASD J. Comput. Sci. Inf. Technol.* **2021**, *2*, 1–12. [CrossRef]
15. Ray, S.; Das, S.S.; Mishra, P.; Al Khatib, A.M.G. Time Series SARIMA Modelling and Forecasting of Monthly Rainfall and Temperature in the South Asian Countries. *Earth Syst. Environ.* **2021**, *5*, 531–546. [CrossRef]
16. Dabral, P.P.; Murry, M.Z. Modelling and Forecasting of Rainfall Time Series Using SARIMA. *Environ. Process.* **2017**, *4*, 399–419. [CrossRef]
17. Szostek, K.; Mazur, D.; Drałus, G.; Kusznier, J. Analysis of the Effectiveness of ARIMA, SARIMA, and SVR Models in Time Series Forecasting: A Case Study of Wind Farm Energy Production. EBSCOhost. Available online: https://openurl.ebsco.com/contentitem/doi:10.3390/en17194803?sid=ebsco:plink:crawler&id=ebsco:doi:10.3390/en17194803 (accessed on 23 January 2025).
18. Hamidi, M.; Roshani, A. Investigation of Climate Change Effects on Iraq Dust Activity Using LSTM. *Atmos. Pollut. Res.* **2023**, *14*, 101874. [CrossRef]
19. Ian, V.-K.; Tang, S.-K.; Pau, G. Assessing the Risk of Extreme Storm Surges from Tropical Cyclones under Climate Change Using Bidirectional Attention-Based LSTM for Improved Prediction. *Atmosphere* **2023**, *14*, 1749. [CrossRef]
20. Gong, Y.; Zhang, Y.; Wang, F.; Lee, C.-H. Deep Learning for Weather Forecasting: A CNN-LSTM Hybrid Model for Predicting Historical Temperature Data. *arXiv* **2024**, arXiv:2410.14963.
21. Bansal, N.; Defo, M.; Lacasse, M.A. Application of Support Vector Regression to the Prediction of the Long-Term Impacts of Climate Change on the Moisture Performance of Wood Frame and Massive Timber Walls. *Buildings* **2021**, *11*, 188. [CrossRef]
22. Jayanthi, S.L.S.V.; Keesara, V.R.; Sridhar, V. Prediction of Future Lake Water Availability Using SWAT and Support Vector Regression (SVR). *Sustainability* **2022**, *14*, 6974. [CrossRef]
23. Kumar, S. A Novel Hybrid Machine Learning Model for Prediction of CO_2 Using Socio-Economic and Energy Attributes for Climate Change Monitoring and Mitigation Policies. *Ecol. Inform.* **2023**, *77*, 102253. [CrossRef]
24. Holsman, K.K.; Aydin, K. Comparative Methods for Evaluating Climate Change Impacts on the Foraging Ecology of Alaskan Groundfish. *Mar. Ecol. Prog. Ser.* **2015**, *521*, 217–235. [CrossRef]
25. Zhang, C.; Ma, Y. *Ensemble Machine Learning: Methods and Applications*; Springer: Cham, Switzerland, 2012.
26. Estrada-Patiño, E.; Castilla-Valdez, G.; Frausto-Solis, J.; Gonzalez-Barbosa, J.J.; Sánchez-Hernández, J.P. HELI: An Ensemble Forecasting Approach for Temperature Prediction in the Context of Climate Change. *Comput. Y Sist.* **2024**, *28*, 1537–1555. [CrossRef]
27. Lu, Y.; Cohen, I.; Zhou, X.S.; Tian, Q. Feature Selection Using Principal Feature Analysis. In Proceedings of the 15th ACM International Conference on Multimedia, Augsburg, Germany, 24–29 September 2007; ACM: New York, NY, USA, 2007; pp. 301–304.
28. Malhi, A.; Gao, R.X. PCA-Based Feature Selection Scheme for Machine Defect Classification. *IEEE Trans. Instrum. Meas.* **2004**, *53*, 1517–1525. [CrossRef]

29. Song, F.; Guo, Z.; Mei, D. Feature Selection Using Principal Component Analysis. In Proceedings of the 2010 International Conference on System Science, Engineering Design and Manufacturing Informatization, Yichang, China, 12–14 November 2010; Volume 1, pp. 27–30.
30. Johnstone, I.M.; Titterington, D.M. Statistical Challenges of High-Dimensional Data. *Phil. Trans. R. Soc. Math. Phys. Eng. Sci.* **2009**, *367*, 4237–4253. [CrossRef]
31. Gers, F.A.; Schmidhuber, J.; Cummins, F. Learning to Forget: Continual Prediction with LSTM. *Neural Comput.* **2000**, *12*, 2451–2471. [CrossRef] [PubMed]
32. Agga, A.; Abbou, A.; Labbadi, M.; Houm, Y.E.; Ou Ali, I.H. CNN-LSTM: An Efficient Hybrid Deep Learning Architecture for Predicting Short-Term Photovoltaic Power Production. *Electr. Power Syst. Res.* **2022**, *208*, 107908. [CrossRef]
33. Frausto-Solís, J.; Galicia-González, J.C.d.J.; González-Barbosa, J.J.; Castilla-Valdez, G.; Sánchez-Hernández, J.P. SSA-Deep Learning Forecasting Methodology with SMA and KF Filters and Residual Analysis. *Math. Comput. Appl.* **2024**, *29*, 19. [CrossRef]
34. Ali, J.; Khan, R.; Ahmad, N.; Maqsood, I. Random Forests and Decision Trees. *Int. J. Comput. Sci. Issues* **2012**, *9*, 272.
35. Breiman, L. Random Forests. *Mach. Learn.* **2001**, *45*, 5–32. [CrossRef]
36. Altman, N.; Krzywinski, M. Ensemble Methods: Bagging and Random Forests. *Nat. Methods* **2017**, *14*, 933–935. [CrossRef]
37. Auret, L.; Aldrich, C. Interpretation of Nonlinear Relationships Between Process Variables by Use of Random Forests. *Miner. Eng.* **2012**, *35*, 27–42. [CrossRef]
38. Balabin, R.M.; Lomakina, E.I. Support Vector Machine Regression (SVR/LS-SVM)—An Alternative to Neural Networks (ANN) for Analytical Chemistry? Comparison of Nonlinear Methods on near Infrared (NIR) Spectroscopy Data. *Analyst* **2011**, *136*, 1703–1712. [CrossRef]
39. Izonin, I.; Tkachenko, R.; Shakhovska, N.; Lotoshynska, N. The Additive Input-Doubling Method Based on the SVR with Nonlinear Kernels: Small Data Approach. *Symmetry* **2021**, *13*, 612. [CrossRef]
40. Victoria, A.H.; Maragatham, G. Automatic Tuning of Hyperparameters Using Bayesian Optimization. *Evol. Syst.* **2021**, *12*, 217–223. [CrossRef]
41. Wu, J.; Chen, X.-Y.; Zhang, H.; Xiong, L.-D.; Lei, H.; Deng, S.-H. Hyperparameter Optimization for Machine Learning Models Based on Bayesian Optimization. *J. Electron. Sci. Technol.* **2019**, *17*, 26–40.
42. Kennedy, J.; Eberhart, R. Particle Swarm Optimization. In Proceedings of the ICNN'95-International Conference on Neural Networks, Perth, Australia, 27 November–1 December 1995; Volume 4, pp. 1942–1948.
43. Wang, D.; Tan, D.; Liu, L. Particle Swarm Optimization Algorithm: An Overview. *Soft Comput.* **2018**, *22*, 387–408. [CrossRef]
44. Hyndman, R.; Koehler, A.B.; Ord, J.K.; Snyder, R.D. *Forecasting with Exponential Smoothing: The State Space Approach*; Springer Science & Business Media: Berlin, Germany, 2008; ISBN 978-3-540-71918-2.
45. Geografía y Medio Ambiente. Climatológicos. Available online: https://www.inegi.org.mx/temas/climatologia/ (accessed on 23 January 2025).

MDPI AG
Grosspeteranlage 5
4052 Basel
Switzerland
Tel.: +41 61 683 77 34

Mathematical and Computational Applications Editorial Office
E-mail: mca@mdpi.com
www.mdpi.com/journal/mca

www.ingramcontent.com/pod-product-compliance
Lightning Source LLC
Chambersburg PA
CBHW041016270926
51850CB00024B/413
* 9 7 8 3 7 2 5 8 5 5 7 5 9 *